NanoScience and Technology

Springer
*Berlin
Heidelberg
New York
Hong Kong
London
Milan
Paris
Tokyo*

Physics and Astronomy ONLINE LIBRARY

http://www.springer.de

NANOSCIENCE AND TECHNOLOGY

Series Editors: P. Avouris   K. von Klitzing   H. Sakaki   R. Wiesendanger

The series NanoScience and Technology is focused on the fascinating nano-world, mesoscopic physics, analysis with atomic resolution, nano and quantum-effect devices, nanomechanics and atomic-scale processes. All the basic aspects and technology-oriented developments in this emerging discipline are covered by comprehensive and timely books. The series constitutes a survey of the relevant special topics, which are presented by leading experts in the field. These books will appeal to researchers, engineers, and advanced students.

Sliding Friction
Physical Principles and Applications
By B.N.J. Persson
2nd Edition

Scanning Probe Microscopy
Analytical Methods
Editor: R. Wiesendanger

Mesoscopic Physics and Electronics
Editors: T. Ando, Y. Arakawa, K. Furuya, S. Komiyama, H. Nakashima

Biological Micro- and Nanotribology
Nature's Solutions
By M. Scherge and S.N. Gorb

**Semiconductor Spintronics
and Quantum Computation**
Editors: D.D. Awschalom, N. Samarth, D. Loss

Semiconductor Quantum Dots
Physics, Spectroscopy and Applications
Editors: Y. Masumoto and T. Takagahara

Nano-Optoelectronics
Concepts, Physics and Devices
Editor: M. Grundmann

Noncontact Atomic Force Microscopy
Editors: S. Morita, R. Wiesendanger, E. Meyer

Nanoelectrodynamics
Electrons and Electromagnetic Fields
in Nanometer-Scale Structures
Editor: H. Nejo

Single Organic Nanoparticles
Editors: H. Masuhara, H. Nakanishi, K. Sasaki

Epitaxy of Nanostructures
By V.A. Shchukin, N.N. Ledentsov, D. Bimberg

Series homepage – http://www.springer.de/phys/books/nst/

Vitaly A. Shchukin
Nikolai N. Ledentsov
Dieter Bimberg

Epitaxy
of Nanostructures

With 192 Figures

 Springer

Dr. V.A. Shchukin
Professor N.N. Ledentsov
Technische Universtität Berlin
Hardenbergstrasse 36
10623 Berlin, Germany
E-mail: shchukin@sol.physik.TU-Berlin.de
leden@sol.physik.TU-Berlin.de
and
A.F. Ioffe Physical Technical Institute
of the Russian Academy of Sciences
Politekhnicheskaya 26
194021 St. Petersburg, Russia

Professor D. Bimberg
Technische Universtität Berlin
Hardenbergstrasse 36
10623 Berlin, Germany
E-mail: bimberg@physik.TU-Berlin.de

Series Editors:
Professor Dr. Phaedon Avouris
IBM Research Division, Nanometer Scale Science & Technology
Thomas J. Watson Research Center, P.O. Box 218
Yorktown Heights, NY 10598, USA

Professor Dr., Dres. h.c. Klaus von Klitzing
Max-Planck-Institut für Festkörperforschung, Heisenbergstrasse 1
70569 Stuttgart, Germany

Professor Hiroyuki Sakaki
University of Tokyo, Institute of Industrial Science, 4-6-1 Komaba, Meguro-ku
Tokyo 153-8505, Japan

Professor Dr. Roland Wiesendanger
Institut für Angewandte Physik, Universität Hamburg, Jungiusstrasse 11
20355 Hamburg, Germany

ISSN 1434-4904
ISBN 3-540-67817-4 Springer-Verlag Berlin Heidelberg New York

Library of Congress Cataloging-in-Publication Data. Shchukin, V.A. (Vitaly A.) 1960– . Epitaxy of nanostructures/V.A. Shchukin, N.N. Ledentsov, D. Bimberg. p.cm. – (Nanoscience and technology) ISBN 3-540-67817-4 (alk. paper). 1. Nanostructures. 2. Epitaxy. I. Ledentsov, N.N. (Nikolai N.), 1959– . II. Bimberg, Dieter, 1942– . III. Title. IV. Series. QC176.8.N35S53 2003   537.6'221–dc22   2003054394

This work is subject to copyright. All rights are reserved, whether the whole or part of the material is concerned, specifically the rights of translation, reprinting, reuse of illustrations, recitation, broadcasting, reproduction on microfilm or in any other way, and storage in data banks. Duplication of this publication or parts thereof is permitted only under the provisions of the German Copyright Law of September 9, 1965, in its current version, and permission for use must always be obtained from Springer-Verlag. Violations are liable for prosecution under the German Copyright Law.

Springer-Verlag Berlin Heidelberg New York
a member of BertelsmannSpringer Science+Business Media GmbH
http://www.springer.de

© Springer-Verlag Berlin Heidelberg 2004   Printed in Germany

The use of general descriptive names, registered names, trademarks, etc. in this publication does not imply, even in the absence of a specific statement, that such names are exempt from the relevant protective laws and regulations and therefore free for general use.

Typesetting: Data conversion by the authors using a Springer TeX macro package
Final processing by LE-TEX, Leipzig
Cover design: *design& production*, Heidelberg

Printed on acid-free paper   57/3141/tr - 5 4 3 2 1 0

Preface

The general trend in modern solid state physics and technology is to make things smaller. The size of key elements in modern devices approaches the nanometer scale, for both vertical and lateral dimensions. Ultrathin layers, or quantum wells, had already gained broad acceptance for applications in micro- and optoelectronics by the 1980s. However, the development of heterostructures with lower dimensionality (quantum wires, where carriers are confined in two directions and move freely in one, and quantum dots, where carriers are confined in all three directions) took longer. It became clear that quantum wire and dot structures constitute the utmost technological challenge, whilst providing enormous advantages.

At the beginning of the 1990s, a few outstanding discoveries concerning self-organization phenomena at crystal surfaces for direct fabrication of nanostructures led to a change in the major paradigms of semiconductor physics and technology. This new approach in epitaxy enables fast parallel fabrication of large densities of quantum dots or wires for almost unlimited material combinations and has become the basis for a powerful new branch of nanotechnology. Quantum dots, coherent inclusions in a semiconductor matrix with zero-dimensional electronic properties persistent up to room temperature, have demonstrated fascinating physical properties and given birth to a novel generation of optoelectronic devices and systems.

The last decade of intense research into quantum dots by a great number of leading laboratories has clearly demonstrated that it is not sufficient to use the effects of self-organization alone when the aim is to achieve practical applications of funny objects. A profound understanding of the physics of nanostructure formation and the development of tools for controlling and tuning geometrical parameters and electronic spectra of the epitaxial systems is a prerequisite for any significant success in nanotechnology. To achieve this goal requires complementary studies including specially designed growth experiments, theoretical modeling of growth, and structural and optical characterization on a nanoscale. Combining effects generously provided by nature with further engineering of complex nanostructures allows significant progress in fabricating nanostructures suitable for devices. Many fundamental phenomena are now qualitatively understood, but no comprehensive survey on this subject exists in the literature. This book attempts to fill the gap.

Chapter 1 provides the motivation for entering the nanoworld, and discusses the fundamental paradigm changes in semiconductor physics. Chapter 2 describes the basic physics of molecular beam epitaxy used for the growth of semiconductor nanostructures, and gives an overview of structural and optical characterization techniques. Chapter 3 presents physical mechanisms governing the fundamental phenomena underlying the spontaneous formation of surface nanostructures, including periodically faceted surfaces and arrays of coherently strained islands. Chapter 4 expounds a large number of techniques and results in the engineering of nanostructures: improving uniformity in island size and spacing by stacking of dots, seeding of quantum dots allowing independent control of dot size and density, engineering exciton wavefunctions and control of photoluminescence polarization via stacking of dots, nanoengineering using a transition between different vertical arrangements of dots in multistacks, shifting quantum dot optical spectra towards longer wavelengths via activated alloy phase separation in the cap layer, and defect reduction techniques using a multiple cycle of partial overgrowth and thermal etching of the dots. For the In(Ga)As/GaAs model system, a combination of the techniques is demonstrated which allows defect-free nanostructures on GaAs substrates emitting at the technologically important wavelength of 1.3–1.55 µm. Chapter 5 gives a brief overview of the development of quantum dot lasers. It is shown that the newcomers are now taking over from conventional quantum well lasers, thereby becoming the first successful example of a laboratory discovery pervading the whole of human society within the short period of one decade. The summary in Chap. 6 presents a prospective view of the future development of semiconductor nanotechnology.

The book results from long-term cooperation between the Instutut für Festkörperphysik, Technische Universität Berlin, and Abraham Ioffe Physical Technical Institute of the Russian Academy of Sciences, St. Petersburg. The work of the two teams would not have been possible without generous support from the Deutsche Forschungsgemeinschaft (Sbf 296), Bundesministerium für Bildung und Forschung (bmb+f), Competence Center for Nanooptoelectronics (NanOp), and the Russian Foundation for Basic Research. In particular, cooperation and exchange of scientists between the teams were supported by the Volkswagen Stiftung, the governments of Germany and Russia within the framework of their general science cooperation agreement, and INTAS. V.A. Shchukin personally acknowledges support from the Alexander von Humboldt Foundation in the form of a research fellowship and from the Russian Foundation for Promotion of Science via a grant for young researchers. N.N. Ledentsov was supported by the Deutsche Akademische Austauschdienst (DAAD) Guest Professorship program, the Alexander von Humboldt Foundation via donation of equipment, the Berlin-Brandenburg Academy of Sciences and the Peregrinus-Stiftung (Rudolf Meimberg). We could not have carried out the research or written this book without their help and are therefore very grateful to the respective administrations and many anonymous referees who approved the project.

Many individuals have helped us through their personal advice and stimulating discussions. We are particularly indebted to Zh.I. Alferov, A. Madhukar, M. Scheffler, and E. Schöll. Valuable comments from D. Jesson, P. Kratzer, E. Pehlke, E. Penev, C. Teichert, and J. Tersoff are also acknowledged. Others have contributed to this book by their work as members of our teams. Particular thanks should go to N.A. Bert, G.E. Cirlin, A.Yu. Egorov, I.P. Ipatova, S.V. Ivanov, P.S. Kop'ev, A.R. Kovsh, I.L. Krestnikov, V.V. Lundin, N.A. Maleev, V.G. Malyshkin, M.V. Maximov, Yu.G. Musikhin, A.V. Sakharov, Yu.M. Shernyakov, I.P. Soshnikov, A.N. Starodubtsev, A.F. Tsatsul'nikov, V.M. Ustinov, B.V. Volovik, and A.E. Zhukov at the Ioffe Institute, to S. Bognár, J. Christen, M. Grundmann, F. Guffarth, F. Heinrichsdorff, R. Heitz, A. Hoffmann, L. Müller-Kirsch, N. Kirstaedter, A. Krost, M. Meixner, U. Pohl, K. Pötschke, S. Rodt, A. Schliwa, R. Sellin, O. Stier, M. Straßburg, A. Strittmatter, and V. Türck, at TU Berlin, to U. Gösele, J. Heydenreich, P. Werner, and N. Zakharov, at the Max-Planck Institut für Mikrostrukturphysik, Halle/Saale, to D. Gerthsen, D. Litvinov, and A. Rosenauer at the Univerität Karlsruhe, and to J.M. Hvam at the Technical University of Denmark, Lyngby.

Berlin, *Vitaly Shchukin*
July 2003 *Nikolai Ledentsov*
  Dieter Bimberg

Contents

1. **Introduction** 1
   1.1 Approaching the End of Moore's Law: What Next? 1
   1.2 Paradigm Changes in Semiconductor Physics and Technology 4
   1.3 Surfing Through Books and Reviews 11

2. **Growth and Characterization Techniques** 15
   2.1 Basics of Molecular Beam Epitaxy 17
       2.1.1 MBE Apparatus 17
       2.1.2 Understanding MBE Growth Processes 19
       2.1.3 Phase Diagrams 22
       2.1.4 Solid–Liquid–Vapor Equilibrium
             for Binary Compounds 26
       2.1.5 Solid–Vapor Equilibrium for Ternary Alloys 28
       2.1.6 Segregation Effects 30
   2.2 Basics of Metalorganic Chemical Vapor Deposition 33
   2.3 Main Characterization Techniques 35
       2.3.1 Direct Imaging Methods 36
       2.3.2 Transmission Electron Microscopy 40
       2.3.3 Diffraction Methods 48
       2.3.4 Optical Methods 54

3. **Self-Organization Phenomena at Crystal Surfaces** 57
   3.1 Periodically Faceted Surfaces 61
       3.1.1 Equilibrium Crystal Shape:
             Two Distinct Formulations of the Problem 61
       3.1.2 Faceting: Analogy with Phase Separation 63
       3.1.3 Intrinsic Surface Stress of a Solid 65
       3.1.4 Thin Strained Epitaxial Film as a Model of a Surface . 67
       3.1.5 Simple Lattice Model for Intrinsic Surface Stress .. 68
       3.1.6 Capillarity Phenomena at Solid Surfaces 71
       3.1.7 Periodically Faceted Surfaces 73
       3.1.8 Faceting Phenomena
             on (311) Surfaces of GaAs and AlAs 76

		3.1.9 Macroscopic Step Bunching and Faceting of Vicinal Surfaces 96

- 3.1.9 Macroscopic Step Bunching and Faceting of Vicinal Surfaces 96
- 3.1.10 Variety of Periodically Faceted Surfaces 101
- 3.1.11 Faceted Surfaces: Understanding and Prospects 103

3.2 Surface Arrays of Two-Dimensional Islands 104
- 3.2.1 Homoepitaxial Systems at Submonolayer Coverage ... 108
- 3.2.2 Energetics of a Heteroepitaxial System at Submonolayer Coverage........................ 110
- 3.2.3 Arrays of 2D Strained Islands at Low Temperatures .. 123
- 3.2.4 Arrays of 2D Strained Islands at Low Coverage 135
- 3.2.5 Equilibrium Distribution of Island Sizes 136
- 3.2.6 Crossover from Kinetically Controlled to Thermodynamically Limited Growth of 2D Strained Islands 140
- 3.2.7 Submonolayer Arrays of InAs/GaAs Islands 142
- 3.2.8 Submonolayer Islands at Work 145

3.3 Arrays of Three-Dimensional Coherently Strained Islands.... 156
- 3.3.1 The In(Ga)As/GaAs System: From Three-Dimensional Islands to Quantum Dots ... 156
- 3.3.2 Coherent vs. Dislocated Islands in Lattice-Mismatched Systems 165
- 3.3.3 Size-Limited Island Growth: Are Islands Stable Against Ripening? 168
- 3.3.4 Energetics of a Lattice-Mismatched Heteroepitaxial System...... 173
- 3.3.5 Dilute Array of 3D Strained Islands................. 175
- 3.3.6 Ordering of Islands in Terms of Shape 178
- 3.3.7 Size Ordering of Islands vs. Ostwald Ripening........ 180
- 3.3.8 Lateral Arrangement of Islands 183
- 3.3.9 Phase Diagram of Arrays of Interacting Strained Islands............. 188
- 3.3.10 Equilibrium Thickness of the Wetting Layer 190
- 3.3.11 Two Exact Theorems on the Shape vs. Volume Dependence of 3D Islands ... 194
- 3.3.12 Kinetic Theories of Size-Limited Island Growth 199
- 3.3.13 Experimental Studies of 3D Island Formation in the In(Ga)As/GaAs System 206
- 3.3.14 Temperature Ramping and Cooling in InAs/GaAs Systems: Evidence of Close-to-Equilibrium Behavior 214
- 3.3.15 Formation of InAs/GaAs Islands at Ultra-Low Temperatures 224
- 3.3.16 3D Islands in Other Material Systems............... 226

| | | 3.3.17 | What Have we Learned about 3D Coherently Strained Islands?............. 231 |

4. **Engineering of Complex Nanostructures: Working Together with Nature** 235
   - 4.1 Multisheet Arrays of Strained Islands 237
     - 4.1.1 Vertical Correlation of Strained Islands.............. 238
     - 4.1.2 Order Enhancement in Multisheet Arrays 239
     - 4.1.3 Electronically Coupled Multisheet Quantum Dots 243
     - 4.1.4 Seeding of Quantum Dots 246
     - 4.1.5 Engineering the Exciton Wave Function by Stacking Quantum Dots 249
     - 4.1.6 Surface Evolution During Overgrowth of Strained Islands 251
     - 4.1.7 Defect-Reduction Techniques 253
   - 4.2 Anticorrelation in Multisheet Arrays of Strained Islands 263
     - 4.2.1 Generalized Rayleigh Waves in Elastically Anisotropic Crystals 264
     - 4.2.2 Formation of Multisheet Arrays in Elastically Anisotropic Crystals 265
     - 4.2.3 Multisheet Arrays of CdSe/ZnSe Submonolayer Islands 269
     - 4.2.4 Highly-Ordered Quantum Dot Superlattices 276
     - 4.2.5 Anticorrelated Multisheet Nanostructures in III–V Semiconductors 280
   - 4.3 Activated Alloy Phase Separation During Overgrowth of Quantum Dots 282
     - 4.3.1 Basic Physics of Phase Separation in Alloys.......... 282
     - 4.3.2 Steady-State Composition-Modulated Structures in Growing Alloy Films 302
     - 4.3.3 Alloy Growth on Stressors: Activated Phase Separation...................... 308

5. **Devices Based on Epitaxial Nanostructures** 315
   - 5.1 Quantum Dot Heterostructure Lasers 316
     - 5.1.1 Basic Advantages of Heterostructure Lasers.......... 317
     - 5.1.2 Development of Heterostructure Lasers............. 318
     - 5.1.3 The Key Breakthrough: Self-Organized Growth 321
     - 5.1.4 State of the Art in Quantum Dot Lasers: Taking an Upper Hand......................... 323
   - 5.2 Quantum Dot Nanostructures for Single-Electron Devices.... 333

6. **Conclusion** 335

A. **Energy of a Strained Disk with Perturbed Shape** 337
    A.1 Energy of the Disk Boundary 338
    A.2 Elastic Relaxation Energy of the Disk 339
    A.3 Evaluation of Integrals 341
    A.4 Stiffness of the Disk against Shape Perturbations 346

B. **Elastic Interaction of Two Strained Disks** 349

C. **Stiffness of a Hexagonal Array of Interacting Strained Disks** 355

References 359

Index 385

1. Introduction

> So perhaps the best thing to do is to stop writing Introductions and get on with the book.
>
> Alan Alexander Milne. *Winnie-The-Pooh*

The quest for information has marked the entire history of mankind. Free access to information is a prerequisite for freedom in the widest sense: personal freedom, freedom of choice, freedom of communication, and indeed everything that is understood under the notion of democracy. Modern information and communication technology has dramatically changed the world, significantly affecting all areas of human existence. This global technology is the driving force changing the industrial society into an information and knowledge-based society.

The highest recognition of the extremely great significance of modern information technology (IT) for mankind was the Nobel Prize for Physics in 2000. The prize was awarded to scientists and inventors whose work laid the foundations of modern information technology, particularly through their invention of rapid transistors, laser diodes, and integrated circuits. Half of the prize was awarded jointly to Zhores I. Alferov and Herbert Kroemer "for developing semiconductor heterostructures used in high-speed- and optoelectronics", whilst the other half went to Jack S. Kilby "for his part in the invention of the integrated circuit" [1.1].

The change to an information-based society occurred over the last few decades of the twentieth century. The development of information technology was made possible by a remarkable advance in microelectronics known as Moore's Law. Now we must ask a further question: what new challenges await us along this road?

1.1 Approaching the End of Moore's Law: What Next?

The general trend in microelectronics towards miniaturization of device features aims to increase operation rates, and reduce power consumption and

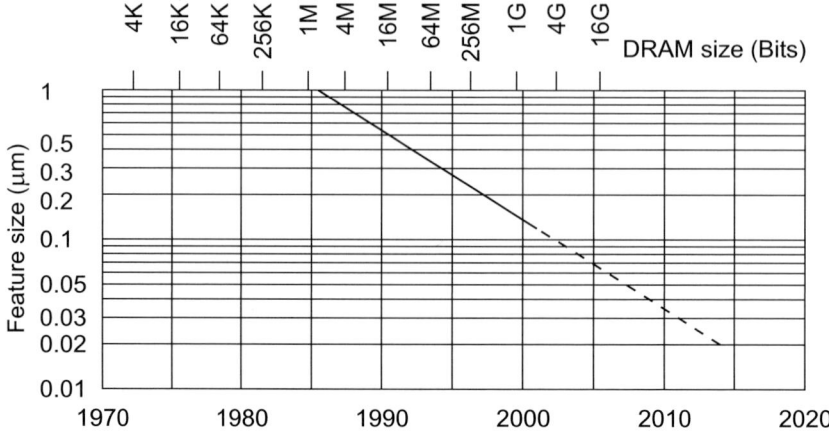

Fig. 1.1. Lithography trend showing miniaturization of device features vs. time. The size of the dynamic random access memory (DRAM) chip is plotted as an example

costs. For the past 50 years or so, i.e., since the invention of the transistor, the exponential progress in microelectronics has been fueled by a 15% yearly miniaturization rate, inducing almost 30% cost decrease and 50% performance improvement in all electronic functions each year. In its turn, this technological progress has caused a 15% growth rate on the semiconductor market, which generated enough revenues to cover the exponential increase (also about 15% per year) in the research and development (R&D) expenses needed to feed such exploding technical progress. Nowadays, the major part of these expenses is devoted to the huge and ever increasing manufacturing R&D needed to optimize chip size, wafer size, defects, and interconnects [1.2]. All these factors are building up to make Moore's Law a reality.

Several times over the last few decades, severe limitations have been identified, predicting the imminent end to Moore's Law. Fortunately, each time, conceptually novel scientific and technological solutions were found allowing further progress to be made. A detailed discussion on the limitations existing in our own time and on future trends may be found in the book [1.2]. Here we focus on just two limitations that may be thought of as fundamental.

Figure 1.1 shows the trend in miniaturization for a typical lateral dimension of a device feature [1.3]. At the present time, sizes are going down from the submicrometer region to the nanometer region, i.e., decreasing below 100 nm. We are thus entering the nanostructure field. Semiconductor nanostructures are structures having characteristic features with a typical size of 1–100 nm in the lateral plane.

As long as the operation of chips or other elements is based on classical physics, a limitation for miniaturization is set by the de Broglie wavelength, related to the thermal energy of electrons:

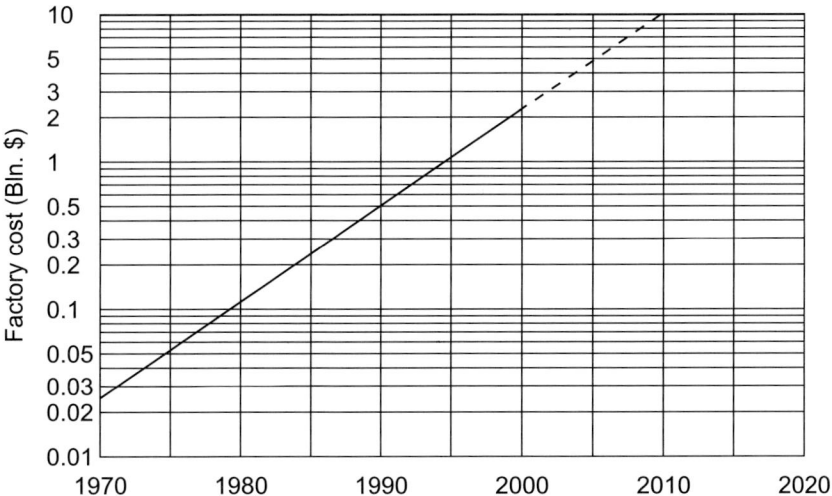

Fig. 1.2. The trend in the cost of building a new microelectronics factory

$$\lambda = \frac{2\pi\hbar}{\sqrt{m^* k_B T}}, \qquad (1.1)$$

where m^* is the effective mass of the electrons, T the absolute temperature, and k_B Boltzmann's constant. At room temperature (1.1) yields $\lambda = 0.03$ μm for GaAs and $\lambda = 0.01$ μm for Si.

If we extrapolate the exponential decrease in the lateral size of device features in Fig. 1.1, the size will become comparable with the de Broglie wavelength within the next 10–15 years. If this size is reached, the operation of such devices will necessarily be governed by quantum physics.

Another limitation may be seen from Fig. 1.2. The cost of building a new microelectronics factory is growing faster than the world's gross domestic product (GDP). This trend will certainly break down at the very latest when the cost is no longer affordable to all the world's leading companies combined. Thus we are already, or at least will be in the very near future, facing the end of Moore's Law, unless a 'leapfrog' strategy is found.

Fortunately, nanostructures have already appeared in the form of quantum wires or quantum dots (Fig. 1.3). We expect the use of such structures to become mainstream throughout semiconductor technology. Our optimism is strongly supported by the successful development of semiconductor quantum dot technology that has occurred over the last 10 years.

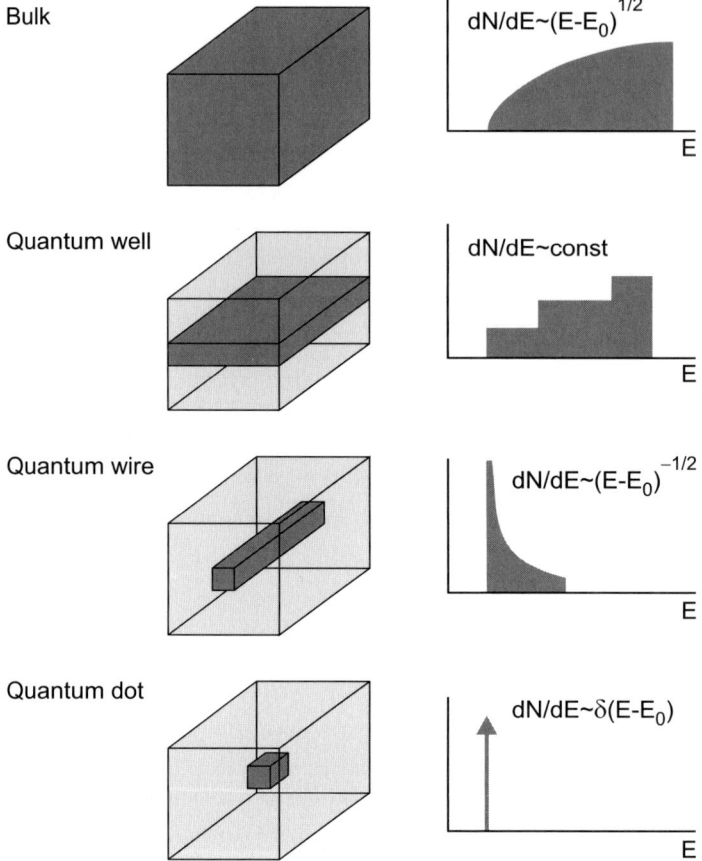

Fig. 1.3. Semiconductor structures and corresponding electronic density of states near the edge of electronic bands (subbands)

1.2 Paradigm Changes in Semiconductor Physics and Technology

The breakthrough that occurred in the field of semiconductor physics and technology at the beginning of 1990s can be viewed as a major change of paradigm.

Paradigm 1: Artificial Atoms vs. Layers. Figure 1.4 compares the electronic levels of a single atom, a bulk semiconductor, and a quantum dot. It is well known that a single atom has discrete energy levels separated by forbidden energy gaps, as shown in Fig. 1.4. When the atom is excited, the electron goes to the higher energy level, and when it relaxes back to the ground state, a photon with strictly defined energy is emitted. The width of the emission or absorption line (ΔE) is defined by a fundamental relation involving the

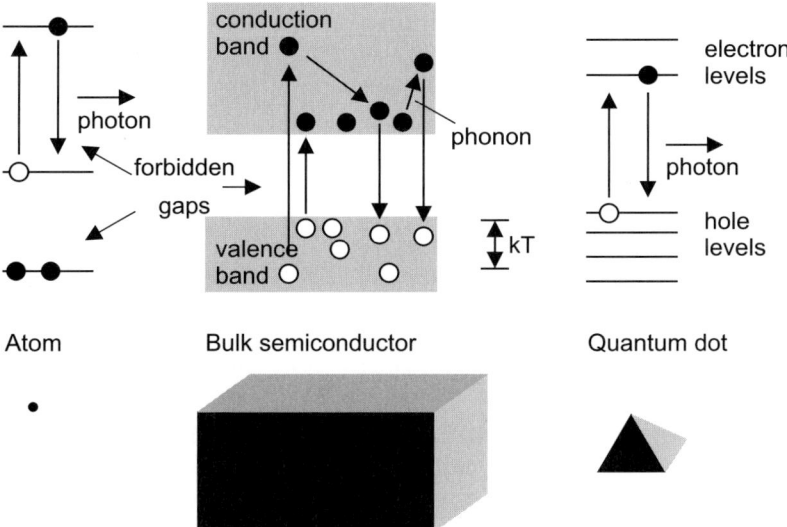

Fig. 1.4. Schematic representation of energy levels in a single atom, a bulk semiconductor and a quantum dot

lifetime of the electron in the upper state. The uncertainty in the emitted energy is

$$\tau \Delta E \geq \hbar \,, \tag{1.2}$$

where τ is the coherence time of the electron in the excited state.

In contrast to the case of a dilute gas of atoms, atoms in crystals are strongly bound to each other. Their high density in crystals plays a very important role in modern solid-state devices. It allows high absorption or (in the case of population inversion) gain coefficients, provides high conductivity, and facilitates high density flows of charged carriers through the crystal. For this reason, a modern semiconductor laser with length 1 mm and cross-section 10^{-4} mm^2 can emit continuous light with a power of a few watts, while the corresponding gas laser is a few meters long. At the same time, small separations between atoms make interactions between their electron levels unavoidable. This interaction results in the formation of wide bands of allowed states, in contrast to the discrete (δ-function-like) energy spectrum of single atoms. In semiconductors, the last filled band of allowed states is called the valence band and the next empty band is called the conduction band. Due to the broad spectrum of allowed states in these bands, a wide range of transition energies exists between electrons from the filled valence band to empty states in the conduction band. The absorption band becomes rather broad, of the order of a few electronvolts, in marked contrast to the sharp line absorption spectrum of single atoms. Excited electrons in the conduction band, as well as empty states in the valence band (so-called holes) can move in the crystal via tunneling between crystal lattice sites. Since the atomic potential profile in a crystal is periodic, electrons and holes can move freely through the crystal, as free carriers do in vacuum. However, the motion of charged carriers in crystals is described by a different mass to that of free electrons, defined by the crystal field. The carriers are thus called quasiparticles. In the widely used optoelectronic III–V materials, e.g., gallium arsenide (GaAs), indium arsenide (InAs), etc., electron effective masses lie in the range 0.01–0.1 of the electron mass in vacuum.

Wide bands of allowed states in the crystal provide ample opportunity for scattering of electrons and holes. Lattice vibrations easily stimulate transitions of charge carriers in the energy range defined by the lattice temperature and/or scatter the direction of motion of the carriers. The tails of the carrier distribution near the bottom of the conduction band and the top of the valence band increase remarkably with temperature. Thus, the concentration of carriers per energy interval near the band edge drops. For the same concentration of injected carriers, a broadening of their energy spectra results in a decrease in maximum gain, and degradation of laser performance, among other disadvantages.

The situation changes remarkably if the motion of the charged carrier in the crystal is limited to a very small volume, e.g., to a three-dimensional rectangular box. Localization of carriers can be provided by a surrounding (matrix) material. For laser applications, it is important for the matrix material to have a larger bandgap than the box material and also for the potential wells to be attractive for both electrons and holes. Since electrons exhibit both particle and wave properties, if the size of the box is small, the electron energy spectrum is quantized rather as it would be in the attractive Coulomb potential of a nucleus. In the simplified case of an infinite barrier at the box–matrix interface, the size quantization energy is described by

$$E(n_x, n_y, n_z) = \frac{\hbar^2}{2m_e^*} \left(\frac{\pi^2 n_x^2}{L_x^2} + \frac{\pi^2 n_y^2}{L_y^2} + \frac{\pi^2 n_z^2}{L_z^2} \right), \qquad (1.3)$$

where m_e^* is the electron effective mass, L_x, L_y, L_z are the dimensions of the box, and n_x, n_y, n_z are integer quantum numbers. Electrons in crystals usually have rather small effective mass, and an already large box size of about 10 nm can result in a large energy separation between electron sublevels (about 100 meV for a GaAs QD). The latter value significantly exceeds the thermal energy at room temperature (26 meV), so that population of excited states can be avoided. In this sense the optical spectrum of such a box will display no temperature dependence over a wide temperature range, and the realization of temperature-insensitive devices becomes possible.

Quantum dots thus combine the advantages of single atoms (discrete energy spectrum) and of solids (a rather large volume density). In view of their discrete spectrum, quantum dots may be called artificial atoms, although they may consist of a few 10^2 to a few 10^5 atoms.

1.2 Paradigm Changes in Semiconductor Physics and Technology

The trend towards studying and fabricating semiconductor heterostructures with reduced effective dimensionality, i.e., quantum wires (1D), followed by quantum dots (0D), has been motivated by the major advantages expected from QDs as compared with quantum wells. Due to the discrete nature of their electronic spectrum, QD-based devices should exhibit a higher temperature stability and allow temperature-insensitive device operation.

In order to realize their potential advantages, quantum dot structures should obey certain requirements [1.4]:

- The QD should not be too small, otherwise it will not have localized states.
- The QD should not be too large, otherwise the spacing between energy levels becomes too small and may hinder the temperature stability of the structure. An estimate made by Ledentsov [1.4] for InAs QDs in a GaAs matrix suggests that the QD size should lie in the interval

$$4 \text{ nm} < L < 20 \text{ nm} . \tag{1.4}$$

- The density of QDs should be rather high to ensure a high modal gain for lasers.
- QDs should be uniform in shape and size.
- A QD heterostructure should contain a low density of defects which produce centers of irradiative recombination.

The traditional approach in semiconductor technology involves planar epitaxial growth followed by batch lithographic processing. Traditional lithographic techniques do not allow fabrication of the required quantum dot structures due to size limitations and the high density of defects generated by lithography.

Advanced lithographic techniques are under development that employ radiation with wavelength shorter than visible light. These methods are electron beam lithography [1.6], X-ray lithography using synchrotron light sources (a review may be found in [1.7]), focused ion beam techniques [1.8], and atom beam lithography [1.9]. Although these techniques yield lateral resolution down to a few 10 nm and offer almost infinite design variations, the disadvantages are the high number of technological processes involved and the high cost. Moreover, these techniques still generate a high density of defects hindering possible applications of nanostructures in optoelectronics.

Paradigm 2: Self-Organized Nanoepitaxy vs. Lithography. Recently, two alternatives to the concept of optical lithography have been introduced. The first is based on scanning probe microscopy (SPM) techniques that have been developed since the early 1980s. Nanostructures down to an atomic scale may be achieved either by manipulating single atoms [1.10] or by using the SPM tip as a pen to 'write' nanoscale structures [1.11]. Although the 'writing speed' of these procedures has been steadily increased, they are still very time-consuming because of their serial nature. To overcome this disadvantage, attempts have been implemented to operate more than one probe

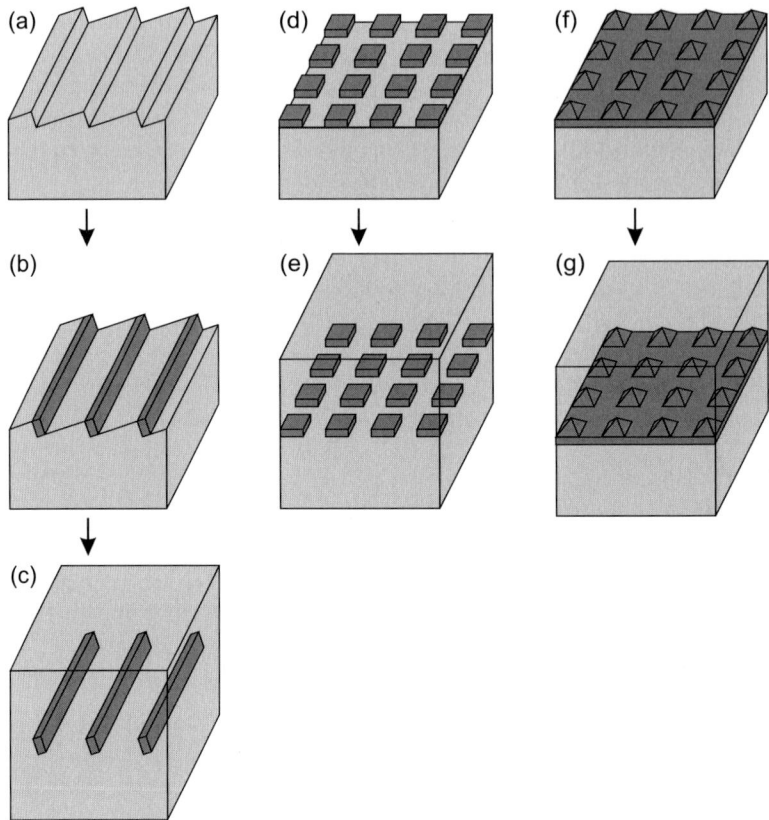

Fig. 1.5. Examples of spontaneously formed nanostructures. (**a**) Periodically faceted surface. (**b**) Wires of a deposited material formed in the grooves of the periodically faceted surface. (**c**) Capped structure (b) forming quantum wires within a matrix. (**d**) Array of two-dimensional islands in submonolayer heteroepitaxy. (**e**) Capped structure (d) forming two-dimensional quantum dots embedded in a matrix. (**f**) Array of three-dimensional coherently strained islands over a wetting layer on a substrate surface. (**g**) Capped structure (f) forming an array of three-dimensional quantum dots embedded in a matrix

in parallel [1.12], resulting in linear arrays [1.13]. Although some 2D arrays have been fabricated [1.14], the prospects for this 'writing' technique are still challenging.

The second, much more elegant and exciting alternative to lithography uses the effects of self-organization. Spontaneous formation of spatial, temporal, or spatio-temporal patterns by self-organization of individual constituents is a common phenomenon in nature [1.15]. It covers a wide range of length scales from atomic to cosmic dimensions (see also [1.16]). Famous examples are lasers and heated fluids [1.15] in physics, the Belousov–Zhabotinsky reaction [1.17] in chemistry, dune patterns [1.18] in earth science, and mor-

1.2 Paradigm Changes in Semiconductor Physics and Technology

phogenes in biology. Pattern formation is even present in human society, where we observe the self-organization of cities and settlements [1.19], and in pedestrian or automotive traffic [1.20, 1.21]. Research in the general area of self-organization helps us to understand interconnections between inorganic and organic matter. It also helps us to learn how complex patterns appear independently of our own designs, how they subsequently behave, and what ultimately is the origin of life.

In the early 1990s the discovery of self-organization phenomena on a nanometer scale for a large variety of material systems marked a breakthrough in the area of quantum dot and quantum wire research.

The use of self-organization phenomena at surfaces allows us to fabricate quantum dot structures within the context of planar technology: just the 'conventional' planar growth is used. The massive parallel process of spontaneous formation of nanostructures makes it possible to produce 10^{10} to 10^{12} quantum dots per cm^2 per second. Hence, the effects of self-organization can result in the formation of ordered nanostructures from an initially random distribution of atoms.

Figure 1.5 shows a few examples of spontaneously forming nanostructures. When the phenomenon occurs in semiconductor materials, further overgrowth of the surface nanostructure (capping) results in the formation of wire-shaped or dot-shaped insertions in the matrix. In the case of narrow bandgap insertions in a wide bandgap matrix, quantum wires or quantum dots are formed.

Paradigm 3: Lattice-Mismatched vs. Lattice-Matched Growth. The classical approach to heteroepitaxial growth focused mainly on lattice-matched or nearly lattice-matched growth. Lattice-mismatched systems grow coherently (dislocation-free) only below a certain critical thickness, beyond which the structure becomes dislocated. This deteriorates device characteristics for both microelectronic and optoelectronic applications. One disadvantage was a severely limited range of materials. Only a few material combinations, e.g., GaAs/Ga$_{1-x}$Al$_x$As, InP/Ga$_{0.52}$Al$_{0.48}$As, GaAs/ZnSe, were possible.

Additionally, in the classical approach, the goal was to stay with a planar morphology on the surface. Islands were highly undesirable, as they were thought to be necessarily dislocated.

In the self-organized growth of quantum dots, lattice-mismatched growth is the main route, and lattice-matched growth is in many cases inappropriate. Islands are widely used as quantum dots, since they are dislocation-free below a certain critical volume. A wide variety of material combinations may thus be used to fabricate quantum dots.

Figure 1.6 illustrates the formation of QDs in a wide variety of highly-mismatched systems. Plan-view transmission electron microscopy (TEM) images of InAs/GaAs QDs [1.22] and GaSb/GaAs QDs [1.23] show the strain contrast representing strained insertions. Cross-sectional high resolution transmission electron microscopy (HRTEM) images of InGaN/GaN

10   1. Introduction

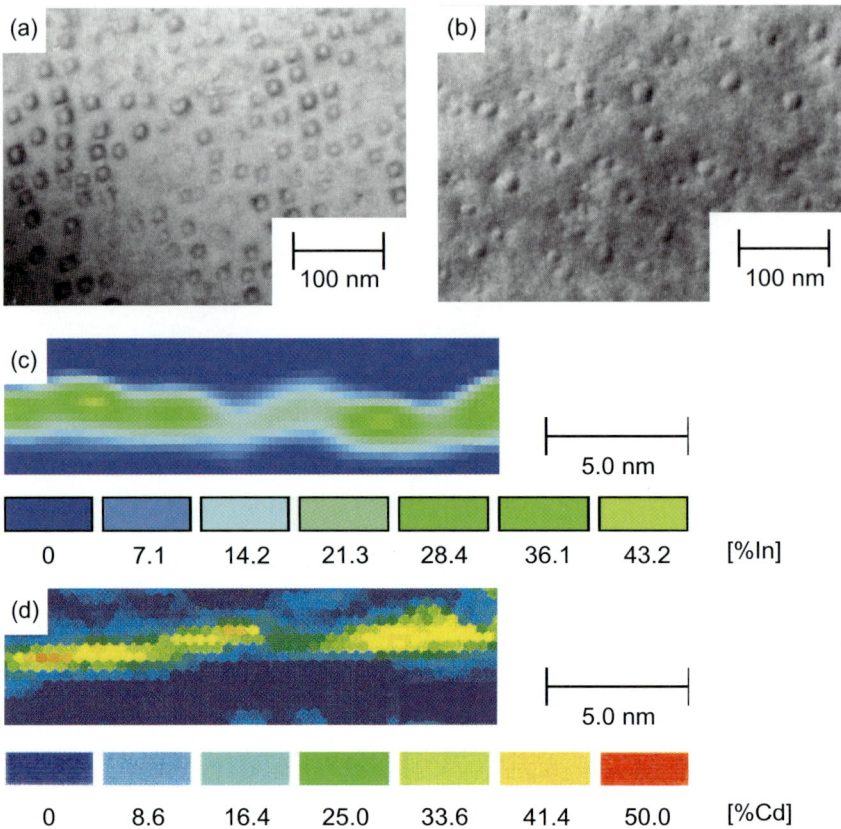

Fig. 1.6. Spontaneously formed arrays of quantum dots (QDs) in various material systems. (**a**) Plan view transmission electron microscopy (TEM) image of InAs/GaAs QDs. (**b**) Plan view TEM image of GaSb/GaAs QDs. (**c**) Cross-sectional high resolution transmission electron microscopy (HRTEM) image of In-GaN/GaN QDs processed by the digital analysis of lattice images (DALI) evaluation technique. (**d**) Cross-sectional HRTEM image of CdSe/ZnSe QDs processed by the DALI technique

[1.24] and CdSe/ZnSe [1.25] structures have been processed by the digital analysis of lattice images (DALI) technique which reveals the local map of the lattice parameter in the vertical direction. This map allows us to estimate the local content of In (Fig. 1.6c) or Cd (Fig. 1.6d).

Progress in the area of epitaxial nanostructures and, in particular, the development of quantum dot and quantum wire semiconductor technology employing self-organization phenomena requires a profound understanding of the basic physics behind the spontaneous formation of nanostructures. Such progress can only be reached by combined efforts to:

- design growth experiments,
- develop a theory of spontaneous nanostructuring,

- perform precise structural and optical characterization of grown objects,
- develop experimental tools for controlling and tuning geometrical parameters and electronic spectra of nanostructures,
- optimize growth techniques to meet device requirements,
- fabricate novel nanostructure-based devices which actually fuel further research.

This, briefly, is the subject of the present book.

1.3 Surfing Through Books and Reviews

It is hardly possible to list all the books and review articles relevant to nanostructure formation. Here we describe those that have significantly helped us to improve our understanding of the subject, to develop our own work and indeed to write the present book, providing new insights into the fields described in our monograph.

The book by Khachaturyan published in 1974 in Russian under the title *Theory of Phase Transformations and the Structure of Solid Solutions* [1.26] and in 1983 in an extended English version under the title *Theory of Structural Transformation in Solids* [1.27] is devoted to the theory of spinodal decomposition and atomic-scale ordering in bulk metal alloys. A theoretical method is developed to solve the problems of strain fields in heterogeneous systems. It uses the concept of a stress-free strain tensor $\varepsilon_{ij}^{(0)}$ which describes the difference in lattice parameters between the two materials. This method has been widely used by ourselves for different physical situations involving inhomogeneously strained epitaxial structures at a crystal surface.

The book by Pimpinelli and Villain entitled *Physics of Crystal Growth* [1.28] and published in 1998 is a general theoretical book on crystal growth, well-suited as a tutorial for anyone interested in the subject. A few issues on strain-related phenomena are addressed: the concept of surface stress, adatom–adatom, adatom–step, and step–step elastic interactions, and Asaro–Tiller–Grinfel'd instability of a flat surface under applied stress. However, spontaneous formation of nanostructures is not the main focus of this book. Key problems such as the final state of developing instability, the role of configuration entropy, and growth of multilayered nanostructures are not discussed here.

The book by Bimberg, Grundmann, and Ledentsov entitled *Quantum Dot Heterostructures* [1.5], published in 1998, was the first book, and is still the only book, covering all major complementary directions in QD research: growth, theoretical modeling of growth, structural characterization, optical studies, calculations of QD electronic spectra, and laser applications.

The book *Growth Processes and Surface Phase Equilibria in Molecular Beam Epitaxy* by Ledentsov [1.29], published in 1999, focuses on the basic processes underlying molecular beam epitaxy (MBE) growth of semiconductor materials. It is shown that the thermodynamic model applies successfully to major processes like deposition, doping, and impurity segregation.

The review by Zinke-Allmang [1.30] published in 1999 focuses on nucleation, coarsening and coalescence kinetics on solid surfaces. The existence of equilibrium structures, the interplay between thermodynamics and kinetics, the relation between structural and optical properties of nanostructures, effects of configuration entropy, and the concept of nanoengineering are beyond the scope of the paper.

The review by Merz et al. [1.31] published in 1999 gives a survey of the mechanisms of self-organization in a single sheet of strained islands. Some control techniques are discussed, including the deposition of QDs on pre-patterned substrates, and fixing of island nucleation sites by scanning probe techniques like scanning tunneling microscopy (STM) or atomic force microscopy (AFM). Complex multi-sheet nanostructures are not considered.

The review by Shchukin and Bimberg [1.32] published in 1999 deals mainly with formation mechanisms in single sheets of nanostructures: periodically faceted surfaces, arrays of two-dimensional islands in submonolayer heteroepitaxy and arrays of three-dimensional coherently strained islands. Transitions between different types of vertical correlation in multi-sheet arrays of islands are discussed as a nanoengineering tool.

The review by Politi et al. [1.33] published in 2000 discusses growth instabilities in molecular beam epitaxy. The main focus is on highly non-equilibrium processes in which the advancing planar surface becomes unstable against surface undulations. Various aspects of elastic interactions, e.g., adatom–adatom, adatom–step, and step–step interactions, and their effect on growth instability are addressed. The formation of ordered nanometer-scale structures and ways of tuning their parameters are not discussed.

The review by Krestnikov et al. [1.34] published in 2001 focuses on submonolayer islands in semiconductors. The emphasis is on demonstrating that, in a wide class of semiconductor systems, these islands are indeed quantum dots, allowing efficient engineering of their geometrical parameters and electronic spectra, and well-suited to laser applications.

The review by Teichert [1.16] published in 2002 discusses self-organization mechanisms mainly in the $Si_{1-x}Ge_x/Si$ system. The improvement of island uniformity in multisheet growth, ripple formation on vicinal surfaces and their self-organization via step bunching, and the interplay between island and ripple formation are the main subjects of the paper. While many effects in lattice-mismatched growth are driven by elastic interactions and are rather general, some key issues like the existence of the equilibrium island volume vs. infinite ripening depend on particular material parameters. Besides that, in optically indirect SiGe/Si materials, it is hardly possible to establish a relationship between structural and optical properties of nano-objects.

The present monograph has only a minor overlap with books and reviews by other authors and largely extends those of our own team. The major focus is given on physical mechanisms for spontaneous formation of surface nanostructures including quantum dot and quantum wire nanostructures. Theoretical concepts underlying the formation of nanostructures are presented in detail. From a vast base of experimental data, key experiments on growth, and structural and optical characterization are discussed which allow one to distinguish between various formation mechanisms and shed light on basic physics. We show how a firm understanding of nanogrowth allows one to combine self-organized phenomena and further engineer nanostructures, fabricating complex systems that should ensure breakthroughs in device applications.

Success in nanophysics and nanotechnology up until 2002 has resulted in QD-based semiconductor diode lasers that surpass conventional quantum-well-based lasers with respect to the major parameters. The efforts of many leading research groups have brought quantum dot nanostructures to a level where they are now beginning to conquer a wide field of industrial applications. The whole of human society is thus entering a new nano-era.

2. Growth and Characterization Techniques

> The days went by, and the wisest little pig's house took shape, brick by brick.
>
> *The Three Little Pigs*

To build nanostructures one has to deal with individual atoms, which are the elementary building blocks for these constructions. Our 'hands' and tools are usually too large for these tiny things. Although some tools have been proposed to manipulate individual atoms brick by brick, as it were, e.g, by using a scanning probe microscopy (SPM) tip [2.1], these approaches seem to be far too expensive, and major success in nanofabrication has been achieved along the main channel of epitaxial growth via self-organization phenomena at surfaces.

A deep understanding of the fundamental physics underlying epitaxial growth techniques is a prerequisite for any significant success in semiconductor nanotechnology. Of great importance is the question of the relative role of kinetics and thermodynamics in modern epitaxial growth techniques like molecular beam epitaxy (MBE) or metalorganic chemical vapor deposition (MOCVD). This issue defines fundamentally different ways of controlling chemical composition, point defect concentration, growth rate, etc., and plays a significant role in optimizing growth parameters to obtain high-quality structures with planar or periodically-modulated interfaces.

There has been a lot of controversy in the past over the possibility of applying thermodynamics to growth and doping in molecular beam epitaxy (see, for example, [2.2] and the references therein). Although these discussions concerning the relative importance of kinetics and thermodynamics were far from over, MBE was very successfully developed on the experimental side in the 1980s, and most of the problems existing in the early stages of this technology were solved empirically. Further success in device applications, particularly for heterostructures with ultrathin layers in GaAs–(Al,Ga)As and (In,Ga,Al)As material systems gave the impression that most of the problems with this technology were either solved, or could be solved without entering too much into the details of growth fundamentals. The dispute concerning the relative importance of thermodynamics and kinetics in MBE was then considered as having a solely academic character.

The situation changed dramatically in the 1990s. To begin with, the range of materials used extensively in MBE growth was dramatically expanded, and there was an explosion of interest in spontaneous formation of ordered nanostructures on crystal surfaces. Approaches to describing self-organization of nanostructures on crystal surfaces can be roughly divided into two groups: kinetic models (e.g., [2.3–2.13]) and thermodynamic models (e.g., [2.14–2.27]).

Thermodynamic approaches are generally based on the assumptions, firstly, that a real system can be approximated as a closed system, and secondly, that all kinetic processes are sufficiently fast to drive the system into its equilibrium (lowest free energy) state. Kinetic approaches, on the contrary, suppose either that approximation by a closed system is inappropriate, or that the equilibrium state cannot be reached on the time scale of real experiments. Then all the main properties of the system, such as concentrations and types of point defects, geometrical size and shape of spontaneously formed nanoislands, etc., are defined by particular kinetic pathways. This separation is, however, rather oversimplified. Firstly, any realistic kinetic model should lead to the same result as thermodynamics, if the growth rate is extrapolated to zero, or a long growth interruption or annealing is introduced. Secondly, the characteristic rates of various elementary processes can differ by a few orders of magnitude, so that we may consider different types of partial equilibrium governed by constraint thermodynamics. (This point will be addressed in a more detailed way in Chap. 3.) In the general case, knowledge of the thermodynamically favorable state is highly important for constructing a proper kinetic pathway. From a different point of view, even if the equilibrium state can be reached by the system under given conditions, one may intentionally increase the growth rate, reduce the substrate temperature, and moderate fluxes, i.e., enhance the 'kinetic' component of the process, to achieve more flexibility in the system.

Increased interest in self-organized growth of nanostructures has stimulated the always existing interest in the interplay between kinetic and thermodynamic effects in MBE growth. Section 2.1 presents a review of a unified thermodynamic model of major growth-related effects in MBE, such as condensation or evaporation of the main elements and impurities, surface segregation of the more volatile amongst the main elements, segregation of impurities, and so on.

Section 2.2 gives a brief description of the MOCVD growth technique, and various general aspects of MBE and MOCVD are compared.

Section 2.3 presents a brief overview of the main experimental methods used in structural and optical studies of nanostructures. The discussion of scanning tunneling micrsocopy (STM), atomic force microscopy (AFM), transmission electron microscopy (TEM), reflection high energy electron diffraction (RHEED), and others, focuses mainly on the relative advantages

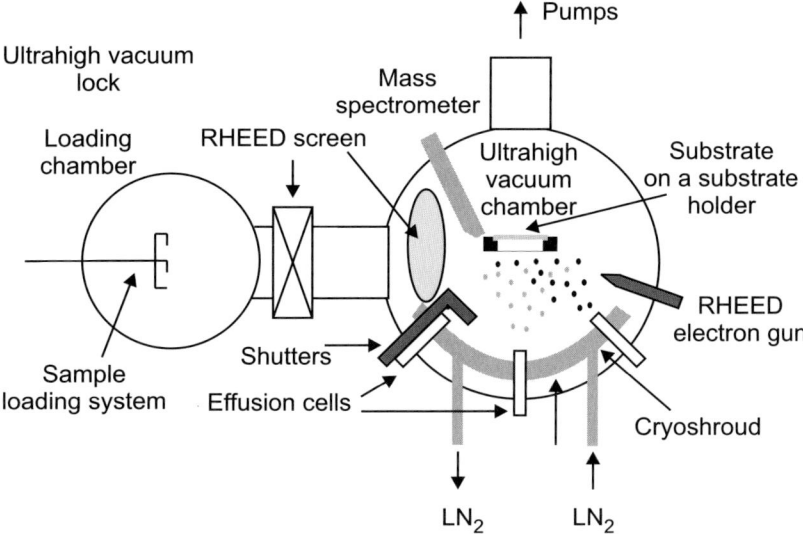

Fig. 2.1. Schematic representation of molecular beam epitaxy apparatus

and shortcomings of these techniques and on measurement accuracy. For further details of the techniques, references are given to specialized books.

2.1 Basics of Molecular Beam Epitaxy

2.1.1 MBE Apparatus

Molecular beam epitaxy (MBE) is a technique for the epitaxial growth of materials. It operates via the chemical interaction of one or several molecular or atomic beams of different intensities and compositions, which occurs on the surface of a heated single-crystalline substrate. A schematic representation of the MBE apparatus is shown in Fig. 2.1

Source materials are placed in evaporation cells composed of a crucible, whose shape and dimensions ensure the required angular distribution of atoms or molecules in the beam, a resistive heater, and thermal screens. The angular distribution of the beam and the distances between the sources and the substrate determine the homogeneity of the parameters of epilayers and heterostructures grown by this technique. A manipulator with a substrate holder is used to ensure the required position of the substrate relative to the cells and to heat it to the required temperature. The homogeneity of the resulting films is often improved by rotating the substrate.

For typical distances between the sources and the substrate, the molecular beam condition is ensured (i.e., the free path of the particle is larger than the geometrical size of the chamber), if the total pressure does not exceed 10^{-4} torr. However, all MBE systems are provided with the means to reach and maintain an ultrahigh vacuum ($\approx 10^{-11}$ torr) and operation is usually oil-free. One of the reasons why MBE systems have to be oil-free is the need to ensure that the substrate is atomically clean before growth. On the other hand, a low level of background doping and control over the properties of the grown materials and structures can be ensured only if uncontrolled fluxes of atoms reaching the substrate surface are as weak as possible. An ultrahigh vacuum is essential for this purpose, but it is not a sufficient condition. Firstly, any vacuum represents an equilibrium between the rate of gas evolution and the rate of pumping, so that construction and crucible materials with the lowest possible rate of gas evolution must be used. The usual crucible material for MBE with III–V compounds is boron nitride, which combines a low rate of gas evolution with a weak chemical activity right up to temperatures of the order of $1\,500°C$. Secondly, it is important to ensure cryogenic screening around the substrate in such a way that it will minimize stray fluxes of atoms and molecules from the walls of the chamber, which are at room temperature, and from the heated components of the apparatus. Not only the total pressure, but also the partial composition of the atmosphere must be monitored and, if necessary, altered. It can be altered, for example, by exposing the substrate surface to a hydrogen flux. A hydrogen flux can be used in order to reduce the partial pressures of the most active components of the background atmosphere and hinder the incorporation of undesirable impurities. Thirdly, only ultra-pure materials can be used as source materials in MBE.

One advantage of MBE technology that has made it very popular among crystal growers and device engineers is the intrinsic feasibility of controlling the profile of the composition and doping of a growing structure at the monolayer level. This feasibility is ensured by a molecular beam regime during growth that excludes any interaction between atoms or molecules in the beam and between different beams, in combination with a relatively low growth rate. An abrupt change in the composition and/or degree and nature of doping are achieved by opening or closing the relevant fluxes using the shutters with which each cell is equipped. The operation time of a shutter (< 0.1 s) is usually considerably less than the time needed to grow one monolayer (typically 1–5 s). Variation of cell temperatures and, consequently, the intensities of molecular fluxes, and the corresponding variation, if necessary, of the growth rate, potentially provide a way of establishing any specified composition and doping profile in the film.

Ultrahigh vacuum conditions and the open growth surface provide extensive opportunities for controlling the technological process at all stages. Preliminary preparation of an atomically clean, defect-free substrate surface is exceptionally important in this technology. As a rule, if non-epi-ready substrates are used, this process includes chemical–mechanical treatment with a polishing etchant, passivation of the surface by oxidation, and removal of a protective oxide film in a vacuum chamber during heating.

The MBE chamber is equipped with a reflection high energy electron diffraction (RHEED) system and mass spectrometers for monitoring the beams, their molecular composition, and the residual atmosphere, and also, ionisation gauges for monitoring the fluxes. A modification of the RHEED technique, which involves the study of oscillations in the intensity of diffraction reflections during growth, makes it possible to monitor not only the reconstruction of the film surface, but also its smoothness at the monolayer level, the surface diffusion length of migrating atoms, and the deposition rate.

2.1.2 Understanding MBE Growth Processes

The early MBE investigations in conventional III–V material systems revealed that at relatively low substrate temperatures all atoms of a group III element have unit sticking coefficient to the substrate surface [2.28]. It has also been found that group V atoms (molecules) do not stick to the surface in the absence of a flux of group III atoms. Therefore, the growth rate is completely governed by the rate of arrival of group III atoms on the substrate surface and the excess group V atoms are desorbed from the surface. In contrast to group III elements arriving on the surface in the form of atoms, group V elements can reach the surface in the form of various molecules. For example, in the case of MBE growth of GaAs, arsenic may arrive on the surface in the form of tetramers As_4 if the beam source is heated metallic arsenic, in the form of dimeric molecules As_2 if the source is crystalline GaAs or a high temperature cracker is used for dissociation of tetrameric molecules, or in the form of As atoms if the source is a high temperature device for dissociating arsine AsH_3. The arsenic molecules reaching the surface become adsorbed, participate in association–dissociation reactions on the surface, interact with Ga atoms to form GaAs, or are desorbed from the surface in the form of one or another molecule.

The details of these processes are actually not very well known at present. When the total flux of arsenic atoms reaching the surface is less than the flux of gallium atoms, droplets of liquid gallium are formed on the surface. The early experiments led to the conclusion that the minimum flux of As_2 molecules (J_{As_2}) needed to maintain growth is $(1/2)J_{Ga}$, i.e., that each As atom from an As_2 molecule is used to form GaAs, whereas studies of beams of As_4 molecules show that at best only half the As atoms interact with gallium to form GaAs and, therefore, the As_2 molecules are more effective in the growth of GaAs by the MBE method [2.29]. On the other hand, other more accurate measurements have shown that the effectiveness of As_2 and As_4 molecules in MBE of InAs, GaAs, and (In,Ga,Al)As is the same [2.30] and exceeds 55% [2.31]. It has also been found that the ratio of the various arsenic species in a flux from the surface of GaAs in the typical MBE range is governed by the substrate temperature and is independent of the type of arsenic molecule incident on the surface. At substrate temperatures T_s below 300°C there are As_4 molecules, whereas at $T_s \geq 400°C$ there are predominantly As_2 molecules [2.29].

During the first stage of growth studies, it has been generally assumed that MBE is a fundamentally nonequilibrium process and that thermodynamic approaches are completely inappropriate, so that information on growth processes can only be obtained by investigating the kinetics of the specific reactions on the surface. Experiments using modulated molecular beams have led to a theory according to which the main role in MBE of III–V compounds is played by elementary adsorption, migration and desorption processes of atoms and molecules [2.29]. However, in the case of binary compounds, the real growth picture is very complex and cannot be analyzed by simple approaches. For example, the probability of jumps of gallium atoms between crystal lattice sites on the surface depends on the number and configuration of bonds with arsenic atoms at each specific site [2.32]. Kinetic models of GaAs MBE usually ignore the processes of atom detachment from islands and surface steps, thermal generation of surface vacancies, concentration of arsenic adatoms on the surface, and desorption of gallium atoms, as well as processes associated with adsorption, segregation, and desorption of impurities, possible reactions between impurities and the main elements, transformation of neutral impurity atoms into ions in the course of their incorporation in the crystal lattice, elastic interactions between steps and facets, etc.

Even simplified models require very complex calculations [2.33]. One should also bear in mind that changes in only one parameter of the system (such as the flux of a group V element) may alter the other parameters (concentrations of adatoms and surface vacancies, density of steps, etc.), so that the experimentally observed transient characteristics in experiments with modulated molecular beams are difficult to interpret. Therefore, in spite of the obvious importance of kinetic models in particular cases for improving our understanding of the growth processes in MBE on the microscopic level, these models have so far been unsuccessful in predicting quantitative dependencies. For example, such models provide only a basic qualitative description of changes in the rates of growth or evaporation of III–V compounds with varying T_s and varying intensity of the flux of group V molecules. Considerable difficulties are encountered in attempts to describe the dependence of the composition of a multicomponent semiconductor alloy on the growth parameters, particularly in the case of compounds containing two group V elements. These kinetic models contain a large number of parameters which can usually only be estimated, so that the authors proposing these models frequently arrive at opposite conclusions. On the whole, kinetic models of composition and growth rate control are well behind the experiments. In view of this situation, there have been many attempts to develop a complete thermodynamic description of MBE [2.34–2.44].

The fact that an MBE process occurs under highly nonequilibrium conditions for the group III element does not justify the conclusion that one cannot use a thermodynamic approach to describe this MBE on the basis of equations of mass action in combination with equations describing mass conservation of the interacting elements. This approach is in fact widely and successfully used to describe many chemical reactions that occur under highly nonequilibrium conditions. In the case of sublimation of binary compounds (which is again a highly nonequilibrium process), the validity of the thermodynamic approach had been confirmed by numerous experimental data well before the appearance of MBE (see, for example, [2.45]). This provided a stimulus for attempts to develop a thermodynamic description of the MBE processes and to utilize the extensive experimental and theoretical data already accumulated in liquid phase epitaxy (LPE) and vapor phase epitaxy (VPE).

At first sight, in the case of MBE, a system cannot be described by thermodynamic representations because its different parts are at different temperatures. However, if we assume that the thermalization times of atoms and molecules reaching the substrate surface are considerably less than the time required to grow one monolayer, then the temperature of the system can be assumed to be that of the substrate. The validity of this assumption is confirmed by the fact that the fluxes of atoms and molecules from the substrate are at the temperature of the substrate, irrespective of the temperature of the fluxes arriving at the surface [2.29]. Besides that, the nature of the arsenic molecules in the flux coming from the surface is independent of the nature of the arsenic molecules reaching the surface [2.29]. The comparable efficiencies of different arsenic molecules used in MBE also point to the quasi-equilibrium nature of the growth process. Furthermore, it has also been shown that the best quality structures, such as nearly ideal quantum wells (QWs), modulation-doped structures with ultrahigh electron mobility (10^7 $cm^2V^{-1}s^{-1}$) [2.46], single QW heterostructure lasers with threshold current density as low as 43 A/cm^2 [2.47] have been obtained using As_4 beams. High quality GaAs layers have reportedly been grown using As_4 beams at temperatures as low as 200°C [2.48], which means that the interaction of As_4 molecules with GaAs substrate is not significantly hindered, even at this temperature. So one can assume that the kinetic limitations for arsenic tetramer decomposition are not very significant, at least in the temperature range of interest.

We consider an equilibrium in a system comprising the gaseous phase and the surrounding volume, the temperature of which is governed by the substrate temperature, and the equilibrium partial pressures are those pressures which represent the fluxes of atoms and molecules leaving the substrate surface. A similar approach has been used in a thermodynamic description of the process of sublimation of binary compounds [2.45]. In view of the fundamental importance of the thermodynamic approach, we shall consider the most important aspects of the thermodynamic description and their interconnection with the main growth-related phenomena.

The key points to be handled by the model are:

- How do MBE growth conditions relate to those in VPE and LPE?
- Which parameters of the growth process define the stoichiometry of the growing compound?
- How is this stoichiometry related to impurity incorporation processes?
- What are the optimal growth regimes for particular compounds?
- In the case of growth of a multicomponent semiconductor alloy, which parameters govern the composition?

When developing any model, the most important question concerns basic growth parameters which determine the final state of the system. In some kinetic models for MBE GaAs growth these are: the substrate temperature, the flux ratio of impinging As and Ga atoms and the growth rate. In thermodynamic model of [2.35], it is the substrate temperature and the As/Ga flux ratio. In the thermodynamic model [2.39], it is the substrate temperature and the effective As pressure. The meaning of the last parameter will be clear from the following description.

2.1.3 Phase Diagrams

According to the Gibbs phase rule [2.49], the sum of the number of phases ϕ and the number of degrees of freedom ν in the system equals the number of components (K) plus the number of parameters which define the state of the system. In the case when the state of the system can be changed only by temperature T and pressure P, and the volumes of the phases are large enough to neglect the surface or interface energy, one can write

$$\nu = K + 2 - \phi \, . \tag{2.1}$$

Inside the field of solidus, the solid phase GaAs(s) is at equilibrium with the gas phase ($K = 2$, $\phi = 2$), so that $\nu = 2$, i.e., the state of the system can be changed by independent adjustment of two parameters: pressure and temperature.

At the boundary of the field of solidus, there are three phases which are at equilibrium, so that $\nu = 1$, and the state of the system is completely defined by only one parameter, P or T. The latter are strictly related,

$$P = f(T) \, . \tag{2.2}$$

We shall now consider the reactions between the main components in the case of GaAs growth:

$$\mathrm{GaAs(s)} \rightleftharpoons \mathrm{GaAs(g)} + \frac{1}{2}\mathrm{As_2(g)} \, , \tag{2.3a}$$

$$2\mathrm{As_2(g)} \rightleftharpoons \mathrm{As_4} \, , \tag{2.3b}$$

where Ga(g) is a Ga atom in the gas phase, As_2(g) and As_4(g) are arsenic dimeric and tetrameric molecules in the gas phase, respectively, and GaAs(s) is a GaAs molecule in the solid phase. We assume that the GaAs concentration in the solid state is very close to unity. (The maximum deviation approaches 10^{-4} at high temperatures.)

The equilibrium constants are defined as

$$K_i = \exp\left(\frac{\Delta S}{k_B}\right) \exp\left(-\frac{\Delta H}{k_B T}\right) = K_i^0 \exp\left(-\frac{\Delta H}{k_B T}\right), \quad (2.4)$$

where ΔS is the change in entropy associated with the reaction, ΔH is the enthalpy of the reaction, and k_B is the Boltzmann constant.

The equation of mass action for the reaction (2.3a) is

$$\frac{P_{Ga} P_{As_2}^{1/2}}{\alpha_{GaAs}} = K_{GaAs}^* = 2.73 \times 10^{11} \exp\left(-\frac{4.72\,\text{eV}}{k_B T}\right), \quad (2.5)$$

where P_{Ga} and P_{As_2} are the partial equilibrium pressures of gallium and arsenic at the substrate surface (in atmospheres), α_{GaAs} is the activity of GaAs in the solid phase, equal to unity for a binary compound, K_{GaAs}^* is the inverse equilibrium constant [2.39], and the value of $k_B T$ is given in electronvolts.

The Ga pressure is maximum over the Ga liquidus of GaAs and corresponds approximately to the Ga pressure over pure Ga at moderate temperatures. When we shift to the As-rich boundary of the field of solidus, the arsenic pressure increases, and according to (2.5), the equilibrium Ga pressure decreases. The total pressure equals

$$P^T = P_{Ga} + P_{As_2} = \frac{K_{GaAs}^*}{P_{As_2}^{1/2}} + P_{As_2}. \quad (2.6)$$

At temperature T_1, there exists a minimum in the total pressure for some particular stoichiometry of the solid phase. The vapor pressure corresponding to this minimum can be derived from the condition $dP^T/dP_{As_2} = dP^T/dP_{Ga} = 0$, which results in

$$P_{As_2} = \frac{1}{2} P_{Ga} = \left(\frac{K_{GaAs}^{*2}}{4}\right)^{1/3}. \quad (2.7)$$

In the case of free sublimation (i.e., sublimation in vacuum), this condition corresponds to congruent decomposition of the material. For example, if the GaAs substrate is heated in vacuum in this regime, no Ga droplets will appear on the substrate surface.

When the substrate temperature is increased further, the Ga and As pressure over the Ga liquidus both increase. At the same time the As pressure increases faster with temperature, and at some temperature $T_2 > T_1$, the arsenic pressure becomes larger than the Ga pressure in the whole field of solidus. In practice, this means that, if the sublimation temperature is below

T_2, the state of the system upon evaporation is described by the condition of minimum total pressure, which corresponds to the case of congruent sublimation ($P_{\text{As}_2}^{\text{subl}} = (1/2) P_{\text{Ga}}^{\text{subl}}$). If the substrate temperature exceeds T_2, the minimum in the total pressure disappears, and the material decomposes in a non-congruent way, when precipitates of the other phase (Ga) are formed on the surface. The critical temperature, which separates these two regimes can be derived from (2.1) and (2.5) and is called the temperature of maximum sublimation. For GaAs this temperature is equal to 630°C, and it is risky to anneal the substrate at higher temperatures in view of the surface morphology degradation.

If one applies an arsenic flux and creates an externally defined arsenic pressure $P_{\text{As}_2}^{\text{ext}}$ over the GaAs surface in such a way that $P_{\text{As}_2}^{\text{ext}} \gg P_{\text{As}_2}^{\text{s}}$, then $P_{\text{Ga}}^{\text{sl}} = K_{\text{GaAs}}^{*}/(P_{\text{As}_2}^{\text{ext}})^{1/2} \ll P_{\text{Ga}}^{\text{s}}$. This means, that the flux of Ga atoms going from the substrate will be much weaker in this case [2.28].

We now consider growth when the external Ga flux is larger than the flux of evaporating Ga atoms. Let us assume that the flux of group III atoms reaching the substrate surface corresponds to pressure P_{III}^0 and that P_{III} is the equilibrium partial pressure of group III vapor at the surface, so that the growth rate V_{gr} is

$$V_{\text{gr}} = \frac{gW \left(P_{\text{III}}^0 - P_{\text{III}}^0\right)}{\sqrt{2\pi m k_{\text{B}} T}} , \qquad (2.8)$$

where g is the sticking coefficient ($g \approx 1$), m is the mass of a molecule, and W is the volume of a molecule in a growing crystal.

Ignoring the weak temperature dependence described by the factor $T^{-1/2}$, we obtain

$$V_{\text{gr}} = t \left(P_{\text{III}}^0 - P_{\text{III}}^0\right) , \qquad (2.9)$$

where t is a constant.

According to [2.39, 2.50], for the chemical reaction of dissociation of the As$_4$ molecule [see (2.3b)], the reaction constant equals

$$P_{\text{As}_4} P_{\text{As}_2}^{-2} = K_{\text{As}} = \frac{27.3 \times 10^3}{T} - 19.8 . \qquad (2.10)$$

Here the temperature T is given in K. The given value of K_{As} is also close to the one reported in [2.51]. In the MBE case, the external arsenic flux fixes the total flux of all arsenic atoms arriving at the surface. Neglecting those As atoms participating in growth, $P_{\text{As}}^{\text{T}} = P_{\text{As}_2} + P_{\text{As}_4}$. The partial pressures of the components can then be derived as

$$P_{\text{As}_2} = \frac{-1 + \sqrt{1 + 4K_{\text{As}} P^{\text{T}}}}{2K_{\text{As}}} , \qquad (2.11\text{a})$$

$$P_{\text{As}_4} = \frac{\left(-1 + \sqrt{1 + 4K_{\text{As}} P^{\text{T}}}\right)^2}{4K_{\text{As}}} . \qquad (2.11\text{b})$$

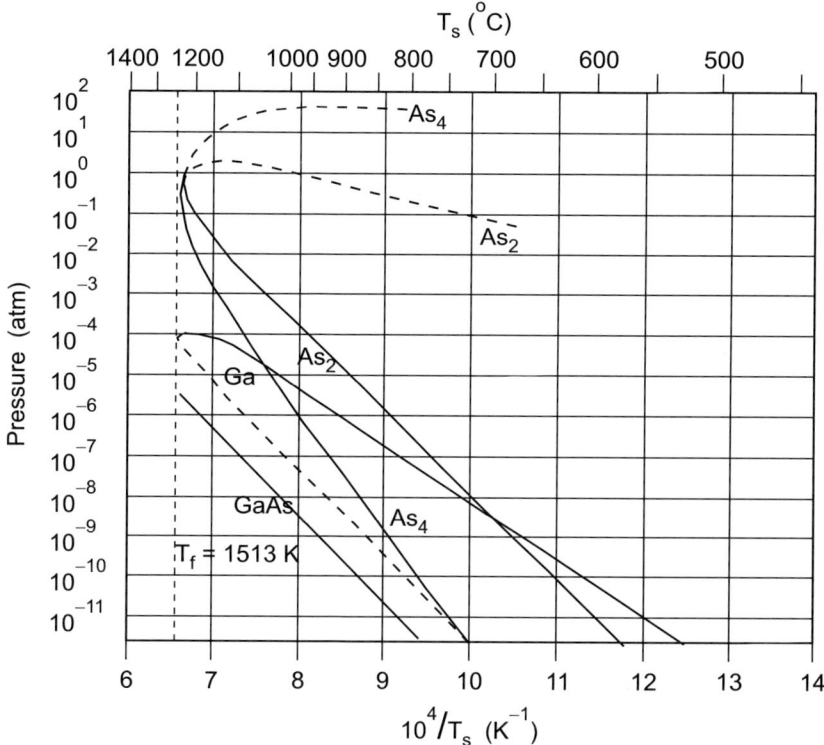

Fig. 2.2. Equilibrium vapor pressures of the components over the liquidus of the Ga–As system

According to thermodynamic predictions, the arsenic tetrameric molecules must dominate in the flux from the substrate surface at temperatures of 300°C and below, irrespective of the type of arsenic molecules impinging on the surface. At temperatures of 450°C and higher, dimeric molecules start to dominate. This behavior was discovered experimentally [2.29].

Figure 2.2 shows the phase diagram of the GaAs–Ga–As system. Along the Ga boundary of the field of solidus, one can see that dimeric molecules dominate at all substrate temperatures. This also follows from Fig. 2.2, which shows the composition of the gaseous phase along the liquidus of the Ga–As system. It is important to note, that in vapor phase epitaxy, even in the case of sufficiently high substrate temperatures, tetrameric molecules can contribute significantly to the total As pressure ($0.04 P^T$ at 800°C, $0.27 P^T$ at 700°C, where $P^T = 10^{-1}$ torr).

From thermodynamic considerations, the efficiency of the arsenic species must be similar if the total flux of arsenic atoms is the same.

In a general case of non-zero growth rate, the mass conservation law reads

$$P_{Ga}^0 - P_{Ga} = P_{As}^0 = 4 P_{As_4} + 2 P_{As_2} \;, \tag{2.12}$$

where P_{Ga}^0 and P_{As}^0 correspond to the arrival rates of arsenic and gallium atoms on the surface, respectively. If the substrate temperature T_s exceeds 400°C and $P_{As}^0 = 10^{-5}$ torr, it follows from (2.10) that $P_{As_2} \gg P_{As_4}$, so that if $P_{As}^0 \gg P_{Ga}^0$, one obtains $P_{As}^0 \approx 2P_{As_2}$, and

$$V_{gr} = t\left(P_{Ga}^0 - P_{Ga}\right) = t\left[P_{Ga} - K_{GaAs}\left(\frac{P_{As}}{2}\right)^{-1/2}\right]. \tag{2.13}$$

This dependence describes well the change in the growth rate of GaAs in molecular beam epitaxy when the substrate temperature is varied [2.37]. The enthalpy of evaporation of GaAs determined experimentally for MBE conditions by studying the reflection high energy electron diffraction (RHEED) intensity oscillations due to layer-by-layer evaporation of GaAs is 4.6 eV [2.52], and the rate of evaporation at a given temperature is inversely proportional to the square root of the arsenic flux to the substrate.

2.1.4 Solid–Liquid–Vapor Equilibrium for Binary Compounds

GaAs. Under Ga-rich conditions, when $P_{As}^0/P_{Ga}^0 \ll 1$, growth occurs close to the appearance of a second phase (liquid gallium), the value of P_{As_2} from (2.12) is governed by the arsenic vapor pressure over the Ga–GaAs liquidus of GaAs [2.50],

$$\left[P_{As_2}^{(Ga_L)}\right]^{1/2} = 9.49 \times 10^5 \exp\left(-\frac{1.98\,\text{eV}}{k_B T}\right). \tag{2.14}$$

Arsenic precipitates do not form on the GaAs surface under typical MBE growth conditions. At typically used substrate temperatures, the equilibrium arsenic pressure over metallic arsenic ranges from a few torr to a few tens of atmospheres, i.e., it is far beyond typical arsenic beam equivalent pressures used in MBE (10^{-6}–10^{-4} torr).

It follows that the growth rate given by the thermodynamic model is mainly governed by the arrival rate of Ga atoms. The excess As atoms that are not used to bind Ga atoms are re-evaporated. However, the excess arsenic flux determines a point on the phase diagram within the homogeneity region, and consequently, the type and concentration of point defects.

The equilibrium constants according to [2.38, 2.39] are given in Table 2.1. The data given in the review article by Kop'ev and Ledentsov [2.39] are summarized from [2.53–2.55]. The interaction parameters in the solid phase according to [2.56] can be found in [2.57].

As has already been noted, at moderate substrate temperatures, III–V materials decompose congruently, i.e., each arsenic atom leaving the surface is accompanied by a gallium atom. However, as the substrate temperature increases, the arsenic flux increases and finally, at some temperature, reaches the value corresponding to $\left(P_{Ga_2}^{[Ga_L]}\right)^{1/2}$. As noted above, this temperature T_{subl} is called the temperature of maximum sublimation [2.45]. A further

Table 2.1. The inverse equilibrium constants $K_{\mathrm{III-V}}$, according to [2.38] (Seki and Koikito SK) and [2.39] (Kop'ev and Ledentsov KL). At low temperatures, III–V materials decompose congruently, so that, for example, each arsenic atom leaving the surface is accompanied by a gallium atom. T_{subl} is the temperature of maximal sublimation. T^*_{subl} is the temperature of non-congruent dissociation of a III–V compound at $P^0_{\mathrm{V}} = 2 \times 10^5$ torr. T^* is the temperature at which the evaporation rate equals 1 ML/s at $P^0_{\mathrm{V}} = 2 \times 10^5$ torr

	$K_{\mathrm{III-V}}$ (SK)	$K_{\mathrm{III-V}}$ (KL)	T_{subl} [°C]	T^*_{subl} [°C]	T^* [°C]
AlP	$1.39 \times 10^{23} e^{-5.85/k_B T}$				
AlAs	$2.90 \times 10^{12} e^{-5.73/k_B T}$	$1.63 \times 10^{10} e^{-5.39/k_B T}$	902	974	900
GaAs	$1.61 \times 10^{10} e^{-4.60/k_B T}$	$2.73 \times 10^{11} e^{-4.72/k_B T}$	630	723	700
GaP	$2.85 \times 10^{10} e^{-4.60/k_B T}$	$2.26 \times 10^{11} e^{-4.71/k_B T}$	571	774	704
InAs	$1.09 \times 10^{11} e^{-4.22/k_B T}$	$7.76 \times 10^{11} e^{-4.34/k_B T}$	508	688	607
InP	$3.76 \times 10^{10} e^{-3.88/k_B T}$	$8.34 \times 10^{11} e^{-4.02/k_B T}$	268	684	613
InSb[a]	$3.76 \times 10^{10} e^{-3.88/k_B T}$				
GaSb[a]	$1.20 \times 10^{11} e^{-4.61/k_B T}$				

[a]For InSb and GaSb the equilibrium species under MBE growth conditions are tetrametric molecules Sb$_4$. Activation energies in the exponentials are given in electronvolts.

increase in the substrate temperature will result in a non-congruent decomposition of GaAs, and Ga droplets will appear on the surface. (T_{subl} is given in Table 2.1 for some of the III–V compounds.) If the substrate is exposed to an external flux of group V elements ($P^0_{\mathrm{V}} = 2 \times 10^5$ torr), the temperature of non-congruent dissociation of a III–V compound can be significantly increased (T^*_{subl} in Table 2.1). The material evaporates congruently and even the total evaporation rate can be rather high. The temperature at which the evaporation rate equals 1 ML/s for this arsenic pressure is given in Table 2.1 as T^*.

InAs. Figure 2.3 shows the partial pressures of As$_2$, As$_4$, In and InAs species over the In–As liquidus [2.55]. The reactions between the main components for InAs growth are

$$\mathrm{InAs(s)} \rightleftharpoons \mathrm{In(g)} + \frac{1}{2}\mathrm{As}_2\mathrm{(g)} \,, \tag{2.15}$$

with equilibrium constant (see also Table 2.1)

$$\frac{P_{\mathrm{In}} P^{1/2}_{\mathrm{As}_2}}{a_{\mathrm{InAs}}} = \frac{1}{K_{\mathrm{InAs}}} = K^*_{\mathrm{InAs}} = 7.76 \times 10^{11} \exp\left(-\frac{4.34\,\mathrm{eV}}{k_B T}\right) \,. \tag{2.16}$$

The indium pressure over the liquidus of the In–InAs(s) system equals

$$P_{\mathrm{In}} = 1.38 \times 10^5 \exp\left(-\frac{2.44\,\mathrm{eV}}{k_B T}\right) \,. \tag{2.17}$$

The arsenic pressure equals

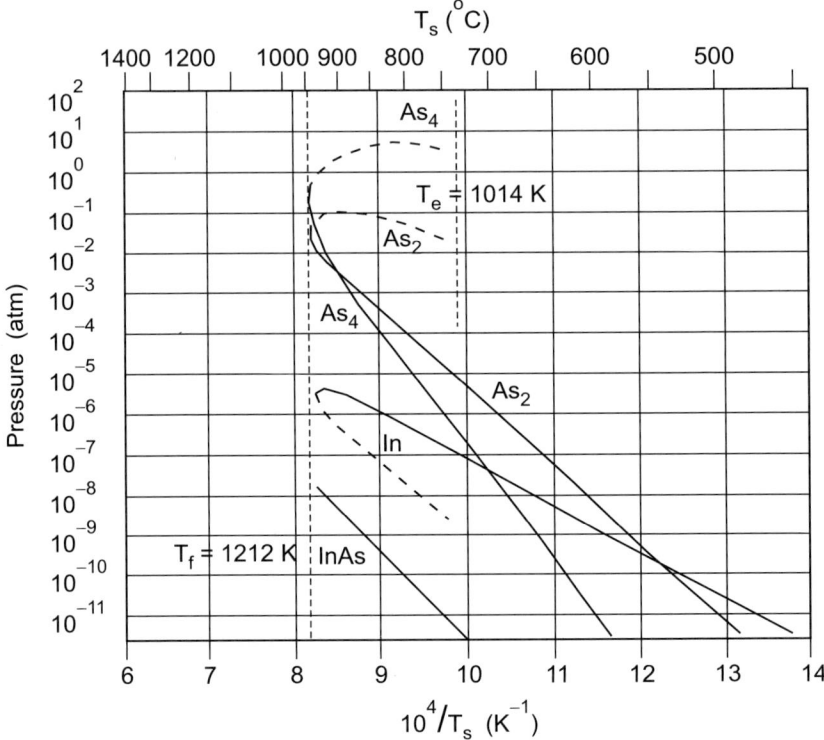

Fig. 2.3. Equilibrium vapor pressures of the components over the liquidus of the In–As system

$$P_{As_2}^{1/2} = 5.62 \times 10^6 \exp\left(-\frac{1.9\,\text{eV}}{k_B T}\right). \qquad (2.18)$$

The temperature of maximum sublimation is 508°C and the arsenic pressure at this temperature is only 1.16×10^{-8} torr. The temperature of non-congruent decomposition of InAs for an external arsenic pressure of 10^{-5} torr is 688°C. The indium pressure approaches 10^{-6} torr (evaporation rate of about 1 monolayer per second) at the same arsenic pressure and 607°C. Thus, for typical growth rates of about one monolayer per second, indium evaporation already plays a role at temperatures of about 550°C. Strain effects in the epilayer can further reduce this temperature.

2.1.5 Solid–Vapor Equilibrium for Ternary Alloys

When a ternary, or pseudobinary semiconductor alloy such as $Ga_x In_{1-x} As$ is grown by MBE, the activities of GaAs and InAs in the solid phase are smaller than unity:

2.1 Basics of Molecular Beam Epitaxy

$$\alpha_{\text{GaAs}} = \gamma_{\text{GaAs}} x = x \exp\left[\frac{\Omega_{\text{InAs-GaAs}}(1-x)^2}{k_B T}\right], \quad (2.19a)$$

$$\alpha_{\text{InAs}} = \gamma_{\text{InAs}}(1-x) = (1-x)\exp\left[\frac{\Omega_{\text{InAs-GaAs}} x^2}{k_B T}\right], \quad (2.19b)$$

where γ_{GaAs} and γ_{InAs} are the activity coefficients of GaAs and InAs in the solid state, respectively, and Ω is the interaction parameter [2.53]. For example, the numerical values of the interaction parameters for common semiconductors are $\Omega_{\text{AlAs-GaAs}} = 0$ and $\Omega_{\text{InAs-GaAs}} = 0.13$ eV.

Combining (2.16) and (2.19a), we obtain

$$\begin{aligned}
P_{\text{In}} &= \gamma_{\text{InAs}}(1-x) K_{\text{InAs}} P_{\text{As}_2}^{1/2} \\
&= \exp\left[\frac{\Omega_{\text{InAs-GaAs}}(1-x)^2}{k_B T}\right](1-x) K_{\text{InAs}} P_{\text{As}_2}^{1/2}.
\end{aligned} \quad (2.20)$$

The characteristic energy of evaporation (enthalpy) of $\Delta H = 4.4$ eV, found experimentally from the temperature dependencies of the rates of evaporation of $\text{Ga}_x\text{In}_{1-x}\text{As}$, $\text{Ga}_x\text{In}_{1-x-y}\text{Al}_y\text{As}$, and $\text{Al}_y\text{In}_{1-y}\text{As}$ under typical MBE conditions in [2.30, 2.58], agrees fairly well with the value 4.34 eV given in Table 2.1. Furthermore, the dependence of the evaporation rate on the growth temperature and on the arsenic pressure [2.59] agree qualitatively with calculations based on (2.20).

For the ternary compound $\text{Al}_x\text{Ga}_{1-x}\text{As}$, the interaction parameter vanishes, i.e., $\Omega_{\text{AlAs-GaAs}} = 0$, and the equation for the Ga evaporation rate takes the simple form

$$P_{\text{Ga}} = (1-x) K_{\text{GaAs}} P_{\text{As}_2}^{-1/2}. \quad (2.21)$$

This leads to the fact that, at the same values of T_s and P_{As_2}, the gallium evaporation rate is higher for GaAs MBE growth than for $\text{Al}_x\text{Ga}_{1-x}\text{As}$ growth, in agreement with experimental results [2.52].

Synthesis of a ternary compound containing two group V elements (e.g., $\text{GaAs}_x\text{P}_{1-x}$) is characterized by the mass action equations

$$P_{\text{Ga}} P_{\text{As}_2}^{1/2} = \gamma_{\text{GaAs}} K_{\text{GaAs}} x, \quad (2.22a)$$

$$P_{\text{Ga}} P_{\text{P}_2}^{1/2} = \gamma_{\text{GaP}} K_{\text{GaP}}(1-x), \quad (2.22b)$$

$$K_{\text{As}} = P_{\text{As}_4} P_{\text{As}_2}^{-2}, \quad (2.22c)$$

$$K_{\text{P}} = P_{\text{P}_4} P_{\text{P}_2}^{-2}, \quad (2.22d)$$

and the mass conservation law

$$P_{\text{Ga}}^0 - P_{\text{Ga}} = P_{\text{As}}^0 + P_{\text{P}}^0 - (4P_{\text{As}_4} + 2P_{\text{As}_2} + 4P_{\text{P}_4} + 2P_{\text{P}_2}). \quad (2.23)$$

Let us assume that $T_s \geq 500°$ C and $P_{\text{As}}^0, P_{\text{P}}^0 \gg P_{\text{Ga}}^0$, which leads to the following relationships: $P_{\text{As}_2} \gg P_{\text{As}_4}$, $P_{\text{P}_2} \gg P_{\text{P}_4}$, $P_{\text{As}_2} = (1/2) P_{\text{As}}^0$, and

$P_{P_2} = (1/2)P_P^0$, so that one obtains the composition x of the ternary compound in the solid phase as

$$x = \frac{1}{\left(\dfrac{\gamma_{\text{GaAs}} K_{\text{GaAs}}}{\gamma_{\text{GaP}} K_{\text{GaP}}}\right) \left(\dfrac{P_P^0}{P_{\text{As}}^0}\right)^{1/2} + 1} . \qquad (2.24)$$

2.1.6 Segregation Effects

Impurity Segregation. To get an understanding of the nature of an impurity segregation process using the thermodynamic approach described in the previous sections, we consider the arsenic equilibrium pressure over the Ga–As–Sn liquid phase, which is at equilibrium with GaAs doped with Sn.

Within the substrate temperature range of interest ($T_s < 800°C$), the arsenic concentration in the liquid phase can be neglected [2.50, 2.60]: $[\text{As}_L] \ll 1$, $[\text{Ga}_L] + [\text{Sn}_L] \ll 1$.

The gallium equilibrium partial pressure over the Ga–Sn–GaAs:Sn liquidus can be written [see (2.19a)]

$$P_{\text{Ga}}^{(\text{Ga-Sn})_L} = \gamma_{\text{Ga}} P_{\text{Ga}}^{\text{Ga}_L} [\text{Ga}_L] , \qquad (2.25)$$

where γ_{Ga} is the Ga activity coefficient in the liquid phase. Calculations using the data provided by [2.53] for $T_s = 700°C$ and an Sn concentration in GaAs of $\approx 10^{18}$ cm^{-3} ($[\text{Sn}_L] \approx 0.8$, $[\text{As}_L] \approx 0.015$, $[\text{Ga}_L] \approx 0.2$ [2.50, 2.60]) give the value $\gamma_{\text{Ga}} = 1.04$. Further, taking into account the fact that $\gamma_{\text{Ga}} \approx 1$, we obtain the following expression for the arsenic equilibrium pressure:

$$P_{\text{As}_2}^{(\text{Ga-Sn})_L} = P_{\text{As}_2}^{\text{Ga}_2} (1 - [\text{Sn}_L])^2 . \qquad (2.26)$$

Figure 2.4 represents the schematic temperature dependence of the arsenic equilibrium partial pressure $P_{\text{As}_2}^{(\text{Ga-Sn})_L}$ over the Ga–As–Sn liquid phase, which is at equilibrium with GaAs doped with Sn grown from an Sn solution with tin concentration [Sn].

This is opposite to the growth of the binary undoped compound, where an As overpressure in excess of $P_{\text{As}_2}^{(\text{Ga})_L}$ automatically results in transition to the field of solidus, where only gaseous and solid phases coexist. In the case of the doped binary, a situation where the liquid phase coexists with the solid and vapor phases is also possible. In practice, this may lead to the formation of a segregated impurity layer. The preconditions for such a process are: higher arrival rate of impurity atoms as compared to the impurity evaporation rate over the impurity liquid phase, and high enough diffusion coefficients to enable significant impurity mass transfer during the growth of one monolayer.

The equilibrium arsenic vapor pressure over the GaAs:Sn solidus as a function of tin concentration in the melt is shown schematically in Fig. 2.4. Liquidus isotherms in the Ga–As–Sn system are shown in the insert. As the As

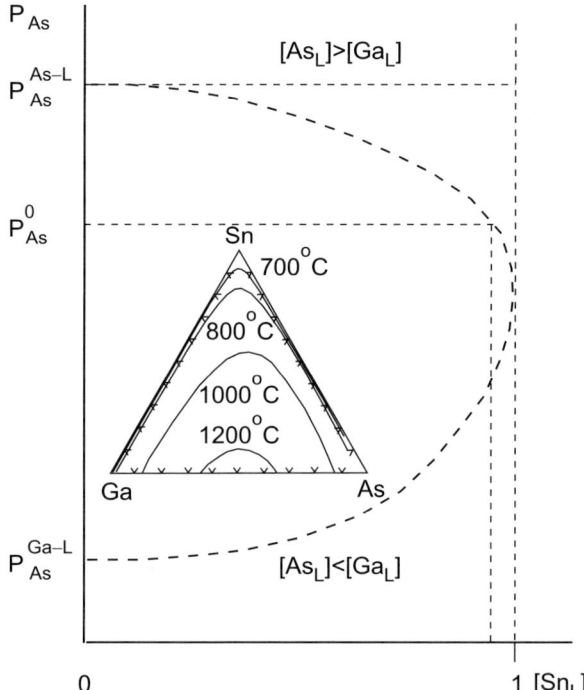

Fig. 2.4. Schematic representation of the equilibrium arsenic vapor pressure over the GaAs:Sn solidus as a function of the tin concentration in the melt. Liquidus isotherms in the Ga–As–Sn system are shown in the *insert*

solubility limits are high for the As–Sn melt, even at moderate temperatures, practically any reasonable As overpressure in MBE and, even MOCVD, will result in a liquid phase enriched in arsenic, so that formation of the liquid phase (Sn segregation) is possible during the growth.

The In–Ga–As ternary system can be qualitatively compared to the Sn–Ga–As system, providing that In and Ga play the roles of Sn and Ga, respectively. The essential difference is that the doping impurity (Sn in our case) has a solubility limit in the solid phase, i.e., a maximum possible dopant concentration which can be introduced in the semiconductor material (GaAs) at a given growth temperature before impurity precipitation occurs, while the InAs molar concentration can reach unity. Another point is that the maximum arsenic pressure is limited by the As overpressure over the InAs liquidus curve.

There exist two possibilities for InGaAs epitaxial growth, on lattice-matched or lattice-mismatched substrates. In the case of pseudomorphic lattice-mismatched growth, the energy of the solid state increases, resulting in a corresponding increase in the enthalpy of the formation reaction. For InAs–GaAs growth, this enthalpy can be estimated as 0.23 eV. For the

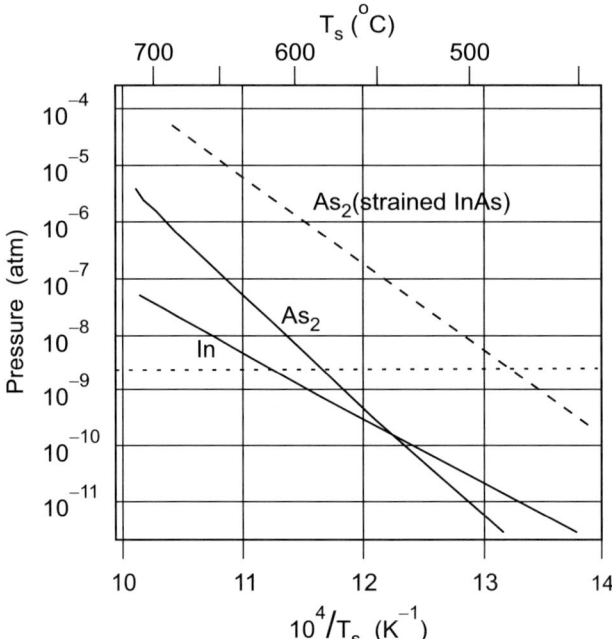

Fig. 2.5. Equilibrium arsenic vapor pressure over the In–Ga–As liquidus for the case of homoepitaxial growth (*solid line*) and for the case of heteroepitaxial growth on a GaAs substrate (*dashed line*). The *dotted line* shows a typical As overpressure for MBE growth

rate equation (2.16), this results in a weaker dependence on the substrate temperature and an increase in K^*_{InAs}. The prefactor governed by the change in entropy is not affected:

$$\frac{P_{\text{In}} P^{1/2}_{\text{As}_2}}{\alpha_{\text{InAs}}} = (K_{\text{InAs}})^{-1} = K^*_{\text{InAs}} = 7.76 \times 10^{11} \exp\left(-\frac{4.11\,\text{eV}}{k_B T}\right). \quad (2.27)$$

As a result, for the same arsenic overpressure and substrate temperature, the indium equilibrium pressure will be higher in the case of strained InAs film growth. In practical situations, this leads to a significant increase in the InAs evaporation rate and a corresponding decrease in the measured In 'sticking coefficient' for the same In input flux.

In the case of the liquidus boundary, the In pressure is fixed by the In pressure over liquid In, as the liquid phase is incompressible. Consequently, the arsenic equilibrium pressure over the liquidus increases. Due to the dimeric nature of the As molecules, the increase in As pressure is much stronger, in fact $\propto \exp(0.46\,\text{eV}/k_B T)$, compared to the increase in the indium equilibrium pressure in the case of As-stabilized growth. The calculated As pressure over the liquidus of a strained InAs film grown on a GaAs surface is shown in Fig. 2.5.

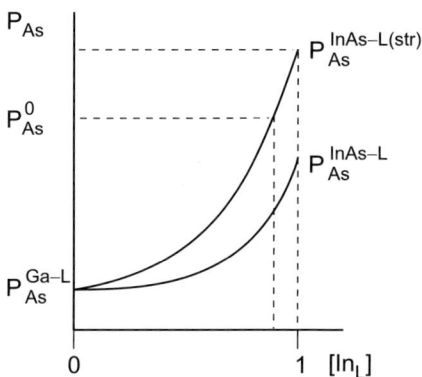

Fig. 2.6. Schematic representation of the equilibrium arsenic vapor pressure over the In–Ga–As liquidus as a function of indium composition in the melt

It follows from Fig. 2.6 that the formation of a strained solid-phase InAs layer on top of GaAs substrates is hardly possible above 500°C. This effect stimulates accumulation of the In liquid phase on the surface of strained InAs or InGaAs heteroepitaxial films, or can favor the formation of 3D elastically relaxed islands of the solid phase, where strain relaxation makes the island crystallization process possible.

2.2 Basics of Metalorganic Chemical Vapor Deposition

Metalorganic chemical vapor deposition (MOCVD), also known as metalorganic vapor phase epitaxy (MOVPE), is another modern growth technique widely applied to grow semiconductor heterostructures including quantum wire and quantum dot (QD) nanostructures. A detailed description of the MOCVD process may be found in the textbook by Stringfellow [2.61]. Here we give a brief survey of the growth technique and compare various general aspects of MBE and MOCVD.

Figure 2.7 shows a schematic view of a typical apparatus used for MOCVD of III–V semiconductor compounds. Particular precursors used in the setup of Fig. 2.7 are trimethylgallium ($Ga(CH_3)_3$, TMG), triethylaluminum ($Al(CH_3)_3$, TMA), and trimethylindium ($In(CH_3)_3$, TMI) for group III elements, and arsine (AsH_3) and phosphine (PH_3) for group V elements. Molecular hydrogen H_2 or molecular nitrogen N_2, and very occasionally Ar, are used as a carrier gas.

If the precursor is in the vapor phase, a defined gas flux mixed with the carrier gas is directed into the reactor. Less volatile liquid or solid precursors are placed in special bubblers, through which the carrier gas flows. These bubblers are, in turn, placed in thermal baths, to stabilize the concentration of the precursors within the carrier gas. The flowing gas, saturated with

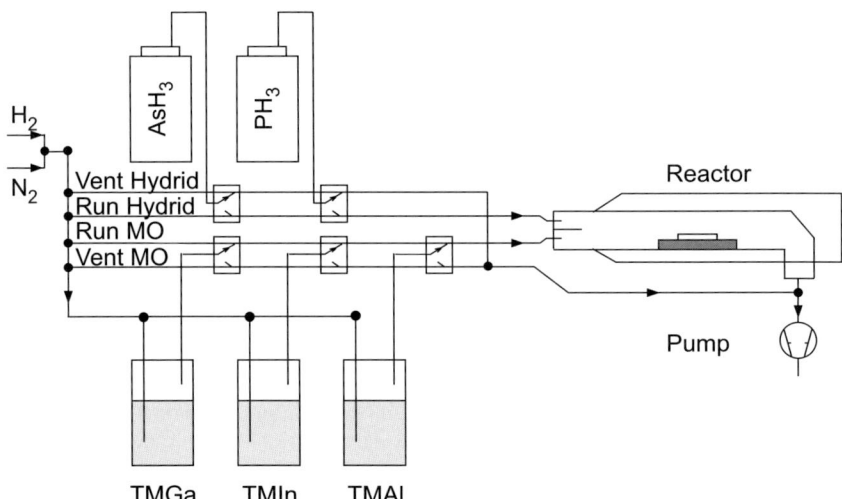

Fig. 2.7. Schematic view of the apparatus for metalorganic chemical vapor deposition (MOCVD)

the metalorganic precursors, flows into the reactor with a well defined flux. The gas flux with group III precursors is only mixed with the gas flux with group V precursors at the reactor entrance, in order to avoid pre-reactions. Additional bubblers (not shown in Fig. 2.7) are used for the precursors of n- and p-doping impurities.

In the reactor shown as an example in Fig. 2.7, the precursors dissolve in the carrier gas and flow with the laminar flux at a typical pressure of 20–100 mbar over the graphite susceptor. The susceptor is subject to heating. The substrate is placed in the middle of the susceptor. The precursors decompose over the susceptor at typical temperatures 450–750°C. Significant oversaturation of the reactants leads to the growth of a crystal over the substrate. In the case of GaAs growth, the basic chemical reaction regarding mass conservation may be written as

$$\text{Ga}(\text{CH}_3)_3 + \text{AsH}_3 \rightleftharpoons \text{GaAs} + 3\text{CH}_4 \; . \tag{2.28}$$

During the growth of heterostructures, the control of growing components is carried out by switches. Switches (not shown in Fig. 2.7 for simplicity) redirect the fluxes between the reactor and the vent, allowing alternating growth of GaAs, AlAs, InAs, GaP, AlP, InP, as well as the growth of alloy films, turning dopant deposition on and off, etc.

Each of the two major modern epitaxial techniques, MBE and MOCVD, has its own advantages and shortcomings. A brief comparison may be summed up in the following list:

- The reflection high energy electron diffraction (RHEED) technique can only be used for in situ monitoring of growth in the ultrahigh vacuum MBE chamber and not in the MOCVD reactor.
- It is easier to control the amount of deposited material at the submonolayer and monolayer level in MBE than in MOCVD, because the diffusion of tiny quantities of MOCVD precursors through the gas phase is essential in MOCVD growth.
- In MBE, growth is possible at lower temperatures (e.g., 200°C), whereas the precursors in MOCVD growth may not decompose at such low temperatures.
- In MOCVD, the growth of GaN and related nitrides is easier, while in MBE growth N_2 molecules do not decompose under typical conditions and require more complicated setups, e.g., plasma source MBE. In MOCVD, NH_3 decomposes reasonably as a precursor above 400°C.
- In MBE growth the partial pressure of components is lower than in MOCVD growth, and this imposes certain limitations at high temperatures. In MOCVD growth, a higher arsenic pressure can be used. This allows the growth of GaAs up to 800°C and above, leading to structures with a lower concentration of defects. It is then much easier to grow samples at high temperatures by MOCVD than by MBE, thus ensuring a higher PL intensity.
- Recharging of crucibles is more difficult in MBE as it involves opening the system and then subsequently reestablishing the ultrahigh vacuum.
- In MOCVD, the reactor suffers from growth on the reactor walls, which therefore require frequent cleaning.
- Servicing of MBE setups may be more expensive, as MBE involves expensive and complicated ultrahigh vacuum systems.
- In MOCVD, higher group III partial pressures can be adjusted to be directly proportional to the growth rate in the transport-limited growth regime. In addition, higher substrate temperatures also lead to higher growth rates, suitable for the mass production of simple structures.

2.3 Main Characterization Techniques

Detailed description of characterization techniques is not a goal of the present book. In this section, we only present a brief overview of the main experimental methods used in structural and optical studies of nanostructures. References to specialized books are given.

Primary methods for structural characterization can be divided into:

- direct imaging methods such as scanning tunneling microscopy (STM) [2.62, 2.63] and atomic force microscopy (AFM),
- transmission electron microscopy (TEM) [2.64–2.70] combining properties of direct imaging and diffraction,

- diffraction methods such as reflection high energy electron diffraction (RHEED) [2.71, 2.72], its ellipsometric equivalent reflectance anisotropy spectroscopy (RAS) [2.73–2.75], and X-ray diffraction [2.76–2.80].

2.3.1 Direct Imaging Methods

Atomic Force Microscopy. Modern scanning probe techniques, like atomic force microscopy (AFM), or scanning tunneling microscopy (STM) are well suited for imaging surface structures on the nanometer scale, even down to atomic resolution. For a detailed discussion of scanning probe techniques, see the review article by Teichert [2.81] and references therein.

AFM measures the surface morphology described by the height function $z = z(x, y)$. According to the AFM imaging principle, in which a sharp probe is scanned across the surface, a typical scan contains $N \times N$ (or $N_x \times N_y$) equidistant pixels. The surface morphology may be described by the height function $z = z(x, y)$, and the scan assigns to each point (x_i, y_j) a height $z(x_i, y_j) = z(\boldsymbol{r})$ in Cartesian coordinates. In addition to observing individual surface features, AFM is well suited to obtaining statistical information about the surface. Given the measured surface morphology $z = z(x, y)$, the two-dimensional (2D) height–height correlation function may be calculated from

$$C(\boldsymbol{r}) = \left\langle \left(z(\boldsymbol{r}_0 + \boldsymbol{r}) - \langle z \rangle\right)\left(z(\boldsymbol{r}_0) - \langle z \rangle\right) \right\rangle, \tag{2.29}$$

where $\langle \ldots \rangle$ is the average over all possible pairs in the matrix that are separated by $\boldsymbol{r} = x\boldsymbol{e}_x + y\boldsymbol{e}_y$. The z values that are taken into account in (2.29) are the deviations from the average height $\langle z \rangle$. The characteristic feature of the surface morphology is the Fourier transform of the height–height correlation function,

$$\widetilde{S}(\boldsymbol{k}) = F\{C(\boldsymbol{r})\} = \frac{1}{A}\int \mathrm{d}x\,\mathrm{d}y\, \exp\left(-\mathrm{i}\boldsymbol{k}\cdot\boldsymbol{r}\right) C(\boldsymbol{r}), \tag{2.30}$$

where A is the scan area. The squared absolute value of $\widetilde{S}(\boldsymbol{k})$,

$$|\widetilde{S}(\boldsymbol{k})|^2 = |F\{C(\boldsymbol{r})\}|^2, \tag{2.31}$$

is regarded as the power spectral density (PSD) or power spectrum of the surface roughness [2.82, 2.83], with the wave vector $\boldsymbol{k} = k_x\boldsymbol{e}_x + k_y\boldsymbol{e}_y$ denoting the spatial frequency. Detailed formalisms for calculating the power spectrum from digitized scanning probe microscopy are given in [2.84].

As the formation of 3-dimensional strained islands in lattice-mismatched heteroepitaxial systems is frequently accompanied by the formation of tilted facets, measuring of local facet orientation becomes an important tool for characterizing the surface. The existence of the facets and their orientation with respect to the nominal sample surface can be derived by analyzing a 1-dimensional cut through an AFM image, z_{ij}, along certain directions. In order to characterize an ensemble of nanostructures with regard to faceting, a 2D histogram of the local surface orientations may be constructed and analyzed [2.85].

The widespread use of AFM for characterizing nanostructures is due to its relative simplicity compared to other techniques like STM or TEM. In particular, AFM does not require conducting or semiconducting surfaces and it may be used even if the surface is covered by an insulating oxide layer.

One of the well known problems with AFM is the finite size of the probe. A 'dilation' with the probe geometry must therefore be taken into account [2.86], and the AFM scan is normally a convolution of a surface nanoscale object and the probe shape. This effect can be neglected for lateral structure sizes that exceed the tip radius (typically of 10–20 nm) and structure slopes less than half the opening angle of the AFM tip. Hence, for an array of 3D islands on the surface, AFM correctly evaluates the island density and typically overestimates the island lateral size.

A more severe restriction on the AFM technique is related to the fact that the grown structure normally has to be cooled down to room temperature before the AFM scan can be taken. For the growth of InAs/GaAs structures, this means cooling from, say, 480°C down to 25°C. It will be shown below (Sect. 3.3) that the cooling of the InAs/GaAs uncovered structure after the formation of 3D islands, e.g., from 480°C to 420°C, already leads to dramatic changes in the surface morphology: a decrease in the wetting layer thickness due to adatom condensation on islands, increased island density, increased average island volume, and an increase in the height-to-width aspect ratio of the islands. Any InAs/GaAs system that is being cooled down after growth for AFM measurements goes through this temperature region. Besides, indium adatoms are mobile even at 325°C, allowing dramatic redistribution of material over the surface. For example, well-separated large-volume lateral associations of islands form at this temperature. These and other experimental data [2.87–2.89] clearly demonstrate that the cooling of the InAs/GaAs system may cause dramatic changes in island density, volume, and shape.

In view of these results, a question arises for any particular system as to whether the surface morphology does or does not change upon cooling. To overcome this uncertainty, it is important to establish under which conditions the surface morphology after cooling corresponds to the surface morphology during growth. A straightforward experimental option for resolving this question is to investigate how the surface morphology measured in AFM or STM depends on the cooling rate. It is worthwhile performing such a study for any material system. This question is of particular importance for the InAs/GaAs system as dramatic changes have been observed upon cooling. However, to the best of our knowledge, no such systematic studies have yet been reported in the literature. On the other hand, surface modification upon cooling may be used intentionally for fabrication of nanostructures.

Scanning Tunneling Microscopy. Scanning tunneling microscopy (STM) has the advantage that it directly reveals the morphology of a surface on an atomic level and can be used to manipulate surface atoms, e.g., to create lines and figures on the surface [2.1]. With respect to the formation of 3D islands (quantum dots, QDs), STM has proved to be very successful (see, for example, [2.90]). In this method, the tunneling current between the surface and the ultra-sharp metal tip for the same voltage, or the voltage for the same tunneling current, are measured as a function of the tip position. Atomic resolution may be achieved with STM.

In contrast to AFM, STM gives the lateral size of nanostructures to greater accuracy, without overestimates. However, STM may underestimate the height (or depth) of surface features, since the tip cannot go deep without running the risk of destroying the surface or the tip itself. Ultrahigh vacuum conditions are normally required for STM, to avoid possible oxide formation on the surface, which may deteriorate conductivity.

Like AFM, STM allows statistical evaluation of surface morphology, e.g., calculation of the power spectral density of the surface. It should be noted that plan-view scanning tunneling techniques, both AFM and STM, cannot generally distinguish between coherent and dislocated islands, if dislocations do not extend to the surface. Besides that, STM is also taken normally at room temperature and requires the sample to be cooled from the growth temperature down to room temperature. The same problem arises for STM as it does for AFM: under which conditions does the surface morphology after cooling correspond to the surface morphology after growth? This is still an open question.

Problems connected with the cooling of a sample are not encountered if STM cross-section experiments are performed on covered samples [2.91–2.93]. However, one should bear in mind that it is the tunneling current that is measured. This current depends strongly on the Fermi level pinning at the metal tip/semiconductor interface. On the one hand, this means a possibility for the tunneling current spectroscopy to distinguish areas with different Schottky barriers. On the other hand, the analysis is complicated by all the problematics related to the physics of Schottky barrier formation and the shape of electron and hole wavefunctions in 'open' or free-standing nanostructures (see, for example, [2.94]). Thus, for an accurate determination of island shape and size, and the distribution of alloy constituents in the case of alloy-based islands, e.g., $In_xGa_{1-x}As/GaAs$ islands, precise modeling is needed.

An accurate approach to the modeling and interpretation of cross-sectional STM images has recently been performed by Eisele [2.95]. It has been demonstrated that strain relaxation of the buried quantum dot upon cleavage plays a major role as a contrast mechanism. At low bias voltage, it is superposed with an additional contrast due to electronic effects. However, the electronic contrast almost vanishes at high bias voltages, since the number of electronic states to tunnel into or out of the constituent materials, and with

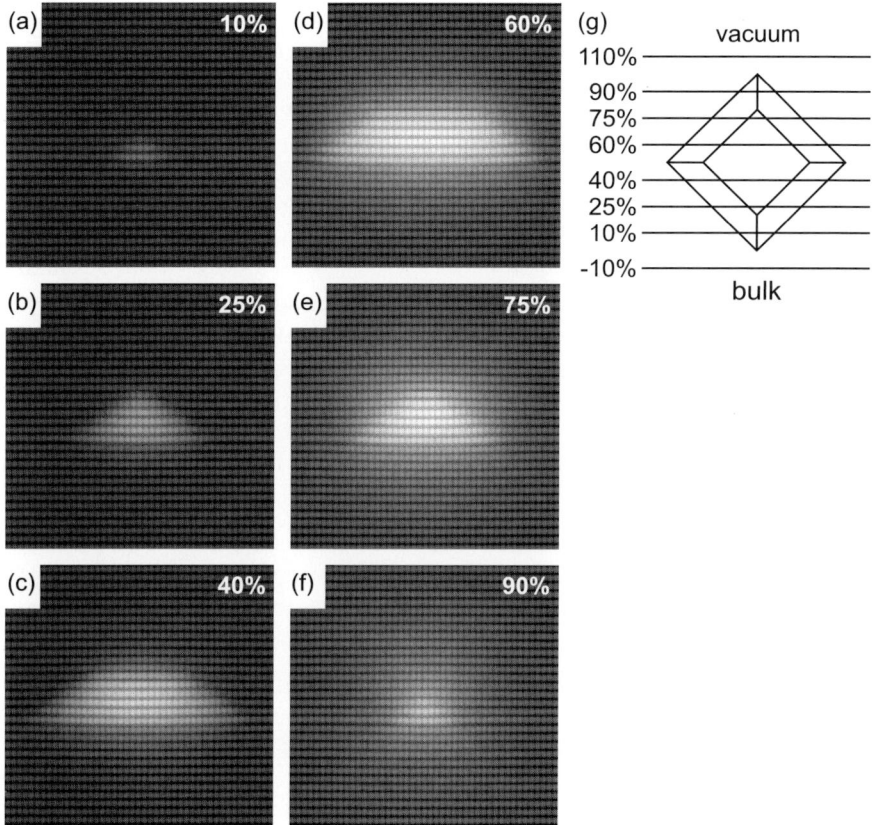

Fig. 2.8. Calculated images of the strain relaxation of a dot with different immersion depths underneath the cleavage surface. (**a**)–(**f**): Immersion depths increase from 10% to 90% of the base diagonal. (**g**) The definition for the immersion depth. With kind permission of H. Eisele [2.95]

it the tunneling probability, then becomes independent of the material. For In(Ga)As/GaAs quantum dots, the optimum bias voltage is $V \geq 3.0$ V.

It has been found that the strain-relaxation-induced contrast dominates for the cases of zero-dimensional structures (quantum dots) and one-dimensional structures (quantum wires), while it is one order of magnitude smaller at the two-dimensional wetting layer. This effect can be assigned to a lacking of opportunity for the quantum dots and wires to relax most of their strain energy in the growth direction, in contrast to the case of quantum wells [2.95].

Figure 2.8 demonstrates a particular example of the modeling. It shows how the strain relaxation depends on the amount of the dot remaining underneath the cleavage surface. The immersion depth d_i describes the position of the cleavage surface relative to the dot, as shown in Fig. 2.8g: $d_i = 0\%$

indicates a dot that is completely cleaved away, while a dot with $d_i = 100\%$ remains completely, but directly underneath the cleavage surface. As can be seen from Fig. 2.8a–c, the cuts below 50%, where the dots are mostly cleaved away, show a much smaller outward relaxation than the complementary ones in Fig. 2.8d–f, in which more material remains underneath the cleavage surface. Furthermore, a larger tailing of the relaxation into the surrounding GaAs matrix can be observed in Fig. 2.8d–f, while in Fig. 2.8a–c, the outward relaxation occurs mostly within the dot region at the cleavage surface.

With this information about the strain relaxation of dots at different positions relative to the cleavage surface, one is able to determine whether the dots observed in actual cross-sectional STM images remain mostly underneath the cleavage surface or whether they are mostly cleaved away.

To conclude, it has been shown by strain-relaxation simulations [2.95] that the following information is available concerning the dot structure underneath the the cleavage surface:

- The orientation of the side facets can be determined.
- The immersion depth of the dot can also be determined by analyzing the tailing of the strain-relaxation profile into the surrounding host material.
- The stoichiometry at the cleavage surface is directly visible.

These results demonstrate that, in order to obtain reliable results on the shape, size, position, and stoichiometry of quantum dots, such detailed modeling should be integrated into any cross-sectional STM investigations.

2.3.2 Transmission Electron Microscopy

Most of the problems that arise when using AFM and STM techniques can be solved by using transmission electron microscopy (TEM) and, particularly, high resolution transmission electron microscopy (HRTEM), which provide information about covered islands. In most cases, islands are covered by the substrate material forming inclusions of material 2 (quantum dots) in the matrix of material 1.

Formation of a TEM Image. Figure 2.9 shows an electron beam passing through the specimen and the objective lens. The incident electron beam is diffracted in the crystalline specimen. The electron wave propagates towards the objective lens, which generates a diffraction pattern in the back focal plane. The latter can be described as the Fourier transform $\mathcal{F}\Psi(r)$ of the object exit plane wave function $\Psi(r)$. The space vector r is assumed to be 2-dimensional and to lie in a plane perpendicular to the electron beam direction. The position and size of the objective aperture in the back focal plane allows the selection of beams contributing to the image. The propagation of the electron beam from the diffraction plane to the image corresponds to an inverse Fourier transform. Assuming an 'ideal' objective lens, the image

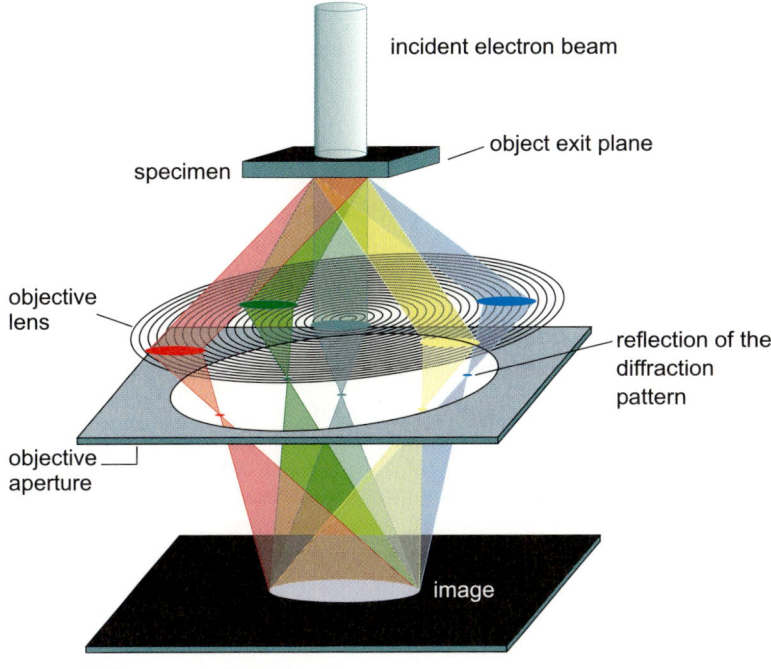

Fig. 2.9. Schematic view of the operating principle of the transmission electron microscope. Each diffracted beam is focused on a spot in the diffraction pattern that occurs in the back focal plane of the objective lens. Beams selected with the objective aperture interfere in the image. With kind permission of A. Rosenauer [2.96]

intensity $|\mathcal{F}^{-\infty}\mathcal{F}\Psi(\boldsymbol{r})|^2 = |\Psi(\boldsymbol{r})|^2$ is given by the intensity of the (magnified) object exit function $\Psi(\boldsymbol{r})$.

The incident electron beam is accelerated up to an energy ΔE of 100–400 keV which is comparable to the electron rest energy ($mc^2 = 511$ keV), so that the electrons in the beam are relativistic. The de Broglie wavelength may then be calculated as

$$\lambda = \frac{2\pi\hbar c}{\sqrt{\Delta E \left(2mc^2 + \Delta E\right)}} \,. \tag{2.32}$$

Equation (2.32) yields $\lambda \approx 3.7 \times 10^{-2}$ Å for an electron beam energy of 100 keV and $\approx 1.6 \times 10^{-2}$ Å for an energy of 400 keV. Thus the de Broglie wavelength of the incident electrons is much smaller than the lattice parameter in the crystal.

Atomic scattering amplitudes calculated for a set of atoms in [2.97, 2.98] show a rapid decrease with the scattering angle for the relevant interval of electron energies. Hence, only scattering at small angles contributes to the TEM image.

An incident electron beam with the wave vector $\boldsymbol{k}$ diffracts at the specimen, giving rise to a set of beams with wave vectors $\boldsymbol{k}'$,

$$\boldsymbol{k}' = \boldsymbol{k} + \boldsymbol{g} , \qquad (2.33)$$

where $\boldsymbol{g}$ is a reciprocal lattice vector of the crystal under investigation. This vector $\boldsymbol{g}_{hkl}$ (where h, k and l are integers) labels different diffracted beams.

By varying the aperture, it is possible to select one or more diffracted beams that will reach the CCD camera of a screen and contribute to the image. In semiconductors having a zinc-blende structure, the main beams are $\{200\}$ and $\{220\}$.

Focusing on these semiconductors, one can calculate the structure factor of a beam $\boldsymbol{g}_{hkl}$. For a binary material AC the structure factor is given by

$$F_\mathrm{S}(\boldsymbol{g}_{hkl}) = 4\left[f_\mathrm{e}^{(\mathrm{A})}(\boldsymbol{g}_{hkl}) + f_\mathrm{e}^{(\mathrm{C})}(\boldsymbol{g}_{hkl}) \exp\frac{2\pi\mathrm{i}(h+k+l)}{4}\right] . \qquad (2.34)$$

In the ternary material $\mathrm{A}_{1-x}\mathrm{B}_x\mathrm{C}$, one may, for example, assume a random distribution of two different kinds of metal atoms (e.g., In and Ga for $\mathrm{Ga}_{1-x}\mathrm{In}_x\mathrm{As}$) in the metal sublattice. The structure factor of the ternary material is then obtained to a good approximation by assuming the atomic scattering amplitude of the metal site to be composed of a linear combination of the scattering amplitudes of the two metal atoms:

$$f_\mathrm{e}^{(\mathrm{A}_{1-x}\mathrm{B}_x)}(\boldsymbol{g}_{hkl}) = (1-x)f_\mathrm{e}^{(\mathrm{A})}(\boldsymbol{g}_{hkl}) + x f_\mathrm{e}^{(\mathrm{B})}(\boldsymbol{g}_{hkl}) . \qquad (2.35)$$

The structure factor of the ternary material is obtained if $f_\mathrm{e}^{(\mathrm{A})}(\boldsymbol{g}_{hkl})$ is replaced by $f_\mathrm{e}^{(\mathrm{A}_{1-x}\mathrm{B}_x)}(\boldsymbol{g}_{hkl})$ in (2.34).

In the TEM study of zinc-blende semiconductors, $\boldsymbol{g}_{200}$ and $\boldsymbol{g}_{220}$ diffracted beams are most frequently used. The structure factor of the $\boldsymbol{g}_{200}$ beam is given by

$$F_\mathrm{S} \propto \left[f_\mathrm{e}^{(\mathrm{A,\,B})}(\boldsymbol{g}_{200}) - f_\mathrm{e}^{(\mathrm{C})}(\boldsymbol{g}_{200})\right] . \qquad (2.36)$$

Cations and anions contribute to $\boldsymbol{g}_{200}$ out of phase, so that the structure factor is proportional to the difference in scattering amplitudes between cations and anions. This difference is particularly small for pure GaAs, because gallium and arsenic have close atomic numbers (31 and 33). The $\{200\}$ beam is thus particularly sensitive to the deviation of the chemical composition from pure GaAs, and this beam is said to be chemically sensitive [2.99].

As regards the $\boldsymbol{g}_{220}$ beam, cations and anions contribute to it in phase. The image is then sensitive to the deviation of the atomic positions from those in an ideal crystal lattice. The $\{220\}$ beam is therefore sensitive to the strain state of the specimen as well as the presence of defects.

Two basic geometries are used in TEM studies of epitaxial structures: plan view and cross-section. Depending on the properties of the incident electron beam, one may distinguish conventional TEM and high resolution TEM (HRTEM). In conventional TEM the energy of electrons in the incident beam is typically 100–200 keV, and the beam has a relatively low degree of coherence. This yields a coarse-scale image with a resolution of a few nanometers. In HRTEM, coherent beams with an energy of 200–400 keV and higher are used to yield atomic resolution.

The shape and size of quantum dots can be obtained to high accuracy, and this technique provides reliable information about the coherent or incoherent nature of insertions in the matrix [2.100, 2.101]. The particular importance of TEM is related to the fact that most applications of quantum dots require covered dots, and the covered dots are the ones whose shape and size should be optimized for device applications.

Modeling of TEM Images in Inhomogeneously Strained Systems. Any inhomogeneous distributions of the chemical composition within a specimen, e.g., quantum wells, wires, or dots, contribute to forming the TEM image. Although TEM is often considered to provide a direct image of the structure, the accurate interpretation of TEM images is rather challenging. Various factors such as aberration by the objective lens and incoherence effects caused by lens current fluctuations or electron energy losses, contribute to the recorded image. Further complications arise due to the phase loss of the electron wave, because only the intensity of the wave can be recorded in the image plane. Apart from rather complicated techniques like electron holography (see, for example, [2.102]), the interpretation of TEM images requires modeling.

In models, distributions of chemical compositions and/or shape and size of insertions within the foil are usually assumed. The strain field is calculated by taking into account the actual geometry of the specimen, including stress-free boundary conditions at the surfaces, etc. The diffraction problem is solved for an electron beam and the model image is calculated. Such calculations are repeated for a set of different model structures, and the actual measured image is compared with model images. The modeling allows, in particular, an optimization of imaging conditions such as foil thickness, defocus, etc. (see, for example, [2.103, 2.104]).

Ruvimov and Scheerschmidt [2.105] have carried out molecular dynamic modelling to check visualization of coherently strained nanometer-scale InAs islands (QDs) embedded in a GaAs matrix. Details of the growth are given in [2.101]. The structure was grown by molecular beam epitaxy. Three monolayers of InAs were deposited at 480°C. A pyramid-shaped InAs QD was used in simulations. It was demonstrated that visualization using conventional TEM is rather complicated owing to the high strain level around the island. Being of pyramidal shape, the InAs island always looks truncated regardless of the usual defocus variation. On the other hand, HRTEM contrast of the island is shown to be rather sensitive to both foil thickness and defoci. Optimum HRTEM imaging conditions (focus and foil thickness less than or roughly equal to twice the base length of the dot) are found to be best suited to revealing the shape of such objects, owing to the difference in structure factors between In and Ga atoms.

Results of simulations coincide with observed TEM and HRTEM images. Plan view TEM indicates that the islands have square base, that the principal axes are close to the two orthogonal $\langle 100 \rangle$ directions, and that the average base length is about (12 ± 1) nm. In cross-section micrographs, the QDs always look like truncated pyramids or lenses, due to specific strain contrast. In the cross-sectional HRTEM image, the pure pyramidal shape of the QDs is fully demonstrated. The height of the pyramid is about 6 nm and the side facets close to $\{101\}$ (see also [2.101]).

Certainly, one should keep in mind that the overgrowth process can in some cases affect the shape, size, and chemical composition of QDs, and measured QDs can in principle differ in shape from free-standing islands on the surface. As regards InAs/GaAs islands that are being overgrown by InAs, intermixing of In and Ga may cause some changes in island size and shape after overgrowth.

Cross-section HRTEM studies of covered InAs/GaAs islands were carried out independently by Xie et al. [2.106] and by Ruvimov et al. [2.101]. Both papers revealed a pyramid-like shape for the QDs. In both papers, the growth temperature was 480°C.

The role of In and Ga intermixing during overgrowth may be estimated by considering experimental results on quantum well structures. The broadening of the compositional profile was estimated for both InGaAs/GaAs quantum well structures [2.107] and InGaAs/GaAs QDs [2.108, 2.109]. These data revealed that the extension of the In profile in samples grown at 500°C exceeds that in samples grown at 450°C by about only 1 nm. We do not therefore expect the capping procedure at temperatures at and below 500°C to significantly affect the shape and volume of the QDs.

On the other hand, overgrowth at higher temperatures (530–580°C) does indeed lead to pronounced modifications in the QD structures. These modifications include the elongation of initially isotropic islands in the [110] direction, a crater-like suppression of the island height in the middle of the islands, and intermixing of Ga and In [2.110].

In order to carefully control such modifications induced by overgrowth, it is especially important to perform structural characterization for both uncovered and covered islands. Consequently, Lian et al. [2.111] controlled the morphology of an InAs/GaAs(001) system by in situ STM and AFM measurements before overgrowth, and capped structures were examined ex situ by TEM and scanning TEM. Comparison of in situ and ex situ results show that overgrowth at 530–580°C results in partial or complete evaporation of InAs islands and a reduction of island density. Incoherent and large coherent islands are shown to evaporate first. This effect can eventually narrow the size distribution of QDs and eliminate incoherent islands, at the expense of reducing the QD density.

To sum up this discussion, we emphasize that problems of strain contrast in conventional TEM can be overcome if HRTEM measurements are combined with HRTEM simulations. The problem of the potential transformation of island morphology during overgrowth can be overcome if islands are completely capped at a moderate temperature, e.g., 480°C for InAs islands on GaAs. Obviously, the temperature of overgrowth can be higher for more thermally stable material systems, such as SiGe/Si.

Digital Analysis of Lattice Images. With high resolution transmission electron microscopy providing atomic resolution images, quantified information can be obtained concerning local and averaged lattice parameters and displacements. Several methods of this kind have been suggested in [2.103, 2.104, 2.112–2.116]. Images presented in this book have been obtained using the technique developed by Rosenauer et al. [2.117] and called digital analysis of lattice images (DALI). A further update of the method may be found in [2.118]. All these methods involve the following four analysis steps, as detailed by Rosenauer et al. [2.96, 2.117]:

1. Noise Reduction. The Wiener filtering technique [2.119] is used in the DALI program. The noise level is locally estimated in the Fourier transformation image $\widetilde{C}$. It consists of the undisturbed signal $\widetilde{S}$ and the noise part $\widetilde{N}$:

$$\widetilde{C} = \widetilde{S} + \widetilde{N} \ . \tag{2.37}$$

   Noise reduction is carried out by applying a filter Φ to the Fourier transformation image $\widetilde{C}$,

$$\widetilde{C}_{\mathrm{Nr}} = \widetilde{C} \times \Phi \ , \tag{2.38}$$

   where $\widetilde{C}_{\mathrm{Nr}}$ is the Fourier transform of the noise-reduced image. The filter Φ is given by

$$\Phi = \begin{cases} \dfrac{|\widetilde{C}|^2 - |\widetilde{N}|^2}{|\widetilde{C}|^2} & \text{if } |\widetilde{C}|^2 > |\widetilde{N}|^2 \ , \\ 0 & \text{otherwise} \ . \end{cases} \tag{2.39}$$

   Other filtering techniques may be found in [2.120].

2. Detection of Lattice Sites and Gridding. This involves:
   - finding the positions of the local brightness maxima $M^{(1)}$;
   - fitting a parabola to the intensity profiles along two lines L_1 and L_2 through $M^{(1)}$, along the x and y directions, to yield a more accurate position $M^{(2)}$;
   - fitting a parabola to the intensity profiles along four lines through $M^{(2)}$, that is, L_1, L_2, and two others, L_3 and L_4 rotated by 45°, then averaging the maximum parabola positions to give a final position $M^{(3)}$;
   - forming grid lines by connecting positions along each of two directions;

- continuously numbering the grid lines, to yield the two-dimensional grid in which a pair of indices is assigned to each position.
3. Calculation of Lattice Base Vectors. The key point is that the lattice base vectors should be deduced in a reference lattice region without deviations from the perfect crystal structure which is located far away from any lattice defects. The reference area is chosen in the image. For each grid line, all lattice positions that lie within the reference area are used to fit a straight line which results in two sets of straight lines. The directions $\hat{a}_1$ and $\hat{a}_2$ of the lattice base vectors are calculated by averaging the gradients of the fitted straight lines. The positions of each grid line formed inside the reference area are projected onto the directions $\hat{a}_1$ and $\hat{a}_2$. The distances between neighboring projected points are averaged for each of the sets 1 and 2 which yields the lengths of the base vectors $\hat{a}_1$ and $\hat{a}_2$.
4. Finally, the lattice base vectors $\hat{a}_1$ and $\hat{a}_2$ are used to calculate the deviations of the actual lattice site positions from the ideal lattice positions. Then the local distribution of strain may be obtained from differences in displacements of neighboring sites.

Figure 2.10 shows an example of a cross-sectional HRTEM image of a multisheet structure of submonolayer CdSe insertions in a ZnSe matrix [2.121]. Figure 2.10a is the image as-taken, whilst Fig. 2.10b shows the local lattice parameter map obtain via the DALI technique.

TEM vs. AFM and STM: Advantages and Shortcomings. None of the above methods for structural characterization, like AFM, STM, and TEM, can be considered as absolute. Each has its own strong and weak points. Here is a brief summary.

- TEM studies overgrown, covered, or capped structures, i.e., those structures that are to be optimized for device applications. At the same time, AFM and plan-view STM deal with uncovered structures.
- AFM and STM studies involve cooling structures from growth temperature down to room temperature. It has been demonstrated by Leon et al. [2.87] and Ledentsov et al. [2.89] that such cooling in the InAs/GaAs system results in severe modifications to the QD volume and shape, the wetting layer thickness, and photoluminescence spectra. In view of these effects, relating observations at room temperature to structures grown at elevated temperature, is quite challenging, particularly for the InAs/GaAs system. To our knowledge, no experimental data has yet been reported in the literature concerning the impact of cooling on morphology in Ge/Si systems.
- Modeling of TEM and HRTEM images is based on the well understood physics of high energy electron diffraction by atoms in a strained crystal. Although complex, it is an achievable task. Modeling of cross-sectional STM requires a detailed understanding of the tunneling current from a structure containing inhomogeneously strained inclusions close to or at the

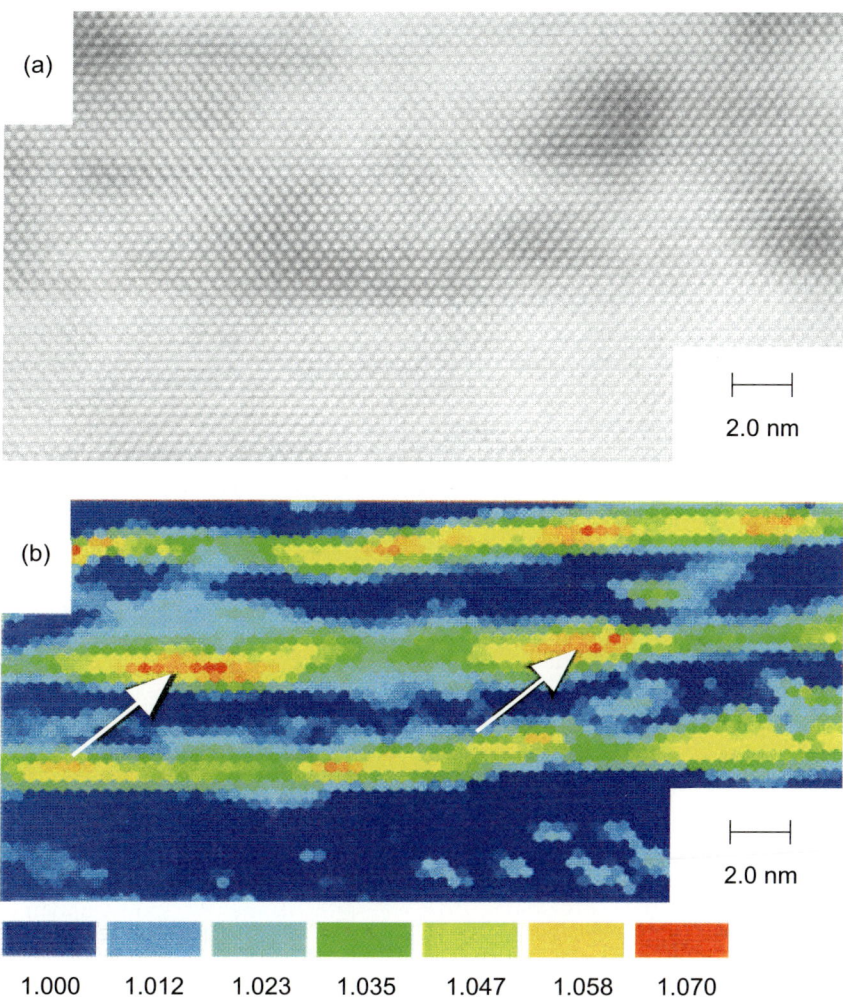

Fig. 2.10. Cross-sectional high resolution transmission electron microscopy (HRTEM) image of a multi-sheet structure of submonolayer CdSe insertions in a ZnSe matrix. (**a**) HRTEM image as-taken. (**b**) Image after processing by the digital analysis of lattice images (DALI) technique. The *color-coded map* represents the local lattice parameter in the vertical direction

surface, of electronic structure near the surface of such systems, etc., which still remains a challenge.
- Overgrowth of quantum dots and quantum wire nanostructures may cause some modifications due to intermixing and segregation effects. However, for InAs/GaAs systems, the problem of possible transformation of island morphology during overgrowth can be overcome if islands are completely capped at a moderate temperature (e.g., 480°C for InAs islands on GaAs).

48   2. Growth and Characterization Techniques

Obviously, the temperature of the overgrowth can be higher for more thermally stable material systems, e.g., for SiGe/Si.
- Cross-sectional HRTEM allows atomic resolution and provides information about the coherent or incoherent nature of inclusions. On the other hand, it typically samples only a small number of islands, which precludes statistical analysis.
- TEM is a much more expensive and time-consuming technique than AFM.

When we compare the advantages and disadvantages of different characterization techniques, research often indicates the complementary nature of TEM, on the one hand, and STM and AFM, on the other. Thus Kamins et al. [2.122] emphasize that AFM provides statistically significant data and accurate height information, but does not provide an unambiguous value for the island diameter, owing to the finite size and shape of the AFM tip. Cross-sectional TEM provides more detailed information about the island diameter and shape, but samples fewer islands. Complementary TEM and AFM data for the same sample make it possible to quantify the apparent enlargement of the island diameter by AFM, so that AFM can provide statistical confirmation of the trends observed by TEM.

2.3.3 Diffraction Methods

Reflection High Energy Electron Diffraction. Reflection high energy electron diffraction (RHEED) is a highly surface-sensitive ultrahigh vacuum technique used to monitor growth in situ in molecular beam epitaxy (MBE) systems. A detailed description may be found in the book [2.123], as well as in the paper by Lagally et al. [2.124].

Briefly, two assumptions are commonly made when analysing RHEED images:

- electrons are scattered elastically from the surface,
- only the topmost atomic layer contributes to scattering.

The condition of elastic scattering read:

$$\boldsymbol{k}_\mathrm{i} + \boldsymbol{q} = \boldsymbol{k}' \,, \tag{2.40a}$$

$$\frac{\hbar^2 k_\mathrm{i}^2}{2m} = \frac{\hbar^2 (k')^2}{2m} \,, \tag{2.40b}$$

where $\boldsymbol{k}_\mathrm{i}$ and $\boldsymbol{k}'$ are initial and final wave vectors of the electrons, $\boldsymbol{q}$ is the scattering wave vector, and m is the mass of a free electron.

If all atoms of the topmost layer are the same, the RHEED signal is proportional to the squared absolute value of the form factor of the surface layer,

$$I(\boldsymbol{q}) \sim \left| \sum_a \exp\left(-\mathrm{i}\boldsymbol{q} \cdot \boldsymbol{R}_a\right) \right|^2 \,, \tag{2.41}$$

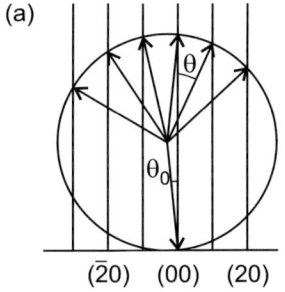 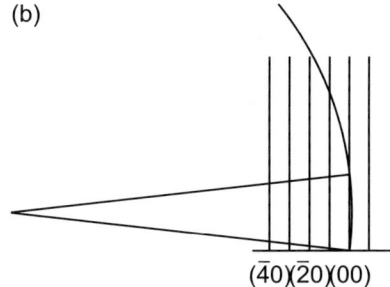

(2̄0) (00) (20)   (4̄0)(2̄0)(00)

Fig. 2.11. Reciprocal lattice for a single plane of atoms and Ewald construction for (**a**) low energy electron diffraction (LEED) and (**b**) reflection high energy electron diffraction (RHEED) geometries. The relationship between rod separations and the lengths of the **k** vectors correspond to a lattice with a 3 Å row spacing, and electron energies of ∼ 150 eV and 12 000 eV, respectively. Note that the widths of the lines for both the Ewald sphere and the reciprocal-lattice rod vastly exceed physically realistic values. With kind permission of M.G. Lagally et al. [2.124]

where the summation is carried out over atoms a in the topmost surface layer, and $\boldsymbol{R}_a$ is the position vector of atom a.

The energy conservation law (2.40b) tells us that the final wave vector of the electron has the same absolute value as the wave vector of the incident electrons. It is then possible to construct the Ewald sphere in the $\boldsymbol{k}$-space with center at the origin and radius $|\boldsymbol{k}_\mathrm{i}|$. In this case, the RHEED signal is non-vanishing only for those values of the scattering vector $\boldsymbol{q}$ for which $\boldsymbol{k}' = \boldsymbol{k}_\mathrm{i} + \boldsymbol{q}$ lies on the Ewald sphere.

Figure 2.11 shows an Ewald construction for both the conventional low energy electron diffraction (LEED) and RHEED geometries, for a crystal consisting of a single plane of atoms arranged in an infinite regular mesh. For such a crystal, the reciprocal lattice consists of a set of infinitely narrow rods normal to the crystal plane and spaced $2\pi/a$, $2\pi/b$ apart, where a and b are the inter-row distances in the x and y directions, respectively.

In the conventional LEED geometry, the Ewald sphere cuts a number of rods nearly parallel to the surface. Such a geometry is favorable for studying 2-dimensional overlayers, because it is possible to measure a number of reflections at varying azimuths with essentially the same cut through the rods. In the RHEED geometry, the Ewald sphere cuts rods at a much more grazing angle. An immediate consequence of this is that the resolution in the plane formed by the exiting wave vector and the surface normal (by the xz plane) is greater, relative to a cut parallel to the surface, by $1/\sin\theta_0$, where θ_0 is the grazing angle the exciting beam makes with the surface [2.124]. Therefore, RHEED is a very sensitive technique for monitoring the surface profile, indicating surface roughness, faceting, islanding, etc.

Figure 2.12 illustrates form factors of several periodic surface configurations. The structure factor of the ideal flat surface in Fig. 2.12a is a 2-

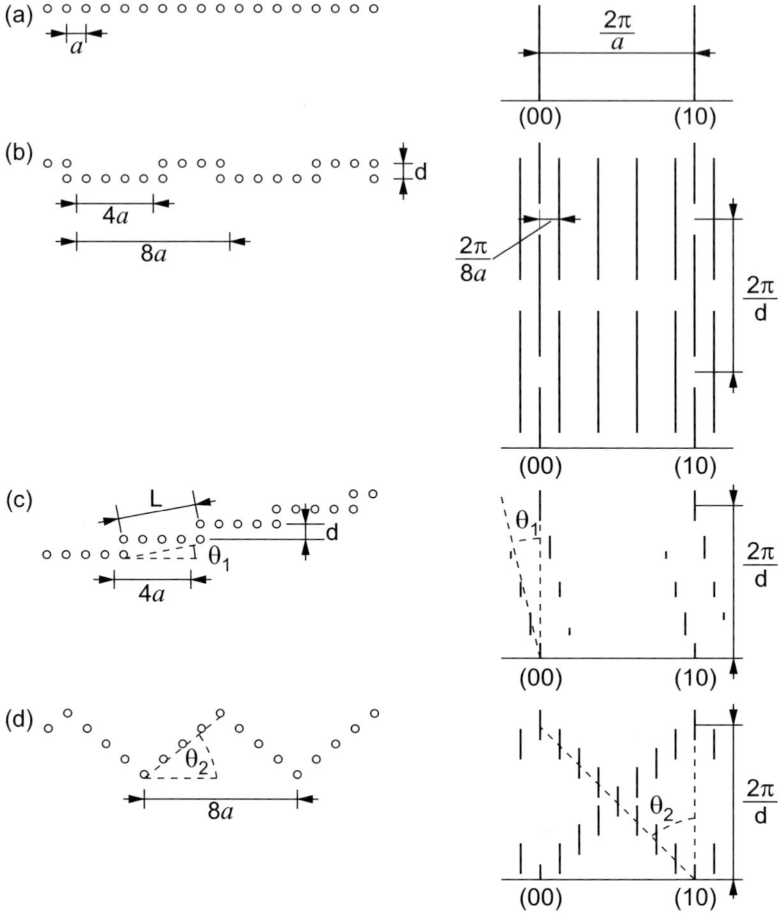

Fig. 2.12. Atomic configurations and structure factors of the topmost surface layer for various perfect surface structures. (**a**) Atomically clean flat surface. (**b**) Two-level surface containing a periodic array of down and up steps. (**c**) Vicinal surface. (**d**) Periodically faceted surface

dimensional reciprocal lattice of infinite vertical rods. In complicated surface patterns, one may find characteristic features that contribute their own structure factors to that of the whole surface. A two-level surface containing a periodic array of down and up steps (Fig. 2.12b) has super-period $8a$ in the lateral direction. The structure factor consists of delta-function rods that are split at periodic positions in G_z defined by a two-level surface. Main rods are separated by $2\pi/a$, and a set of satellite rods appears according to a reciprocal lattice constant $2\pi/(8a)$. Some satellite rods vanish due to destructive interference.

2.3 Main Characterization Techniques

A vicinal surface (Fig. 2.12c) can be considered as a special case of a two-level system. The observed reflections are the product of the terrace structure factor and the reciprocal lattice rods corresponding to the repeat unit L of the vicinal surface. The sequence of vertical rods is tilted with respect to the z-direction, at an angle θ_1 equal to the misorientation angle of the initial vicinal surface.

A periodically faceted surface (Fig. 2.12d) has super-period $8a$ in the lateral direction, and characteristic tilt facets. The structure factor consists of the two sequences of vertical rods each of which is tilted with respect to the z-direction, at angle θ_2 equal to the tilt angle of the facets.

Roughness on the surface results in broadening of the structure factors, and hence of the RHEED patterns. Figure 2.13 illustrates transmission electron diffraction through surface objects.

RHEED is widely used for in situ monitoring of MBE growth, since it allows:

- observation of intensity oscillations at the first stage of layer-by-layer growth;
- transition from layer-by-layer to step-flow growth on vicinal substrates;
- determination of the misorientation angle and step roughness at vicinal surfaces;
- determination of the facet tilt angle on faceted surfaces;
- observation of transitions between 2D and 3D growth in lattice-mismatched epitaxy.

In the latter case, before the onset of 3D islands, the surface is rough, as a high density of flat features form. It is represented schematically by Fig. 2.13d. The onset of 3D islands corresponds rather to Fig. 2.13a, although islands normally have a flatter shape. The 2D → 3D transition in heteroepitaxy is thus monitored in RHEED as a transition from a streaky to a spotty pattern.

Reflectance Anisotropy Spectroscopy. The development and use of reflectance anisotropy spectroscopy has been of particular importance for in situ monitoring of metalorganic chemical vapor deposition (MOCVD). One of the problems associated with MOCVD is the experimental difficulty involved in studying processes on the growing surface. This situation is in contrast to molecular beam epitaxy (MBE), where considerable knowledge about growth processes on the surface has been gained. The main reason is that all the ultrahigh-vacuum-based classical surface science tools (using electrons and ions), especially reflection high energy electron diffraction (RHEED), can be applied in vacuum-based MBE but not under the gas-phase conditions of MOCVD.

This situation has changed in the last decade, since optical surface science tools have been developed. In particular, with linear optical techniques such as reflectance anisotropy spectroscopy (RAS) and spectroscopic ellipsometry, there are now quasi-standard tools at hand to investigate all kinds of pregrowth and growth in MOCVD (and, of course, in MBE).

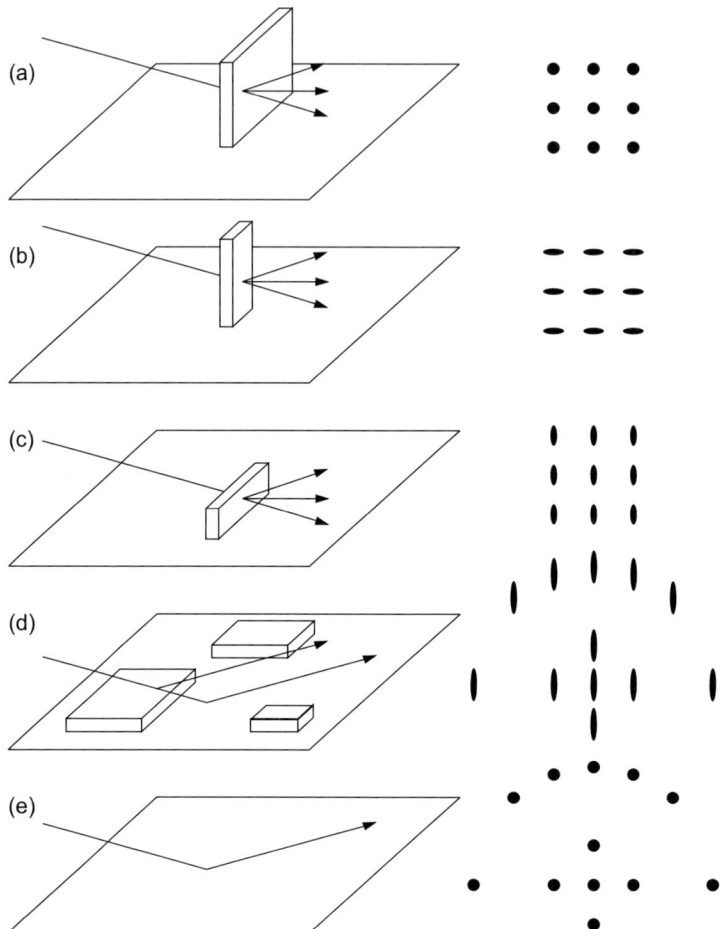

Fig. 2.13. Schematic illustration of transmission electron diffraction through asperities, using glancing incidence geometry. (**a**) A high, wide and thin crystal is sitting on the flat surface. An incident beam illuminates this crystal at nearly normal incidence, giving a sharp transmission pattern. (**b**) If the crystal is made narrow, broadening of the diffracted beams occurs parallel to the surface. (**c**) If the crystal is made less tall, broadening of the diffracted beams occurs normal to the surface (streaking). (**d**) As the density of asperities decreases, the reflection diffraction pattern of a multilevel surface become visible as long streaks centered at positions on Laue circles. (**e**) As the surface becomes flat, the diffraction pattern becomes a series of spots lying on Laue circles. With kind permission of M.G. Lagally et al. [2.124]

A detailed description of these methods is presented in the review article by Richter [2.75] (see also [2.74]). Briefly, reflectance anisotropy spectroscopy is based to a large extent on the high symmetry of bulk materials, like Si and Ge with diamond structure, or III–V and II–VI zinc-blende semiconductors.

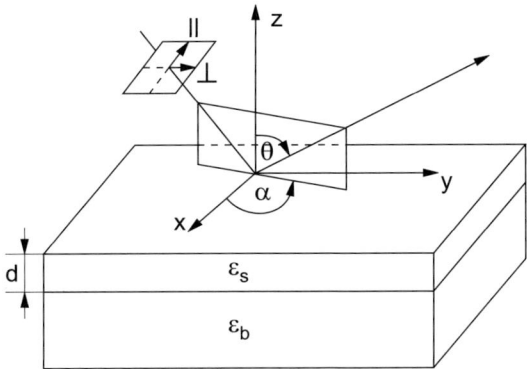

Fig. 2.14. Schematic sketch of a surface reflectance experiment with polarized light. With kind permission of W. Richter [2.75]

The reduction of symmetry at surfaces can be exploited to concentrate the signal generation on the surface. If the surface techniques are applied in high symmetry materials, the bulk does not or only marginally contributes to the total signal.

The scenario for the reflection of polarized light is sketched in Fig. 2.14. The surface layer is modeled by a thin layer of thickness d, which is of the order of a few lattice constants and much smaller than the light wavelength λ. The dielectric constant ε_s is assumed to be anisotropic, while the bulk is described by an isotropic dielectric function ε_b:

$$\varepsilon_s = \begin{pmatrix} \varepsilon_{xx} & 0 & 0 \\ 0 & \varepsilon_{yy} & 0 \\ 0 & 0 & \varepsilon_{zz} \end{pmatrix}, \quad \varepsilon_b = \begin{pmatrix} \varepsilon_b & 0 & 0 \\ 0 & \varepsilon_b & 0 \\ 0 & 0 & \varepsilon_b \end{pmatrix}, \tag{2.42}$$

where x and y are the anisotropic eigenvectors of the surface and z is the surface normal. For a typical example of growth on the (001) surface of zincblende semiconductors, x and y eigenvectors are directed along the $[\bar{1}10]$ and $[110]$ axes. For normal incidence ($\theta = 0$) one obtains the result for the complex reflectance anisotropy with $\delta r = r_{xx} - r_{yy}$ [2.125]:

$$\left.\frac{\Delta \tilde{r}}{\tilde{r}}\right|_{\text{RAS}} = \frac{\Delta r}{r} + i\Delta\Theta = \frac{4\pi i d n_a (\varepsilon_{xx} - \varepsilon_{yy})}{\lambda (\varepsilon_b - \varepsilon_a)}, \tag{2.43}$$

where ε_a, n_a denote the ambient dielectric function and refractive index.

For interpretation and comparison with theory, it is especially useful to eliminate the bulk dielectric function ε_b from this expression, because it introduces an additional spectral dependence. Then, in order to discuss just the desired surface dielectric anisotropy $\Delta\varepsilon = \varepsilon_{xx} - \varepsilon_{yy}$, one needs the dielectric function of the bulk ε_b, which can be obtained from an ellipsometric measurement at oblique incidence [2.75].

A more general description of the method can be found in [2.126–2.128], where experimental details are discussed. A general approach to RAS involves:

- combining RHEED and RAS in an MBE apparatus for the analysis of surface structures,
- assigning particular RAS spectra to particular surface reconstructions and stoichiometry measured by RHEED, and recording RAS 'fingerprints' of various surface reconstructions,
- employing RAS in an MOCVD growth reactor and comparing the observed spectra with the 'fingerprints'.

One should bear in mind, however, that the state of the surface in MOCVD growth differs from that in MBE growth due, for example, to adsorbed hydrogen, etc., which may alter the RAS spectra. However, RAS spectra of GaAs(001) obtained under MBE and MOCVD conditions show a strong correspondence [2.75], which allows us to use reflectance anisotropy spectroscopy as a surface science tool.

RAS thus allows in situ monitoring of MOCVD growth. The method identifies surface reconstructions [2.129], oscillations due to monolayer growth [2.130], and the formation of larger objects like 3D islands [2.131]. To conclude, reflectance anisotropy spectroscopy has today become a powerful tool for in situ observation of MOCVD growth.

X-Ray Diffraction. X-ray diffraction techniques are useful for structural investigation after growth. Results for single-sheet arrays of quantum dots [2.80] and multi-sheet arrays [2.132, 2.133] have been reported for InAs/GaAs and Ge/Si systems. The diffraction signal due to dots is rather weak since the dots are much larger (~ 10 nm) than the probing wavelength (~ 0.1 nm).

In particular, X-ray diffraction has proved to be a powerful tool for measuring the alloy composition of InGaAs in the wetting layer [2.80]. The observed interference pattern is caused by the phase shift between diffracted waves from the upper GaAs cap layer and the lower GaAs substrate, the phase shift depending on the strain–thickness product. The latter allows us to determine the In composition in the InGaAs wetting layer.

2.3.4 Optical Methods

For quantum dots based on direct bandgap semiconductors, e.g., InAs QDs in GaAs matrix, photoluminescence (PL) spectroscopy is a powerful additional tool for characterizing QDs and quantum wires.

- The position of the PL line contains information about the volume of QDs and the depth of the confinement potential. An increase in the QD volume and the depth of the confinement potential results in a shift of the PL line towards lower photon energies, or longer wavelengths (the so-called redshift). The broadening of the PL spectrum is related to the width of the distribution of QD volumes.

- If the wavefunction of an exciton is extended in one spatial direction, as in quantum wires or in strongly anisotropic QDs, the PL line is polarized in the same direction. Thus, the polarization of the PL spectrum carries information about the spatial shape of electronic states.
- The loss of coherence and the onset of misfit dislocations create a high concentration of non-radiative recombination centers and significantly reduce the integral intensity of the PL spectrum.

Other optical techniques include photoluminescence excitation (PLE) spectroscopy, absorption measurements, Raman scattering, spatially selective cathodoluminescence (CL) allowing resolution of emission lines from individual QDs, various kinds of time-resolved spectroscopy, and others. A detailed description of optical techniques and their applications to QD heterostructures may be found in an earlier book by our team [2.134].

In the present book, the main focus is on the physics of nanostructure formation, on theoretical models of formation, on experimental verification of formation mechanisms, and on the engineering of complex nanostructures. Experimental results on the structural and optical characterization of nanostructures will be discussed in connection with processes of self-organized formation and with subsequent nanoengineering techniques.

3. Self-Organization Phenomena at Crystal Surfaces

> Erkennest dann der Sterne Lauf,
> Und wenn Natur dich unterweist,
> Dann geht die Seelenkraft dir auf,
> Wie spricht ein Geist zum andern Geist.
>
> Johann Wolfgang von Goethe. *Faust: Der Tragödie erster Teil. Nacht.*
>
>
> Stars' orbits you will know; and bold,
> You learn what nature has to teach;
> Your soul is freed, and you behold
> The spirits' words, the spirits' speech.
>
> Johann Wolfgang von Goethe. *Faust: The First Part of the Tragedy. Night.*
> Translated from the German
> by Walter Kaufmann

The discovery of the spontaneous formation of nanometer-scale structures on crystal surfaces marked a turning point in the physics and technology of low-dimensional (1D and 0D) systems. It is hard to overestimate the significance of these phenomena. Nature indeed offers us great assistance, providing small scale and high quality systems that cannot be fabricated by conventional lithography.

The present chapter focuses on the basic physical principles of spontaneous formation of nanostructures. From a large number of different classes of nanostructures, three are chosen for detailed consideration, as illustrated in Fig. 3.1. These are periodically faceted surfaces (Fig. 3.1a), arrays of two-dimensional islands in heteroepitaxial systems at submonolayer coverage (Fig. 3.1b), and arrays of essentially three-dimensional strained islands in heteroepitaxial systems (Fig. 3.1c). We emphasize two reasons for this choice. Firstly, elastic capillarity effects on surfaces, i.e., effects involving in-

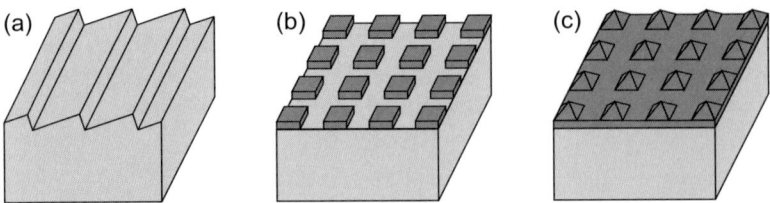

Fig. 3.1. Nanostructures spontaneously formed on crystal surfaces. *Light grey* marks the substrate material, and *dark grey* is used for the deposited material. (**a**) Periodically faceted surface. (**b**) Periodically ordered array of two-dimensional submonolayer islands on a substrate. (**c**) Periodically ordered array of three-dimensional coherently strained islands over a wetting layer

trinsic surface stress, are the driving force for formation of nanostructures in all these classes. Secondly, these are the main types of nanostructure generally employed for fabricating semiconductor quantum wires and quantum dots.

The aim of this chapter is to elucidate the mechanisms underlying the self-organization of the nanostructures in question. The term 'self-organization' is commonly used for the spontaneous formation of spatial, temporal, or spatio-temporal patterns from initially random distributions (see, for example, [3.1, 3.2]). This term was initially applied to kinetically-driven spontaneous ordering in open systems far from equilibrium [3.1].

In the last decade, when spontaneous spatial nanometer-scale ordering in epitaxial systems became a top research area in solid state physics, the term 'self-organization' was widely applied to both kinetically-dominated and thermodynamically-dominated processes. We follow this line throughout our book for the following two reasons.

Firstly, the spontaneous formation of epitaxial nanostructures covers a wide class of physical phenomena from nearly equilibrium phenomena like nanofaceting upon annealing to far-from-equilibrium phenomena like the onset of island formation during deposition. We intend to elucidate the common physics underlying these phenomena.

Secondly, thermodynamics and kinetics play complementary roles to a certain extent. If a system can be approximated as closed and evolving towards thermodynamic equilibrium, its final state can only be reached in experiment if the kinetics is sufficient, e.g., diffusivity of atoms is not too low. On the other hand, the evolution kinetics in a closed system is directed towards its equilibrium state determined by thermodynamics, i.e., towards the state which meets the conditions of the Helmholtz free energy minimum.

Another term, 'self-assembly', is also widely used in the literature. The formation of epitaxial structures, e.g., arrays of strained islands with a narrow size distribution is referred to as self-assembly. If these structures show a tendency to form ordered arrays, this process is called self-organization. In this connection we note that self-assembly can be regarded as partial, or incomplete self-organization.

In this chapter, key theoretical concepts related to self-organization of nanostructures are presented and compared with decisive experimental results. Discussing experimental observations of self-organized nanostructures, we focus on particular experimental designs that can distinguish the relative role of thermodynamic and kinetic effects in nanostructure formation.

It has been emphasized by Madhukar [3.3] that a careful and precise definition must be made every time one discusses thermodynamic vs. kinetic effects. The same goes for the concept of 'equilibrium' in epitaxial systems, particularly in relation to molecular beam epitaxy (MBE). A distinction needs to be made between:

- the equilibrium state of the entire system, the substrate, the growing film, and the vapor,
- equilibrium between the substrate and the growing film,
- thermodynamic equilibrium on a local spatial scale between the growing film and the substrate.

In [3.3] these are referred to as global, partial and local equilibrium, respectively.

Much emphasis in [3.3] is placed on the complexity of MBE growth, which involves a few key conceptual steps: impingement, physisorption (for molecular species), chemisorption (dissociative for molecular species), surface migration, evaporation, and incorporation into the growing film. Characteristic rates and time scales for individual steps can differ by a few orders of magnitude. This allows us to distinguish fast and slow kinetics and thus define partial equilibrium.

In the present chapter, when we discuss the relative role of thermodynamic and kinetic effects, we invoke experiments on either annealing or growth interruption introduced in the growth process. It is assumed that evaporation is negligible. Under these conditions, an epitaxial film on a substrate behaves as a closed system with a fixed number of atoms of every species. In addition, we usually neglect intermixing between the deposited film and the substrate. Under these assumptions, the system evolves towards a constrained equilibrium which may also be called a partial equilibrium according to [3.3].

However, the kinetics of equilibration may be rather slow, and the system may still not reach its final state on an accessible time scale. In this connection, a key question arises for any experiment: if a system is subject to annealing or growth interruption, is the observed state the equilibrium state or just an intermediate state on a slow pathway towards equilibrium? We describe experiments specifically designed to address this question and discuss their results.

In Sect. 3.1, we describe in detail the origin of the surface stress in solids and its role in the formation of periodically faceted surfaces. Among the discussed experimental data, we focus in particular on faceting on the

GaAs(311)A surface. Indeed, this is the first experimental system in which the spontaneous formation of a periodically faceted structure was observed. A second reason is that the interpretation of experimental results for this system has long been controversial in the literature. We also discuss our recent experiments specially designed to prove the formation of nanofaceted structures on GaAs(311) and AlAs(311) surfaces.

The other two sections are devoted to self-organization in heteroepitaxy. In Sect. 3.2 the focus is made on submonolayer heteroepitaxial systems. The first motivation is that such systems are intermediate states in any heteroepitaxial growth. Secondly, they are model objects for studying the effects of configuration entropy. We show that entropy effects govern the temperature dependence of the volume and density of submonolayer islands. On the basis of our results, we formulate an experimental tool that allows us to distinguish arrays of islands whose formation is dominated by thermodynamic effects, and arrays of islands whose formation is mainly kinetically controlled. Concerning experimental data on submonolayer arrays of InAs/GaAs(001) islands, we demonstrate that the formation of these arrays is governed mainly by thermodynamics.

Section 3.3 focuses on arrays of essentially 3-dimensional coherently strained islands which form on a bare substrate in Volmer–Weber growth or on a flat wetting layer of the adsorbate in Stranski–Krastanow growth. We demonstrate that in such a system, due to the dependence of surface energies on strain, an array of 3D islands may correspond to a stable state of the heteroepitaxial system. This explains the formation of arrays of islands having the same shape, a narrow size distribution, and stable against Ostwald ripening. We also discuss kinetic theories of the formation of metastable arrays that are also apparently stable against ripening. Experimental studies of In(Ga)As/GaAs(001) islands are discussed in detail. We present key experimental results on reversible changes in the morphology of the InAs/GaAs system upon variations in arsenic pressure or substrate temperature. These strongly support the quasi-equilibrium picture of island formation in this system.

It would not be possible to cover all major classes of self-organized nanostructures, even in a book. Besides those presented in Fig. 3.1, we list a few others, directing the reader to the original papers or review articles. The following are widely discussed in the literature:

- domain structures on the singular and vicinal Si(001) surfaces (see, for example, [3.4] and references therein);
- faceted structures on vicinal Si(111) surfaces (see, for example, [3.5, 3.6] and references therein);
- self-organization of surface ripples and strained islands during lattice-mismatched growth of $Si_{1-x}Ge_x$ on vicinal Si(001) substrates (see, for example, the review article by Teichert [3.7] and references therein);

- self-organized growth of quantum dots and quantum wires on nonplanar patterned substrates (see, for example, [3.8, 3.9] and references therein);
- the onset of composition modulation during the growth of short-period superlattices (see, for example, [3.10] and references therein).

3.1 Periodically Faceted Surfaces

3.1.1 Equilibrium Crystal Shape: Two Distinct Formulations of the Problem

The phenomenon of equilibrium faceting plays a key role in determining the equilibrium crystal shape (ECS). We emphasize here two distinct problems related to the ECS, corresponding to two different physical situations. The first problem is the ECS of a single crystal, which dates back at least to Wulff's paper of 1901 [3.11], and further developments of the ECS theory can be found in papers by Herring [3.12], Chernov [3.13], and Mullins [3.14].

The exact thermodynamic formulation may be found, for example, in the review by Rottman and Wortis [3.15], who discussed the shape of a single solid inclusion of fixed volume ω in equilibrium with the liquid or vapor phase. The surface free energy of the inclusion is the integral over the surface of ω,

$$F_{\text{surf}}(T,\omega) = \oint_{\partial \omega} \gamma(\widehat{\bm{m}}; T) \mathrm{d}A, \qquad (3.1)$$

where $\gamma(\widehat{\bm{m}}; T)$ is the surface free energy per unit surface area, depending on the orientation $\widehat{\bm{m}}$ of the surface element $\mathrm{d}A$ relative to the crystal axes. The thermodynamics of the ECS states that, at equilibrium, a macroscopic inclusion of fixed volume

$$V(\omega) = \int_{\omega} \mathrm{d}V \qquad (3.2)$$

takes that shape which minimizes the surface free energy (3.1) subject to the constraint (3.2).

A brief summary of ECS theory may be given as follows [3.15]. The orientational dependence of the surface free energy $\gamma(\widehat{\bm{m}}; T)$ is expected to have cusps in symmetry directions leading to facets in the crystal shape at sufficiently low temperatures. These cusps represent discontinuities in the angular derivatives of the surface free energy, the discontinuities being associated with the free energy of the steps on a given facet. With increasing temperature, step free energies decrease, cusps blunt, and the corresponding facets shrink. A given facet finally disappears at the roughening transition temperature T_{R} of the corresponding infinite planar surface. T_{R} is different for different symmetry directions. Above each T_{R}, the corresponding region of the ECS becomes smoothly rounded.

References to experimental data on ECSs in micrometer-scale metal clusters are presented in [3.15]. Transmission electron microscopy (TEM) studies of the equilibrium shape of voids in Si revealed the orientational dependence of the surface free energy of Si [3.16].

Statistical mechanics is the appropriate tool for microscopic evaluation of the orientational dependence of the surface free energy $\gamma(\widehat{\boldsymbol{m}};T)$, and for determining the ECS. An overview of theoretical results may be found in [3.15].

The important issue in ECS theory is that there exist surface orientations which are not present in the crystal shape at a given temperature. At $T = 0$, only a couple of high-symmetry surface orientations are present in the ECS, and all others are passive in the sense that they do not contribute to the ECS. With increasing temperature, the domain of passive orientations shrinks as the crystal becomes more rounded.

The importance of this issue becomes even more striking when one considers the second problem of the ECS. This concerns a crystal which is in equilibrium with the liquid or vapor phase and has fixed volume and all surfaces fixed except for the top one. This formulation of the problem is relevant to the experimental situation in which only the top crystal surface is studied, e.g., thermal annealing of the crystal or growth interruptions introduced in crystal growth experiments.

The top crystal surface is not fixed and is allowed to rearrange into a hill-and-valley structure. The question here is: when can the free energy of a plane surface be lowered by rearranging the atoms into hills and valleys? Since the hills and valleys under consideration are large compared to the lattice parameter, new tilted facets can be defined, and the free energy of the hill-and-valley structure can be written as a surface integral over tilted facets:

$$F_{\text{surf}} = \int \frac{\gamma(\widehat{\boldsymbol{m}};T)}{(\widehat{\boldsymbol{m}}\cdot\widehat{\boldsymbol{n}})}\mathrm{d}A, \qquad (3.3)$$

where $\widehat{\boldsymbol{m}}$ is the coordinate-dependent unit vector locally normal to the surface at each point, and $\widehat{\boldsymbol{n}}$ is the constant unit vector normal to the initially planar surface. The scalar product $(\widehat{\boldsymbol{m}}\cdot\widehat{\boldsymbol{n}})$ in the denominator of the integrand means that the surface element $\mathrm{d}A$ of a tilted facet is normalized per unit projected area of the nominally planar surface.

The fixed side surfaces of the crystal imply that the average normal to the top surface coincides with the normal to the nominally planar surface, i.e.,

$$\frac{1}{A}\int \widehat{\boldsymbol{m}}\,\mathrm{d}A = \widehat{\boldsymbol{n}}, \qquad (3.4)$$

where A is the total area of the nominally planar surface.

The theorem proved exactly by Herring [3.12] reads: "If a given macroscopic surface of a crystal does not coincide in orientation with some portion of the boundary of the equilibrium shape, there will always exist a hill-and-valley structure which has a lower free energy than a flat surface, while if the given surface does occur in the equilibrium shape, no hill-and-valley structure can be more stable."

In the case where the planar surface is unstable, the resulting hill-and-valley structure is determined by the minimum of the surface free energy (3.3), subject to the constraint (3.4). This minimization will yield the orientation of tilted facets as well as the fraction of the nominal planar surface onto which each facet is projected. Microscopic theory based on statistical mechanics can yield the orientational dependence of the surface free energy $\gamma(\widehat{\boldsymbol{m}}; T)$ and thus allow us to determine the ECS. Recent developments in this area have been made by Mukherjee et al. [3.4] for surfaces vicinal to Si(001) and by Williams et al. [3.5] for surfaces vicinal to Si(111).

3.1.2 Faceting: Analogy with Phase Separation

In general, the relationship between the orientational variation of the surface free energy and the surface stability is cumbersome to apply quantitatively. An easier approach is possible for the 2-dimensional case, as described by Williams et al. [3.5]. For a surface with orientation $\widehat{\boldsymbol{n}}_0$, where $\widehat{\boldsymbol{n}}$ is the unit surface normal, the requirements for phase separation to two new orientations $\widehat{\boldsymbol{n}}_a$ and $\widehat{\boldsymbol{n}}_b$, as illustrated in Fig. 3.2, are simply that the net orientation should be conserved and the total surface energy reduced:

$$A_0 \widehat{\boldsymbol{n}}_0 = A_a \widehat{\boldsymbol{n}}_a + A_b \widehat{\boldsymbol{n}}_b \ , \tag{3.5a}$$

$$A_0 \gamma(\widehat{\boldsymbol{n}}_0) > A_a \gamma(\widehat{\boldsymbol{n}}_a) + A_b \gamma(\widehat{\boldsymbol{n}}_b) \ , \tag{3.5b}$$

where A_i is the area of the surface with orientation $\widehat{\boldsymbol{n}}_i$. A straightforward approach to evaluating the stability of a surface is to construct a Wulff plot.

A more easily applied approach to evaluating the conditions for faceting is to define a free energy for which the standard convexity arguments familiar to phase separation in fluids apply [3.13, 3.17–3.19]. To make this analogy with phase separation, one needs to define a 'specific free energy' f, an extensive parameter A' (analogous to volume) related to the surface area, and the corresponding 'density', ρ. The definition of these parameters may be chosen such that the convexity requirement for phase separation in terms of these variables is [3.5]

$$A'_0 = A'_a + A'_b \ , \tag{3.6a}$$

$$A'_0 f(\rho_0) > A'_a f(\rho_a) + A'_b f(\rho_b) \ , \tag{3.6b}$$

$$\frac{A'_b}{A'_a} = \frac{\rho_a - \rho_0}{\rho_0 - \rho_b} \ . \tag{3.6c}$$

The geometric considerations illustrated in Fig. 3.2 immediately show that the definition of the extensive 'area' parameter and the 'specific surface free

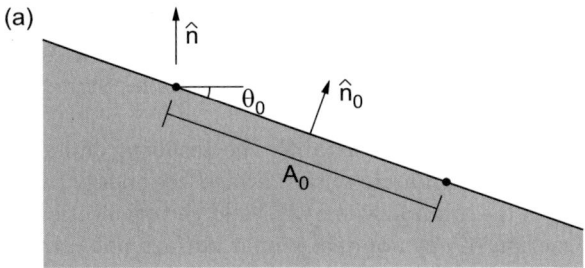
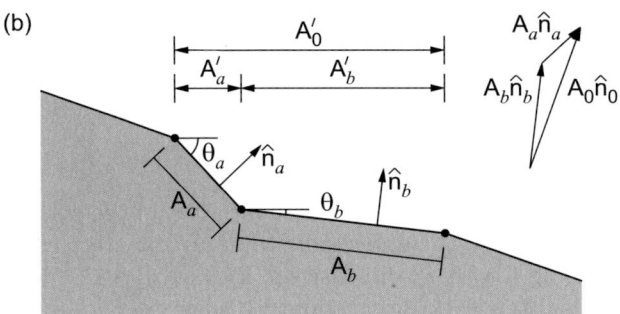

Fig. 3.2. A surface will be unstable with respect to faceting if the total surface energy decreases in going from (**a**) to (**b**). Note that projected areas A'_i are additive (analogous to volume in fluids, and in contrast to total areas). The requirement of conservation of macroscopic orientation, given by (3.5a), is illustrated in the *insert*. With kind permission of E.D. Williams et al. [3.5]

energy' which satisfy (3.6a) and (3.6b) are the projections of the true area and the surface energy onto a reference plane of orientation $\widehat{\boldsymbol{n}}$:

$$A'_i = A_i \widehat{\boldsymbol{n}}_i \cdot \widehat{\boldsymbol{n}} , \tag{3.7a}$$

$$f_i = \frac{\gamma(\widehat{\boldsymbol{n}}_i)}{\widehat{\boldsymbol{n}}_i \cdot \widehat{\boldsymbol{n}}} . \tag{3.7b}$$

We can refer to f as the reduced surface energy and A' as the projected area. Chernov [3.13] noted that the appropriate thermodynamic density should be the step density $\rho = \tan\theta$. If we define the surface orientation in spherical coordinates by defining the polar angle θ, illustrated in Fig. 3.2, and an azimuthal angle φ, we find by manipulating the vector components of (3.5a) that the step density is a vector, and that either of its two components can satisfy (3.6c):

$$\rho^i_x = \tan\theta_i \cos\varphi_i , \tag{3.8a}$$

$$\rho^i_y = \tan\theta_i \sin\varphi_i , \tag{3.8b}$$

if we choose the reference plane as a low index surface for which the polar angle $\theta_{\widehat{\boldsymbol{n}}}$ is zero [such that $\widehat{\boldsymbol{n}}_i \cdot \widehat{\boldsymbol{n}} = \cos\theta$, and (3.7a) and (3.7b) have corre-

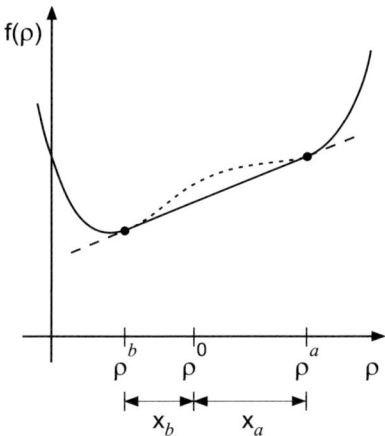

Fig. 3.3. Orientational phase separation occurs when a hill-and-valley structure has a lower total surface energy than a flat surface, as in (3.5b) and (3.6b). This translates into a convexity requirement on the 'reduced surface energy' (which is analogous to a Helmholtz free energy) vs. either component [(3.8a), (3.8b)] of the step density $\widehat{\rho}$. The figure illustrates a non-convex surface energy curve which would lead to faceting. The illustrated curve is schematic only: the form shown would occur physically only if there were attractive interactions between steps strong enough to overcome entropic and elastic interactions. The phase separation is indicated by the tie-bar connecting points a and b. For a macroscopic orientation $\widehat{\rho}_0$, the relative areas of the orientations $\widehat{\rho}_a$ and $\widehat{\rho}_b$ are found from (3.6c), as illustrated in the figure with $x_a/x_b = (\rho_a - \rho_0)/(\rho_0 - \rho_b) = A'_b/A'_a$. With kind permission of E.D. Williams et al. [3.5]

spondingly simple forms]. We may thus apply the convexity arguments to orientational phase separation between two orientations $\widehat{n}_a$ and $\widehat{n}_b$ by considering the reduced surface energy as a function of the vector density defined in (3.8a) and (3.8b), as illustrated in Fig. 3.3. We can construct the tie-bars relating the phases in equilibrium according to the geometric requirement that they be tangent to the reduced surface energy at the two points $\widehat{n}_a$ and $\widehat{n}_b$.

It should be noted that the theory considered above does not give the linear scale of the equilibrium faceted structure. The formation of periodic structures was not therefore discussed in [3.11–3.15]. In order to address the problem of periodically faceted surfaces, we require additional concepts of intrinsic surface stress and capillarity effects on solid surfaces, which are introduced below.

3.1.3 Intrinsic Surface Stress of a Solid

Since atoms in the surface layer of any material are in a different environment to the bulk, the surface layer energetically favors a lattice parameter differing from the bulk value in directions parallel to the surface. Being adjusted to the bulk lattice parameter, the surface layer is intrinsically stretched or compressed. The surface is therefore characterized by intrinsic surface stress.

The intrinsic surface stress of a solid is analogous to the surface tension of a liquid. However, there exists a fundamental difference between the thermodynamic properties of a liquid surface and those of a solid surface, as pointed out long ago by Gibbs [3.20]. An explanation may also be found in papers by Marchenko and Parshin [3.21] and by Needs [3.22]. The basic reason is that any liquid is considered to be incompressible. When a liquid film is stretched, atoms or molecules move out from the bulk to form a new surface which is structurally identical to the existing surface. Thus, an attempt to stretch a liquid film by 1% say, by means of a thought experiment, results in a 1% increase in the number of surface atoms or molecules, whereas the spacing between surface atoms remains unchanged. Thus the processes of creation and deformation of a liquid surface are identical and are described by a single parameter γ, the energy required to create unit area of the surface.

However, when a crystal surface is stretched, the distance between atoms increases and the nature of the surface itself changes. This process is quite different from the creation of a new surface by the cutting of bonds. The energy to create unit area of the surface of a given orientation is characterized by the scalar quantity γ, termed the surface energy, but the energy change due to deformation of the crystal surface is described by the intrinsic surface stress tensor $\tau_{\alpha\beta}$. The concept of intrinsic surface stress tensor was proposed by Gibbs [3.20] and discussed later by Shuttleworth [3.23], Herring [3.24], and Marchenko and Parshin [3.21].

The linear change in the surface energy with respect to the strain may be written as the following integral over the surface:

$$\int \tau_{\alpha\beta}(\widehat{\boldsymbol{m}})\varepsilon_{\alpha\beta}\mathrm{d}A \,, \tag{3.9}$$

where the intrinsic surface stress tensor $\tau_{\alpha\beta}$ has non-vanishing components only in the surface plane, $\alpha, \beta = 1, 2$ [3.21]. The principal values of the intrinsic surface stress tensor can be either positive (tensile) or negative (compressive). A tensile surface stress is associated with the surface which favors contraction while a compressive surface stress favors expansion.

Values of the intrinsic surface stress for solids are known either from first-principles calculations, or from comparison with indirect experimental data on the parameters of surface domain structures (see also Sect. 3.2). No direct experimental method for determining the intrinsic surface stress of a solid has been proposed so far. For most solid surfaces, the intrinsic surface stress is tensile (see, for example, [3.22, 3.25]), whereas a compressive surface stress is known for the Si(001) surface in the direction perpendicular to the dimers [3.26, 3.27]. The order of magnitude of τ is 100 meVÅ^{-2} for both tensile and compressive surface stresses.

At this point it is worth noting the following. Since a liquid surface is characterized by the single quantity γ, which is both the energy for creation of a unit surface area (i.e., the surface energy), and the quantity responsible for capillarity effects of Laplace-pressure type, the term 'surface tension' is widely used for the surface energy γ. For mainly historical reasons, the use of this term has been extended to surfaces and interfaces of solids. However, the use of 'surface tension' for solids turns out to be totally ambiguous, since it may cause confusion between the surface energy γ and the intrinsic surface stress $\tau_{\alpha\beta}$. For this reason, we shall consistently distinguish the 'surface energy' and 'intrinsic surface stress' and will not use the term 'surface tension' for solids.

The change in surface energy due to strain (3.9) indicates the interconnection which exists between surface effects and strain-related phenomena. Within the framework of the linear theory of elasticity, where the bulk strain energy is a quadratic function of the strain, it is possible to expand the surface energy up to second-order terms in strain. The thermodynamics of solid surfaces and interfaces up to second-order terms in strain was studied by Andreev and Kosevich [3.28]. Further developments were made by Nozières and Wolf [3.29] (see also the review article by Kosevich [3.30]). The dependence of the surface or interface free energy on strain may be written as the following expansion:

$$\gamma(\widehat{\boldsymbol{m}};\varepsilon_{\alpha\beta}) = \gamma_0(\widehat{\boldsymbol{m}}) + \tau_{\alpha\beta}(\widehat{\boldsymbol{m}})\varepsilon_{\alpha\beta} \qquad (3.10)$$
$$+ \frac{1}{2}S_{\alpha\beta\varphi\psi}(\widehat{\boldsymbol{m}})\varepsilon_{\alpha\beta}\varepsilon_{\varphi\psi} + \frac{1}{2}h_{\alpha\beta i}(\widehat{\boldsymbol{m}})\varepsilon_{\alpha\beta}\sigma_{ij}m_j + \cdots,$$

where $\widehat{\boldsymbol{m}}$ is the normal to the surface or interface. Greek characters label 2-dimensional indices in the surface plane, whereas Roman indices are 3-dimensional. Quadratic coefficients $S_{\alpha\beta\varphi\psi}$ and $h_{\alpha\beta i}$ correspond to surface excess elastic moduli and can be either positive or negative. The tensor σ_{ij} is the bulk elastic stress tensor, and the fourth term in (3.10) exists on solid–solid interfaces but vanishes on stress-free surfaces where the bulk stress tensor obeys the boundary conditions $\sigma_{ij}m_j = 0$.

Following the conventional approach to elasticity theory [3.31], all surface quantities will hereafter be defined per unit area of the undeformed surface. The dependence of the surface free energy on the strain (3.10) can be interpreted as the strain-induced renormalization of the surface free energy.

3.1.4 Thin Strained Epitaxial Film as a Model of a Surface

A key difference between the elastic properties of a surface and those of the bulk is that the expansion (3.10) of the surface free energy in powers of strain contains a linear term, while the expansion in the bulk starts from the quadratic term. The origin of the linear term is the physical difference between the environment of the surface atoms and that of the bulk atoms. In this section, we introduce a simple model illustrating the effect.

Figure 3.4a shows an epitaxial film on a substrate. For a mismatch ε_0 between film and substrate, the elastic energy is

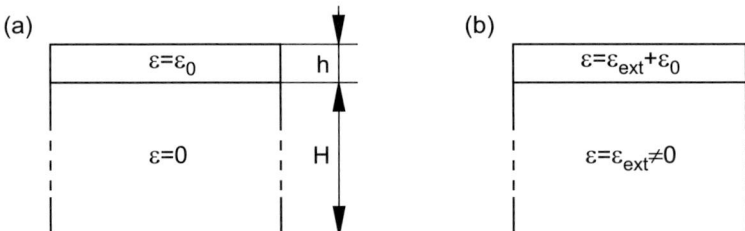

Fig. 3.4. An epitaxial film lattice-mismatched to the substrate as a model for the surface. (**a**) The substrate is unstrained. The film is strained to maintain a coherent conjugation with the substrate. (**b**) The system substrate+film under external strain ε_{ext}. The total strain in the film is $\varepsilon_{\text{ext}} + \varepsilon_0$

$$E^1_{\text{elast}} = \frac{Y_{\text{film}}}{1 - \nu_{\text{film}}} \varepsilon_0^2 h A , \qquad (3.11)$$

where Y_{film} is the Young's modulus of the film, ν_{film} the Poisson ratio, h the film thickness, and A the substrate area. Now let the whole system (substrate and film) be biaxially strained by an external strain $\varepsilon_{xx} = \varepsilon_{yy} = \varepsilon_{\text{ext}}$. The subsequent strain in the film is $\varepsilon_{\text{ext}} + \varepsilon_0$, and the elastic energy of the whole system is

$$E^{(2)}_{\text{elast}} = \frac{Y_{\text{sub}}}{1 - \nu_{\text{sub}}} \varepsilon_{\text{ext}}^2 H A + \frac{Y_{\text{film}}}{1 - \nu_{\text{film}}} (\varepsilon_{\text{ext}} + \varepsilon_0)^2 h A . \qquad (3.12)$$

If we consider the whole system (substrate and film) as a bulk crystal with height $H + h$ and neglect surface effects, we obtain the elastic energy in the form

$$E^{(0)}_{\text{elast}} = \frac{Y_{\text{sub}}}{1 - \nu_{\text{sub}}} \varepsilon_{\text{ext}}^2 (H + h) A . \qquad (3.13)$$

The difference between the two elastic energies (3.12) and (3.13) gives the energy of the film, viz.,

$$E^{(2)}_{\text{elast}} - E^{(0)}_{\text{elast}} = \frac{Y_{\text{film}}}{1 - \nu_{\text{film}}} \varepsilon_0^2 h A + 2 \frac{Y_{\text{film}}}{1 - \nu_{\text{film}}} \varepsilon_0 \varepsilon_{\text{ext}} h A \qquad (3.14)$$
$$+ \left[\frac{Y_{\text{film}}}{1 - \nu_{\text{film}}} - \frac{Y_{\text{sub}}}{1 - \nu_{\text{sub}}} \right] \varepsilon_{\text{ext}}^2 h A .$$

The expansion (3.14) of the film energy in powers of the strain ε_{ext} contains a non-vanishing linear term, the linear coefficient being proportional to the mismatch ε_0 between the film and the substrate. The quadratic coefficient in (3.14) is due to the difference in the Young's modulus and/or the Poisson ratio. The quadratic coefficient has the physical meaning of the excess elastic modulus of the film.

3.1.5 Simple Lattice Model for Intrinsic Surface Stress

To elucidate the origin of the surface stress from a microscopic point of view, we consider here a simple 2-dimensional crystal with square lattice. We model

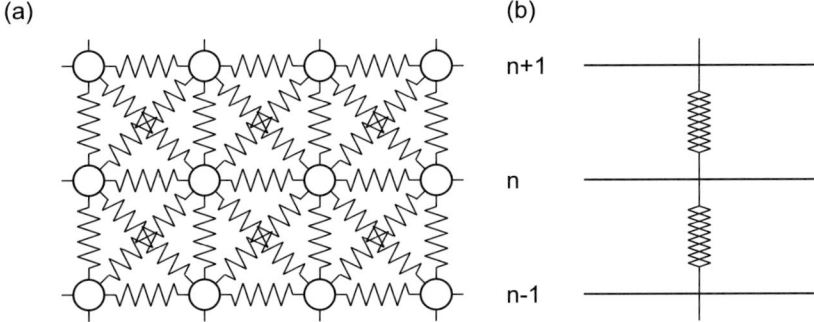

Fig. 3.5. A model 2-dimensional crystal with a square lattice. (**a**) A ball-and-spring model illustrates a model crystal where every atom interacts with its nearest and next-nearest (diagonal) neighbors. (**b**) A vertical chain model of interacting objects. For the unstrained or uniformly strained crystal of (**a**), every atomic layer can be considered as a single object. These layers form a vertical chain where every layer interacts only with nearest neighbors

this crystal as a system of balls connected by springs, as shown in Fig. 3.5. Every atom interacts with its nearest and next-nearest neighbors (diagonal), with potentials $V_1(|r|)$ and $V_2(|r|)$, respectively. Such a model has been considered by Saito et al. [3.32] to study the adatom–adatom, adatom–step, and step–step interactions. Here we apply this model to illustrate the origin of the surface stress. The lattice constant of the square lattice is determined so as to minimize the total energy of the rigid structure,

$$U(a) = 2N \left[V_1(a) + V_2(a) \right] , \tag{3.15}$$

for a system of N atoms. The minimum total energy condition $\mathrm{d}U(a)/\mathrm{d}a = 0$ implies that

$$-V_1'(a) = \sqrt{2} V_2'(\sqrt{2}a) . \tag{3.16}$$

The nearest-neighbor bond energy given by $-J_1 = V_1(a)$ should be attractive to ensure atomic cohesion, whereas the next-nearest neighbor bond energy, $-J_2 = V_2\left(\sqrt{2}a\right)$ can be repulsive or attractive with moderate strength. Equation (3.16) implies that the length of every bond connecting neighboring atoms is not the equilibrium bond length, and the bonds are stresses. Saito et al. [3.32] introduced the term spontaneous stress $\sigma_0 = -V_1'(a) = \sqrt{2} V_2'\left(\sqrt{2}a\right)$. Figure 3.6a depicts the forces acting on a given atom from the nearest and next-nearest neighbors. Equation (3.16) gives the following relation between the forces:

$$|F_1| = |F_2| = |F_3| = |F_4| = \sigma_0 , \tag{3.17a}$$

$$|F_5| = |F_6| = |F_7| = |F_8| = \frac{\sqrt{2}}{2}\sigma_0 . \tag{3.17b}$$

Now let us consider the change in the total energy of a slab subject to the strain ε_{xx} and relaxing in the z-direction. An important point is that a

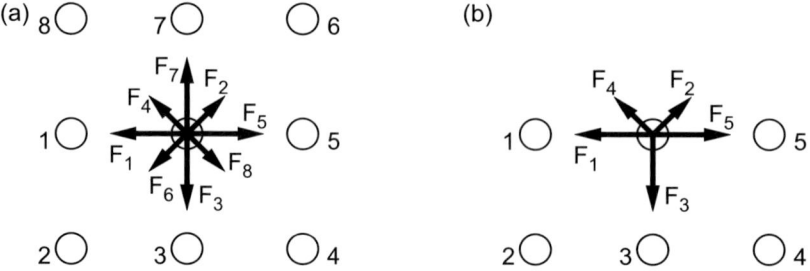

Fig. 3.6. Forces acting on a given atom from nearest and next-nearest (diagonal) neighbors. (**a**) Forces acting on a bulk atom. The values of these forces are related by (3.17a) and (3.17b). (**b**) Forces acting on a surface atom. Due to (3.17a) and (3.17b), the total force acting on a surface atom vanishes

uniformly strained horizontal layer of atoms can be considered as a single object, and the interaction between atoms can be reduced to an interaction between layers. Then every nth layer interacts with one layer below [the $(n-1)$th layer] and with one layer above [the $(n+1)$th layer]. Thus, while atoms interact with the nearest and next-nearest (diagonal) neighbors, the layers interact only with the nearest neighboring layers. A slab consisting of, say, $(2N_z+1)$ horizontal layers $(-N_z \leq n \leq N_z)$ of atoms is then equivalent to a vertical chain of $(2N_z+1)$ objects with interaction only between nearest neighbors. Cheng et al. [3.33] have shown that the surface relaxation changing the spacing between neighboring layers near the surface occurs only if a layer interacts with the second neighboring layer in the vertical direction. Thus, for the system described in Fig. 3.5, the spacing between the surface (N_zth) and the subsurface $[(N_z-1)$th] layer is exactly the same as the spacing between two neighboring bulk layers.

For an unstrained slab, this feature is also illustrated in Fig. 3.6b. Let us (by a thought experiment) create the '(001) surface' of the crystal by cutting the bonds, as shown in Fig. 3.6b, and keeping atoms at their bulk positions. Then (3.17a) and (3.17b) ensure that the force acting on a surface atom vanishes (see Fig. 3.6b). Thus, no surface relaxation occurs at the '(001) surface'. The arguments given above show that the same remains valid for a uniformly strained slab. (It should be noted, that relaxation occurs at surfaces with other orientations.)

This example allows us to consider the surface stress without the complications which could be caused by surface relaxation. If a slab is subject to the strain ε_{xx}, it relaxes in the vertical direction, and the ε_{zz} component of the strain tensor comes into play. The surface energy is due to the lack of strained bonds. The surface energy can be calculated up to second-order terms in strain and is given in terms of ε_{xx}. Tedious but straightforward calculations yield the surface energy per unit length of the surface:

$$\gamma(\varepsilon_{xx}) = \frac{1}{a}\left(\frac{J_1}{2} + J_2\right) + \tau\varepsilon_{xx} + \frac{1}{2}S\varepsilon_{xx}^2, \tag{3.18}$$

where

$$\tau = -\frac{1}{2a}\sigma_0, \tag{3.19a}$$

$$S = -\frac{a}{4}\left[V_2'' + \frac{1}{2a}\sigma_0 - \frac{\left(V_2'' - \frac{1}{2a}\sigma_0\right)^2}{V_1'' + \frac{1}{2a}\sigma_0 + V_2''}\right]. \tag{3.19b}$$

It follows from (3.19a) that the surface stress is proportional to the 'spontaneous stress' σ_0. Figure 3.6a corresponds to attractive forces between nearest neighbors, i.e., to $V'(a) > 0$ and $\sigma_0 < 0$. Then (3.19a) tells us that $\tau > 0$, i.e., the surface stress is tensile. The surface excess elastic modulus S from (3.19b) can be either positive or negative.

3.1.6 Capillarity Phenomena at Solid Surfaces

Despite the difference between solid and liquid surfaces emphasized in Sect. 3.1.3, there exists a basic similarity between these two types of surface. Indeed they are both intrinsically stressed. Hence solid surfaces exhibit capillarity (or surface stress-induced) phenomena similar to the familiar Laplace capillarity pressure under a curved liquid surface, and a strain field is generated at the curved surface of a solid. The existence of a surface stress-induced strain field was pointed out by Marchenko and Parshin [3.21], who considered the elastic energy of a solid including both bulk and surface contributions:

$$E_{\text{elast}} = \frac{1}{2}\int \lambda_{ijlm}\varepsilon_{ij}\varepsilon_{lm}\mathrm{d}V + \int \tau_{\alpha\beta}(\widehat{\boldsymbol{m}})\varepsilon_{\alpha\beta}\mathrm{d}A. \tag{3.20}$$

Equation (3.20) yields the elastic energy as a function of strain for a given configuration of the surface. The term linear in the strain means that the unstrained state $\varepsilon_{ij}(\boldsymbol{r}) = 0$ does not correspond to mechanical equilibrium. Indeed, it implies effective forces applied to the crystal edges and these give rise to elastic relaxation and non-zero strain at mechanical equilibrium.

To explain the origin of these forces, we note that the components of the tensor $\tau_{\alpha\beta}$ depend on the surface orientation $\widehat{\boldsymbol{m}}$. (Even the orientations of the axes where the tensor components do not vanish change with $\widehat{\boldsymbol{m}}$.) A divergence of the surface stress tensor thus appears, yielding an effective elastic force applied to the crystal,

$$F_i = \frac{\partial \tau_{i\beta}}{\partial r_\beta}. \tag{3.21}$$

The force F_i creates the elastic strain field in the crystal. It should be noted here that although the elastic forces in (3.21), in Fig. 3.7 and below are

72   3. Self-Organization Phenomena at Crystal Surfaces

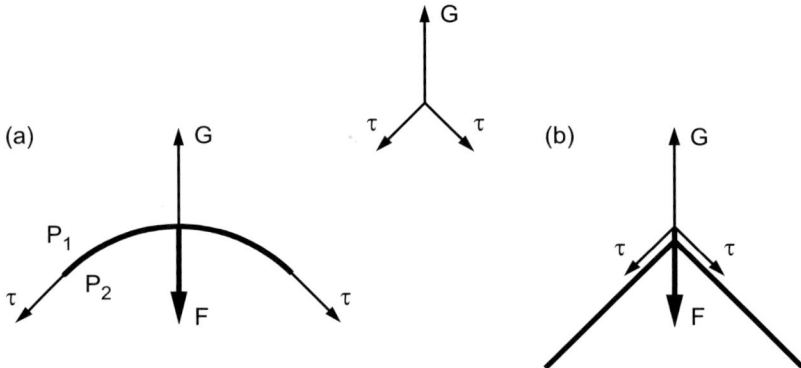

Fig. 3.7. Capillarity effects at liquid and solid surfaces. (**a**) The balance of forces acting on an element of the curved surface layer of a liquid. Forces caused by the surface tension τ are balanced by the force $\boldsymbol{G}$ acting from the bulk of the liquid. According to Newton's third law, a reaction $\boldsymbol{F} = -\boldsymbol{G}$ acts from the surface layer on the bulk of the liquid. This force results in the excess Laplace pressure $\Delta P = P_2 - P_1$ below the curved surface of the liquid, where P_1 and P_2 are the values of the pressure below and above the curved liquid surface. (**b**) The balance of forces acting on a crystal edge. Forces caused by the intrinsic surface stress τ are balanced by the force $\boldsymbol{G}$ acting from the bulk of the crystal. According to Newton's third law, the reaction $\boldsymbol{F} = -\boldsymbol{G}$ acts from the surface layer on the bulk of the crystal, resulting in a strain field. The *central inset* depicts the balance of forces, similar for a liquid and a crystal. All forces in the figure are defined per unit length in the direction perpendicular to the plane of the figure

denoted $\boldsymbol{F}$, the Helmholtz free energy is denoted F, so there should be no confusion on this point. The force is a vector quantity $\boldsymbol{F}$ denoted in bold type throughout the text, or with Cartesian subscripts F_i, whereas the free energy is a scalar quantity, denoted by F.

The ultimate case of a curved solid surface is a sharp edge between neighboring facets. Figure 3.7 depicts the force balance on the curved surface of a liquid and at the crystal edge, and thereby shows the similarity between capillarity effects in liquids and solids. The surface tension on a curved liquid surface results in an excess pressure below the curved surface. In a similar manner, the intrinsic surface stress of crystal surfaces generates an effective force applied to the edge of the crystal.

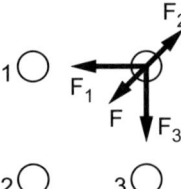

Fig. 3.8. Forces acting on a corner atom of a 2-dimensional square lattice. Forces from the nearest neighbors and from the next-nearest diagonal neighbor are not balanced, resulting in a total force $\boldsymbol{F}$ applied to a corner atom

Figure 3.8 shows the forces acting on a corner atom of the 2-dimensional square lattice considered above. The forces from the nearest neighbors and from the single next-nearest diagonal neighbor no longer cancel. An effective total force is then applied to the corner atom. Its value is $|\boldsymbol{F}| = \sqrt{2}\sigma_0/2$.

The effective forces acting at the crystal edges generate a strain field which significantly affects the energetics of faceted surfaces and promotes the formation of a periodic structure of facets.

3.1.7 Periodically Faceted Surfaces

If a planar crystal surface is unstable and breaks up into a system of tilted facets, the conservation of the average orientation of the normal to the surface (3.4) implies coexistence of alternating facets (Fig. 3.9a). At the intersection of neighboring facets there appear either sharp crystal edges or narrow rounded parts of the surface. Both these types of intersection may be described as linear defects at the surface. Such linear defects give a short-range contribution to the surface free energy and a long-range contribution due to elastic strain energy.

It was shown by Andreev [3.34] that if the order parameter related to the phase transition at the surface is linearly coupled to the strain field, i.e., if linear striction effects exist, they favor the formation of a periodic structure whose period can be macroscopically large. For the faceting phase transition, Andreev pointed out that a faceted surface with a macroscopic period could form [3.35].

The theory of periodically faceted surfaces was developed by Marchenko [3.36] and the period of the equilibrium structure was found. Following [3.36], we consider the faceted surface with a one-dimensional periodic saw-tooth profile depicted in Fig. 3.9. The free energy per unit projected area equals

$$F = F_{\text{surf}} + E_{\text{edges}} + \Delta E_{\text{elast}} , \qquad (3.22)$$

where F_{surf} is the free energy of tilted facets, E_{edges} is the short-range energy of the edges, and ΔE_{elast} is the elastic energy due to the discontinuity of the surface stress tensor τ_{ij} at the crystal edges.

The free energy of the tilted facets per unit projected area is a sum of contributions from the two types of facet,

$$F_{\text{surf}} = \gamma(\theta_1)\frac{L_1}{D} + \gamma(\theta_2)\frac{L_2}{D} . \qquad (3.23)$$

From Fig. 3.9a, one may write down the following relationship between the widths of the tilted facets:

$$L_1 \sin\theta_1 = L_2 \sin\theta_2 , \qquad (3.24\text{a})$$
$$L_1 \cos\theta_1 + L_2 \cos\theta_2 = D . \qquad (3.24\text{b})$$

Extracting L_1 and L_2 from (3.24b) and substituting them into (3.23), one obtains the surface energy in the form

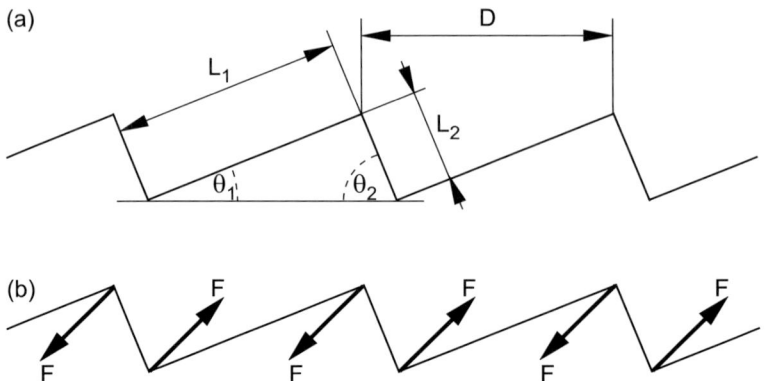

Fig. 3.9. Periodically faceted surface. (a) Geometry of a faceted surface. (b) Alternating forces of opposite sign applied at the edges

$$F_{\text{surf}} = \gamma(\theta_1)\frac{\sin\theta_2}{\sin(\theta_1+\theta_2)} + \gamma(\theta_2)\frac{\sin\theta_1}{\sin(\theta_1+\theta_2)}, \quad (3.25)$$

which does not depend on the period D of the faceted structure. The short-range energy of the edges per unit projected area equals

$$E_{\text{edges}} = \frac{\eta}{D} \equiv \frac{\eta^+ + \eta^-}{D}, \quad (3.26)$$

where η^+ is the short-range energy of the convex edge per unit length of the edge, and η^- denotes the same energy for the concave edge.

While discussing the elastic strain energy, we note the following. Firstly, E_{elast} is given by the general formulae (3.20), where the energy is zero in the absence of strain, and contains both linear and quadratic terms in the strain. Therefore, the elastic strain energy at equilibrium is negative, which corresponds to relaxation of the surface stress at crystal edges. Below, this negative elastic energy will be called the elastic relaxation energy due to crystal edges. Secondly, effective elastic forces acting at the edges are displayed in Fig. 3.9. Force monopoles act at each edge, and forces applied to neighboring edges are counterbalanced so that the total force applied to the system vanishes.

Elastic strains generated by linear crystal edges propagate into the crystal over a distance of the order of D, and decay at larger distances from the surface. Since the strain field is generated by linear defects at the surface, namely by the linear crystal edges, the elastic relaxation energy depends logarithmically on the period D of the structure [3.36]:

$$\Delta E_{\text{elast}} = -\frac{\widetilde{C}F^2}{YD}\ln\left(\frac{D}{a}\right) = -\frac{C\tau^2}{YD}\ln\left(\frac{D}{a}\right). \quad (3.27)$$

Here τ is the characteristic value of the intrinsic surface stress tensor, Y is the Young's modulus, a is the lattice parameter, and C is the geometric factor

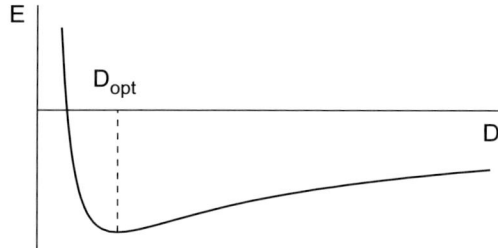

Fig. 3.10. The energy of a periodically faceted surface per unit surface area vs. the period D. There always exists an optimum period of faceting D_{opt} due to the logarithmic dependence of the elastic relaxation energy on D

accounting for a particular symmetry of the tensor τ_{ij}, elastic anisotropy of the crystal, etc.

Substituting (3.26) and (3.27) into (3.22), one obtains the following expression for the free energy of the faceted surface per unit projected area:

$$F = \left[\gamma(\theta_1)\frac{\sin\theta_2}{\sin(\theta_1+\theta_2)} + \gamma(\theta_2)\frac{\sin\theta_1}{\sin(\theta_1+\theta_2)}\right] + \frac{\eta}{D} - \frac{C\tau^2}{YD}\ln\left(\frac{D}{a}\right). \tag{3.28}$$

The dependence of the free energy vs. the period of the faceted surface D is displayed in Fig. 3.10. Due to the logarithmic dependence of the elastic relaxation energy on the period D, there always exists an optimum period of faceting D_{opt} equal to

$$D_{\text{opt}} = a\,\exp\left(\frac{\eta Y}{C\tau^2} + 1\right). \tag{3.29}$$

All material parameters entering the exponential in (3.29) have typical atomic values. Therefore the combination that appears as the argument of the exponential is of the order of 1. Since the exponential function is steep, and the argument can eventually be, say, greater than or equal to 3, the period D can exceed the lattice parameter a by at least an order of magnitude, i.e., it can be macroscopically large. In this case the macroscopic approach is justified.

It should be noted that the free energy of facets, i.e., the first term in (3.22) and (3.28), contains the entropy contribution which describes the dependence of the faceting itself and of the facet orientation on temperature. In addition, there exists another entropy contribution to F, associated with possible deviations of the actual structure from perfect periodicity (the configuration entropy). For periodically faceted surfaces, the configuration entropy has not yet been considered in the literature. Therefore the present discussion refers, strictly speaking, to the case $T=0$. In the rest of this section, we will omit the entropy contribution to F and discuss only the total energy of the system. Entropy effects in spontaneous formation of nanostructures will be considered in detail in the next section.

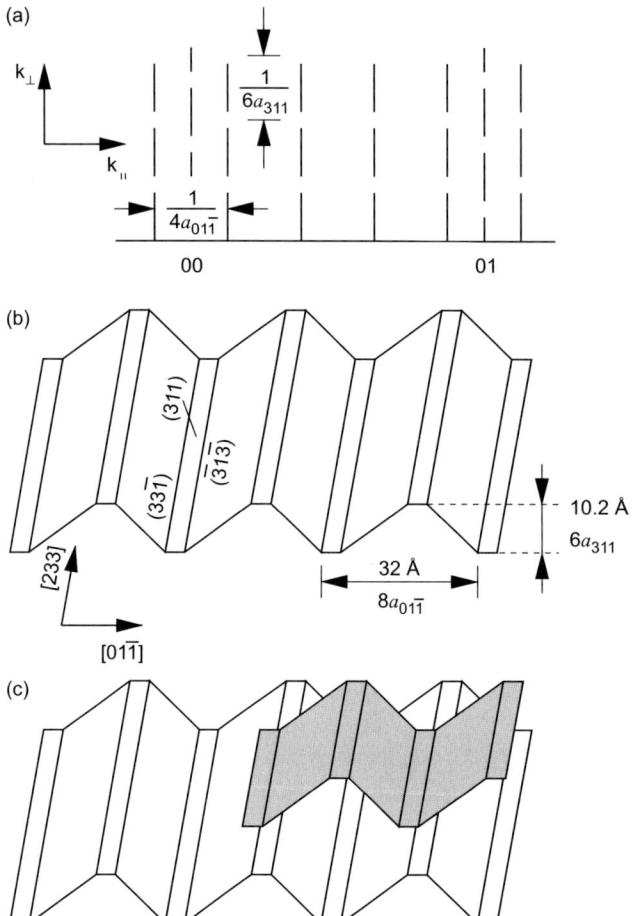

Fig. 3.11. Schematic reflection high energy electron diffraction (RHEED) pattern from the (311)A GaAs surface and the model of the surface structure. (**a**) RHEED pattern from the faceted (311)A GaAs surface. (**b**) Schematics of the faceted (311)A GaAs surface. (**c**) Schematics of the AlAs/GaAs(311) structure exhibiting out-of-phase corrugation of the two surfaces

3.1.8 Faceting Phenomena on (311) Surfaces of GaAs and AlAs

First Observations. The high-index (311) surface of GaAs and related systems, such as the (311) surface of AlAs and corrugated superlattices formed in the heteroepitaxial AlAs/GaAs(311) system, were historically the first examples of spontaneous formation of macroscopically ordered (with respect to atomic distance) semiconductor nanostructures from initially random distributions of species in atomic and molecular beams (i.e., self-organized growth process). The first observation of a periodically faceted surface of GaAs was reported by Nötzel et al. in 1991 [3.37]. A more detailed analysis is presented in [3.38].

Figure 3.11a depicts the reflection high energy electron diffraction (RHEED) pattern from the (311)A surface of GaAs. The RHEED pattern images the reciprocal lattice of the stepped surface.

- With the electron beam in the $[01\bar{1}]$ direction, the RHEED pattern shows pronounced streaking, indicating a high density of steps oriented along the $[\bar{2}33]$ direction.
- Depending on the wave vector $k_\perp$, the main streaks (marked 00 and 01) are split into sharp satellites or unsplit. The separation of the satellites gives the lateral periodicity as 32 Å ($\Delta = 8a_{110}$).
- The splitting of the pattern along the main streak with varying $k_\perp$ indicates the presence of a two-level system with a step height of 10.2 Å $= 6a_{311}$.
- The high degree of ordering indicates the presence of an almost perfect two-level system.

Based on the RHEED data, Nötzel et al. [3.37, 3.38] proposed the model of a faceted surface shown in Fig. 3.11b. The surface is composed of (311) terraces of width 4 Å ($1a_{110}$) and two sets of $(33\bar{1})$ and $(\bar{3}13)$ facets corresponding to upward and downward steps of height 10.2 Å ($6a_{311}$).

It should be emphasized that the RHEED pattern corresponding to the faceted surfaces is observed after oxide removal from the GaAs substrate surface at 580°C in the molecular beam epitaxy (MBE) growth chamber and remains stable during growth of GaAs. This shows the equilibrium nature of faceting on the (311) GaAs surface.

The RHEED dynamics inspected during GaAs/AlAs growth shows pronounced oscillations at the onset of GaAs or AlAs growth. After deposition of six monolayers, the RHEED pattern of the surface approaches the pattern corresponding to the stable faceted surface during growth. It was concluded by Nötzel et al. [3.37, 3.38] that the surface corrugation changes phase by π after deposition of six monolayers of AlAs over the faceted GaAs surface, and vice versa. In this case, if the phase of corrugation were preserved during overgrowth, no oscillations would occur in the RHEED pattern. In addition, the out-of-phase corrugation of GaAs/AlAs and AlAs/GaAs interfaces in a multilayered structure was confirmed by cross-section high resolution transmission electron microscopy (HRTEM) data [3.37].

Figure 3.12a shows a GaAs/AlAs(311) quantum well wire superlattice which is separated electronically from the GaAs substrate by thick AlAs barriers. Alferov et al. [3.39] investigated the effect of thin insertions (less than six monolayers) of GaAs into an AlAs barrier (see Fig. 3.12c). In this case a new faceted surface of GaAs(311) cannot be completed, and a more complicated morphology occurs. Photoluminescence (PL) spectroscopy data showed [3.39] a new line shifted to smaller photon energies (redshift) compared with the PL maximum observed for the structure of Fig. 3.12b. This redshift occurs for very thin insertions of GaAs (1 monolayer or less) and

78   3. Self-Organization Phenomena at Crystal Surfaces

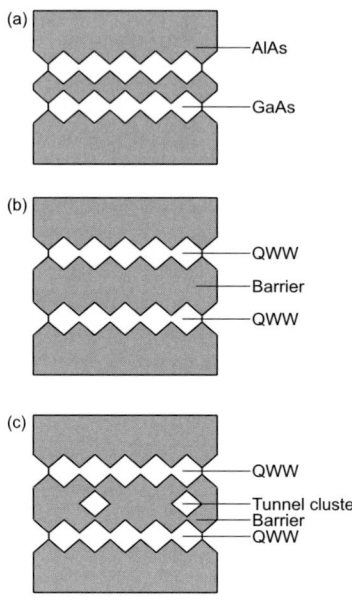

Fig. 3.12. GaAs/AlAs(311) quantum well wire superlattices with corrugated interfaces. (**a**) Schematic cross-sectional view of a GaAs/AlAs quantum well wire superlattice with corrugated interfaces. (**b**) Two layers of GaAs (*white*) separated by a layer of AlAs (*grey*) and surrounded by thick AlAs barriers. (**c**) Insertions of GaAs inside an AlAs barrier form clusters providing some tunneling of carriers between the two neighboring quantum wells of GaAs and redshifting the photoluminescence spectrum

does not change for thicker insertions. Based on PL data, a growth model was proposed in [3.39]: GaAs deposited on a faceted surface of AlAs first forms isolated clusters (Fig. 3.12c), where the top surface already reproduces locally a faceted surface with an out-of-phase corrugation with respect to the AlAs surface. With further GaAs deposition, the length of clusters increases in the [$\bar{2}33$] direction, new clusters form in neighboring grooves, and finally, the complete faceted surface restores itself after deposition of six monolayers.

Thus, insertion of GaAs in thick AlAs barriers may lead to the formation of isolated quantum wires if the thickness of the insertion is about 1 ML, for groups of a few quantum wires placed in the neighboring grooves, and for the complete quantum well wire (QWW) if the thickness exceeds six monolayers. Isolated clusters between the two quantum well wires may induce tunneling of carriers between the QWWs.

Physics of Quantum Wire Formation on Faceted Surfaces. The theory of quasi-equilibrium heteroepitaxial growth on periodically corrugated substrates was developed by Shchukin et al. [3.40]. They focussed on a system possessing two features at the same time: firstly, the two materials were lattice-matched like AlAs and GaAs, and secondly, the flat surface of both materials was unstable against faceting. This is the case for the AlAs/GaAs(311) system, for example.

The total energy of the heteroepitaxial system considered in [3.40] equals

$$E = E_{\text{surf}} + E_{\text{interface}} + E_{\text{edges}} + \Delta E_{\text{elast}} \,. \tag{3.30}$$

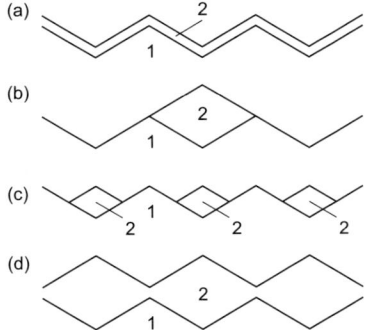

Fig. 3.13. Possible structures of the heteroepitaxial system where material 2 is deposited on a periodically faceted surface of material 1. (**a**) Homogeneous coverage. (**b**) System with separated 'thick' clusters. (**c**) System of 'thin' clusters. (**d**) Heteroepitaxial system at high coverage, where the periodic surface corrugation is restored and the hills of the top surface of the heteroepitaxial system appear over the valleys of the substrate, and vice versa. The heteroepitaxial system contains a continuous layer of material 2 with periodically modulated thickness

Here, in addition to the three contributions to the energy of the faceted surface [see (3.22)], one more contribution enters into account, namely the interface energy. The comparison of total energies for the distinct types of heteroepitaxial structure depicted in Fig. 3.13 was carried out in [3.40]. It yields the following conclusions.

The selection between two possible growth modes is determined by whether or not the deposited material wets the substrate. If the deposited material wets the substrate, then homogeneous coverage of the periodically corrugated substrate occurs (Fig. 3.13a). An example is the growth of AlAs on a periodically corrugated vicinal surface of GaAs(001) 3°-off towards [1$\bar{1}$0] [3.41], which will be considered in the next section.

If the deposited material does not wet the substrate, then isolated clusters of the deposited material form on the periodically corrugated substrate (Fig. 3.13b). This situation is likely to be realized for the growth of GaAs on the vicinal surface of AlAs(001) 3°-off towards [1$\bar{1}$0] [3.41], and for both GaAs/AlAs and AlAs/GaAs heteroepitaxial growth on the (311)A surface [3.37].

In the case of inhomogeneous cluster coverage, the periodic surface corrugation is restored after deposition of the first few monolayers. Then the hills of the top surface of the heteroepitaxial system appear over the valleys of the substrate, and vice versa, and a continuous layer of the deposited material is formed with periodically modulated thickness (Fig. 3.13d). Hence, the formation of clusters allows direct fabrication of quantum wires and quantum wire superlattices in heteroepitaxial semiconductor systems.

Since any periodically faceted surface is a structure of elastic domains, its geometrical parameters (e.g., the period) can be tuned in a controlled way by applying external stress. Detailed considerations may be found in the paper by Shchukin et al. [3.42].

Controversy Over the Structure of the GaAs(311) Surface. Since the discovery of corrugated superlattices in 1991 by Nötzel et al. [3.37], there has been a lot of interest in GaAs–AlAs structures grown on (311)A surfaces. A 3.2 nm in-plane periodicity was confirmed by the observation by Popovic et al. of folded acoustic phonons from lateral thickness modulation [3.43]. Similar lateral periodicity was also demonstrated for confined LO phonons by Shields et al. [3.44]. One should note, however, that the same authors making experiments on different samples grown under nominally similar conditions observed different phonon line widths, and the features due to lateral periodicity were not revealed in some cases [3.45]. It was also observed by Tournié et al. [3.46] that the surface corrugation is very sensitive to surface strain. For example, Ilg et al. [3.47] observed suppression of faceting after the deposition of only 0.05 ML of In or Si on the corrugated surface. This may be attributed to the impact of external strain on the surface corrugation studied theoretically in [3.40].

More recently it was shown that atomic hydrogen may also affect corrugation of the (311)A surface, resulting in facet bunching and a much larger (30–100 nm) lateral periodicity of $[\bar{2}33]$-oriented grooves with increased height [3.48]. This instability of nanofaceted high-index surfaces with respect to step-bunching, reported first by Nötzel et al. [3.49], or suppression of faceting by very dilute concentrations of impurities [3.47], may result in different surface and interface structures for nominally similar deposition conditions, and as a consequence, different optical and electrical properties.

Direct studies using scanning tunneling microscopy (STM) performed under different experimental conditions also gave different results. Using an ultrahigh vacuum scanning tunneling microscope attached to an MBE chamber, Wassermeyer et al. [3.50] found a well-periodic array of grooves with a periodicity of 0.32 nm, in agreement with the model proposed by Nötzel et al. [3.37]. At the same time the corrugation height was found in [3.50] to be only two monolayers. One should note, however, that STM studies are carried out after cooling the surface to room temperature. Cooling may completely suppress surface corrugation, as is likely to be the case in [3.37]. STM studies performed on samples capped with arsenic and characterized in a separate STM chamber after the As decapping procedure have shown an absence of long-range order on the (311)A GaAs surface [3.51]. More recent STM studies confirmed the lateral periodicity of the (311)A GaAs surface and the two-monolayer step height at room temperature [3.52].

Once the two-monolayer corrugation height had been observed by Wassermeyer et al. [3.50], most researchers came to the conclusion that the interface corrugation of GaAs–AlAs superlattices on a (311)A GaAs surface is either absent [3.53] or very weak (not more than 2 ML) [3.54]. Furthermore, it was generally agreed that only the normal GaAs–AlAs interface is corrugated, while the inverted AlAs–GaAs interface is intermixed [3.55]. Optical anisotropy in (311)A-grown GaAs–AlAs superlattices was attributed to intrinsic anisotropy of the (311) surface [3.56] and it was proposed that "corrugation modifies the density of states only slightly, giving no evidence of a quantum-wire behavior" [3.57]. More recently, Raman studies have demonstrated a very significant polarization splitting of the TO phonon lines in GaAs–AlAs superlattices grown on (311)A substrates in the case of thin ($\approx$ 5–8 ML) GaAs layers [3.58]. The value of the splitting was close to that calculated in [3.59] for the case where both interfaces are corrugated, and the corrugation height is 1.02 nm. No such splitting was observed in [3.59] for multilayer structures grown side-by-side on (311)B substrates. It was also observed in that paper that intentional doping of the (311)A-grown superlattices results in a systematically smaller polarization splitting for any studied thickness of the GaAs layer.

Surprisingly, very few works have been carried out for structural characterization of (311)A-grown superlattices using high resolution transmission electron microscopy (HRTEM). This may be due to the difficulties involved in proper sample preparation along the [$\bar{2}33$] zone axis, the relatively small lattice fringe distances of the (311) and (220) planes, fast AlAs oxidation, the change in image pattern induced by small sample thickness, and interface non-uniformity along the thickness of the HRTEM sample, which may mask the interface structure. HRTEM studies were recently performed by Vorob'ev et al. on (311)A GaAs–AlAs superlattices with beryllium-doped AlAs layers [3.60]. These studies demonstrated a corrugation height of about 1 nm for each interface and a GaAs layer thickness modulation of 2 nm, in general agreement with the original results of Nötzel et al. [3.37]. At the same time an imperfect lateral periodicity was observed with a characteristic waviness in the GaAs layer of about 10–20 nm. Furthermore, RHEED intensity modulations pointing to a monolayer-by-monolayer growth mode were observed during AlAs–GaAs overgrowth, contradicting the model proposed in [3.37]. On the other hand, the experiment of [3.60] is not very convincing. It was shown that doping may, in general, dramatically suppress surface [3.47] and interface [3.58] corrugations which might also occur in the case of beryllium doping. Volodin et al. [3.58] also found that, in their samples, only the GaAs layers are modulated in thickness, while no phase shift of the surface corrugation occurs during AlAs growth. The evident diversity of experimental data made it possible for some research groups to doubt the concept of corrugated superlattices proposed by Nötzel et al. [3.37, 3.38, 3.49] in a general sense. Other explanations were invoked [3.50, 3.51, 3.55–3.57, 3.60–3.63] to interpret experimental data which did not agree with the concept of corrugated superlattices.

Interface corrugation was also claimed by Nötzel et al. [3.37] to be observed in high-resolution transmission electron microscopy (HRTEM), but imperfect imaging conditions regarding the objective lens defocus and sample thickness as well as surface oxidation of the TEM samples (in particular of the AlAs) prevented conclusive statements concerning the degree of ordering and the actual strength of lateral compositional modulation.

Cross-Sectional TEM Studies of Multilayered AlAs/GaAs(311) Structures. Aiming specifically to solve this discrepancy between different interpretations of AlAs/GaAs(311) heterostructures, Ledentsov et al. [3.64] recently undertook a detailed study of structural and optical properties of GaAs–AlAs multilayered structures grown on the (311)A surface of GaAs. The interface structure of these systems was studied using high resolution transmission electron microscopy (HRTEM) and a contrast-enhancing image processing technique. The structures were grown in the regime where the RHEED pattern indicated interface corrugation with a height of 1.02 nm and a lateral period of 3.2 nm along [$\bar{2}$33]. Two distinctly different growth regimes were chosen. In the first case the thickness of GaAs and AlAs layers was quite large (11 ML each). Then, if the surface corrugation model proposed by Nötzel et al. [3.37] applies, the GaAs or AlAs surface will be completely covered by AlAs or GaAs, respectively. In the second case, the AlAs layer thickness was 11 MLs, but the GaAs thickness was just 1 ML, resulting in only partial coverage of the AlAs surface and in local formation of GaAs clusters (quantum dots), as first proposed by Alferov et al. [3.39] and Kop'ev and Ledentsov [3.65].

In [3.64] Ledentsov et al. revealed corrugation of the GaAs and AlAs layers in both cases. For multilayer deposition with GaAs layer thickness exceeding 1 nm, the lateral periodicity is 3.2 nm along the [0$\bar{1}$1] direction and the corrugation height is about 1 nm. The corrugation is symmetric for both upper and lower GaAs–AlAs interfaces. Thicker parts in the thickness-modulated AlAs and GaAs layers of the corrugated superlattice (CSL) are shifted by half a period with respect to each other, in agreement with the phase shift of the surface corrugation during heteroepitaxy proposed by Nötzel et al. [3.37]. In the regime where the GaAs thickness is only 1 ML (0.17 nm), an additional lateral periodicity of 1.5–2 nm is revealed, and the degree of order is much weaker. We attribute this effect to local formation of GaAs clusters in AlAs grooves, resulting in a local phase reversal of the AlAs surface corrugation. Phase-shifted and unshifted AlAs surface domains coexist, resulting in an additional periodicity. HRTEM studies also confirm the AlAs–GaAs interface inclination angles of about 36° and 144° with respect to the flat (311) surface, in agreement with the facet geometry model proposed by Nötzel et al. [3.37].

The samples were grown using conventional solid-source molecular beam epitaxy at 580°C. The growth rate was about 1 µm/h for GaAs and 0.3 µm/h for AlAs, and the As_4/Ga flux ratio was 3. The growth was monitored by RHEED at 1° glancing angle. After oxide removal at the GaAs substrate at 600°C, a thin 0.1 µm GaAs buffer was grown. The RHEED pattern recorded along the [$\bar{2}$33] and [0$\bar{1}$1] directions manifested the break-up of the flat (311) surface into an ordered array of grooves oriented along the [$\bar{2}$33] direction (see, for example, [3.66]). Observing the [$\bar{2}$33] azimuth parallel to the steps, the streaks were found to be split into sharp satellites or unsplit depending on the position along the height of the streak. The intensity maximum of the satellites corresponds to an intensity minimum of the main streak, in agreement with the reciprocal lattice of the stepped two-level surface, as revealed by Nötzel et al. [3.37, 3.38]. The intensity of the diffraction pattern was modulated according to the intersection of the reciprocal lattice of the two-level system with the Ewald sphere. This points to the uniformity of the surface in the imaging direction.

In contrast, with the electron beam along the [0$\bar{1}$1] direction, the diffraction pattern showed a pronounced streaking, inconsistent with the Ewald construction for the flat surface. This pattern indicates that the uncertainty of the reciprocal lattice in the imaging direction is about $1/a_{[0\bar{1}1]}$, where $a_{[0\bar{1}1]}$ is the lattice parameter in the [0$\bar{1}$1] direction. Thus the diffracting element in this case must have a very narrow top surface, in agreement with the surface model proposed by Nötzel et al. [3.37]. The interpretation of the streaky RHEED pattern as evidencing a smooth surface, given in [3.61], is not correct.

In [3.64] cross-section samples for TEM were prepared along the [$\bar{2}\bar{3}$3] direction. The conventional procedure was applied, which consists in gluing two stripes of the wafer against each other, embedding them in a brass holder (further details are given in [3.67]), plane grinding, dimpling, and Ar^+-ion milling. TEM was performed in a Philips CM200 FEG/ST microscope with a Scherzer resolution of 0.24 nm and an information limit of 0.16 nm. High-resolution images along the [$\bar{2}$33] zone axis were recorded with a CCD camera. The Wiener filtering technique was applied for noise reduction. (Details of the method are given in Chap. 2. See also [3.68]). To enhance the chemical contrast, only the region around the transmitted beam with an amplitude A_{000} was selected in Fourier-transformed images (diffractograms). The back transformation of the diffractograms reveals a stronger contrast difference between the AlAs and GaAs than the original image. The contrast is based on the different extinction lengths of A_{000} in GaAs and AlAs. Unfortunately, the more favorable chemically sensitive (002) reflection is not available in the [$\bar{2}$33] zone axis.

Figure 3.14 shows an HRTEM image of a sample with nominal GaAs and AlAs layer thicknesses of 1 and 11 ML, respectively. The amplitude of the transmitted beam is plotted in a color-coded map, which qualitatively reveals the local composition of the superlattice. The red color on the image corresponds to AlAs regions, as was checked on AlAs buffer layers in the vicinity. The deep blue regions correspond to GaAs-rich regions. It should be noted that TEM sample thickness variations can also induce a modulation of A_{000}. However, no significant contrast variation was observed in the thick GaAs layers neighboring the corrugated superlattice (CSL). Furthermore, a sample thickness variation with regular well-defined periodicity is quite

84   3. Self-Organization Phenomena at Crystal Surfaces

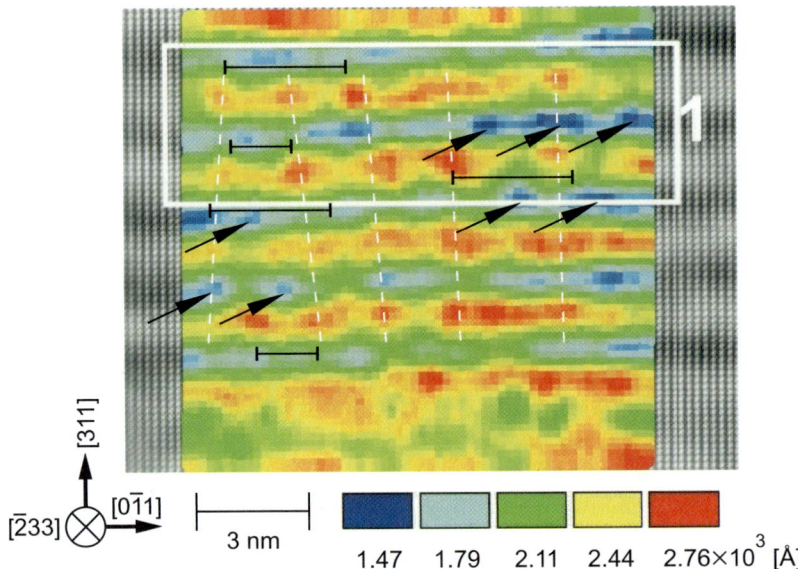

Fig. 3.14. Processed cross-section HRTEM image of the sample with nominal GaAs and AlAs layer thicknesses of 1 and 11 monolayers, respectively. The amplitude of the transmitted beam is plotted in a color-coded map. The *red color* on the image corresponds to AlAs regions, and *deep blue* to GaAs-rich regions

unlikely. Therefore, we may reasonably assume that the modulation of A_{000} shows a compositional periodicity of 3.2 nm and 1.6 nm and that both GaAs and AlAs layers are modulated.

The interface corrugation can also be revealed and is shown to be symmetric close to 1 nm, indicating that both GaAs and AlAs thickness are modulated. Intermediate colors indicate intermediate compositions which are, however, not necessarily linearly correlated with A_{000}. It is nevertheless worth noting that the GaAs-rich layer contrast modulation has comparable strength in both vertical and lateral directions. Strictly speaking, A_{000} reflection cannot be used for numerical evaluation of the composition because the intensity also depends on the thickness of the HRTEM sample and so may not be linearly related to composition. At the same time, the buffer AlAs or GaAs regions demonstrated negligibly weak A_{000} modulations and, taking into account the A_{000} intensities in these regions and assuming that the degree of signal nonlinearity is not very high, one can estimate the strength of the modulation as being up to roughly 40–50% in the lateral direction and 50–80% in the vertical direction for both samples. Significant nonlinearity would mean that either the AlAs or GaAs layers were much more strongly modulated than is revealed in the image. This would indicate that either the GaAs or AlAs layers were only weakly modulated in thickness, and so must be wavy when grown on top of a thickness-modulated layer. In contrast, the corrugation is

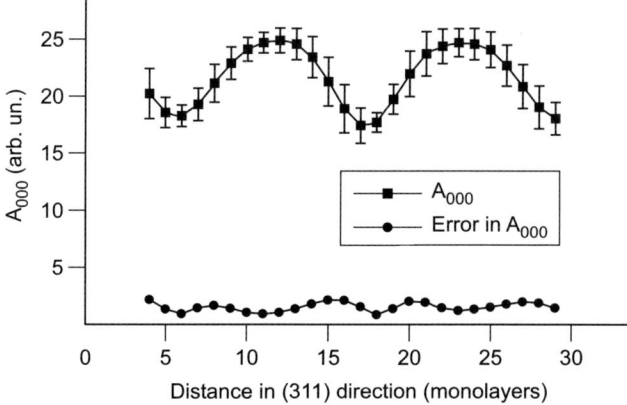

Fig. 3.15. Amplitude of the transmitted beam A_{000} (*upper curve*) as a function of lattice spacing along the (311) direction, averaged for each monolayer along the $[0\bar{1}1]$ direction in the region within the white frame in Fig. 3.14. The standard deviation ΔA_{000} of the amplitude of the transmitted beam along each monolayer is displayed in the *lower curve*

symmetric for both interfaces of the CSL layers and each layer is parallel to the (311) plane. Thus both AlAs and GaAs layers demonstrate comparable thickness modulation, and no significant nonlinearity can be expected.

Figure 3.15 shows the intensity of the A_{000} transmitted signal (upper curve) averaged for each monolayer along the $[0\bar{1}1]$ direction in the region within the white frame in Fig. 3.14. The standard deviation ΔA_{000} of the amplitude of the transmitted beam along each monolayer is displayed in the lower curve. ΔA_{000} is particularly high at intermediate values of A_{000}, which occur at the corrugated interfaces between GaAs and AlAs. This can be taken as another indicator that modulation of the composition is indeed present along the (311) monolayers at the interfaces.

A color-coded map of the sample with 2nm GaAs/2nm AlAs SL is shown in Fig. 3.16. It is clear that the GaAs-rich and AlAs-rich regions are laterally periodic with period 3.2 nm and are vertically correlated and phase shifted as originally proposed by Nötzel et al. [3.37]. The probability of facet irregularities is relatively low in this case (see such an example in the lower right part of Fig. 3.16). We note again that the in-plane GaAs-rich layer contrast modulation corresponds to compositional modulation of up to 40–50%. The angles of the corrugation (black arrows) are revealed in many parts of the image and fit [≈ 40° and ≈ 140° with respect to the flat (311) surface] with the interface structure originally proposed by Nötzel et al. [3.37].

A new and important point observed by Ledentsov et al. [3.64] is the local increase in the lateral frequency of compositional modulation in the sample with 1 ML GaAs insertions. In this case only 1/6 of the surface is covered by GaAs clusters and the growth mode must be significantly different

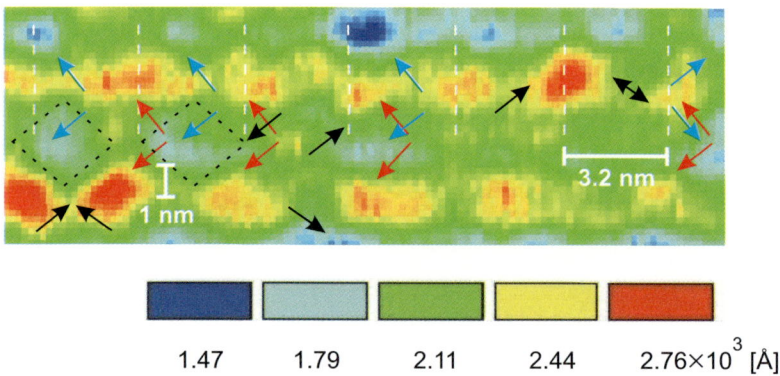

Fig. 3.16. Amplitude of the transmitted beam plotted in a color-coded map for the corrugated superlattice with 11-ML-thick GaAs and 11-ML-thick AlAs layers. The *red color* on the image corresponds to AlAs regions, and *deep blue* to GaAs-rich regions

from the one proposed by Nötzel et al. [3.37, 3.38, 3.49]. The increase in spatial frequency may be explained by the coexistence of two types of AlAs surface domains with phase-shifted corrugation of the surface, an effect whose probability is small in the case of conventional homoepitaxial growth. Since in HRTEM studies both regions can be trapped in the HRTEM foil, one can also see the approximately doubled lateral frequency for the neighboring GaAs clusters. The facet angles and thickness modulation periodicity may also be directly observed in HRTEM images of both structures (Fig. 3.17), even though the contrast is fairly weak in both vertical and lateral directions, as already discussed above.

We note that for the structure with 1 ML GaAs insertions (Fig. 3.17a), an approximate half periodicity can be revealed. The average width of the GaAs-related contrast modulation is much larger than the nominal thickness (1 ML). The average height of the GaAs-related contrast is close to 2 nm, as expected for the model of a single GaAs cluster proposed by Alferov et al. [3.39] (see also [3.65]). At the same time the relatively large surface area corresponding to the GaAs-related contrast indicates that the total projection of all GaAs clusters along the $[\overline{2}33]$ direction gives a much larger cross-section surface (by a factor of about 4), as one would expect for 1D-like 'quantum wire-like' GaAs domains filling 1/6 of the surface. This may be due to the fact that the effective length of the cluster along the $[\overline{2}33]$ direction can be small (e.g., 4–5 nm) compared with the total thickness of the foil used in HRTEM studies ($\approx$ 15–20 nm). Such a length would also explain the RHEED intensity oscillation corresponding to 6 ML GaAs deposition on the AlAs (311)A surface and vice versa [3.38], which was attributed to the filling of AlAs grooves with GaAs and re-establishment of the initial (but phase-shifted) surface morphology.

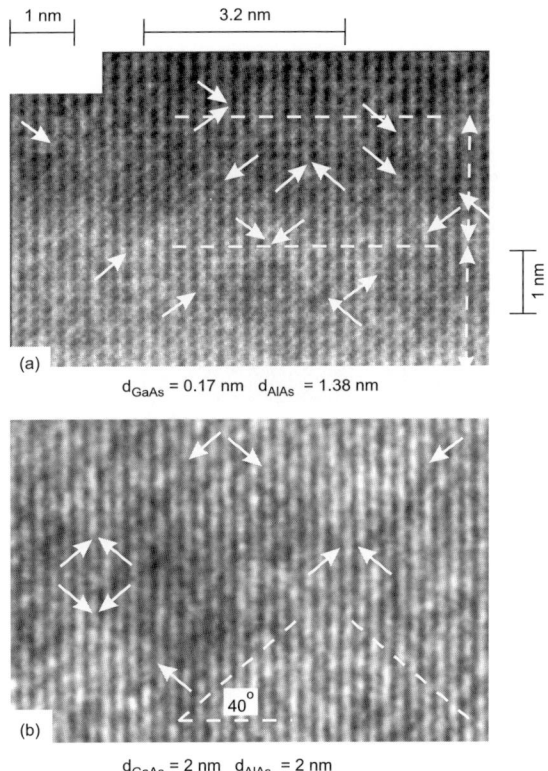

Fig. 3.17. Cross-section HRTEM images of corrugated superlattices (CSLs) (**a**) with 1-ML-thick GaAs and 11-ML-thick AlAs layers and (**b**) with 11-ML-thick GaAs and 11-ML-thick AlAs layers

Short clusters result in an uncertainty of the surface lattice rods in the reciprocal space along the imaging direction and in a reduced intensity of the corresponding diffraction features. More studies are required (e.g., plan-view HRTEM investigations) to characterize the length of the GaAs clusters (quantum dots) quantitatively. The contrast in the HRTEM image of the sample with 11-ML-thick GaAs and AlAs layers (Fig. 3.17b) is clearer, and both the facet spacing and facet angles can be unambiguously distinguished. This is most probably due to the fact that the GaAs–AlAs interfaces are more uniform along the imaging direction in the latter case. One may conclude from the HRTEM images that the interface corrugation height is about 1 nm, and that both normal (GaAs–AlAs) and inverted (AlAs–GaAs) interfaces are corrugated.

Fourier transform images of CSLs with 3.7 nm vertical periodicity (11 ML GaAs/11 ML AlAs) are shown in Fig. 3.18. Figures 3.18a and c are the Fourier transformations of the two idealized model structures. The first structure refers to the model due to Nötzel et al. [3.37]: ideal corrugated SL, where

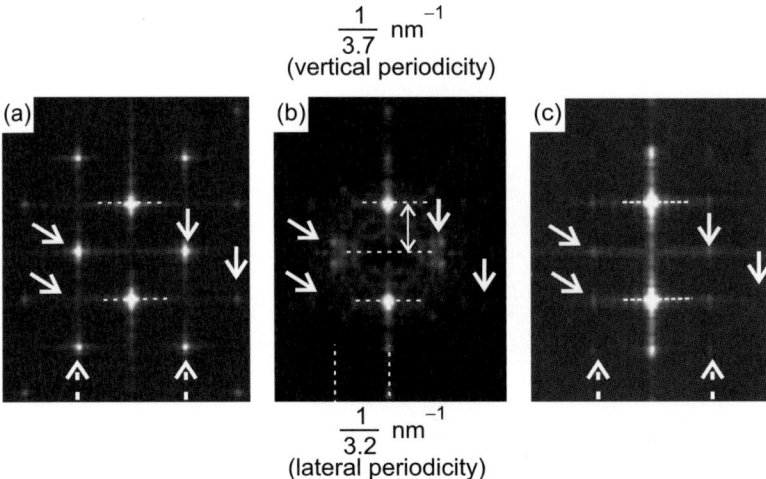

Fig. 3.18. Fourier transform images of corrugated superlattices (CSLs) with 3.7 nm vertical periodicity. (**a**) Fourier transform of the ideal model structure proposed by Nötzel [3.37]. This is an ideal CSL having both interfaces corrugated with height 1 nm, perfect 3.2 nm lateral periodicity, and {331} facets. (**c**) Fourier transform of the model structure proposed by Lüerßen et al. [3.55] and Langbein et al. [3.57]. In this model only the AlAs-on-GaAs interface is corrugated and the corrugation height is 0.34 nm. The compositional gradient at the interface is set in both models to one atomic plane along the (311) direction (0.17 nm). The image in (**c**) was intensified to show the satellites due to lateral periodicity, which are not otherwise resolved. (**b**) Fourier transformation of the processed experimental image of a randomly chosen part (20 nm in the vertical direction, 30 nm in the lateral direction) of the (311A) CSL with 3.7 nm vertical period (11 ML GaAs/11 ML AlAs). Note the clear streaks indicating a lateral periodicity of 3.2 nm in the experimental image (**b**) and a checkerboard arrangement of the pattern in the images (**a**) and (**b**). This disagrees with image (**c**), as indicated by *arrows* for clarity. The high frequency noise is an artifact of the finite size of the images used for Fourier transformations

both interfaces are corrugated with height 1 nm and there is perfect 3.2 nm lateral periodicity. The second structure is the one proposed by Lüerßen et al. [3.55] and Langbein et al. [3.57]: only the AlAs-on-GaAs interface is corrugated and the corrugation height is 0.34 nm. In both models the interface compositional gradient region was set to one atomic plane along the [311] direction (0.17 nm), so as to avoid having contrast sharper than the interplane distance in the crystal. The image in Fig. 3.18c was intensified to show the satellites due to lateral periodicity, which could not be resolved otherwise. The Fourier transformation of the processed experimental image of the randomly chosen part (20 nm in the vertical direction, 30 nm in the lateral direction) of the (311A) CSL with 3.7 nm vertical period (11 ML GaAs/11 ML AlAs) is shown in Fig. 3.18b. Note that the clear streaks indicating a lateral periodicity of 3.2 nm^{-1} in the experimental image of Fig. 3.18b and the checkerboard arrangement of the pattern in Figs. 3.18a and b are in disagreement with

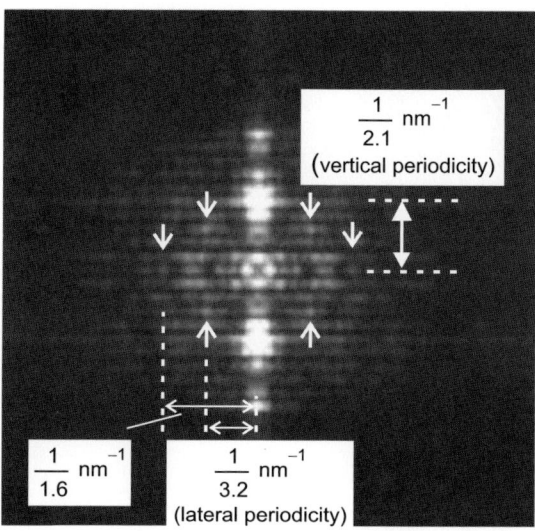

Fig. 3.19. Fourier transformation of the processed experimental image of a randomly chosen part (20 nm in the vertical direction, 30 nm in the lateral direction) of the (311A) corrugated superlattice with vertical period 1.9 nm (1 ML GaAs/10 ML AlAs). Streaks due to the lateral periodicity of 3.2 nm^{-1} are resolved. The characteristic lateral periodicity giving most of the integrated intensity is around 1.5–2 nm. Note the checkerboard arrangement of the pattern, and a higher degree of disorder than in Fig. 3.18b

the image in Fig. 3.18c, as indicated by arrows for clarity. The experimental image also contains disorder-induced weakly tilted streaks, corresponding to the facet angle of the CSL (40°), in agreement with the model by Nötzel et al. [3.37].

Figure 3.19 shows the Fourier transformation of the processed experimental image of the randomly chosen part (20 nm in the vertical direction, 30 nm in the lateral direction) of the (311A) CSL with 2.1 nm vertical period (1 ML GaAs/11 ML AlAs). The streaks due to the lateral periodicity of 3.2 nm are also resolved in this case. However, the characteristic lateral periodicity range, giving most of the lateral streak intensity, is around 1.5–2 nm. This indicates a significant degree of disorder. Again, one can resolve the checkerboard arrangement of the pattern.

To explain the difference in structural properties of the structures with complete and partial heteroepitaxial filling of the corrugated (311)A surface, Ledentsov et al. [3.64] proposed the growth model represented schematically in Fig. 3.20. The growth mode depicted in Fig. 3.20a corresponds to the model proposed by Nötzel et al. [3.37]. Both GaAs and AlAs coverage of the AlAs or GaAs surface, respectively, is complete. The growth mode depicted in Fig. 3.20b, corresponding to the case where only a fraction of the AlAs (311) surface is covered with GaAs clusters, should be different. If the surface

90   3. Self-Organization Phenomena at Crystal Surfaces

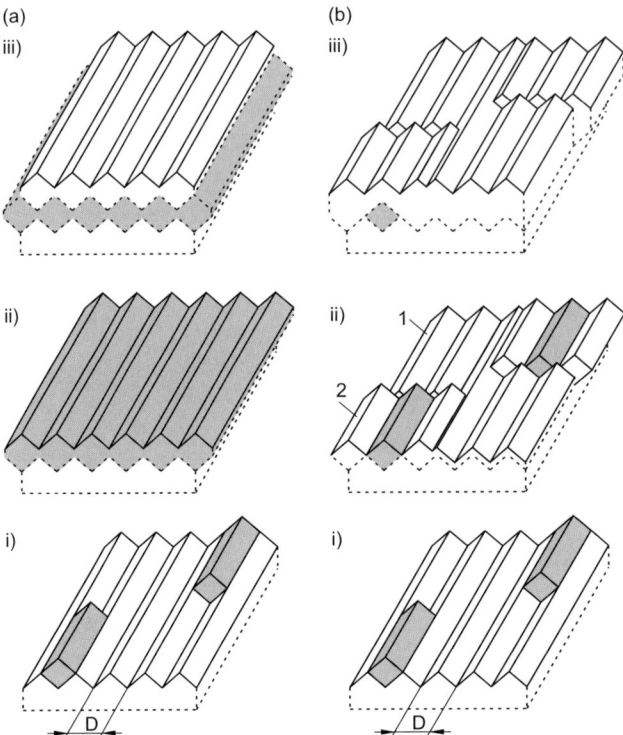

Fig. 3.20. Schematic representation of heteroepitaxial growth on corrugated surfaces. (**a**) AlAs surface completely covered by GaAs clusters. (i) Initial GaAs clusters (*grey*) on the AlAs corrugated surface (*white*). (ii) AlAs surface covered completely by GaAs. (iii) GaAs surface covered by AlAs. The hills on the top AlAs surface are positioned above the hills of the initial AlAs surface. The structure has lateral period D. (**b**) AlAs surface only partially covered by GaAs clusters. (i) Initial GaAs clusters on the AlAs surface. (ii) AlAs surface (1) far from GaAs clusters reproduces the initial AlAs surface. In the vicinity of GaAs clusters, due to non-wetting effects, AlAs forms clusters (2) having the opposite phase with respect to the initial AlAs surface, i.e., hills of the clusters are positioned above valleys of the initial surface. (iii) When GaAs clusters are completely covered by AlAs, the top surface of AlAs contains two types of domains. In additional to the periodicity D, the projected structure also has a periodicity of $D/2$

is only partly covered by GaAs clusters (see Fig. 3.20), AlAs growth may only proceed in a phase-correlated way in regions some distance away from the clusters, as also in the case of conventional AlAs homoepitaxy on corrugated surfaces. In the vicinity of the GaAs cluster, however, AlAs growth should terminate, since AlAs does not wet the GaAs (311)A surface as in the model explaining the corrugation phase shift [3.40]. Thus, AlAs will not grow on top of GaAs clusters unless other scenarios with a lower surface energy are possible. If the supply of Al atoms from the atomic beam continues, AlAs hillocks

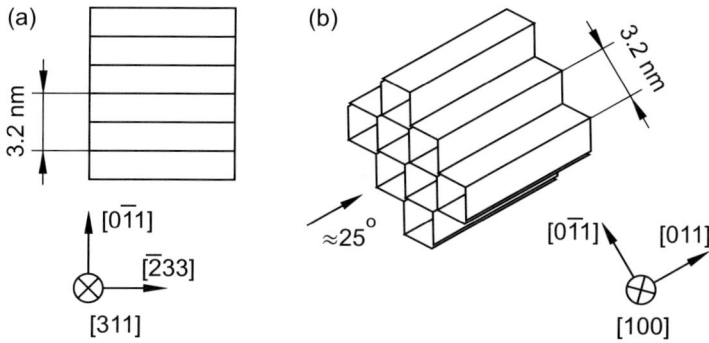

Fig. 3.21. Possible geometry for plan-view transmission electron microscopy (TEM) of superlattices grown on an (311)-oriented surface. (**a**) The direction of the electron beam is parallel to the normal to the sample surface. (**b**) The sample is tilted to adjust the electron beam direction along a [100] crystal direction. The projection of GaAs quantum wire arrays along both directions should give 3.2 nm periodicity, if the growth model proposed by Nötzel et al. [3.37] is valid

may be formed in the vicinity of the GaAs cluster, in regions where surface grooves were initially positioned (see Fig. 3.20). This stimulates the local surface phase-reversal effect. This process does not require any additional energy from the system, as both phase-shifted AlAs and normal homoepitaxy AlAs surfaces are equivalent. However, this process causes the appearance of different kinds of boundary between domains with two different phases of AlAs surface faceting, resulting in disorder within the system.

Plan-View TEM Studies of Multilayered AlAs/GaAs(311) Structures. Litvinov et al. [3.69] studied a short-period GaAs–AlAs superlattice grown on the GaAs (311)A surface using plan-view transmission electron microscopy (TEM) with two-beam dark-field (DF) and bright-field (BF) imaging and chemically sensitive {200}-type reflection, and high-resolution transmission electron microscopy (HRTEM). There is clear in-plane compositional modulation with a period of 3.2 nm in the [$\bar{2}33$]-direction. Our results confirm the formation of well-ordered vertically-aligned arrays of GaAs and AlAs quantum wires via self-organized growth on a corrugated (311)A surface, in full agreement with the model proposed by Nötzel et al. [3.37].

The short-period superlattice (SPSL) studied by Litvinov et al. [3.69] comprised 200 periods of alternating GaAs and AlAs layers with an average thickness of 1.8 nm for each layer. A 10 nm thick GaAs layer was grown on top. Samples for plan-view TEM were prepared by dimpling from the substrate side and rear side chemical etching to electron transparency using a mixture of NaOH (1 mol/l) and H_2O_2 (30%) at a proportion of 5:1. TEM was performed in a Philips CM200 FEG/ST microscope with a Scherzer resolution of 0.24 nm and an information limit of 0.16 nm. High-resolution images were recorded with a CCD camera.

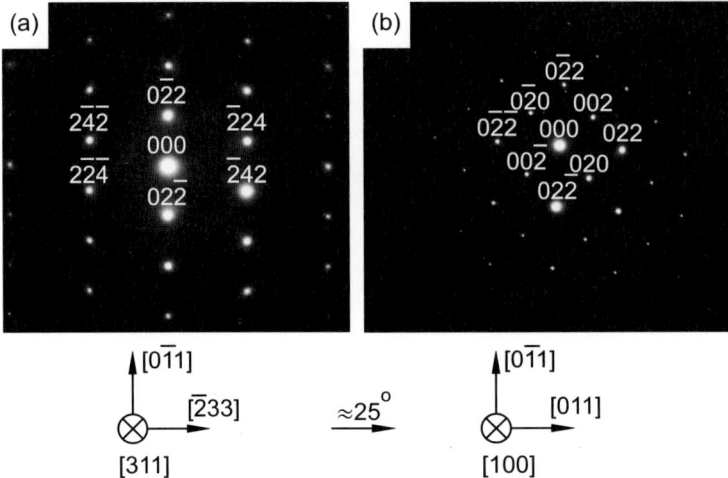

Fig. 3.22. Electron diffraction pattern for the two different sample orientations presented in Fig. 3.21. (**a**) No chemically sensitive {200}-type reflection exists. (**b**) Two pairs of {200} reflections sensitive to chemical composition modulations are available

The basic idea of the experiment is illustrated in Fig. 3.21. In TEM plan-view imaging, the [311] direction can be chosen for the incident electron beam corresponding to the normal to the sample surface (Fig. 3.21a). However, no chemically sensitive reflections exist in the [311] zone axis (see Fig. 3.21a). Tilting the sample from the [311] zone axis by about 25° around the [0$\bar{1}$1] direction (see Figs. 3.21b and 3.22) allows us to perform studies in the [100] zone axis, which contains chemically sensitive {200}-type reflections, as shown in Fig. 3.22b. Since the projection of the expected GaAs and AlAs wires along the electron beam direction are not affected by [111] tilting, one may expect to reveal the quantum-wire-related chemical contrast, if the latter is indeed present.

We note that approximately 10 alternating GaAs and AlAs layers are contained in the TEM sample, with an estimated maximum thickness of approximately 30 nm. Hence, any irregularities such as different local surface corrugation, disorder in different layers, tilting of compositional domains in the vertical direction due to step bunching, etc., would affect the modulation contrast. Sample bending and strain non-uniformities may also adversely affect the chances of observing lateral compositional modulation.

Figures 3.23a and b show {200} two-beam dark-field (DF) and bright-field (BF) TEM images of the same sample area. A stripe-like pattern is clearly revealed in the images. The characteristic width of the feature is close to 3.2 nm, as expected for the growth model described in [3.37]. The high contrast already revealed in the unprocessed DF TEM image supports a strong compositional modulation due to the formation of vertically-correlated

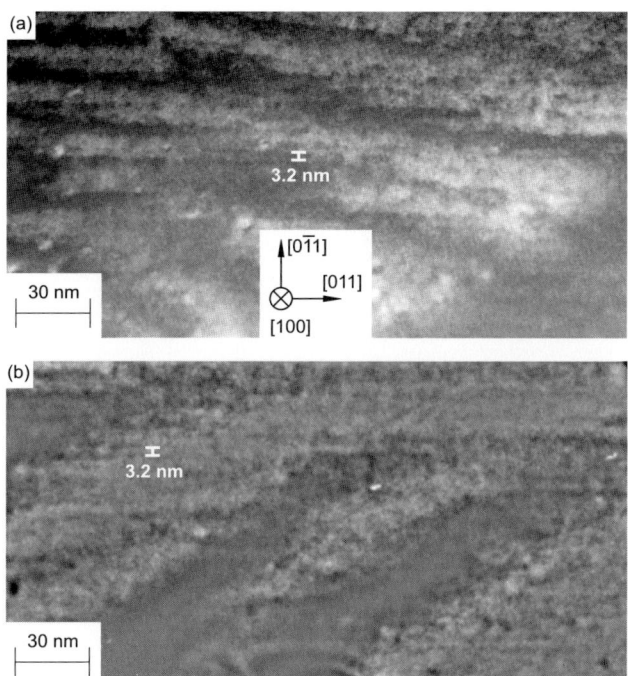

Fig. 3.23. Plan view transmission electron microscopy (TEM) images of the same sample area taken with a {200}-type reflection. (**a**) Two-beam dark field image. (**b**) Two-beam bright field image. A stripe-like pattern is clearly revealed in both images

GaAs and AlAs regions. We note that the compositional modulation in lateral superlattices is weaker ($\approx 50\%$ in the present case) than in conventional GaAs–AlAs vertical superlattices. In addition, the existence of the top 10 nm thick protective GaAs layer reduces the maximum contrast modulation to only 30% in the ideal case. The small period and effective grading of the composition by the lateral period further complicate imaging with TEM. We note that the assumption [3.55] of only one interface corrugated by 0.34 nm in SPSL, with the other being intermixed, would result in a compositional modulation by only 6%. Even more important, a vertical correlation of quantum wires can hardly be expected in the latter case, as no physical reason for this correlation exists if one of the interfaces is intermixed.

We note that the 3.2 nm characteristic feature size revealed in Fig. 3.23 is already at the limit of the resolution of conventional TEM imaging. To obtain more information on the uniformity of the quantum wire array, HRTEM studies were performed. An HRTEM image of the sample is shown in Fig. 3.24a. Stripes with periodicity 3.2 nm can also be revealed, even in the raw image. After the procedure of noise reduction and smoothing, these 3.2-nm-wide stripes become clearly visible along the whole imaging area (Fig. 3.24b). This

94   3. Self-Organization Phenomena at Crystal Surfaces

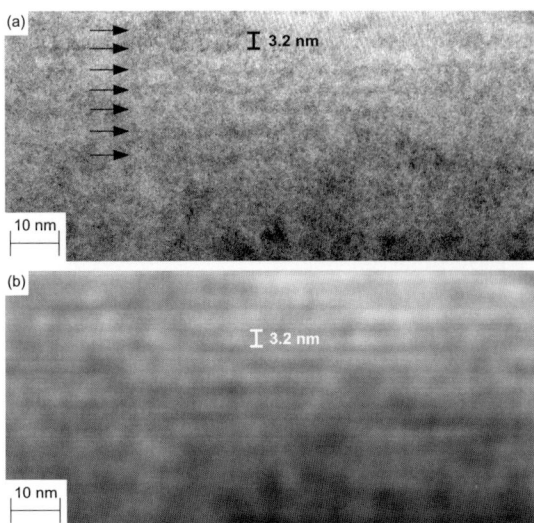

Fig. 3.24. Plan-view HRTEM image (**a**) of the (311)A-grown short-period superlattice. Note the striped pattern with 3.2 nm periodicity. The processed HRTEM image (**b**) more clearly evidences the 3.2 nm periodicity

illustrates the good uniformity of the stacked, vertically-correlated quantum wire arrays. Improved resolution in HRTEM studies seems to be important for adequate evaluation of the lateral superlattice periodicity and uniformity.

To evaluate the uniformity of the quantum wire array more quantitatively, Litvinov et al. [3.69] performed a Fourier transformation (FT) of the original HRTEM image depicted in Fig. 3.24a. The resulting FT image is shown in Fig. 3.25a. Clear periodic satellites are observed in the FT image, evidencing a highly periodic compositional modulation in the direction perpendicular to the quantum wires (indicated by the arrow in Fig. 3.25a). The period of the streaks corresponds to 3.2 nm. Even streaks corresponding to a 6.4 nm period are revealed. According to the results obtained for FTI of idealized lattices, the 6.4 nm streaks should always have high intensity if the contrast of neighboring stripes in the direct image differs significantly. The intensity of the corresponding reciprocal lattice streaks is depicted in Fig. 3.25b. One can see the high intensity of the first streak and a high peak-to-valley ratio for the periodic intensity profile, which also supports a high level of periodicity and uniformity.

To conclude this discussion about the structure of the GaAs(311) surface and GaAs/AlAs(311) interface, one should emphasize the significant progress that has been made in understanding this complicated structure. High resolution transmission electron microscopy (HRTEM) followed by mathematical processing of images allowed Ledentsov et al. [3.64] to make a clear reconstruction of the structure of the GaAs(311)A surface and confirm periodic

3.1 Periodically Faceted Surfaces

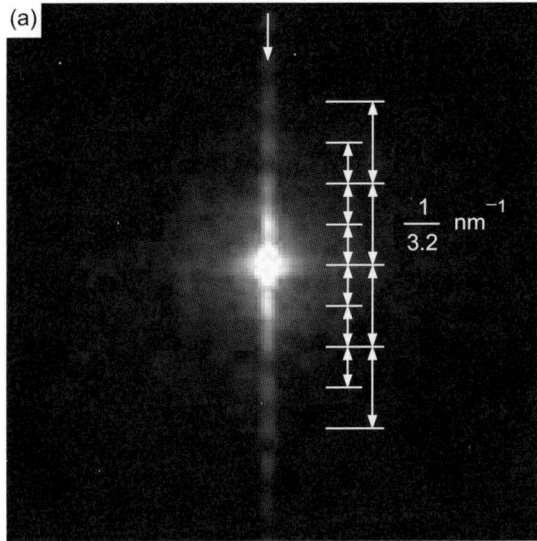

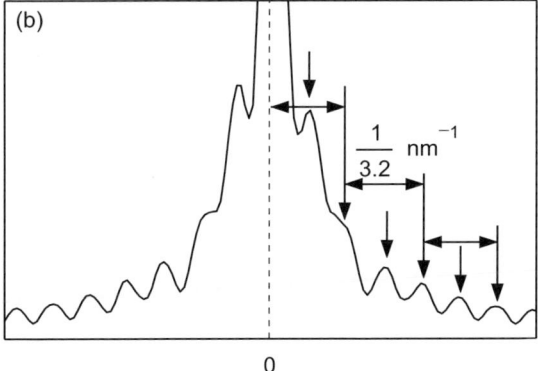

Fig. 3.25. (a) Fourier transform of the original HRTEM micrograph presented in Fig. 3.24a. (b) Intensity distribution along the direction of compositional modulation, indicated by an *arrow* in (a). Note the clear streaks in the reciprocal lattice, evidencing good vertical and lateral ordering of the quantum wire array

faceting leading to 1-nm-deep surface corrugation, according to the model originally proposed by Nötzel et al. [3.37]. Study of the superlattice (11 ML GaAs/11 ML AlAs) confirmed that both AlAs and GaAs layers are modulated in thickness. The corrugation height is about 1 nm in both cases and it is symmetric for both normal GaAs–AlAs and inverted AlAs–GaAs interfaces. Thicker sections of the thickness-modulated CSL with 2-nm-thick GaAs and AlAs layers are shifted by a half period with respect to each other, in agreement with a phase shift of the surface corrugation in heteroepitaxial growth on (311)A.

In the case of the superlattice (1 ML GaAs/11 ML AlAs) where GaAs only partially covers the AlAs surface, in addition to the 3.2 nm periodicity, a high probability of much smaller periods (1.5–2 nm) is revealed. The appearance of the smaller period in this case is attributed to the local phase reversal of the AlAs surface in the vicinity of GaAs clusters. HRTEM also indicates that the AlAs–GaAs interface inclination angles in both regimes are 40° and 140° with respect to the flat (311) surface.

The diversity of the results published in the literature, often contradicting the growth mode presented by Nötzel et al. [3.37], may be attributed to the non-optimized growth conditions chosen in those papers. For example, the system studied appears to be highly sensitive to intentional and background impurities. Impurity segregation may affect the growth mode and leads to a wavy surface with a corrugation scale of 10–40 nm, which was also observed in some samples grown under non-optimized vacuum conditions.

The recent HRTEM studies undertaken by Ledentsov et al. [3.64] and Litvinov et al. [3.69] confirm the model of GaAs(311) surface faceting and multilayered growth of AlAs/GaAs(311) superlattices first proposed by Nötzel et al. [3.37] and Alferov et al. [3.39]. GaAs–AlAs superlattices grown on a GaAs (311)A surface do indeed form arrays of vertically-correlated GaAs and AlAs quantum wires with a lateral periodicity of 3.2 nm and high uniformity.

3.1.9 Macroscopic Step Bunching and Faceting of Vicinal Surfaces

An important particular example of surface faceting is the faceting of a vicinal surface. A perfect vicinal surface of a crystal consists of flat terraces with low Miller indices, where neighboring terraces are separated by equidistant steps of monolayer height (Fig. 3.26a). Eventually, steps can form step bunches. Step bunching has long been known as one possible kinetic instability for crystal growth on vicinal surfaces (see, for example, [3.70]). Another possibility is equilibrium step bunching due to the faceting of a vicinal surface. The vicinal surface of Fig. 3.26a can break up to form an array of alternating vicinal facets with different tilt angles (or with different terrace widths) (Fig. 3.26b), an array of alternating singular facets (terraces) and vicinal facets (Fig. 3.26c), or an array of alternating singular facets (Fig. 3.26d).

Figure 3.27 depicts a periodic array of step bunches of macroscopic height. The discontinuity in the intrinsic surface stress tensor τ_{ij} at the edges results in effective force monopoles acting at the edges. Then the elastic interaction between the two edges of the same step bunch is the monopole–monopole interaction. Force monopoles acting at the two edges of the same bunch compensate each other, and the strain field due to a single-step bunch behaves at large distances like the strain field of an elastic dipole. Therefore, the elastic interaction between step bunches is the dipole–dipole interaction, which decreases as L^{-2} with the separation L, similarly to the elastic interaction

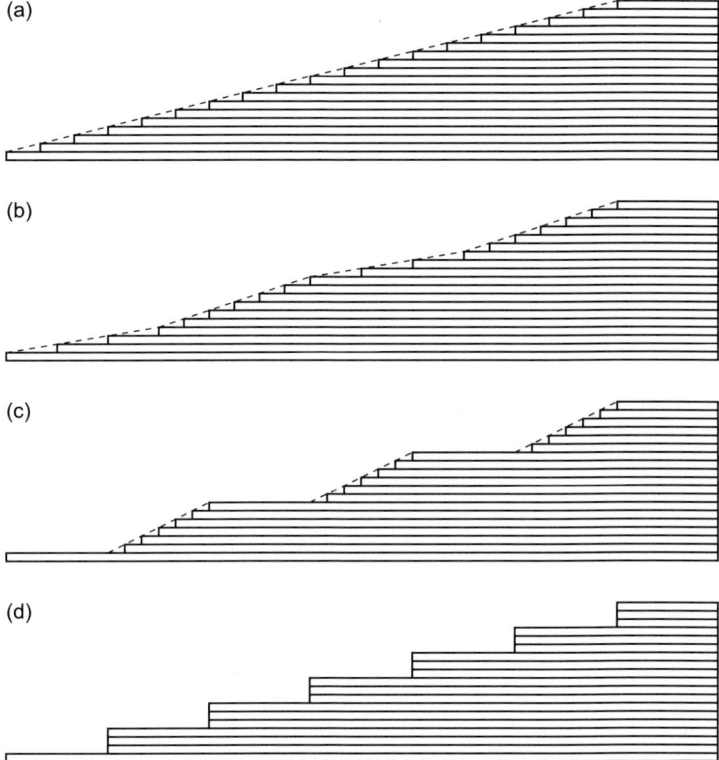

Fig. 3.26. Possible faceting configurations on a stepped vicinal surface. (**a**) Stepped vicinal surface comprising equidistant steps of monolayer height. (**b**) Alternating array of vicinal facets with two different tilt angles. (**c**) Alternating array of singular facets (terraces) separated by vicinal facets. *Dashed lines* in (**a**)–(**c**) indicate the average facet orientation of a stepped vicinal surface. (**d**) Alternating array of singular facets of different orientation

between steps of microscopic height [3.21]. When the dipole–dipole interaction and higher terms are truncated, the total energy per unit surface area equals

$$E = \gamma_0 + \gamma_1 \frac{h}{D} + \frac{C_1 \eta}{D} - \frac{C_2 \tau^2}{YD} \ln\left(\frac{h}{a}\right), \qquad (3.31)$$

where γ_0 is the surface energy of a flat terrace, γ_1 is the surface energy of a step bunch, and η is the energy of two edges, one convex and one concave, per unit length of the edges, similar to the coefficient in the second term in (3.28). C_1 and C_2 are geometric factors.

Since the average orientation of the faceted surface in Fig. 3.27 coincides with the orientation of the initially homogeneous vicinal surface, there exists a relation between the height h of macroscopic step bunches and the period D of the structure, viz., $h = D\theta$, where θ is the miscut angle of the initially

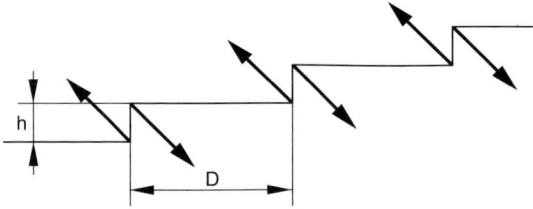

Fig. 3.27. Periodic array of macroscopic step bunches resulting from the faceting of a vicinal surface. Force monopoles act at the edges of the structure. The elastic interaction between the two edges of the same step bunch is the monopole–monopole interaction, whereas the interaction between different step bunches is the dipole–dipole interaction

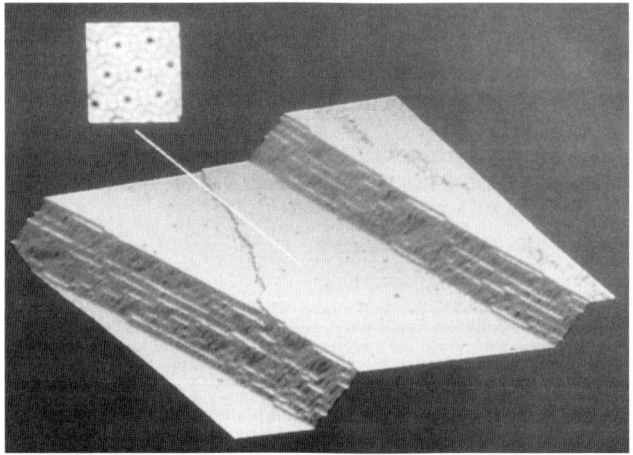

Fig. 3.28. 2000 Å × 2000 Å scanning tunneling microscopy image of the phase separation of a vicinal Si surface. At high temperatures, this surface displays uniform step density. (The net surface orientation is 4° from (111) towards $[\bar{2}11]$.) The two 'phases' which appear at low temperatures are the (7 × 7) reconstructed (111) facets and unreconstructed step bunches. There are 10 steps in each step bunch. The surface normal of the step bunches is temperature dependent. With kind permission of E.D. Williams et al. [3.5]

homogeneous vicinal surface. Using this relation, we can write (3.31) in a form similar to (3.28):

$$E = \gamma_0 + \gamma_1 \theta + \theta \left[\frac{C_1 \eta}{h} - \frac{C_2 \tau^2}{Y h} \ln \left(\frac{h}{a} \right) \right] .$$

Due to the logarithmic dependence of the elastic relaxation energy on the height of the step bunch, there always exists an optimum equilibrium height for step bunches.

Figure 3.28 depicts the scanning tunneling microscopy image of a vicinal Si(111) surface. It breaks up into an array of flat Si(111) terraces separated by step bunches (as in Fig. 3.26c). A detailed review of both experimental and theoretical studies of vicinal Si(111) surfaces can be found in the review article by Williams et al. [3.5], for example.

Figure 3.29 depicts an atomic force microscopy image of the vicinal GaAs(001) surface studied by Kasu and Kobayashi [3.41]. The vicinal GaAs(001) surface, misoriented by 2° towards [$\bar{1}$10], was grown by metalorganic chemical vapor deposition (MOCVD) and reveals the formation of multisteps. The height of multisteps, deduced from the distance between multisteps and the misorientation angle of the substrate, is about 17 ML. The height of the steps is found to be homogeneous throughout the sample.

Kasu and Kobayashi [3.41] discuss, in particular, the equilibrium vs. kinetic nature of step bunching. Triethylgallium (TEG), triethylaluminum (TEA), and arsine (AsH$_3$) were used as source materials. They grew a GaAs buffer layer and an (AlAs)$_{1/2}$(GaAs)$_{1/2}$ fractional layer superlattice (FLS) under an AsH$_3$ partial pressure of 95 Pa. For the final GaAs layer, the AsH$_3$ partial pressure was reduced to 6 Pa. After growing the FLS, the surface had a staircase of monolayer steps with almost the same separation. The height resolution of AFM for GaAs is 3 ML, and it was confirmed that there were no multisteps more than 3 ML high. Reducing the arsine pressure to 6 Pa leads to a longer diffusion distance for Ga atoms, thereby allowing the step structure transition.

Just after growth, the multistep edges are wavy (Fig. 3.29a). As annealing proceeds, the multistep separation remains almost unchanged and the step edges become straighter (Fig. 3.29b). These observations very much favor the equilibrium nature of the multisteps. Vicinal GaAs(001) surfaces misoriented in the [110] direction reveal similar step bunching. The bunches have a smaller height of about 7 ML.

The observation of equal-width terraces (or equidistant step bunches) was reported for vicinal GaAs(001) surfaces grown using molecular beam epitaxy

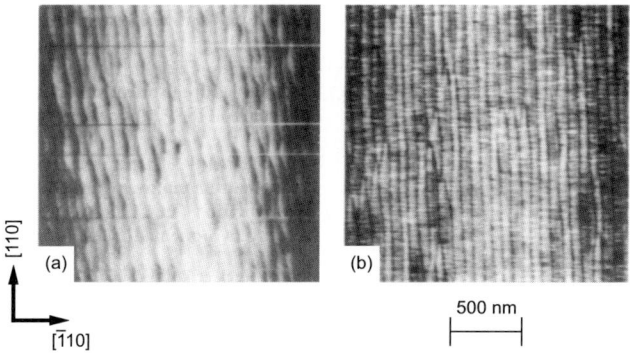

Fig. 3.29. Atomic force microscopy image of GaAs(001) vicinal surfaces (**a**) just after growth and (**b**) after 30 min annealing. The surface steps go down from left to right. With kind permission of M. Kasu and N. Kobayashi [3.41]

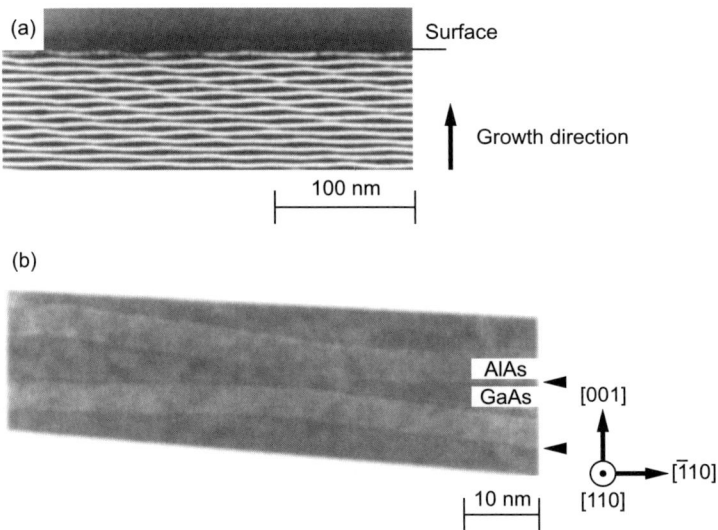

Fig. 3.30. Cross-sectional TEM images of an AlAs/GaAs superlattice on a GaAs(001) vicinal substrate. The *light region* corresponds to AlAs and the *dark region* to GaAs. (**a**) (002) dark-field image and (**b**) a higher-magnification lattice image. *Arrows* indicate the interface of GaAs under AlAs. With kind permission of M. Kasu and N. Kobayashi [3.41]

by Golubok et al. [3.71] and Ledentsov et al. [3.72]. The formation of an array of equidistant step bunches on vicinal surfaces of a semiconductor opens up the remarkable possibility of growing quantum wires by alternating deposition of two semiconductor materials.

Figure 3.30 shows an example of a structure obtained by alternating MOCVD growth of GaAs and AlAs. Kasu and Kobayashi [3.41] found a vicinal GaAs(001) substrate misoriented by 5° towards [$\bar{1}$10] to be the optimum case for studying the multistep structure. During growth, 14 ML amounts of AlAs and GaAs were supplied. Growth was interrupted for 1 min between the layers. AlAs was grown uniformly all over the surfaces and was used as a marker to show GaAs step bunching. If the AlAs surface structure follows the structure of the layer beneath, the top surface corresponds to the surface structure after annealing for 40 min. In Fig. 3.30a, the interface seems to consist of facets with two different orientations. Kasu and Kobayashi [3.41] reported the (001) and (117)B surfaces. Figure 3.30b shows with large magnification that the GaAs layer becomes narrow near the angle between the facets. This narrowing means that the GaAs layer consists of channels (wires) running along the [110] direction, with weak coupling between the wires. Such a structure can be considered as a modulated quantum well, or as an array of quantum wires, similar to the one grown on the faceted GaAs(311) surface by Nötzel et al. [3.37].

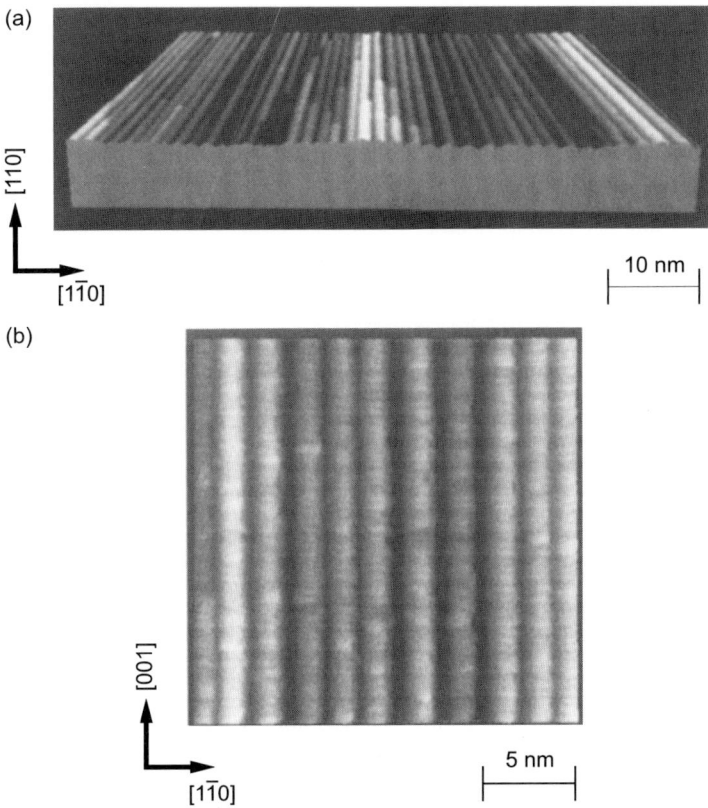

Fig. 3.31. STM image of the faceted TaC(110) surface. (**a**) 700 × 700 Å² height-mode STM prospective view from the clean TaC(110) surface following a slow cooling rate. Height is indicated by the *grey scale*, with *white* corresponding to the highest position. (**b**) Close-up of a ∼ 210 × 210 Å² area from the top of image (**a**). With kind permission of J.-K. Zuo et al. [3.74]

3.1.10 Variety of Periodically Faceted Surfaces

The range of surfaces which undergo spontaneous periodic nanofaceting is not limited to vicinal surfaces and the (311) family of GaAs and AlAs surfaces. Periodic faceting with nanometer-scale period has been observed on various surfaces of different materials. Low-index surfaces, like {100}, {110}, and {111}, are stable in most solids. Even more remarkable are situations where a low-index surface is unstable. One of the best studied cases is the faceting of TaC(110).

Zuo et al. [3.73] observed a rippled structure consisting of alternating (100) and (010) facets with an average periodicity of ∼ 6 lattice spacings from a clean transition-metal carbide surface TaC(110). High resolution low-energy electron diffraction (HRLEED) experiments were performed in an ultra-high vacuum (UHV) chamber with a bare pressure $< 10^{-10}$ torr. Scanning tunneling microscopy (STM) measurements were conducted in a separate UHV chamber, operated at a similar base pressure. The TaC(110) surface was cleaned by flashing to a temperature between 1 500 and 2 000°C, using electron bombardment, and then cooled down to room temperature. To assess the kinetic effects on ordering, a variety of cooling rates from the flash temperature were examined. The STM images show that the surface has a more clearly rippled formation and better long-range order along the unreconstructed direction [001] when a slow cooling rate is employed. This indicates that (100) faceting is an energetically favored structure. In these experiments, slow cooling rates occurred when the heating power was gradually reduced over a few minutes, resulting in a cooling rate of $\sim 250°$C/min. The sample, whose color was still red ($\sim 750°$C), was then cooled to room temperature in situ. In contrast, rapid cooling was accomplished by immediately turning off the heating power after flashing and allowing the sample to cool in situ.

From the HRLEED measurements, Zuo et al. [3.73] concluded that TaC(110) breaks up into an array of alternating (100) and (010) facets, the average period of the array being ~ 6 lattice spacings along the $[1\bar{1}0]$ direction, i.e., $La = 6a \sim 19$ Å. The STM image (Fig. 3.31) shows a ridge-and-valley grating structure, with ridges running along the [001] direction. The stability of the faceted structure was examined by comparing STM images obtained for different cooling rates. In the STM image obtained with a rapid cooling rate, the faceted structure exists, but in a quite disordered form, and the inter-range spacings are about the same as in the case of slow cooling. Along the [001] direction, ridges appear as beads with diameters of 10–15 Å, a length of two to three times the Ta–C basis spacing in this direction, and some ridges are displaced and terminated randomly. By comparing the two different STM images, Zuo et al. [3.73] drew the following conclusions. Firstly, the faceted surface is an energetically favored structure. If it were induced by high-temperature flashing and were frozen during the cooling process, then the slower cooling would result in a less pronounced facet structure. Secondly, the observed long bead chains may be a frozen-in morphology in the middle of the ordering process, which implies a short-range order of the facets at high temperatures.

The faceting of a low-index surface has also been observed on Ir(110) by Koch et al. [3.75], although no periodicity was revealed.

High-index surfaces are usually less stable than low-index surfaces, and we are more likely to observe faceting on a high-index surface. Nötzel et al. [3.37, 3.38, 3.76] observed faceting of several high-index surfaces of GaAs. They used molecular beam epitaxy to grow GaAs on substrates with various orientations. RHEED measurements were performed in situ. The RHEED pattern of the GaAs(211) surface observed along the $[\bar{1}11]$ azimuth shows an array of regular facets. The surface undergoes a reversible transition from the regular faceted surface to the flat surface below 590°C. The transition occurs continuously over a temperature range between 590°C and 550°C, indicating the equilibrium nature of the faceted structure. Faceting was also observed by Nötzel et al. for the (331) and (210) surfaces of GaAs.

Faceting has been observed on other high-index surfaces of GaAs. Higashiwaki et al. [3.77] reported faceting of the GaAs(775)B surface, which exhibits extremely straight step edges along [$\bar{1}$10], and a lateral corrugation period of about 12 nm.

3.1.11 Faceted Surfaces: Understanding and Prospects

To conclude the present section, we emphasize that the spontaneous periodic faceting of semiconductor surfaces and cluster growth in grooves provides the possibility of directly fabricating isolated quantum wires, quantum wire superlattices, and quantum well superlattices with modulated quantum well thicknesses.

So far, most applications have been connected with the family of (311) faceted surfaces of GaAs and AlAs and corresponding corrugated superlattices. Self-organized growth of ordered nanostructures on corrugated surfaces may have particularly important applications in opto- and microelectronics, as demonstrated in the patent by Nötzel et al. [3.78]. Such applications include lateral superlattices for infrared intersubband photodetectors with normal incidence, high-frequency two- and three-terminal devices based on lateral Esaki–Tsu superlattices, lasers, such as cascade lasers, and others.

Figure 3.32 shows photoluminescence (PL) spectra of two SPSL samples with different average thicknesses of GaAs and AlAs layers grown on (311)A GaAs substrates. The sample with 0.7 nm GaAs and 1.7 nm AlAs layer thicknesses exhibits broad PL emission in the green–orange spectral range, and the sample with 1.7 nm GaAs and 1.7 nm AlAs layer thicknesses exhibits PL in the orange–red spectral range. No luminescence in this spectral range is found for SPSLs grown side-by-side on (311)B and (100) substrates. As the (311)A structures studied in this work are type II SPSLs [3.38], we attribute the observed PL to giant mixing between Γ and X minima in the conduction band due to interface corrugation.

To conclude, it has been established that GaAs–AlAs superlattices grown on a GaAs (311)A surface form arrays of vertically-aligned GaAs and AlAs quantum wires with lateral periodicity 3.2 nm and high uniformity. Only the SLs grown on (311)A substrates demonstrate room-temperature luminescence in the green–yellow spectral range, evidencing a giant impact of interface corrugation on optical properties. It is proposed that vertical-cavity surface-emitting lasers for the yellow–green spectral range could be realized on (311)A GaAs substrates [3.79]. The structures discussed may also find numerous applications in diode lasers, Esaki–Tsu lateral superlattices, including three-terminal devices, normal-incidence middle- and far-infrared photodetectors and emitters, including cascade lasers, and so on.

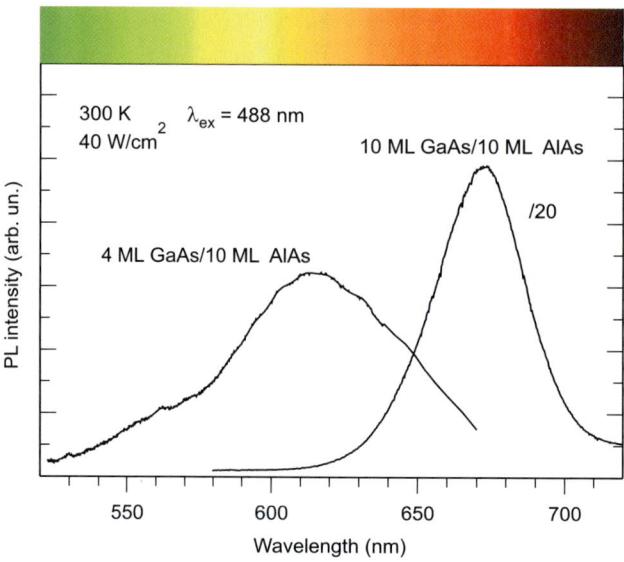

Fig. 3.32. Photoluminescence (PL) spectra of short-period superlattices with different numbers of GaAs and AlAs monolayers (4 and 10, or 10 and 10, respectively) grown on GaAs(311)A substrates. Note the spectral range of the PL. No PL in this spectral range is found in SPSLs grown side-by-side on (311)B and (100) substrates

The considerations in the present section have focused on single-crystal surfaces or lattice-matched AlAs/GaAs systems. Lattice-mismatched heteroepitaxy gives a much wider variety of self-organized nanostructures and this will be the subject of the following sections.

3.2 Surface Arrays of Two-Dimensional Islands

Heteroepitaxial semiconductor systems in which material 2 is deposited on a substrate of material 1 provide a large variety of self-organized nanostructures. In the equilibrium theory of heteroepitaxial growth, three growth modes are traditionally distinguished [3.80]. They are the Frank–van der Merwe (FM) [3.81], Volmer–Weber (VW) [3.82], and Stranski–Krastanow (SK) [3.83] growth modes. They may be described as layer-by-layer growth (2D), island growth (3D), and layer-by-layer growth plus islands (Fig. 3.33). The particular growth mode for a given system depends on the interface energies and lattice mismatch.

In lattice-matched systems, the growth mode is governed by the interface and surface energies only. If the sum of the epilayer surface energy γ_2 and the interface energy γ_{12} is lower than the energy of the substrate surface, $\gamma_2 + \gamma_{12} < \gamma_1$, i.e., if the deposited material wets the substrate, the FM mode occurs. A change in $\gamma_2 + \gamma_{12}$ alone may drive a transition from the FM to

3.2 Surface Arrays of Two-Dimensional Islands

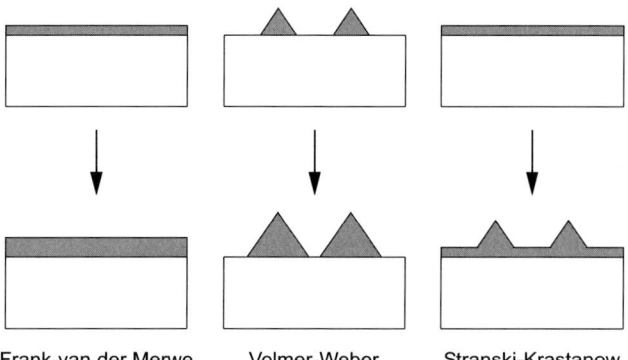

Frank-van der Merwe   Volmer-Weber   Stranski-Krastanow

Fig. 3.33. Schematic diagrams of the three growth modes for heteroepitaxial systems: Frank–van der Merwe (FM), Volmer–Weber (VW), and Stranski–Krastanow (SK)

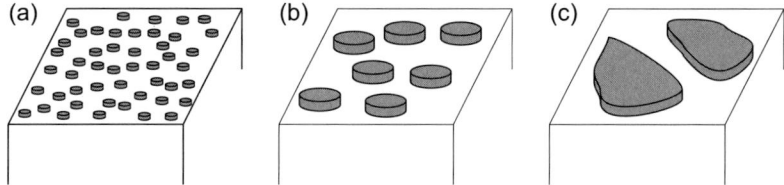

Fig. 3.34. Possible structures of a heteroepitaxial system with a submonolayer deposit. (**a**) Random distribution of deposited atoms over the surface. (**b**) Surface array of 2-dimensional islands. (**c**) Large random islands formed via ripening

the VW growth mode. For a strained epilayer with small interface energy, initial growth may occur layer-by-layer, but a thicker layer has a large strain energy and can lower its energy by forming isolated islands in which strain is relaxed. Thus the SK growth mode occurs.

If a given heteroepitaxial system grows according to FM or SK mode and the amount of deposited material is below one monolayer, one may distinguish the following possible arrangements of the deposit (Fig. 3.34):

- atoms of the deposit may be distributed randomly over the surface (Fig. 3.34a),
- atoms may form islands (Fig. 3.34b) determined by the intrinsic properties of the system,
- islands may undergo ripening and form large islands limited by some surface inhomogeneities, extended defects, or slow kinetics (Fig. 3.34c).

Atoms of the deposit may of course form some surface reconstruction on an atomic scale, or a combination of the above patterns.

Strictly speaking, the transition from a 2D to 3D morphology may also occur below one monolayer (ML) coverage. However, in the present section, we focus on systems in which the 2-dimensional morphology persists at least up to 1 ML.

In semiconductor systems, a submonolayer insertion may be overgrown by a substrate material, thus forming an attractive potential for electrons and holes in a semiconductor matrix. In the case of random distributions of atoms (Fig. 3.34a), the layer is a quantum well characterized by an alloy composition. In the case of large islands (Fig. 3.34c) with characteristic size exceeding the exciton Bohr radius, the layer may be regarded as consisting of fragments of a quantum well. Finally, in the case of islands (Fig. 3.34b) smaller than or of similar size to the exciton Bohr radius, the islands are quantum dots localizing excitons in the lateral plane.

The existence of an optimum island size follows from the theory of capillarity effects on a crystal surface, where the formation of equilibrium domain structures has been predicted theoretically by Andreev [3.35], Marchenko [3.84], Alerhand et al. [3.85] and Vanderbilt [3.86]. The first observations seem to be those in metallic systems O/Cu(110) by Kern et al. [3.87].

The first experimental studies of a submonolayer semiconductor system were those by Wang et al. [3.88]. Optical reflectance spectroscopy studies of a submonolayer (0.3 ML) insertion of InAs in a GaAs matrix revealed a high optical anisotropy. Such an anisotropy is not consistent with the picture of a homogeneous quantum well or with large fragments of a quantum well. It can only be explained by the formation of anisotropic islands, where anisotropy of island shape and/or the strain pattern leads to optical anisotropy. By scanning tunneling microscopy studies of an uncovered submonolayer InAs/GaAs system, Bressler-Hill et al. [3.89] confirmed the formation of islands with nearly equal width (4 nm) in the [$\bar{1}$10] direction and highly elongated in the [110] direction. Ledentsov et al. [3.90] showed that submonolayer InAs/GaAs systems remain stable, if an ultra-long (up to 1 000 s) growth interruption is introduced after deposition of InAs and before capping by GaAs. Such stability suggests the equilibrium nature of submonolayer islands of InAs on GaAs. Belousov et al. [3.91] revealed an ultra-high exciton oscillator strength for systems with submonolayer InAs insertions in a GaAs matrix. This shows the efficient localization of excitons in the lateral plane by InAs nanometer-scale islands.

Optical studies of CdSe submonolayer and monolayer insertions in a ZnSe matrix by Ledentsov et al. [3.92] also demonstrated a high modulation of the optical reflectance spectrum and no-phonon lasing at an energy close to the heavy-hole exciton. This type of lasing is not the same as that from quantum well-type structures and is likely to be caused by excitons that are strongly localized in the lateral plane. Cross-sectional high resolution transmission electron microscopy (HRTEM) studies of 1 ML CdSe insertions in a ZnSe matrix [3.93] revealed the formation of flat nanoscale islands.

For a direct proof of quantum dot-like behavior of submonolayer islands in semiconductor systems, an observation of an atom-like energy spectrum

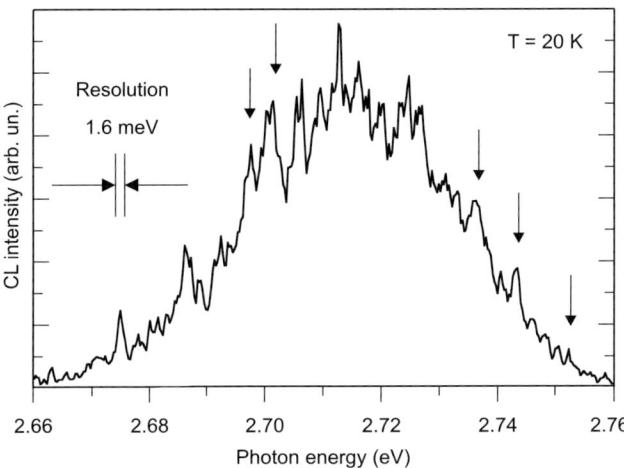

Fig. 3.35. Low temperature (20 K) cathodoluminescence (CL) spectra of a single 1 ML CdSe insertion in a ZnSSe matrix. The sharp luminescence lines in the spectrum originate from single quantum dots

of excitons is needed. For very dense arrays of QDs, e.g., as formed by submonolayer (SML) insertions, it may be difficult to resolve luminescence lines originating from individual QDs. Single QD luminescence can be revealed using spot-focus cathodoluminescence (CL) studies. The CL spectrum of an MOCVD-grown sample with a 1 ML CdSe insertion in a ZnSSe matrix obtained by Engelhardt et al. [3.94] is depicted in Fig. 3.35. The full width at half maximum of the sharp emission lines is limited by the spectral resolution. The lines originate from transitions in single QDs. Since the density of QDs, and hence also the density of single sharp lines, is still too high in the case of CL excitation conditions, a broad luminescence background is also detected.

To resolve luminescence from only a few individual QDs, a further reduction in the investigated area is necessary. One possible way to perform such measurements is to use small etched mesas. Ultrasharp luminescence lines due to single QDs and a high density of nanoscale QDs formed by 1–2 ML CdSe deposition in a ZnSe matrix using MBE growth have been observed by Kümmell et al. [3.95]. Measurements on mesas, however, suffer from pronounced non-radiative recombination at the mesa side-walls, particularly at elevated temperatures. Damaged regions near the mesa side-walls also result in an underestimation of the number of islands per mesa.

To avoid non-radiative leakage of nonequilibrium carriers at mesa sidewalls while keeping the investigated area small, one may use the approach of ultrasmall openings in mesa masks, as proposed by Türck et al. [3.96]. The technique has been used to resolve single QD emission lines up to high observation temperatures and to calculate the density of the QDs. The results appear to be in good agreement with values obtained from HRTEM studies.

The current section focuses on formation mechanisms for submonolayer arrays of islands in heteroepitaxial systems. There are several reasons why these are so very important:

- A submonolayer array of islands is a relatively simple model system allowing detailed study of the mechanisms underlying self-organized formation of arrays of nanometer-scale islands.
- The entropy effects in nanostructure formation and the interplay between thermodynamic and kinetic effects in island formation can be studied theoretically in great detail for SML systems and the results extended to arrays of 3D islands.
- Submonolayer islands can indeed exhibit electronic properties of quantum dots.
- The possibility of forming arrays of ultra-small islands with ultra-high density creates a broad field of applications for these systems.

3.2.1 Homoepitaxial Systems at Submonolayer Coverage

A fraction $q < 1$ of a monolayer is deposited on a flat surface and growth is interrupted. We assume that the desorption rate is very low. Then it is possible to define a 'surface equilibrium' in which the substrate works as a thermal bath with a given temperature T, the number of adsorbed atoms on the surface is fixed, atoms do not desorb, and no exchange processes occur between adatoms and substrate atoms. A necessary condition for the very existence of this surface equilibrium is that the time needed to reach this state be considerably shorter than the desorption time, i.e.,

$$t_{\text{eq}} \ll t_{\text{des}} \,. \tag{3.32}$$

A basic assumption in the present and following sections is that the strong inequality (3.32) holds. Then it is possible to introduce the growth interruption on a time scale t_{GI} such that

$$t_{\text{eq}} \ll t_{\text{GI}} \ll t_{\text{des}} \,. \tag{3.33}$$

Growth interruption during the time interval t_{GI} will drive the system into the surface equilibrium defined above.

To describe such an equilibrium, we consider a simple model of a lattice gas of adatoms on a square lattice with lattice parameter a. Two adatoms on two neighboring sites form a chemical bond with energy $(-w/2) < 0$. In large 2-dimensional islands, if boundary effects are neglected, the energy per atom equals $(-w)$. The chemical potential of atoms in large islands is

$$\mu_{\text{isl}} = -w \,. \tag{3.34}$$

The chemical potential of a dilute gas of adatoms is

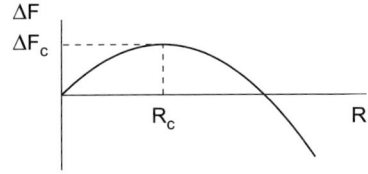

Fig. 3.36. Change in the free energy due to formation of a 2-dimensional circular island

$$\mu_{\mathrm{ad}} = k_{\mathrm{B}} T \ln q , \tag{3.35}$$

where q is the adatom density per site. If the gas of adatoms is in equilibrium with large islands, the two chemical potentials (3.34) and (3.35) must be equal, and the equilibrium density of adatoms is

$$q_{\mathrm{eq}} = \exp\left(-\frac{w}{k_{\mathrm{B}} T}\right) . \tag{3.36}$$

Deposited atoms at a higher surface coverage $q > q_{\mathrm{eq}}$ lead to supersaturation in the surface gas of adatoms:

$$\Delta \mu = k_{\mathrm{B}} T \ln \left(\frac{q}{q_{\mathrm{eq}}}\right) . \tag{3.37}$$

During growth interruption, adatoms tend to form islands. Consider the net change in the free energy due to the formation of a circular island of radius $R \gg a$,

$$\Delta F = 2\pi \eta_{\mathrm{b}} R - \frac{\pi R^2}{a^2} \Delta \mu , \tag{3.38}$$

where η_{b} is the boundary energy per unit length of the boundary (see Fig. 3.36). The quantity vanishes at $R = 0$, is negative at large R, and has a maximum

$$\Delta F_{\mathrm{c}} = \frac{\pi \eta_{\mathrm{b}}^2 a^2}{\Delta \mu} , \tag{3.39}$$

at a critical radius $R = R_{\mathrm{c}}$ equal to

$$R_{\mathrm{c}} = \frac{\eta_{\mathrm{b}} a^2}{\Delta \mu} . \tag{3.40}$$

To estimate the order of magnitude of the critical radius R_{c}, we note that the boundary energy per unit length η_{b} is due to the lack of a chemical bond for boundary atoms and is related to the bond energy by

$$\eta_{\mathrm{b}} = \frac{w}{2a} . \tag{3.41}$$

Substituting η_{b} from (3.41) into (3.40) yields

$$R_{\mathrm{c}} = \frac{a}{2} \frac{w}{w + k_{\mathrm{B}} T \ln q} = \frac{a}{2} \frac{w}{k_{\mathrm{B}} T \ln \left(\frac{q}{q_{\mathrm{eq}}}\right)} . \tag{3.42}$$

Equation (3.42) is valid only if the critical radius R_c is macroscopically large. Since typical values of the binding energy are about 1 eV, and typical growth temperatures are about 300–700°C, the equilibrium density of adatoms satisfies $q_{eq} < 10^{-4}$. Then (3.42) yields a macroscopic critical radius only if $q < 10^{-2}$. Otherwise, a critical island consists of just a few (e.g., 2) atoms.

The classical picture of island growth considers three stages in the evolution of an array of islands. In the first stage, critical islands nucleate. In the second stage, the islands grow due to incorporation of atoms from the dilute gas of adatoms. Supersaturation decreases until island growth comes to a halt. In the last stage, small islands become unstable and dissolve, providing atoms for the growth of large islands. This process is known as Ostwald ripening. The above picture lies behind the classical theory of Ostwald ripening developed by Lifshits and Slyozov [3.97, 3.98] and Wagner [3.99], and extended in a great number of subsequent papers.

When discussing Ostwald ripening here and throughout the book, we focus only on the thermodynamic force that drives the system towards ripening, or the absence of this driving force. Consider the free energy of an island normalized per atom in the island:

$$\frac{\Delta F}{N} = \frac{2\eta_b a^2}{R} - \Delta\mu \ . \tag{3.43}$$

The minimum of the free energy per atom corresponds to the size of islands $R \to \infty$. Thus, the evolution of an array of islands is the process where the average size of the islands increases, and the number of islands decreases. In real systems, large islands form which may be regarded as terraces for the next atomic layer. As the average size of the islands increases, the kinetics slows down, and the system contains a low density of large islands.

3.2.2 Energetics of a Heteroepitaxial System at Submonolayer Coverage

The energetics of a heteroepitaxial system at submonolayer coverage differs drastically from that of a homoepitaxial system. A heteroepitaxial system at submonolayer coverage can be considered as a system of two coexisting phases on the surface. The two phases are a bare substrate surface and islands formed by the deposit. The two coexisting phases generally have different surface energies and different intrinsic surface stress tensors τ_{ij}. Then effective force monopoles are applied to the crystal at the interphase boundary (Fig. 3.37),

$$F_\alpha = (\Delta\tau_{\alpha\beta}) m_\beta \equiv \left(\tau^{(2)}_{\alpha\beta} - \tau^{(1)}_{\alpha\beta}\right) m_\beta \ , \tag{3.44}$$

where m_β is the 2-dimensional vector normal to the domain boundaries. The effective force from (3.44) gives rise to elastic relaxation. The elastic energy of such a system can be written in a form similar to (3.20). A term linear in the strain indicates the existence of linear striction effects which favor

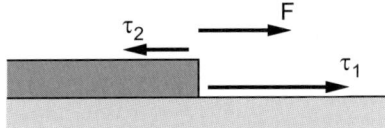

Fig. 3.37. Forces acting at the boundary of a monolayer-high island. *Light grey* corresponds to the substrate and *dark grey* denotes the island

the formation of periodic domain structures with macroscopic periods, as shown by Andreev [3.35]. The theory of surface structures of planar domains governed by the discontinuity of the intrinsic surface stress tensor τ_{ij} was developed by Marchenko [3.84]. Although the geometry of these structures is very different from that of periodically faceted surfaces, the energetics is basically the same. The total energy of the domain structure per unit surface area equals

$$E = E_{\text{surf}} + E_{\text{bound}} + \Delta E_{\text{elast}} \,. \tag{3.45}$$

The surface energy E_{surf} is determined by the fraction q of the surface covered by the adsorbate,

$$E_{\text{surf}} = (1-q)\gamma_{\text{sub}} + q\gamma_{\text{ads}} \,. \tag{3.46}$$

The energy of the domain boundaries equals

$$E_{\text{bound}} = \eta_{\text{b}} n_{\text{b}} \,, \tag{3.47}$$

where η_{b} is the short-range energy of the domain boundaries defined per unit length of the boundary, and n_{b} is the domain-boundary density (length per unit area).

The elastic relaxation energy can be written as a double integral over the domain boundaries,

$$\Delta E_{\text{elast}} = -\frac{1}{2} \oint \oint dl\, dl'\, (\Delta\tau)_{\alpha\gamma}\, m_\gamma(\boldsymbol{r}) G_{\alpha\beta}\left(\boldsymbol{r}-\boldsymbol{r}';z,z'\right)\bigg|_{\substack{z=0\\z'=0}} (\Delta\tau)_{\beta\delta}\, m_\delta(\boldsymbol{r}') \,. \tag{3.48}$$

where $G_{\alpha\beta}\left(\boldsymbol{r}-\boldsymbol{r}';z,z'\right)\big|_{\substack{z=0\\z'=0}}$ is the static Green tensor of the semi-infinite medium bounded by a flat stress-free surface $z=0$, and $\boldsymbol{m}(\boldsymbol{r})$ is the unit vector of the outer normal to phase 2. It is also useful to introduce the shape function $\Theta(\boldsymbol{r})$ of surface domains which equals 0 or 1 for phases 1 and 2, respectively. Then $\boldsymbol{m}(\boldsymbol{r}) = -\nabla\Theta(\boldsymbol{r})$. By applying the Gauss theorem twice to the integral (3.48), we obtain the elastic relaxation energy in the equivalent form

$$\Delta E_{\text{elast}} = \frac{1}{2} (\Delta\tau)_{\alpha\gamma} (\Delta\tau)_{\beta\delta} \int\int d^2r\, d^2r' \tag{3.49}$$

$$\times \Theta(\boldsymbol{r})\nabla_\gamma \nabla'_\delta \left[G_{\alpha\beta}\left(\boldsymbol{r}-\boldsymbol{r}';z,z'\right)\bigg|_{\substack{z=0\\z'=0}}\right] \Theta(\boldsymbol{r}') \,.$$

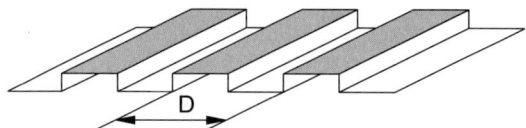

Fig. 3.38. An example of a striped domain structure with surface coverage $q = 1/2$

Striped Domain Structure. The simplest structure of surface stress domains is a striped domain structure (see Fig. 3.38). Let stripes run in the y direction, forming a periodic superlattice, and let the surface of both phases have at least two-fold symmetry, where the y and $(-y)$ directions are equivalent. Then the tensor $(\Delta\tau)_{\alpha\beta}$ has diagonal form with respect to the axes (x, y). For a periodic structure, it is convenient to define the reciprocal lattice. Then the elastic relaxation energy can be written as a sum over reciprocal lattice vectors $\boldsymbol{g}$, viz.,

$$\Delta E_{\text{elast}} = -\frac{(\Delta\tau)_{xx}^2}{2} \sum_{\boldsymbol{g}} g_x^2 \widetilde{G}_{xx}(g_x, g_y = 0; z, z') \Big|_{\substack{z=0 \\ z'=0}} \left|\widetilde{\Theta}(g_x)\right|^2 . \tag{3.50}$$

Below, we consider the elastically isotropic medium. The surface Green tensor in coordinate space is given in the textbook by Landau and Lifshits [3.31] as

$$G_{\alpha\beta}(\boldsymbol{r} - \boldsymbol{r}', z, z') \Big|_{\substack{z=0 \\ z'=0}} \tag{3.51}$$

$$= \frac{(1+\nu)}{\pi Y} \left[(1-\nu)\frac{\delta_{\alpha\beta}}{|\boldsymbol{r} - \boldsymbol{r}'|} + \nu \frac{(r_\alpha - r'_\alpha)(r_\beta - r'_\beta)}{|\boldsymbol{r} - \boldsymbol{r}'|^3} \right],$$

where Y is the Young's modulus, ν is Poisson's ratio, and $\delta_{\alpha\beta}=1$ if $\alpha = \beta$ and 0 otherwise. The Fourier transform of the Green tensor equals

$$\widetilde{G}_{\alpha\beta}(\boldsymbol{k}, z, z') \Big|_{\substack{z=0 \\ z'=0}} = \frac{2(1+\nu)}{Y} \left[\frac{\delta_{\alpha\beta}}{k} - \nu \frac{k_\alpha k_\beta}{k^3} \right] . \tag{3.52}$$

In each of (3.48), (3.49), and (3.50), logarithmic divergences occur at small length scales. They may be removed, for example, by introducing a cutoff a in (3.50) in the form of a damping factor $\exp(-2ga)$, where the cutoff length a is of the order of the lattice parameter.

Let us consider a striped domain structure in which phases 1 and 2 are present with fractions $(1 - q)$ and q, respectively. Then, evaluation of the elastic relaxation energy from (3.50) yields [3.85]

$$\Delta E_{\text{elast}} = -\frac{2\eta_d}{D} \ln \frac{D\sin(\pi q)}{2\pi a} , \tag{3.53}$$

where the coefficient

$$\eta_d \equiv \frac{(\Delta\tau)_{xx}^2 (1-\nu^2)}{\pi Y} \tag{3.54}$$

can be interpreted as the elastic relaxation energy per unit length of the domain boundary. We note, however, that such an interpretation is not rigorous, since ΔE_{elast} from (3.53) contains a logarithmic factor due to the long-range nature of elastic forces. By adding up the boundary energy and the elastic relaxation energy from (3.53) and substituting them into (3.45), one obtains the total energy of a domain structure per unit surface area as a function of the period D,

$$E_{\text{total}} = (1-q)\gamma_0 + q\gamma_1 + \frac{2\eta_{\text{b}}}{D} - \frac{2\eta_{\text{d}}}{D} \ln \frac{D \sin(\pi q)}{2\pi a} . \qquad (3.55)$$

Both coefficients η_{b} and η_{d} in (3.55) are positive, so that the energy of domain boundaries suppresses domain formation while the elastic relaxation energy promotes it. Because of the logarithmic dependence of E_{total} on the period D, there always exists a total energy minimum at a certain period D_{opt} given by

$$D_{\text{opt}} = \frac{2\pi a}{\sin(\pi q)} \exp\left(\frac{\eta_{\text{b}}}{\eta_{\text{d}}} + 1\right) . \qquad (3.56)$$

The period of the domain structure at coverage $q = 1/2$ equals

$$l_0 = 2\pi a \exp\left(\frac{\eta_{\text{b}}}{\eta_{\text{d}}} + 1\right) . \qquad (3.57)$$

It will be shown below that l_0 defines a scale of domain sizes for various coverages and different domain geometries.

Substituting D_{opt} from (3.56) into (3.45) gives the total energy of the domain structure with the optimum period,

$$E_{\text{total}} = (1-q)\gamma_0 + q\gamma_1 - \frac{2\eta_{\text{d}}}{l_0} I(q) , \qquad (3.58)$$

where the dimensionless function $-I(q)$ for striped domains equals

$$-I(q) = -I_{\text{strip}}(q) = -\sin(\pi q) . \qquad (3.59)$$

The third term in (3.58) is negative and gives the reduction in the energy of the system due to the formation of a periodic domain structure, compared with the energy of a system which has two completely separated surface phases.

Equations (3.55) and (3.57) were first obtained by Marchenko [3.84] for structures where domains of the two phases have equal widths ($q = 1/2$). The explosion in interest in surface stress domain structures is connected with the domain pattern on the Si(001) surface first observed by Men et al. [3.100] using scanning tunneling microscopy. These observations were explained by Alerhand et al. [3.85], who also obtained (3.56) for arbitrary q. The (001) surface of silicon has tetragonal symmetry and, for typical Si growth conditions, the surface structure is a (2×1) reconstruction, the surface atoms forming dimers. Since bulk silicon has two atoms per unit cell, its lattice is not a Bravais lattice. If one atomic layer is added to the surface, the surface

114   3. Self-Organization Phenomena at Crystal Surfaces

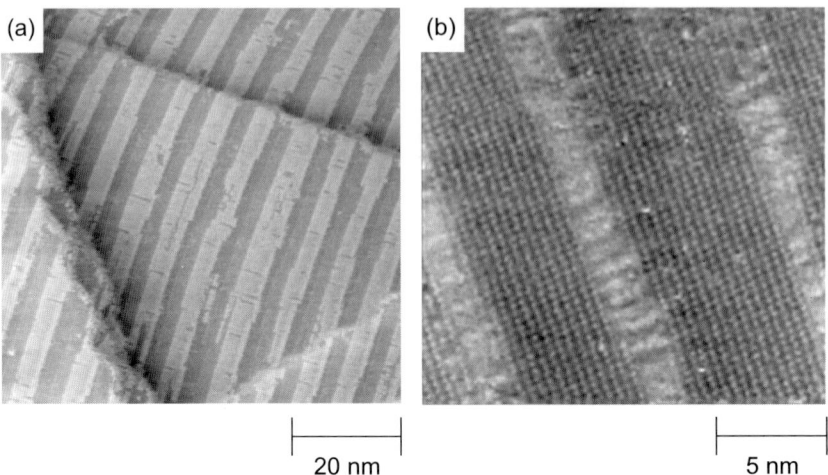

Fig. 3.39. Scanning tunneling microscopy (STM) image of the Cu(110)–(2 × 1)O surface. (**a**) Image at an oxygen coverage of $q = 0.26$. Annealing temperature ~ 550 K. *Dark stripes* are Cu(110)–(2 × 1)O islands and *bright stripes* are clean Cu(110)–(1 × 1) areas. (**b**) Image at an oxygen coverage of $q = 0.38$. Annealing temperature ~ 600 K. With kind permission of K. Kern et al. [3.87]

structure is rotated by 90° and we have the (1 × 2) reconstruction. If a second atomic layer is added, the reconstruction is again (2 × 1). Thus, if two neighboring terraces on the Si(001) surface are separated by a single-height atomic step, the atomic structures of these terraces are rotated by 90° with respect to each other. Hence, the intrinsic surface stress tensors of the two domains are

$$\tau_{ij}^{(1)} = \begin{pmatrix} \tau_\perp & 0 \\ 0 & \tau_\| \end{pmatrix}, \quad \tau_{ij}^{(2)} = \begin{pmatrix} \tau_\| & 0 \\ 0 & \tau_\perp \end{pmatrix}, \tag{3.60}$$

where $\tau_\|$ and $\tau_\perp$ are surface stress components in directions parallel and perpendicular to the dimers, respectively. The discontinuity in the surface stress tensor on domain boundaries,

$$\Delta \tau_{ij} = (\tau_\| - \tau_\perp) \begin{pmatrix} 1 & 0 \\ 0 & -1 \end{pmatrix}, \tag{3.61}$$

gives rise to the effective force monopoles applied at the steps and thus to a domain structure. For a double-height step, the two neighboring terraces have the same structure and no force monopole is applied to the step. The force monopoles at the steps play a key role in forming a wide variety of vicinal Si(001) surface structures, as considered theoretically by Alerhand et al. [3.101] and Poon et al. [3.102]. A review of both experimental and theoretical works on singular and vicinal Si(001) may be found in the paper by Mukherjee et al. [3.4], for example.

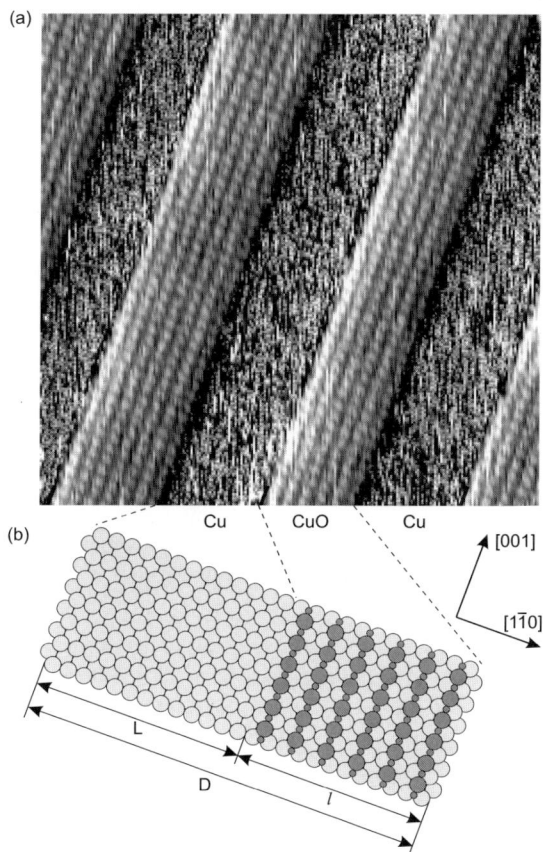

Fig. 3.40. Cu–CuO striped phase. (a) Scanning tunneling microscopy (STM) image of the Cu–CuO striped phase (oxygen coverage $q_0 = 0.25$). Scanned area 154 Å × 154 Å. (b) Schematic atomic structure of the surface. *Light grey* marks Cu atoms whereas *dark grey* is used for oxygen. With kind permission of G. Boishin et al. [3.103]

A classical example of a heteroepitaxial system showing the spontaneous formation of a surface-stress domain structure is submonolayer coverage of oxygen on Cu(110), studied by Kern et al. [3.87]. Oxygen was adsorbed by exposing the Cu(110) surface to an O_2 background pressure of 2×10^{-9} mbar. When the desired coverage was reached, the oxygen atmosphere was pumped off. The surface structure was examined by He diffraction and scanning tunneling microscopy (STM).

Oxygen on the Cu(110) surface forms the (2×1) reconstructed phase, and the coverage $q = 0.5$ ML refers to the full (2×1) coverage. Because of the strongly attractive Cu–O interactions, long Cu–O strings are formed along the [001] direction, on top of the substrate. In a wide average range

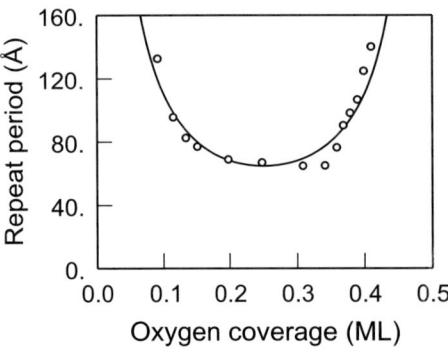

Fig. 3.41. Period of striped domain structures vs. oxygen coverage q. Points are data from Kern et al. [3.87]. The *curve* is (3.62) with $a = 2.55$ Å and $\eta_b/\eta_d = 0.40$. With kind permission of D. Vanderbilt [3.86]

($0.05 < q < 0.45$), Kern et al. [3.87] found that the (2×1)-reconstructed Cu–O island stripes consisting of 8 to 14 Cu–O strings arrange themselves in the form of a one-dimensional periodic grating (see Fig. 3.39). The periodic grating forms properly upon oxygen exposure while keeping the Cu surface at $T > 450$ K, but also under room-temperature exposure and subsequent annealing at $T > 450$ K. Accurate measurements of the fraction of bare Cu(110) surface, and hence the oxygen coverage, were performed by He diffraction. The same method revealed the period of the domain structure. Figure 3.40 shows an STM image with higher resolution, obtained from a similar structure by Boishin et al. [3.103].

The dependence of the period of the domain structure on the oxygen coverage measured in [3.87] has attracted a great deal of theoretical interest. Marchenko [3.104] and Vanderbilt [3.86] independently explained the observed dependence. Marchenko [3.104] described the observed structure as a system of surface stress domains. Vanderbilt [3.86] assumed that the structure is formed due to electrostatic interaction where the two phases differ in work function. Both approaches give the same result, where the period D_opt depends on the coverage q according to (3.56). For the particular (2×1) reconstructed phase of O/Cu(110), the value $q = 0.5$ corresponds to complete coverage, and one should replace $q \to 2q$ in (3.56). The dependence of the structure period on coverage thus takes the form

$$D_\text{opt} = 2\pi a \frac{1}{\sin(2\pi q)} \exp\left(\frac{\eta_b}{\eta_d} + 1\right). \tag{3.62}$$

Figure 3.41 compares Kern's observations [3.87] with (3.62). The figure illustrates the very good agreement between theory and experiment.

Disk-to-Stripe Transition. A 1-dimensional periodic array of stripes is the simplest possible structure for surface stress domains. Vanderbilt [3.86] considered a variety of domain structure geometries formed through electro-

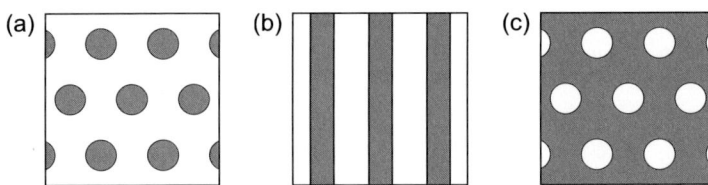

Fig. 3.42. The minimum-energy domain structures for the isotropic dipolar model. (a) Hexagonal array of droplets. (b) Striped structure. (c) Hexagonal array of inverted droplets. With kind permission of K.-O. Ng and D. Vanderbilt [3.105]

static interactions, where the two phases differ in work function. Numerical calculations of the total energy predicted a phase transition sequence as a function of the surface coverage q. For $0 < q < 0.286$, the minority phase forms a hexagonal array of droplets (Fig. 3.42a). For intermediate coverage $0.286 < q < 0.714$, the striped structure of Fig. 3.42b is the most favorable. For $0.714 < q < 1$, a hexagonal array of inverted droplets (Fig. 3.42c) is preferred energetically.

Later, Ng and Vanderbilt [3.105] recognized that a wide variety of physically different domain structures exhibit similar properties and can be described in a general form within the isotropic 2-dimensional dipolar model with $1/r^3$ interaction. Surface stress domain structures belong to this model if we assume an elastically isotropic medium and orientation-independent boundary energy. Then $(\Delta\tau)_{\alpha\beta} = (\Delta\tau)\delta_{\alpha\beta}$.

For a hexagonal array of droplets, we reproduce here the results due to Vanderbilt [3.86]. However, in contrast to [3.86], we will calculate the elastic relaxation energy by summation in real space. For an isotropic system, the elastic relaxation energy (3.48) reduces to

$$\Delta E_{\text{elast}} = \frac{\eta_\text{d}}{2} \int d^2 r \int d^2 r' \Theta(r)\Theta(r') \frac{1}{|r - r'|^3} \ . \tag{3.63}$$

The total energy of the hexagonal array of droplets defined per unit surface area is given by

$$E_{\text{total}} = \frac{1}{A_0} \left[\widetilde{E}_{\text{bound}} + \widetilde{\Delta E}_{\text{elast}} + \frac{1}{2\mathcal{N}} \sum_p \sum_{q \neq p} U(p,q) \right] \ , \tag{3.64}$$

where A_0 is the unit cell area of the hexagonal superlattice of droplets. The quantities $\widetilde{E}_{\text{bound}}$ and $\widetilde{\Delta E}_{\text{elast}}$ are the boundary energy and elastic relaxation energy of a single circular island. They are calculated in (A.50) of Appendix A. For a circular island of radius ρ,

$$\widetilde{E}_{\text{bound}} = 2\pi\rho\eta_\text{b} \ , \tag{3.65a}$$

$$\widetilde{\Delta E}_{\text{elast}} = -2\pi\rho\eta_\text{d} \ln\left(\frac{4\rho}{\text{e}^2 a}\right) \ . \tag{3.65b}$$

The interaction energy between the two circular islands p and q is calculated in (B.15) of Appendix B, and equals

$$U(p,q) = \eta_d \left(\pi\rho^2\right)^2 \frac{1}{R(p,q)^3} F\left[\left(\frac{\rho}{R(p,q)}\right)^2, \left(\frac{\rho}{R(p,q)}\right)^2\right], \quad (3.66)$$

where $R(p,q)$ is the distance between the centers of the two circular islands and $F(\xi_1, \xi_2)$ is the correction factor due to the deviation of the exact interaction energy from that in the dipole–dipole approximation. The factor F is obtained from (B.16).

The unit cell area A_0 of a hexagonal superlattice equals $A_0 = \sqrt{3}R_0^2/2$, where R_0 is the distance between neighboring droplets. Since the coverage q is defined as the ratio of the area of a single disk to the area A_0,

$$q = \frac{\pi\rho^2}{\sqrt{3}R_0^2/2}, \quad (3.67)$$

this allows us to relate the distance R_0 between islands to their radius. When calculating the contribution of the interaction energy between the islands to the total energy, it is useful to express R_0 in terms of the radius ρ of the islands and the coverage q. Thus, the interaction energy of the two neighboring islands is of the order of $\eta_d \rho q^{3/2}$. Substituting (3.65a), (3.65b), (3.66) and (3.67) into (3.64) yields the energy per unit surface area as a function of the disk radius ρ, viz.,

$$E_{\text{total}} = (1-q)\gamma_0 + q\gamma_1 + q\left[\frac{2\eta_b}{\rho} - \frac{2\eta_d}{\rho}\ln\left(\frac{4\rho}{e^2 a}\right) + \frac{\eta_d}{2\sqrt{\pi}\rho}\left(\frac{\sqrt{3}}{2}q\right)^{3/2} S(q)\right]. \quad (3.68)$$

The last term in (3.68) is the interaction energy, and the function $S(q)$ equals

$$S(q) = \sum_{r_j \neq 0} \frac{1}{r_j^3} F\left(\frac{\sqrt{3q}}{2\pi r_j^2}, \frac{\sqrt{3q}}{2\pi r_j^2}\right). \quad (3.69)$$

The summation in (3.69) is taken over the hexagonal superlattice, where the nearest neighbor distance equals 1, r_j is the dimensionless distance between a given disk and the jth disk, and the function F is defined in (B.16). Minimizing the total energy over the disk radius ρ yields

$$\rho_{\text{opt}}(q) = \frac{e^2}{8\pi} l_0 \exp\left[\frac{1}{4\sqrt{\pi}}\left(\frac{\sqrt{3}}{2}q\right)^{3/2} S(q)\right]. \quad (3.70)$$

Figure 3.43 depicts the optimum radius of the island and the optimum number of atoms in the island $N_{\text{opt}} = \pi\rho_{\text{opt}}^2/a^2$ as a function of the surface coverage q. Substituting ρ_{opt} from (3.70) into (3.68) yields the total energy of the hexagonal superlattice of droplets as

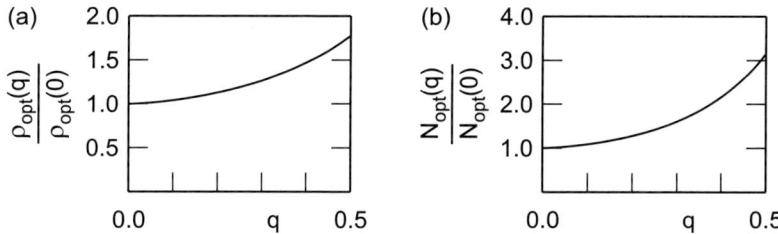

Fig. 3.43. Optimum size of a disk-shape island vs. surface coverage. (**a**) Disk radius ρ as a function of q. (**b**) Number of atoms in the island as a function of q

$$E_{\text{total}}(q) = (1-q)\gamma_0 + q\gamma_1 - \frac{2\eta_d}{l_0} I_{\text{drop}}(q), \qquad (3.71)$$

where the function $-I_{\text{drop}}(q)$ for the hexagonal superlattice of droplets is

$$-I_{\text{drop}}(q) = -q\frac{8\pi}{e^2} \exp\left[-\frac{1}{4\sqrt{\pi}}\left(\frac{\sqrt{3}}{2}q\right)^{3/2} S(q)\right]. \qquad (3.72)$$

Figure 3.44 compares the total energies of the striped domain structure (3.59), the hexagonal superlattice of circular droplets (3.72), and the superlattice of inverted droplets, where $I_{\text{inv}}(q) = I_{\text{drop}}(1-q)$. Our calculations reproduce the results of Vanderbilt [3.86] and Ng and Vanderbilt [3.105], performed earlier by calculations in the $\boldsymbol{k}$ space. The transitions between different structures correspond to coverages $q = 0.286$ and $q = 0.714$. Ng and Vanderbilt [3.105] showed that, in a narrow interval of coverage around transition points, striped domains and droplets may coexist, forming a structure of superdomains.

Real crystal surfaces are always anisotropic. Firstly, the energy of a domain boundary depends on the orientation of the boundary. Secondly, elastic

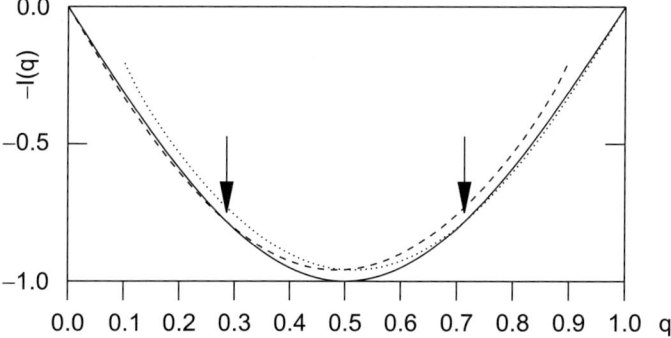

Fig. 3.44. Comparison of dimensionless energies per unit area, optimized with respect to scale, for striped (*solid curve*), droplet (*dashed curve*), and inverted-droplet (*dotted curve*) domain structures, as a function of area coverage q. With kind permission of K.-O. Ng and D. Vanderbilt [3.105]

120   3. Self-Organization Phenomena at Crystal Surfaces

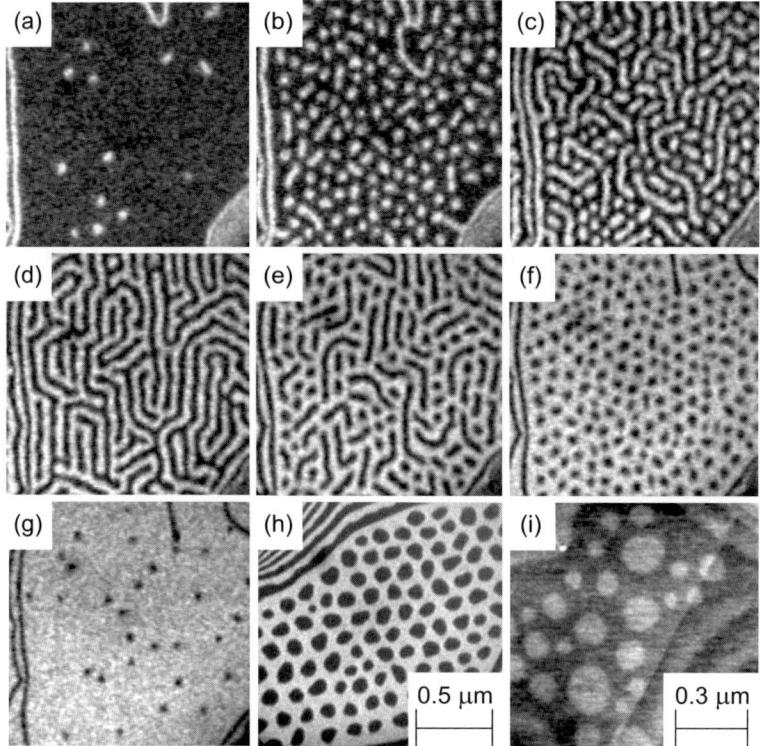

Fig. 3.45. Self-assembly of Pb on Cu(111). Low-energy electron micrographs of the Cu(111) surface at 673 K with different area fractions of the lead overlayer phase (*bright*) in the surface alloy phase (*dark*). (**a**)–(**g**) Area fractions 0.03, 0.28, 0.35, 0.50, 0.65, 0.73 and 0.95, respectively. The domain pattern evolves from circular islands (droplets) to stripes, and then to vacancy islands (inverted droplets) with increasing lead coverage. (**h**) Ordered droplet configuration at 623 K. Scale bar 0.5 μm. (**i**) Atomic force micrograph of a droplet pattern after cooling to room temperature and 2 hr exposure to air. Scale bar 0.3 μm. With kind permission of R. Plass et al. [3.106]

interaction is mediated by an elastically anistropic substrate. Ng and Vanderbilt [3.105] argued that anisotropy should favor striped domains rather than disk-shaped domains.

Theoretical prediction by Vanderbilt [3.86] and Ng and Vanderbilt [3.105] of transitions from droplets to stripes and then to inverted droplets have received remarkable experimental confirmation by Plass et al. [3.106]. Figure 3.45a–g shows a sequence of low energy electron microscope images of the Pb/Cu(111) system at different Pb coverages. The most striking feature of this sequence is the evolution of the pattern from circular islands (average diameter 67 nm) to stripes and then to circular holes within the lead overlayer matrix. The transition from droplets to stripes occurs at a coverage between

0.28 (Fig. 3.45b) and 0.35 (Fig. 3.45c), whereas the transition from stripes to inverted droplets occurs at a coverage between 0.65 (Fig. 3.45e) and 0.73 (Fig. 3.45f), which agrees perfectly with theoretical predictions [3.105].

Stability of Submonolayer Islands Against Ripening. The fact that the equilibrium state of a submonolayer system is a periodic domain structure has a very important consequence in the low coverage limit $q \to 0$. The minority phase forms a hexagonal array of droplets with a negligibly small elastic interaction. The radius of the droplet is given by (3.70) at $q = 0$ and equals

$$\rho_{\rm opt}(q) = \frac{e^2}{8\pi} l_0 \,. \tag{3.73}$$

The value $\rho_{\rm opt}$ is $l_0/3.4$, roughly 3 times smaller than the period l_0 of the domain structure at coverage $q = 1/2$. It has been emphasized by Zeppenfeld et al. [3.107] that a similar relation between the size of an isolated island in the dilute limit and the period of the structure at coverage 0.5 is rather general and holds for a wide variety of surface domain structures.

The existence of an optimum radius (3.73) for islands demonstrates that, if the system evolves towards equilibrium, islands tend to reach this optimum radius and will not undergo Ostwald ripening. The stability of the islands against Ostwald ripening is a crucial difference between heteroepitaxial and homoepitaxial submonolayer systems.

Three-Dimensional Islands with Fixed Height. Some other physically different systems exhibit similar behavior to 2-dimensional submonolayer islands. An example are islands whose height h exceeds 1 monolayer and is fixed by energetics or by kinetic restrictions, but whose lateral size can change. An example of a 3-dimensional island is shown in Fig. 3.46, where

$$a \ll h \ll L \,. \tag{3.74}$$

Here L is the lateral dimension of the island and h its height.

If a uniform flat film of material 2 is coherently conjugated to a substrate of material 1, the stress tensor in the film has non-vanishing in-plane components $\sigma_{\alpha\beta}$ ($\alpha, \beta = 1, 2$). In an island with finite lateral size L, side facets lead to elastic relaxation. A non-uniform strain field can be described as the strain

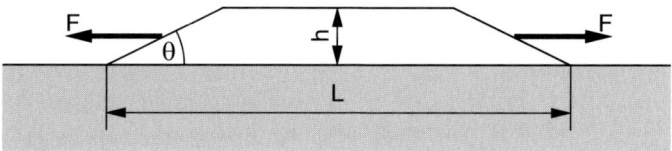

Fig. 3.46. A 3-dimensional island with low aspect ratio $h/L \ll 1$. Effective forces $\boldsymbol{F}$ applied to the side facets result in elastic relaxation. *Light grey* denotes the substrate

field created by effective forces with density $P_\alpha = -\sigma_{\alpha\beta} n_\beta$ applied to the side facets of the island, where n_β is the 3-dimensional unit vector normal to the island surface (see, for example, [3.108, 3.109]). If one considers the boundary of the island base, the total force per unit length of the boundary is then

$$F_\alpha = \sigma_{\alpha\beta} h m_\beta, \qquad (3.75)$$

where m_β is the 2-dimensional unit vector normal to the boundary of the island base, as in (3.44).

The stress tensor $\sigma_{\alpha\beta}$ in the flat uniform film is directly related to the lattice mismatch ε_0. In a film of cubic material on a cubic substrate, the stress tensor is

$$\sigma_{\alpha\beta} = -\frac{(c_{11} + 2c_{12})(c_{11} - c_{12})}{c_{11}} \varepsilon_0 \delta_{\alpha\beta}, \qquad (3.76)$$

where c_{11} and c_{12} are the elastic moduli of the film in the Voigt notations. In the approximation of an elastically isotropic medium, the stress tensor (3.76) reduces to

$$\sigma_{\alpha\beta} = -\frac{Y}{1-\nu} \varepsilon_0 \delta_{\alpha\beta}. \qquad (3.77)$$

There exists a certain similarity between 3-dimensional islands with low aspect ratio and submonolayer islands. Equation (3.75) formally coincides with (3.44) if one sets

$$\Delta \tau_{\alpha\beta} = \sigma_{\alpha\beta} h. \qquad (3.78)$$

However, although the intrinsic surface stress $\tau_{\alpha\beta}$ can be qualitatively modeled as bulk stress in an epitaxial layer, (3.78) does not hold quantitatively for monolayer-high islands. A substrate surface covered by a monolayer-thick epitaxial layer is, strictly speaking, a completely new surface, distinct from both the substrate surface and the surface of the deposited material. This complex surface has its own surface energy and its own intrinsic surface stress tensor.

For example, AlAs and GaAs are nearly lattice-matched materials, with lattice mismatch $\varepsilon_0 < 0.1\%$. However, the difference in intrinsic surface stress tensors between, e.g., AlAs(001) and GaAs(001) is not necessarily small. The opposite situation occurs for Ag/Pt(111), where Grossmann et al. [3.110] reported a giant heteroepitaxial stress, exceeding that of (3.78) by more than an order of magnitude.

Nevertheless, apart from special cases where the discontinuity in the intrinsic surface stress tensor $\Delta \tau_{\alpha\beta}$ differs significantly from (3.78), in many lattice-mismatched heteroepitaxial systems, (3.78) yields the correct sign and the correct order of magnitude of $\Delta \tau_{\alpha\beta}$, even for monolayer-high islands. To clarify the physical reason for this, we make a rough estimate of the right-hand side of (3.78) by substituting $\sigma_{\alpha\beta}$ from (3.77) and setting $h = a$. Putting in $Y \approx 500$ meV/Å^3, $\varepsilon_0 \approx 0.07$, and $a = 3$ Å yields $\Delta \tau \approx 100$ meV/Å^2 which is of the order of the characteristic value of τ for surfaces of pure crystals.

Tersoff and Tromp [3.109] considered a dilute array of 3D islands with low aspect ratio and a fixed height that is kinetically limited to a value considerably smaller than the lateral size, $h \ll L$. They showed that there exists an optimum lateral size for the islands in such an array, and that islands do not undergo Ostwald ripening.

3.2.3 Arrays of 2D Strained Islands at Low Temperatures

Perfect periodic structures of stripes (quantum wires) or disks (quantum dots or antidots) correspond to the minimum of the total energy E of a heteroepitaxial system and are thus related to the equilibrium state at $T = 0$ K. At finite temperatures, a perfect structure will be altered by thermal fluctuations. In our treatment of the finite-temperature behavior of heteroepitaxial submonolayer systems, we focus on an array of droplets or circular disks. At $T = 0$, an array of disks is thermodynamically preferred at the surface coverage $0 \le q \le 0.286$. One should bear in mind that the interval of surface coverage q corresponding to an array of disks can vary with temperature.

Characteristic Energies in an Array of Strained Disks. To study finite-temperature effects in arrays of 2D islands, it is necessary to estimate characteristic temperatures at which these effects become important. To address this question, let us consider characteristic energies for an array of islands at $T = 0$.

Firstly, the energy of an optimum-size island in a dilute array ($q \to 0$) is given in Appendix A by (A.55),

$$-\widetilde{E_0} = -2\sqrt{\pi}\eta_\mathrm{d} a \sqrt{N_\mathrm{opt}} \ . \tag{3.79}$$

Secondly, the energy normalized per atom in an optimum island is

$$-E_0 = -\frac{2\sqrt{\pi}\eta_\mathrm{d} a}{\sqrt{N_\mathrm{opt}}} \ . \tag{3.80}$$

The value E_0 is the energy gain per atom in an optimum island with respect to the energy per atom in a large ($N \to \infty$) ripened island. To estimate E_0 for a heteroepitaxial system, we write the constant η_d in terms of the surface stress discontinuity $\Delta\tau$ at the island boundaries from (3.54). Then, for the value of $\Delta\tau$, we use an approximate expression in terms of the lattice mismatch ε_0 from (3.78). This yields

$$E_0 = \frac{2(1+\nu)}{\sqrt{\pi}(1-\nu)} Y \varepsilon_0^2 a h^2 \frac{1}{\sqrt{N_\mathrm{opt}}} \ , \tag{3.81}$$

where Y and ν are the Young's modulus and the Poisson ratio of the island material, a is the in-plane lattice parameter, and h is the monolayer thickness. To make an estimate for InAs/GaAs(001) submonolayer islands, we substitute $Y \approx 320$ meV/Å^2, $\nu = 0.35$, $h = 2.83$ Å, $a = 4.0$ Å, $\varepsilon_0 = 0.07$, $N_\mathrm{opt} = 1\,000$, and obtain $E_0 \approx 3.7$ meV. A characteristic temperature

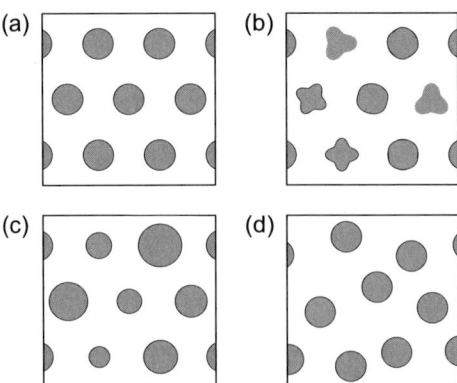

Fig. 3.47. Fluctuations in a 2-dimensional array of submonolayer islands. (**a**) A perfect periodic hexagonal array of circular disks. (**b**) Fluctuations in shape. (**c**) Fluctuations in the number of atoms (in volume). (**d**) Fluctuations in relative arrangement

$T = E_0/k_B$ is roughly 43 K. This temperature is an order of magnitude smaller than typical growth temperatures 300–600°C, or 600–900 K. Such a small energy E_0 does not therefore give the correct energy scale for finite temperature effects at realistic temperatures.

The energy of an optimum island is about 3.7 eV, giving a characteristic temperature $\approx 43\,000$ K which once again cannot be directly related to the temperature dependence at realistic experimental temperatures. To obtain a characteristic temperature for an array of 2D islands, we must consider fluctuations in the system.

Fluctuations in an Array of Strained Disks. Figure 3.47 depicts three different types of fluctuation. Firstly, islands fluctuate in shape (Fig. 3.47b), and their boundary can be described by the following function of the polar angle φ,

$$\rho^{(p)}(\varphi) = \rho^{(p)} + \sum_{m=1}^{\infty} \left[a_m^{(p)} \cos(m\varphi) + b_m^{(p)} \sin(m\varphi) \right] , \qquad (3.82)$$

where the superscript p labels islands. Secondly, islands fluctuate in the number of atoms, or in volume (Fig. 3.47c),

$$N(p) = N_{\text{opt}} + \delta N(p) . \qquad (3.83)$$

Thirdly, islands fluctuate in their relative arrangement (Fig. 3.47d),

$$\boldsymbol{R}(p) = \boldsymbol{R}_0(p) + \delta \boldsymbol{R}(p) , \qquad (3.84)$$

where $\boldsymbol{R}_0(p)$ denotes the position of the center of the pth island in the non-perturbed hexagonal lattice, and $\delta \boldsymbol{R}(p)$ is its displacement. In a real situations, all types of fluctuation are present together.

To account for finite-temperature effects on an array of islands, we consider the partition function of the system. This includes summation over all possible fluctuations,

$$Z = \prod_p \int \frac{\mathrm{d}X(p)}{a} \frac{\mathrm{d}Y(p)}{a} \sum_{N(p)} \int \frac{\mathrm{d}a_m^{(p)}}{a} \frac{\mathrm{d}b_m^{(p)}}{a} \exp\left\{\frac{1}{k_\mathrm{B}T}\left[\sum_p \left[\mu N(p)\right.\right.\right.$$
$$\left.\left.\left. - E\left(N(p), \{a_m^{(p)}\}, \{b_m^{(p)}\}\right)\right] + \frac{1}{2}\sum_p\sum_{q\neq p} U(p,q)\right]\right\}, \qquad (3.85)$$

where μ is the chemical potential of the system, $E(N(p))$ is the energy of the isolated pth island, and $U(p,q)$ is the energy of the elastic interaction between the pth and qth islands.

At low temperatures, fluctuations in shape, volume, and position of the islands are small, and the integrand in (3.85) is an exponentially sharp function. From symmetry arguments, the maximum of the integrand corresponds to a hexagonal array of disks. However, the radius ρ of the disks at finite T may differ from that at $T = 0$ K. Since the coverage q is fixed, the period of the superlattice R_0 changes correspondingly, according to (3.67). The integration in (3.85) can then be performed by the steepest descent technique and yields

$$Z = \left\{\exp\left[\frac{1}{k_\mathrm{B}T}\left(\mu(T)N(T) - \widetilde{E}(T)\right)\right]\exp\left(S_\mathrm{fluct}\right)\right\}^\mathcal{N}, \qquad (3.86)$$

where $\mathcal{N}$ is the number of islands in the system. The quantity $\widetilde{E}$ is the energy per island in the hexagonal superlattice. To obtain the energy per island, we use the last term in (3.68), which gives the energy per unit area, multiply it by the area A_0 of a superlattice unit cell, and express the radius ρ of the islands in terms of the number of atoms N. This yields

$$\widetilde{E} = 2\sqrt{\pi}\eta_\mathrm{d} a \left[-\sqrt{N}\ln\left(\mathrm{e}\sqrt{\frac{N}{N_\mathrm{opt}}}\right) + \frac{\sqrt{N}}{4\sqrt{\pi}}\left(\frac{\sqrt{3}}{2}q\right)^{3/2} S(q)\right]. \qquad (3.87)$$

The entropy S_fluct is defined per island and is related to fluctuations in shape, volume, and position of the islands.

Before carrying out rigorous calculations of the entropy S_fluct, it is worth estimating qualitatively the stiffness of an array of interacting islands against different types of perturbation, and hence elucidating the role of the different types of fluctuation. The stiffness against perturbation of island shape can be estimated as the change in energy of an island due to a perturbation of type (3.82), with an amplitude of one lattice parameter. Such a stiffness is of order

$$\widetilde{E_0}\frac{a^2}{\rho^2} = \frac{\widetilde{E_0}}{N}. \qquad (3.88)$$

126   3. Self-Organization Phenomena at Crystal Surfaces

The stiffness against fluctuations in island volume can be estimated as the second derivative of the energy $\widetilde{E}$ from (3.87) vs. the number of atoms in the island, viz.,

$$\frac{\partial^2 \widetilde{E}(N)}{\partial N^2} \approx \frac{\widetilde{E_0}}{N^2} . \tag{3.89}$$

The stiffness against fluctuations in island position can be estimated as a change in the energy of a given island due to displacement over one lattice parameter in the elastic field of neighboring islands. This yields $\left(\partial^2 U/\partial R^2\right) a^2$, where the interaction energy U is given in (3.66). Hence, the stiffness is of the order of

$$\frac{\widetilde{E_0}}{N} q^{5/2} . \tag{3.90}$$

For nanometer-scale islands, a typical number of atoms is $N > 100$. It then follows from (3.88)–(3.90) that the stiffness of the array of islands against shape fluctuations is larger than the stiffness against the other two types of fluctuation.

Entropy of Shape Fluctuations. The main part of the interaction between two islands is the dipole–dipole interaction, which depends only on the volumes of the two islands and the distance between them, and is independent of the island shape. Therefore fluctuations in island shape can only alter higher order terms in the multipole expansion of $U(p,q)$. It should be noted that we will consider only moderate surface coverages $0 \leq q \leq 0.286$. Ng and Vanderbilt [3.105] have shown that the impact of the elastic interaction between the islands on their shape at $T = 0$ is negligible, even at a coverage $q = 0.286$. Based on their results, we can argue that shape-dependent contributions to the island–island interaction energy are rather small for moderate coverage. We shall neglect these contributions in what follows. Hence, we will only take into account the effect of shape fluctuations on the energy of individual islands and neglect their effect on the interaction energy $U(p,q)$.

In this approximation, the exact partition function Z from (3.85) splits into a product

$$Z = Z_{\text{shape}} \times Z_{\text{circ}} , \tag{3.91}$$

where Z_{shape} accounts only for fluctuations in island shape, whilst Z_{circ} takes into account fluctuations in island volume and position, the islands being considered as circular.

Since fluctuations in island shape are decoupled from fluctuations in island volumes and positions, the partition function for shape fluctuations Z_{shape} (3.85) can be calculated independently for every island. In Appendix A, the energy (3.82) of an isolated island with the shape of a perturbed circular disk is obtained as

$$\widetilde{E}\left(N, \{a_m\}, \{b_m\}\right) = \widetilde{E}_{\text{circ}}(N) + \widetilde{\Delta E}\left(N, \{a_m\}, \{b_m\}\right) , \tag{3.92}$$

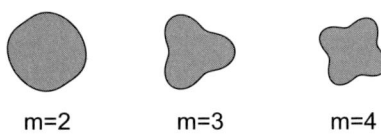

m=2   m=3   m=4

Fig. 3.48. Perturbations in the shape of a circular island. Lowest modes corresponding to angular harmonics with $m = 2$, 3, and 4 are shown

where the energy of a circular island $\widetilde{E}_{\text{circ}}$ and correction $\widetilde{\Delta E}$ are

$$\widetilde{E}_{\text{circ}}(N) = -2\sqrt{\pi} a \eta_{\text{d}} \sqrt{N} \ln\left(e\sqrt{\frac{N}{N_{\text{opt}}}}\right), \quad (3.93\text{a})$$

$$\widetilde{\Delta E}\left(N, \{a_m\}, \{b_m\}\right) = \frac{\pi^{3/2} \eta_{\text{d}}}{a\sqrt{N}} \sum_{m=2}^{\infty} \Lambda_m \left(a_m^2 + b_m^2\right). \quad (3.93\text{b})$$

Stiffness coefficients Λ_m are given by (A.53). It is worth noting that the terms with $m = 1$ correspond to translation of an island and do not alter the energy. Thus the expansion (3.93b) starts with the term $m = 2$. Figure 3.48 illustrates perturbations of circular shape with different m. Figure 3.49 shows values of the stiffness coefficients Λ_m for several of the lowest angular harmonics of the shape perturbation. At large m, the stiffness coefficients increase as $m^2 \ln m$.

Substituting the expansion of the energy of an island (3.92) into the formula for the partition function (3.85), all shape fluctuations of the given mth island yield a factor $Z_{\text{shape}}(N)$

$$Z_{\text{shape}}(N) \equiv \exp\left(S_{\text{shape}}\right) \quad (3.94)$$

$$= \int\int \frac{\mathrm{d}a_m}{a} \frac{\mathrm{d}b_m}{a} \exp\left[-\frac{\pi E_0}{2k_{\text{B}}T}\sqrt{\frac{N_{\text{opt}}}{N}} \sum_{m=2}^{\infty} \Lambda_m \left(\frac{a_m^2}{a^2} + \frac{b_m^2}{a^2}\right)\right].$$

To calculate the entropy of shape fluctuations, we must consider a discrete atomistic structure for the islands. We mimic the atomistic structure by al-

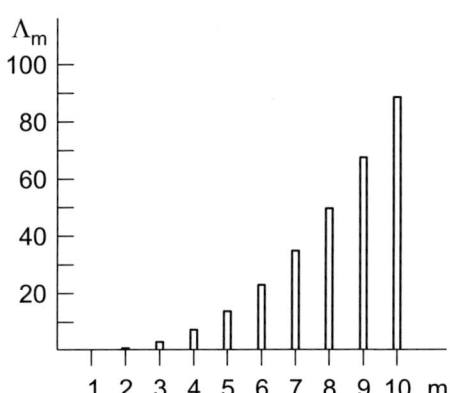

Fig. 3.49. Stiffness Λ_m of the circular island with optimum radius $\rho = \rho_{\text{opt}}$ against the mth angular harmonic of the shape perturbation

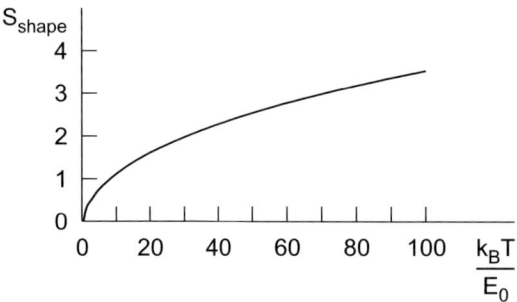

Fig. 3.50. Entropy of shape fluctuations as a function of $X = k_\text{B}T/E_0$

lowing the amplitude of every m th harmonic of the shape perturbation to equal an integer number of lattice parameters a. Hence, the entropy of shape fluctuations of a given island is

$$S_\text{shape} = \sum_{m=2}^{\infty} 2\ln\left[\sum_{n=-\infty}^{\infty} \exp\left(-\frac{\pi E_0}{2k_\text{B}T}\sqrt{\frac{N_\text{opt}}{N}}\Lambda_m n^2\right)\right]. \quad (3.95)$$

At low temperatures, $N \approx N_\text{opt}$. Then Λ_m defined in (A.53) and entering (3.95) are just numbers, and the entropy of shape fluctuations is a function of a single variable $X = k_\text{B}T/E_0$. Evaluation of (3.95) yields the function $S_\text{shape}(k_\text{B}T/E_0)$ plotted in Fig. 3.50. Over a wide range of variables, $X = k_\text{B}T/E_0$ ($10 \leq X \leq 100$) and N/N_opt ($0.4 \leq N/N_\text{opt} \leq 1.6$), the entropy S_shape can be fitted by

$$S_\text{shape} = C\left(\frac{k_\text{B}T}{E_0}\right)^p, \quad (3.96)$$

where

$$C = 0.1510 + 0.2480\frac{N}{N_\text{opt}} - 0.0292\left(\frac{N}{N_\text{opt}}\right)^2, \quad (3.97\text{a})$$

$$p = 0.6019 - 0.1445\frac{N}{N_\text{opt}} + 0.0327\left(\frac{N}{N_\text{opt}}\right)^2. \quad (3.97\text{b})$$

Fluctuations in Island Volumes and Positions. The perfect array of submonolayer islands is a 2-dimensional hexagonal superlattice of equal-sized circular disks (Fig. 3.51a). At low temperatures, fluctuations in island volumes and positions are small. To obtain the partition function Z from (3.85), it is sufficient to expand the total energy up to second-order terms in fluctuation amplitudes and to perform the Gaussian integration.

Since the expansion coefficients reflect the symmetries of an unperturbed structure, in particular the translational symmetry, it is convenient to use the Fourier transformation. Perturbations $\delta N(p)$ in the number of atoms in the islands and $\delta \boldsymbol{R}(p)$ in the positions of the islands can be written as a linear combination of static plane waves with different wave vectors $\boldsymbol{k}$, viz.,

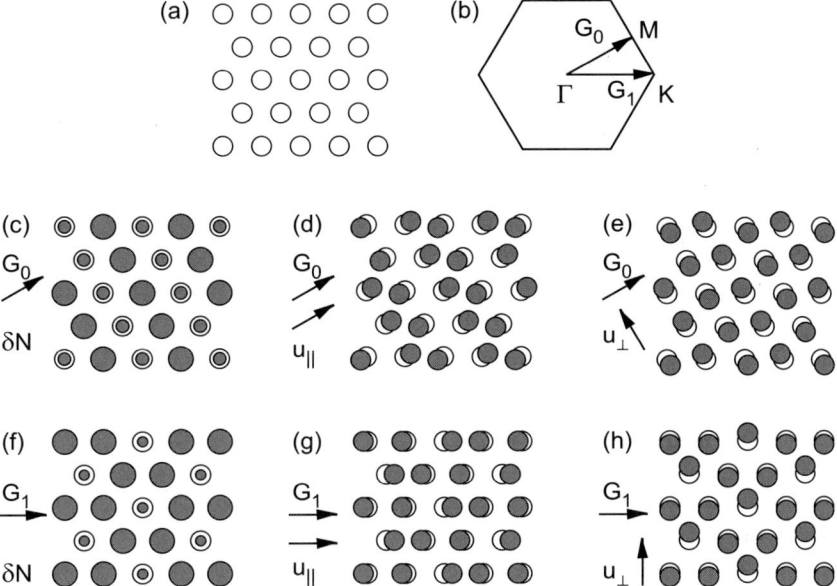

Fig. 3.51. Fluctuations in island volumes and positions for a hexagonal superlattice of circular disks. (**a**) Perfect array of islands: hexagonal superlattice of equal-sized disks. (**b**) First Brillouin zone of the reciprocal lattice. (**c**) Fluctuations in the number of atoms with wave vector G_0. (**d**) Longitudinal displacements of islands with wave vector G_0. (**e**) Transverse displacements of islands with wave vector G_0. (**f**) Fluctuations in the number of atoms with wave vector G_1. (**g**) Longitudinal displacements of islands with wave vector G_1. (**h**) Transverse displacements of islands with wave vector G_1. In (**c**)–(**h**), *open circles* denote unperturbed islands, while *shaded circles* correspond to perturbed islands

$$\delta N(p) = \frac{1}{\mathcal{N}} \sum_{k \in \mathrm{BZ}} \widetilde{\delta N}(k) \exp\left[i k \cdot R(p)\right] , \tag{3.98a}$$

$$\delta R_\alpha(p) = \frac{1}{\mathcal{N}} \sum_{k \in \mathrm{BZ}} \widetilde{\delta R}_\alpha(k) \exp\left[i k \cdot R(p)\right] , \tag{3.98b}$$

where the summation is taken over the wave vectors from the first Brillouin zone (BZ) (Fig. 3.51b) of the reciprocal lattice, and $\mathcal{N}$ is the number of vectors in the BZ equal to the number of islands on the surface. Figures 3.51c–h illustrate perturbations δN in the number of atoms in the islands and displacements δR for two different vectors at the BZ boundary.

To calculate the partition function, we expand the total energy of the system in powers of $\widetilde{\delta N}(k)$ and $\widetilde{\delta R}(k)$ up to quadratic terms,

$$E_{\text{total}} = E_{\text{total}}^{(0)} + \frac{1}{2N} \sum_{\boldsymbol{k} \in \text{BZ}} \left[D_{00}(\boldsymbol{k}) |\widetilde{\delta N}(\boldsymbol{k})|^2 + \frac{1}{2} \sum_{\alpha=1}^{2} D_{\alpha 0}(\boldsymbol{k}) \frac{\widetilde{\delta R}_{\alpha}^{*}(\boldsymbol{k})}{a} \widetilde{\delta N}(\boldsymbol{k}) \right.$$
$$\left. + \frac{1}{2} \sum_{\beta=1}^{2} D_{0\beta}(\boldsymbol{k}) \widetilde{\delta N}^{*}(\boldsymbol{k}) \frac{\widetilde{\delta R}_{\beta}(\boldsymbol{k})}{a} + \frac{1}{2} \sum_{\alpha,\beta=1}^{2} D_{\alpha\beta}(\boldsymbol{k}) \frac{\widetilde{\delta R}_{\alpha}^{*}(\boldsymbol{k})}{a} \frac{\widetilde{\delta R}_{\beta}(\boldsymbol{k})}{a} \right],$$
(3.99)

where $\alpha, \beta = 1, 2$ denote in-plane Cartesian coordinates. The exact equations for the (3×3) stiffness matrix $D_{ij}(\boldsymbol{k})$ are derived in Appendix C. The order of magnitude of the components of the matrix D can be obtained by differentiating the total energy of interacting islands. Hence,

$$D_{00}(\boldsymbol{k}) \propto \frac{\eta_\text{d} a}{4 N^{3/2}}, \qquad (3.100\text{a})$$

$$D_{\alpha 0}(\boldsymbol{k}) \propto \frac{\eta_\text{d} N a^5}{R_0^4}, \qquad (3.100\text{b})$$

$$D_{\alpha\beta}(\boldsymbol{k}) \propto \frac{\eta_\text{d} N^2 a^6}{R_0^5}, \qquad (3.100\text{c})$$

where R_0 is the nearest-neighbor distance in the hexagonal superlattice of the islands. The component D_{00} corresponds to the stiffness against fluctuations in island volumes, whereas $D_{\alpha\beta}$ corresponds to stiffness against displacements of the islands from their equilibrium positions. Since R_0 is directly connected to the surface coverage q, the relation between the two stiffnesses depends crucially on the coverage.

Figure 3.52 displays the eigenvalues of the stiffness matrix $D_{ij}(\boldsymbol{k})$ for an optimum number of atoms $N_{\text{opt}} = 1\,000$ and for different surface coverages. Values are given in units of $(1/4)\eta_\text{d} a / N_{\text{opt}}^{3/2}$ so that, at low coverage $q \to 0$, the stiffness against fluctuations in island volume equals 1. At low coverage $q = 0.01$ (Fig. 3.52a), the upper branch (with hardly any dispersion) corresponds mainly to the stiffness of the system against change in the number of atoms in islands. The lower two branches are related to the stiffness against displacement of the islands. These are similar to acoustic branches of the phonon spectrum, since the stiffness vanishes as $\boldsymbol{k} \to 0$. With increasing coverage, the stiffness against displacement increases approximately as $q^{3/2}$ and the eigenvalue spectrum exhibits anti-crossing of the branches (Fig. 3.52b). At higher coverage $q = 0.286$, the stiffness against displacement is 3 orders of magnitude larger than the stiffness against change in island volume, and anti-crossing occurs in the vicinity of the Γ point of the Brillouin zone (Fig. 3.52c). Apart from the anti-crossing shown with large magnification in Fig. 3.52e, the dispersion of the lower branch (Fig. 3.52d) is still very small ($\approx 4\%$).

Two important quantities are related to the stiffness matrix $D_{ij}(\boldsymbol{k})$. Firstly, at low temperatures, fluctuations in island volumes and island positions are small and the root mean square of the fluctuations can be obtained

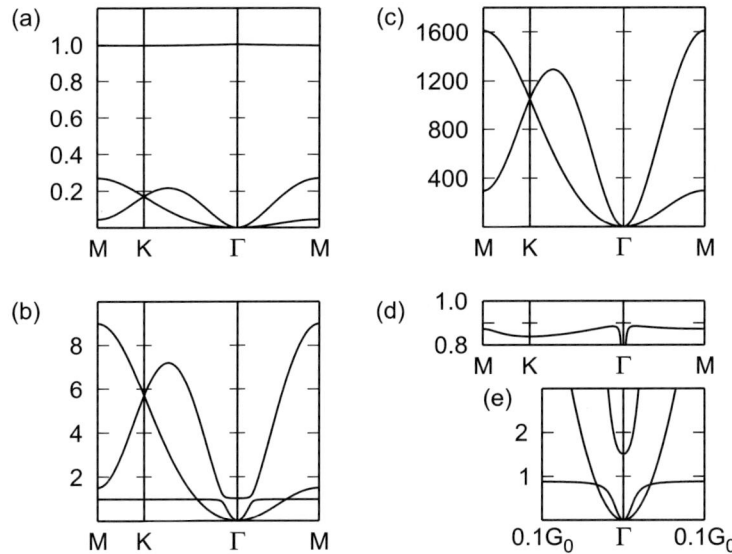

Fig. 3.52. Eigenvalues of the stiffness matrix of a hexagonal superlattice of sub-monolayer islands. Eigenvalues vs. wave vectors are depicted for symmetric directions in the Brillouin zone. (**a**) Coverage $q = 0.01$. The stiffness against fluctuations in island volumes is larger than the stiffness against displacements. (**b**) Coverage $q = 0.04$. Anti-crossing of two branches occurs. (**c**) Coverage $q = 0.286$. Stiffness against displacements. (**d**) Coverage $q = 0.286$. Stiffness against fluctuations in island volumes. (**e**) Coverage $q = 0.286$. The region of the Brillouin zone close to the Γ point. Anti-crossing of two branches is shown. Stiffness is given in units of $(1/4)\eta_\mathrm{d} a/N_\mathrm{opt}^{3/2}$

in terms of the stiffness matrix. Fluctuations in the number of atoms are given by

$$\langle (\delta N)^2 \rangle = \frac{1}{\mathcal{N}} \sum_{\mathbf{k} \in \mathrm{BZ}} k_\mathrm{B} T \left[D^{-1} \right]_{00} (\mathbf{k}) , \qquad (3.101)$$

where $D_{ij}^{-1}(\mathbf{k})$ is the inverse matrix of $D_{lm}(\mathbf{k})$. Evaluation of (3.101) yields

$$\left[\frac{\langle (\delta N)^2 \rangle}{N^2} \right]^{1/2} = 2 \left(\frac{k_\mathrm{B} T}{\eta_\mathrm{d} a N} \right)^{1/2} f_\mathrm{N}(q) . \qquad (3.102)$$

The function $f_\mathrm{N}(q)$ accounting for the effect of the elastic interactions between islands on the fluctuation in island volume is plotted in Fig. 3.53a.

According to a general property of all 2-dimensional systems, the root mean square displacement of a single island increases with the size of the substrate L_sub as $\ln(L_\mathrm{sub}/a)$, due to the contribution of acoustic-type modes of island displacements with small $|\mathbf{k}|$. This value diverges as $L_\mathrm{sub} \to \infty$. However, the root mean square fluctuation in the distance between two neighboring islands is finite since the contributions from modes with small $|\mathbf{k}|$ nearly

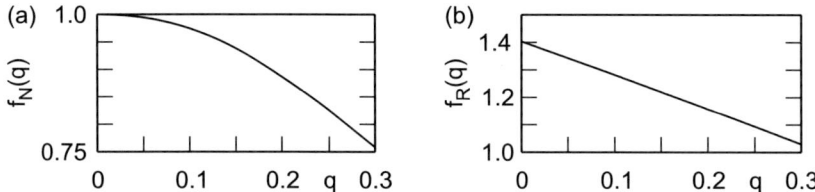

Fig. 3.53. Effect of the elastic interaction between islands on the fluctuations in the system. (**a**) The correction factor $f_N(q)$ from (3.102) to the root mean square fluctuation in the number of atoms in the island. (**b**) The correction factor $f_R(q)$ from (3.104) to the root mean square fluctuation in the distance between neighboring islands

cancel out. This quantity is given by

$$\langle [\delta R_\alpha(p) - \delta R_\alpha(q)]^2 \rangle = \qquad (3.103)$$
$$\frac{1}{\mathcal{N}} \sum_{k \in BZ} 2k_B T \Big[1 - \cos\big[k \cdot (R(p) - R(q))\big] \Big] \big[D_{11}^{-1}(k) + D_{22}^{-1}(k) \big] \ .$$

Evaluation of (3.103) for two neighboring islands yields

$$\left[\frac{\langle (\delta R)^2 \rangle}{R^2} \right]^{1/2} = \left(\frac{k_B T}{\eta_d a N} \right)^{1/2} \frac{2\sqrt{\pi}}{(\sqrt{3}q/2)^{3/4}} f_R(q) \ . \qquad (3.104)$$

The factor $q^{-3/4}$ is due to the dipole–dipole elastic interaction between islands. The q-dependent factor $f_R(q)$ plotted in Fig. 3.53b is due to the deviation of the exact interaction energy from the dipole–dipole approximation.

At sufficiently low temperatures, the relative fluctuation in the number of atoms in the islands,

$$\left[\frac{\langle (\delta N)^2 \rangle}{N^2} \right]^{1/2} \ll 1 \ , \qquad (3.105)$$

and the relative fluctuation in the distance between islands

$$\left[\frac{\langle (\delta R)^2 \rangle}{R^2} \right]^{1/2} \ll 1 \ , \qquad (3.106)$$

are both small. Then all significant contributions to the partition function come from configurations close to the perfect hexagonal superlattice of identical circular islands. To obtain the partition function Z_{circ} describing fluctuations in volumes and positions in an array of circular islands, we substitute the expansion of the total energy (3.99) into (3.85) and perform the Gaussian integration over $\widetilde{\delta N}(k)$ and $\widetilde{\delta R_\alpha}(k)$. This yields

$$Z_{\text{circ}} = [\exp(S_{\text{circ}})]^{\mathcal{N}} \left[\exp\left\{ \frac{1}{k_B T} \left[\mu N_{\text{opt}}(T) - \widetilde{E}(N_{\text{opt}}) \right] \right\} \right]^{\mathcal{N}}, \qquad (3.107)$$

where the entropy of fluctuations in volumes and positions may be written in terms of the stiffness matrix $D_{ij}(k)$ as

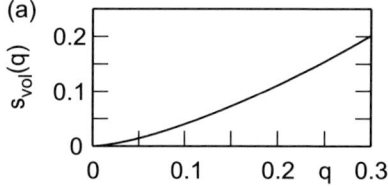 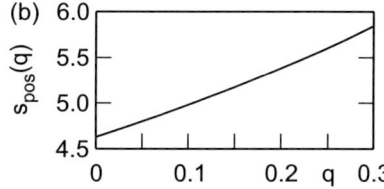

Fig. 3.54. Corrections to the entropy of volume fluctuations and position fluctuations due to elastic interaction between islands. Corrections are given as correction factors in the average determinants of the stiffness matrix. (**a**) Correction factor $s_{\text{vol}}(q)$ entering (3.111) for the entropy of volume fluctuations. (**b**) Correction factor $s_{\text{pos}}(q)$ entering (3.112) for the entropy of fluctuations in island positions

$$S_{\text{circ}} = \frac{1}{2N} \sum_{\mathbf{k} \in \text{BZ}} \ln \left[\frac{(2\pi)^3 (k_B T)^3}{\det D_{ij}(\mathbf{k})} \right] . \tag{3.108}$$

It follows from Fig. 3.52 that the coupling between fluctuations in island volumes and fluctuations in island positions occurs only in a rather narrow region of the Brillouin zone. This means that fluctuations in island volumes and fluctuations in island positions are only weakly coupled. Then the entropy S_{circ} from (3.108) may be written as a sum of two contributions:

$$S_{\text{circ}} = S_{\text{vol}} + S_{\text{pos}} , \tag{3.109}$$

where the entropy of volume fluctuations is given by

$$S_{\text{vol}} = \frac{1}{2N} \sum_{\mathbf{k} \in \text{BZ}} \ln \left[\frac{2\pi k_B T}{D_{00}(\mathbf{k})} \right] . \tag{3.110}$$

Calculating the stiffness matrix $D_{ij}(\mathbf{k})$ and averaging (3.110) over the Brillouin zone gives us the entropy of volume fluctuations in the form

$$S_{\text{vol}} = \frac{1}{2} \ln \left[\frac{8\pi k_B T N}{E_0} s_{\text{vol}}(q) \right] , \tag{3.111}$$

where the function $s_{\text{vol}}(q)$ due to the influence of island–island interactions is plotted in Fig. 3.54a.

The entropy of fluctuations in island positions is

$$S_{\text{pos}} = \frac{1}{2N} \sum_{\mathbf{k} \in \text{BZ}} \ln \left[\frac{(2\pi)^2 (k_B T)^2}{\det D_{\alpha\beta}(\mathbf{k})} \right] , \tag{3.112}$$

where $\alpha, \beta = 1, 2$. Averaging (3.112) over the Brillouin zone, one obtains the entropy of fluctuations in island positions in the form

$$S_{\text{pos}} = \frac{1}{2} \ln \left[\frac{(2\pi)^2 (k_B T)^2 N^4}{(1/4\pi)(\sqrt{3}/2)^5 q^5 E_0^2} s_{\text{pos}}(q) \right] , \tag{3.113}$$

where the q-dependent function $s_{\text{pos}}(q)$ is plotted in Fig. 3.54b.

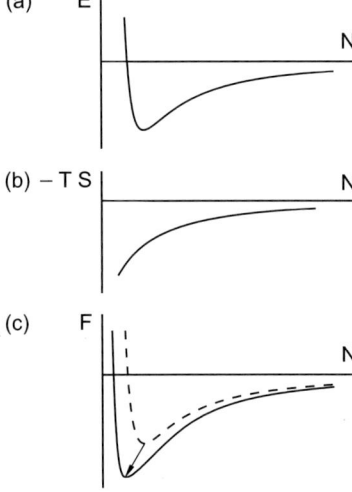

Fig. 3.55. Effect of the entropy on the optimum volume of the island. (**a**) Total energy per atom E in the island. (**b**) The entropy contribution $-TS$ to the Helmholtz free energy per atom. (**c**) The free energy per atom F at a finite temperature (*solid line*) and the total energy E (*dashed line*). The entropy contribution results in a decrease in the optimum number of atoms N in the island (shown by the *arrow*)

It should be noted that there exist two restrictions under which equations (3.111) and (3.113) are valid. Firstly, the temperature should not be so low that a large number of configurations contribute to the partition function Z_{circ} and to the entropy. This is valid at any temperatures of experimental interest. Secondly, the relative root mean square fluctuations in the number of atoms in the islands (3.102) and that for the distance between islands (3.104) must be small.

Adding up the two contributions to the entropy, one obtains the total entropy per island as

$$S_{\text{circ}} = \frac{1}{2}\ln\left[\frac{2^{10}\pi^4}{3^{5/2}}\frac{(k_BT)^3}{E_0^3}\frac{N^5}{q^5}s_{\text{vol}}(q)s_{\text{pos}}(q)\right]. \tag{3.114}$$

Now, once the equation for the entropy (3.114) has been obtained, we can write down the total Helmholtz free energy of the system. By summing up the total energy E and the term $-TS$, and dividing it by the number of atoms in the island N, we obtain the free energy per atom,

$$F = -\frac{E_0\sqrt{N_{\text{opt}}}}{\sqrt{N}}\ln\left(e\sqrt{\frac{N}{N_0(q)}}\right) \tag{3.115}$$

$$-\frac{3}{2}\frac{k_BT}{N}\ln\left[\frac{2^{10/3}\pi^{4/3}}{3^{5/6}}\frac{k_BT}{E_0}\frac{N^{5/3}}{q^{5/3}}[s_{\text{vol}}(q)s_{\text{pos}}(q)]^{1/3}\right].$$

Figure 3.55 illustrates the effect of the entropy on the optimum size of islands. Figure 3.55a displays the total energy of the system and Fig. 3.55b the entropy term $(-TS)$. Since the entropy term is an increasing function of the number of atoms N in the island, the minimum in the free energy $F = E - TS$ shifts with temperature to smaller values of N. This leads to a shrinkage in the number of atoms in each island (the island volume) with temperature, due to entropy effects. To quantify this shrinkage, we differentiate F from (3.115) with respect to N and set $dF/dN = 0$. This yields the temperature dependence of the number of atoms in the islands,

$$\frac{N(q;T)}{N_0(q)} = 1 - \frac{6k_\mathrm{B}T}{E_0} [N_\mathrm{opt} N_0(q)]^{1/2} \quad (3.116)$$

$$\times \ln \left[\frac{2^{10/3} \pi^{4/3}}{3^{5/6} \mathrm{e}^{5/3}} \frac{k_\mathrm{B}T}{E_0} \frac{N^{5/3}}{q^{5/3}} [s_\mathrm{vol}(q) s_\mathrm{pos}(q)]^{1/3} \right].$$

It is worth noting here that (3.116) is valid only at rather low temperatures. It nevertheless gives a semi-quantitative estimate for the characteristic temperature T_char at which islands change significantly, e.g., the number of atoms N decreases by a factor of 2,

$$T_\mathrm{char} = \frac{\frac{\Theta}{12 k_\mathrm{B}} \sqrt{\frac{N_0(q)}{N_\mathrm{opt}}}}{\ln \left\{ \frac{2^{4/3} \pi^{4/3}}{3^{11/6} \mathrm{e}^{5/3}} [s_\mathrm{vol}(q) s_\mathrm{pos}(q)]^{1/3} \frac{N_\mathrm{opt}^{8/3}}{q^{5/3}} \left[\frac{N_0(q)}{N_\mathrm{opt}} \right]^{13/6} \right\}}. \quad (3.117)$$

This equation solves the problem of a characteristic temperature. It follows from (3.117) that, besides two characteristic energies, the energy per atom in the optimum island E_0 and the energy per island Θ, there exists an intermediate energy such that

$$E_0 \ll k_\mathrm{B} T_\mathrm{char} \ll \Theta. \quad (3.118)$$

For an array of submonolayer islands in a typical semiconductor system like InAs/GaAs, the characteristic temperature from (3.117) may be of the order of several hundred K, i.e., of the order of typical formation temperatures.

3.2.4 Arrays of 2D Strained Islands at Low Coverage

The relative fluctuations in the distance between islands, $\left[\frac{\langle (\delta R)^2 \rangle}{R^2} \right]^{1/2}$ from (3.104), and the relative fluctuations in the number of atoms in the islands, $\left[\frac{\langle (\delta N)^2 \rangle}{N^2} \right]^{1/2}$ from (3.102) have the same temperature dependence $\sim T^{1/2}$. However, due to the factor $q^{-3/4}$, the relative fluctuations in the distance between islands are already large at relatively low temperatures when islands still exhibit a rather sharp distribution in the number of atoms. This means a short-range disorder in island positions.

To evaluate the entropy of island positions, we consider a dilute array and neglect elastic interactions between islands. The system then reduces to an array of identical disks, each of which has N atoms and can occupy an arbitrary position on the surface under the constraint that disks are not allowed to overlap. The entropy per island is then

$$\widetilde{S}_{\text{pos}} = k_{\text{B}} \ln\left(\frac{\alpha N}{q}\right), \tag{3.119}$$

where α is a numerical factor of the order of unity.

Adding up the contributions (3.111) and (3.119) to the entropy and dividing by the number of atoms in the island N, one obtains the entropy per atom. Then adding up the energy E and the entropy term $(-TS)$, one gets the Helmholtz free energy per atom as

$$F = -W + E_{\text{elast}}^{(0)} + \frac{C_1}{\sqrt{N}} - \frac{C_2}{\sqrt{N}} \ln(\sqrt{N}) \tag{3.120}$$

$$- \frac{k_{\text{B}}T}{N} \left[\ln\left(\frac{\alpha N}{q}\right) + \ln\left(2N\sqrt{\frac{2\pi k_{\text{B}}T}{\Theta}}\right) \right].$$

The entropy term in the free energy (3.120) is negative and its absolute value decreases monotonically with N. It therefore shifts the free energy minimum towards smaller values of N. This implies an entropy-driven shrinkage of islands with temperature.

To estimate the characteristic temperature at which entropy effects become essential, we equate the entropy term in (3.120) with the energy per atom E_0 in optimum islands. This yields the value of the characteristic temperature as

$$T_{\text{char}} \sim \frac{\Theta}{k_{\text{B}} \ln(N^2/q)}. \tag{3.121}$$

This estimated value of T_{char} agrees with our earlier results [3.111]. Remarkably, the characteristic temperature has an intermediate value between the energy per atom in an optimum island and the energy per island, i.e.,

$$E_0 \ll k_{\text{B}} T_{\text{char}} \ll \Theta. \tag{3.122}$$

A typical value of the characteristic temperature is 800–1 000 K, which lies in the range of typical growth temperatures. It confirms the decisive role played by entropy effects in the formation of arrays of strained submonolayer islands in real experimental systems.

3.2.5 Equilibrium Distribution of Island Sizes

At elevated temperatures the distribution of island sizes is broadened due to thermal fluctuations. To address effects of thermal broadening, we consider the Boltzmann–Gibbs distribution function. For a dilute array, where elastic interaction between islands is neglected, the distribution function can be written as

3.2 Surface Arrays of Two-Dimensional Islands

$$P(N) = \exp\left[\frac{\mu N - \widetilde{E}(N)}{k_\text{B}T}\right] = \exp\left[\frac{[\mu - E(N)]N}{k_\text{B}T}\right]. \tag{3.123}$$

The chemical potential μ can be found from the constraint that the total number of atoms in the islands is fixed by the total coverage q,

$$q = \sum_N NP(N). \tag{3.124}$$

We extend the approach of (3.123) and (3.124) to dense arrays by taking into account the elastic interaction between islands in the mean-field approximation and accounting for individual adatoms. First, we rewrite the matter conservation condition (3.124) as

$$q = q_\text{ad} + \sum_N NP(N), \tag{3.125}$$

where the sum is taken over the macroscopic volumes of the islands, $N \gg 1$, single adatoms are taken into account separately, and islands containing a small number of atoms $N = 2, 3, \ldots$, are neglected. Then, in order to take into account the elastic interaction between islands, we assume that the elastic interaction energy of a given island with other islands is equal to the same energy in a perfect hexagonal array with coverage $q - q_\text{ad}$. Here we neglect the contribution of adatoms to the elastic interaction energy, since adatoms are strongly relaxed. Then, expressing the energy $E(N; q - q_\text{ad})$ from (3.71) in terms of the number of atoms in the islands and substituting it into (3.123), we may write down the distribution function as

$$P(N) = \exp\left\{\frac{[\mu - E(N; q - q_\text{ad})]N}{k_\text{B}T}\right\}. \tag{3.126}$$

The adatom density is related to the chemical potential in a simple way,

$$q_\text{ad} = \exp\left(\frac{\mu}{k_\text{B}T}\right). \tag{3.127}$$

After substituting (3.126) into (3.125), the latter and (3.127) form a set of two equations in the two unknowns q_ad and μ.

For a numerical solution to this set of equations, we have in mind, in particular, a system of InAs/GaAs(001) islands. Although we do not consider a specific strongly anisotropic reconstructed surface (described, for example, in [3.112]), we use numerical values for the system parameters that are of the right order of magnitude. This will give us a reasonable qualitative picture of the temperature dependence of arrays of islands in heteroepitaxial systems.

Following the work by Shitara et al. [3.113], we choose the same value for the binding energy of the adsorbate atoms in a 2-dimensional island, namely, $W = 369$ meV. The strain-induced reduction in the binding energy is taken to be equal to the elastic energy in a flat strained film, i.e., $E_\text{elast}^{(0)} = Y(1 - \nu)^{-1}a^2h\varepsilon_0^2$. To model the energy of the island boundary $C_1\sqrt{N}$, we

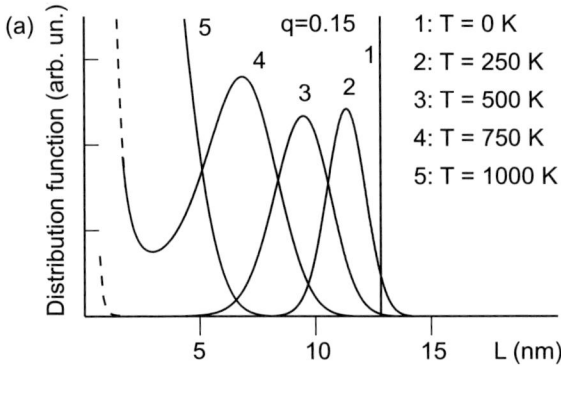

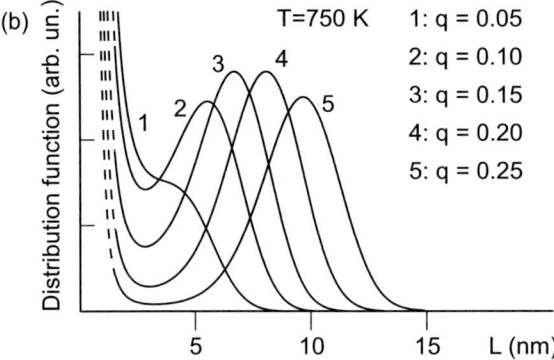

Fig. 3.56. Equilibrium distribution of the sizes of 2-dimensional islands. (a) Distribution at a given surface coverage $q = 0.15$ and different temperatures. (b) Distribution at a given temperature $T = 750$ K and different coverages

equate it to the energy of the broken bonds at the boundary of a circular island, namely, $\beta(W/2)2\pi\sqrt{N/\pi}$. Hence $C_1 = \beta\sqrt{\pi}W$, where the coefficient β refers to a reduction in the energy of the island boundary due to some short-range relaxation at the island boundary. We assume $\beta = 0.6$. For the coefficient C_2 in the elastic relaxation energy, we use the estimate from (3.81).

Substituting these model parameters in (3.125) and (3.126), taking $\varepsilon_0 = 0.07$, and solving the equations numerically, we obtain the distribution function for the number of atoms in the island. This can also be rewritten in terms of the distribution function of island sizes, $P(N)\mathrm{d}N = 2P(L)\pi(4a^2)^{-1}L\mathrm{d}L$, where L is the diameter of a circular island. Figure 3.56a depicts the distribution function of island sizes at a given coverage $q = 0.15$ and different temperatures. At $T = 0$ K, the distribution function is an infinitely sharp peak at some optimum size L_opt. With increasing temperature, the distribution of island sizes broadens and its maximum shifts to smaller sizes. The second maximum corresponding to individual adatoms evolves. The islands have a bimodal size distribution where one local maximum corresponds to single

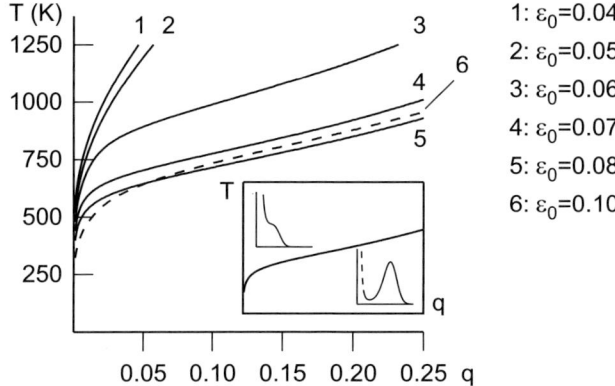

Fig. 3.57. Phase diagrams of an array of 2-dimensional islands showing domain regions of unimodal and bimodal size distribution. The *inset* shows qualitatively a general feature of this phase diagram containing regions of unimodal and bimodal size distributions

adatoms, and the other one refers to nanometer-scale islands. With further temperature decrease, the local maximum corresponding to nanoscale islands disappears and the bimodal size distribution transforms to a unimodal one. Dashed lines refer to relatively small island sizes, say $L < 2$ nm, where the continuum approach does not apply quantitatively but may give a reasonable qualitative description. It should be noted that a bimodal size distribution of strained 2-dimensional islands was obtained earlier from kinetic Monte Carlo simulations in [3.114]. However, no discussion of the temperature dependence of such a distribution was given there.

Figure 3.56b shows the change in the island size distribution at a given temperature as a function of island coverage. The effect of coverage is similar to the effect of temperature. At higher coverages, islands have a bimodal size distribution with some optimum size at the nanometer scale, whereas at lower coverages, the size distribution becomes unimodal and an optimum size can no longer be resolved.

Figure 3.57 represents a phase diagram in the temperature–coverage variables. It depicts two regions, a region of unimodal distribution of island sizes and a region of bimodal distribution of island sizes. The figure shows the line separating these two regions, calculated for different values of the lattice mismatch ε_0.

The main result of our thermodynamic treatment is the shrinkage of island volumes, and hence of island sizes, with temperature. This effect provides an efficient experimental tool for distinguishing between thermodynamically controlled and kinetically controlled arrays of islands. In homoepitaxial systems, where equilibrium corresponds to the ripening of islands, dense arrays of islands with some finite size are kinetically controlled arrays, referring to some intermediate state on the pathway towards ripening. In such arrays, the

140   3. Self-Organization Phenomena at Crystal Surfaces

average island size increases with substrate temperature (see, for example, the experimental results for Fe/Fe islands in [3.115]). The same is expected for heteroepitaxial systems in the case where the size of kinetically controlled islands is far below the equilibrium value. On the other hand, a decrease in the island size with temperature indicates the equilibrium nature of an array of islands in a given heteroepitaxial system.

3.2.6 Crossover from Kinetically Controlled to Thermodynamically Limited Growth of 2D Strained Islands

Aiming particularly to develop one common description of heteroepitaxial submonolayer systems which would cover both kinetically and thermodynamically controlled growth stages, Meixner et al. [3.116] carried out a kinetic Monte Carlo (MC) simulation of the formation of 2-dimensional strained islands upon growth interruption.

In contrast to Metropolis-type algorithms, an event-based algorithm [3.117] was used for the MC simulations, independent of any particular time scale. Every Monte Carlo step corresponds to a diffusion process that is chosen according to its overall probability. The simulated time interval is then given in terms of the probabilities of all possible processes p_i by $\Delta t = 1/\sum_i p_i$. The acceptance ratio for this continuous time scheme is always unity. It is therefore very efficient in simulating low temperature systems.

The growth simulations by Meixner et al. [3.116] are based on a solid-on-solid model with deposition and diffusion as the relevant processes. Diffusion of adatoms occurs on a square lattice by nearest neighbor hopping. Atoms can cross island edges by surmounting a Schwoebel barrier. The relevant energies in our simulations are the binding energy to the surface $E_s = 0.7$ eV and the strength of the $n \leq 4$ nearest-neighbor bonds $E_b = 0.3$ eV, which influence the time scale for diffusion and island formation, respectively. Existing islands generate an elastic strain field caused by the lattice mismatch. This strain field influences detachment from island boundaries and the motion of adatoms in the vicinity of islands through a position-dependent energy correction term E_{str}. The hopping rate for a single atom is then given by an Arrhenius law:

$$p = \nu \exp\left(-\frac{E_s + n E_b - E_{\mathrm{str}}}{k_B T}\right), \tag{3.128}$$

with attempt frequency $\nu = 10^{13}$ s^{-1}.

The strain field is treated within the framework of the continuum theory of elastic media using a Green function formalism. The elastic displacements $u_i(\mathbf{r})$ are calculated using the static Green tensor of elasticity theory $G_{ij}(\mathbf{r}, \mathbf{r}')$, so that

$$u_i(\mathbf{r}) = -\oint_S \mathrm{d}^2 r' G_{ij}(\mathbf{r}, \mathbf{r}') P_j(\mathbf{r}'). \tag{3.129}$$

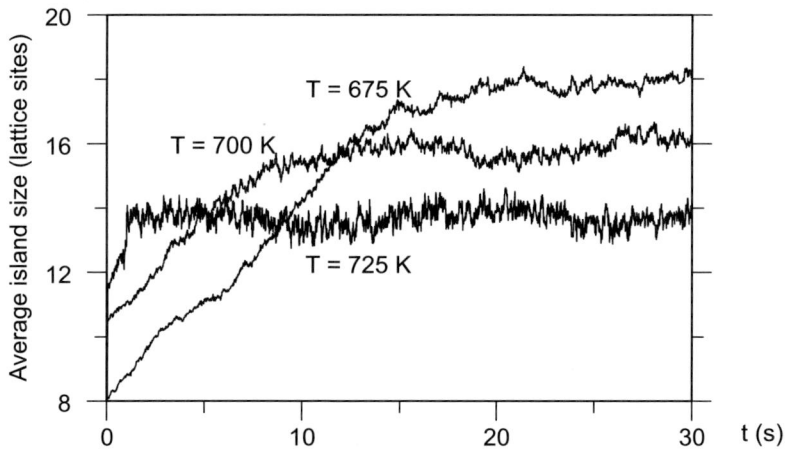

Fig. 3.58. Temporal evolution of average island size for $T = 675$ K, $T = 700$ K, and $T = 725$ K. Monte Carlo simulations were performed on a 250×250 grid and averaged over ten runs with the same set of parameters

The line forces $P_i(\mathbf{r})$ appear at island edges and act as sources for the strain. The integration is thus carried out along all island boundaries S. The strain energy is then calculated from the displacements in the isotropic approximation.

Since [3.116] focused on a dilute array of islands, it was assumed that the strain field extended only five lattice constants away from the island boundaries, to speed up computations. This implies a varying lower energy cut-off for islands of different size. However, due to the steep decrease of the strain energy, this procedure still captures about 85% of the total energy, even for the largest islands.

Simulations were performed on a lattice of 250×250 atomic sites. As an initial step, a coverage of 4% was deposited randomly on the surface at a flux of 1 ML/s. Every 0.01 s, a histogram of the island size distribution was recorded. For higher temperatures, in particular, the fluctuations in the size distributions were considerable. To reduce noise, ten simulations with different initial conditions were used to calculate an average.

To display the temporal evolution of the average island size, the average diameter $\langle \sqrt{N} \rangle$ of the islands was calculated from the histograms. Islands comprising fewer than four atoms were not considered in the averaging. Figure 3.58 displays the simulation results for temperatures of $T = 675$ K, 700 K, and 725 K.

From Fig. 3.58, it is evident that in the initial stages of island growth the size distribution is kinetically controlled. At lower temperatures many small islands are formed, whereas at higher temperatures fewer and larger islands emerge.

On short time scales of a few seconds, islands do not grow very much and the scaling of island size with temperature is still kinetically controlled.

At lower temperatures, nucleation of islands is the dominant process. Since adatom mobility is low, the density of single adatoms increases fast during deposition and pairs of atoms are formed randomly. Those act as nuclei for islands. Consequently, one observes many small islands for low temperatures.

With increasing temperature, adatoms become more and more mobile. A single adatom in a hot system can travel a long way before it finds an existing island to which it can attach. The adatom density therefore decreases and nucleation of new islands is suppressed. The final spatial configuration in the kinetically controlled regime exhibits few and large islands.

Right after deposition, however, the islands begin to equilibrate. The system is now in an intermediate state between kinetically and thermodynamically controlled growth conditions. A slow increase in island sizes and crossover of the average island size for systems of different temperatures is characteristic for this regime.

For low temperatures, the growth process is the slowest. The higher the temperature, the faster the islands approach their average equilibrium size. Once the equilibrium size distribution has been reached, the average island diameter remains constant. In the course of equilibration, islands in low temperature systems continue to grow until they reach their equilibrium size at an average diameter above that of islands in the hotter systems, as is expected for islands grown under equilibrium conditions.

From the results of the thermodynamic theory and kinetic simulations, an experimental tool emerges that allows one to distinguish between kinetically and thermodynamically controlled islands. If, with increasing substrate temperature, the average number of atoms in the islands, or the average island volume, increases, island formation is predominantly controlled by growth kinetics. If, with increasing substrate temperature, the average island volume decreases, island formation is predominantly controlled by thermodynamics. For submonolayer islands, the height is fixed, and the island volume is proportional to the square of the island lateral size. The above arguments therefore apply to the dependence of lateral size on temperature.

3.2.7 Submonolayer Arrays of InAs/GaAs Islands

Using the above tool, we can shed light on the physical nature of arrays of strained islands in any material system. Figure 3.59 shows experimental results for the temperature dependence of submonolayer arrays of InAs/GaAs islands, discussed briefly in [3.118]. A heteroepitaxial InAs/GaAs(001) system containing a submonolayer (0.3 ML) insertion of InAs in a GaAs matrix was grown at different temperatures. InAs was deposited on a GaAs(001) surface by molecular beam epitaxy, 10 s growth interruption was introduced, and the system was overgrown by GaAs. Figure 3.59a displays cross-sectional high

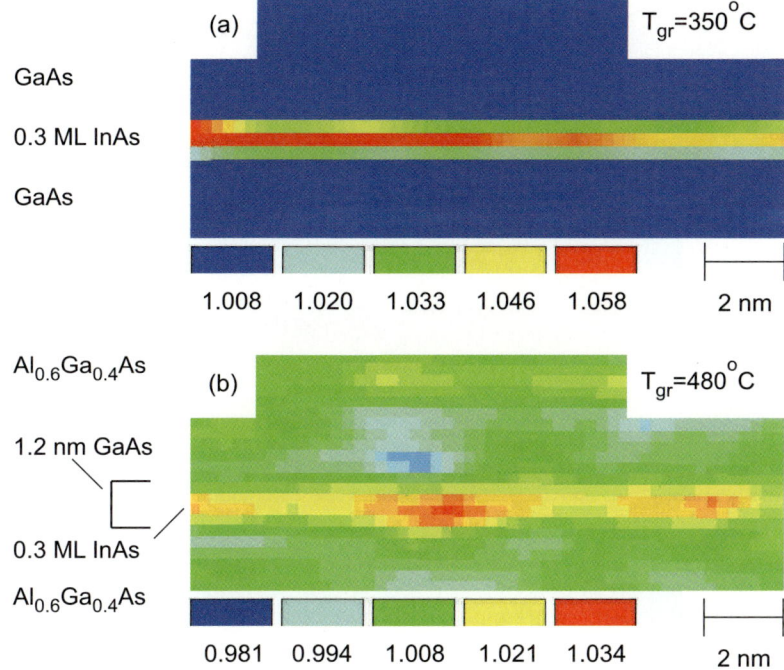

Fig. 3.59. Dependence of a submonolayer (0.3 ML) array of InAs/GaAs islands on growth temperatures. (**a**) DALI processed cross-sectional high resolution transmission electron microscopy (HRTEM) image of the submonolayer array of InAs/GaAs islands grown at 350°C. The *color code* displays the local map of the lattice parameter in the vertical direction. The relative lattice parameter is plotted. The value 1 refers to GaAs. Variation of the color from blue to red corresponds to increasing indium content. (**b**) DALI-processed cross-sectional HRTEM image of the submonolayer array of InAs islands grown within a thin (1.2 nm) GaAs insertion in a $Ga_{0.4}Al_{0.6}As$ matrix at 480°C

resolution transmission electron microscopy (HRTEM) images, processed by the DALI evaluation program, which yields a map of the local lattice parameter a_z in the vertical direction, normalized by the lattice parameter of GaAs. With the given color code, blue refers to GaAs. Variation of the color from blue to red corresponds to increasing indium content. For a sample grown at 350°C, rather large islands (> 8 nm) are revealed by HRTEM. For a sample grown at 480°C, the islands shrink, and the typical size is less than 4 nm.

Some questions arise concerning comparison of the two HRTEM images in Fig. 3.59, because the InAs islands in Fig. 3.59b are grown within a rather thin GaAs insertion surrounded by a $Ga_{0.4}Al_{0.6}As$ matrix. Although GaAs and GaAlAs are lattice-matched, the two materials differ in elastic moduli. Therefore, the GaAs/GaAlAs interface affects the strain profile induced by InAs islands and this makes image processing much more complicated.

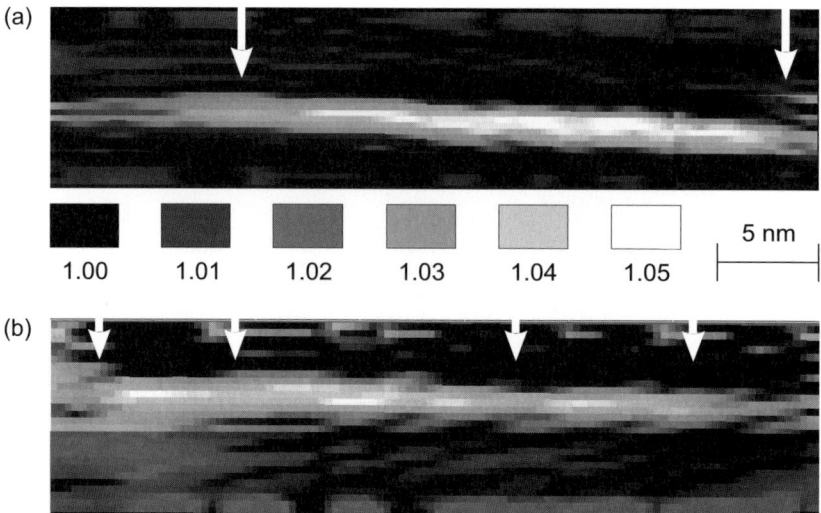

Fig. 3.60. Dependence of a submonolayer (0.3 ML) array of InAs/GaAs islands on growth temperatures. (**a**) DALI processed cross-sectional high resolution transmission electron microscopy (HRTEM) image of the submonolayer array of InAs/GaAs islands grown at 350°C. The *grey scale code* displays the local map of the lattice parameter in the vertical direction. The relative lattice parameter is plotted. The value 1 refers to GaAs. Variation of the grey scale from black to white corresponds to increasing indium content. (**b**) DALI processed cross-sectional HRTEM image of the submonolayer array of InAs islands grown at 480°C. *White arrows* mark island edges

This problem is overcome in Fig. 3.60. The figure compares two HRTEM images of InAs submonolayer (0.3 ML) islands grown at different temperatures, but surrounded by the same GaAs matrix. DALI processed images are shown on a grey scale. Black regions refer to the local lattice parameter of GaAs. Variation of the grey color from black to white corresponds to an increase in the local lattice parameter in the vertical direction, and hence to an increase in indium content. By changing the growth temperature from 350°C (Fig. 3.60a) to 480°C (Fig. 3.60b), the lateral size of islands decreases, similarly to the situation in Fig. 3.59.

As HRTEM images show only a small part of the structure, a complementary study revealing statistical information about the whole structure is needed for an unambiguous conclusion. Photoluminescence (PL) spectra from the submonolayer InAs/GaAs systems grown at 350°C, 450°C, and 480°C are displayed in Fig. 3.61. Each of the spectra contains one peak at 1.514 eV, which refers to the GaAs matrix and does not depend on the growth temperature, and one more peak at lower photon energies, corresponding to PL from excitons localized by InAs insertions. Upon an increase in the growth temperature, the latter PL peak shifts towards higher energies. This corresponds to smaller volumes of submonolayer islands localizing excitons.

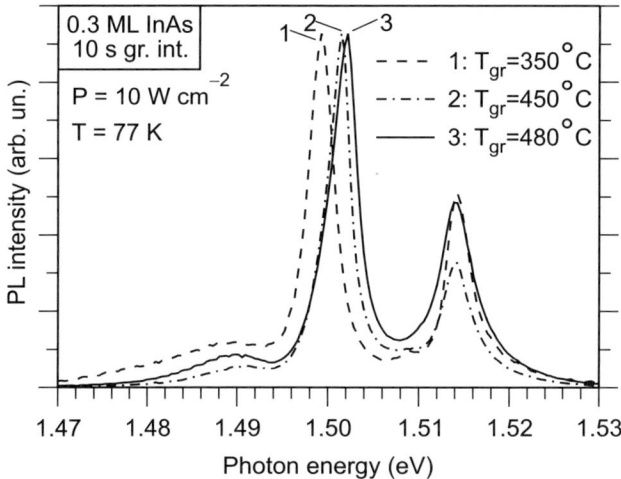

Fig. 3.61. Photoluminescence (PL) spectra of submonolayer arrays of InAs/GaAs islands grown at different temperatures

In conclusion, the HRTEM data of Figs. 3.59 and 3.60 and the PL spectra of Fig. 3.61 demonstrate a decrease in the volume of InAs islands upon an increase in substrate temperature. The shrinkage of the island volume agrees with the thermodynamic theory and confirms the equilibrium nature of arrays of InAs/GaAs islands.

3.2.8 Submonolayer Islands at Work

Ultra-High Island Densities. A major advantage of submonolayer islands in heteroepitaxial systems is their relatively small dimensions. The typical height of the islands may be strictly one monolayer, or weakly smeared up to 2–3 ML. The lateral dimensions may be about 5 nm, and the lateral spacing between centers of neighboring islands may be about 10 nm. Thus, the area density of the islands may be extremely high, up to 10^{12} cm^{-2} and more.

Growth of a multilayered system: substrate of material 1/submonolayer deposition of material 2/spacer of material 1/submonolayer deposition of material 2, etc., causes a system of ultrathin insertions to form, as shown in Fig. 3.62. The various possible arrangements of islands in neighboring layers (random, correlated, anticorrelated) are governed by the strain-mediated interaction of buried islands and surface islands during the growth of each subsequent layer. This will be discussed in detail in the next chapter. The key point is the possibility of forming a very dense multilayered array of submonolayer islands, with volume density higher than 10^{18} cm^{-3}.

The next major advantage of submonolayer islands is the possibility of growing a multi-layered structure with a large number of layers, while keeping the whole structure coherent, i.e., dislocation-free. Since the volume of every

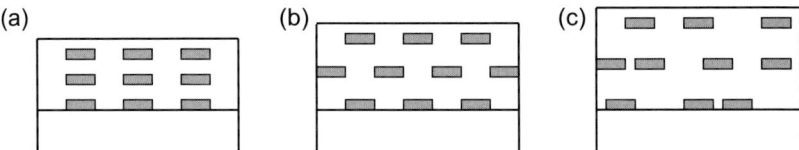

Fig. 3.62. Multi-sheet arrays of submonolayer islands. (**a**) In-phase, or vertically correlated arrangement. (**b**) Out-of-phase, or anticorrelated arrangement. (**c**) Random, or uncorrelated arrangement

single submonolayer island is significantly smaller than that of a typical 3-dimensional island, the onset of misfit dislocations will occur later.

Ultra-Strong Optical Response from Submonolayer Islands. Initial optical studies of systems with submonolayer (0.08–1 ML) InAs insertions in the GaAs matrix already revealed the extremely narrow linewidth of photoluminescence (PL) and photoluminescence excitation (PLE) spectra (0.15 meV at $T = 5$ K) [3.91]. Optical reflection measurements showed remarkably pronounced modulation features related to InAs insertions. The width of the InAs-related exciton peak decreased from 7 meV in a 0.9 ML sample to 0.6 meV in a 0.08 ML sample. The InAs submonolayer-related exciton oscillator strength was deduced from a model of the dielectric function:

$$\varepsilon(E) = \varepsilon_{\mathrm{bcgr}}(E) + \sum_j \frac{S_j E_j}{E_j - E - \mathrm{i}\Gamma_j}, \tag{3.130}$$

where S_j, E_j, and Γ_j are oscillator strength, energy, and damping of the jth exciton transition, and $\varepsilon_{\mathrm{bcgr}}(E)$ is the background dielectric function resulting from GaAs bandgap transitions. The oscillator strength S_j deduced from optical reflection measurements is of order 0.5×10^{-2}–10^{-1}, and the lifetime of excitons is about 50 ns. Hence, large oscillator strengths are of the same order as in GaAs/GaAlAs quantum wells of width 50–100 Å, and considerably larger than expected for ultrathin quantum wells of thickness 1 ML. Correspondingly, the radiative lifetime of excitons is ≈ 20 times shorter than expected for ultrathin quantum wells. These results clearly demonstrate a strong localization of excitons in the lateral plane. This dramatically changes the optical characteristics of the system compared with those of ultrathin quantum wells.

Such an ultra-strong optical response from a single-sheet submonolayer array suggests growing multi-sheet systems. The repetition of a sheet of submonolayer islands will create a medium with a high volume density of oscillators, each having a high oscillator strength. One may thus obtain a 3-dimensional medium with finite thickness exhibiting an ultra-strong optical response near the exciton resonance.

Resonant Waveguide Effect and Laser Without External Waveguide. A medium with a strong optical response may be employed to

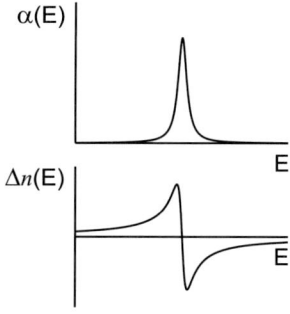

Fig. 3.63. Optical absorption coefficient $\alpha(E)$ and variation of the refractive index $\Delta n(E)$ in the vicinity of the resonant absorption peak

realize the resonant waveguide effect proposed by Alferov et al. [3.119]. Since the real and imaginary parts of the dielectric function $\varepsilon(E) = \varepsilon'(E) + i\varepsilon''(E)$ are related via the Kramers–Kronig relation,

$$\varepsilon'(E) = \frac{1}{\pi} \text{PP} \int dE' \frac{\varepsilon''(E')}{E' - E}, \qquad (3.131)$$

where E is the photon energy and PP denotes the principal part, the refractive index

$$n(E) = \text{Re}\left(\sqrt{\varepsilon'(E) + i\varepsilon''(E)}\right) \qquad (3.132)$$

and the absorption coefficient

$$\alpha(E) = \frac{2E}{\hbar c} \text{Im}\left(\sqrt{\varepsilon'(E) + i\varepsilon''(E)}\right) \qquad (3.133)$$

must also be related. In the vicinity of a resonant absorption peak, the refractive index is modulated as shown in Fig. 3.63. The refractive index increases on the low energy side of the absorption peak and decreases on the high energy side. Enhancement of the refractive index results in optical confinement in the medium. If the medium exhibiting a strong absorption peak works as the active medium for a laser, the refractive index modulation provides confinement of the emitted light, in addition to confinement by a conventional waveguide, as was shown experimentally for CdSe quantum wells in a ZnSe matrix [3.119].

Furthermore, if the absorption peak is very strong, it is possible to achieve lasing based entirely on the resonant waveguide effect, without a conventional waveguide. However, in conventional II–VI quantum well structures, it is usually difficult to use this waveguiding effect because lasing starts at an energy shifted by 30–90 meV down towards lower energies from the heavy hole exciton resonance, where refractive index modulation is rather weak.

For a CdSe/ZnSe submonolayer superlattice, Ledentsov et al. [3.92] showed that lasing occurs at energies in the vicinity of the heavy hole exciton resonance, fully within the region of strongly enhanced exciton-induced modulation of the refractive index. The structure comprised a 20-nm-thick ZnSe layer grown on a GaAs substrate, a 35-nm-thick $ZnS_{0.06}Se_{0.94}$ layer, and a 500-nm-thick lattice-matched ZnMgSSe lower cladding layer. The active region was symmetrically confined by ZnSSe layers. The total thickness of the optical cavity region was equal to 100 nm. The upper cladding ZnMgSSe layer had a thickness of 50 nm and the protective 3-nm-thick ZnSe layer was grown on the top. The active region in one case represented a single 70-Å-thick $Zn_{0.75}Cd_{0.25}Se$ quantum well and, in the other case, an eight-period short-period submonolayer superlattice composed of 1/3 ML CdSe insertions separated by 30-Å-thick ZnSe layers.

The exciton resonance in the optical reflectance spectra was found to be much more pronounced in the case of a submonolayer (SML) superlattice than in the case of a single quantum well. Furthermore, in the sample with the quantum well, lasing occurred at energies $\approx \hbar\omega_{LO}$ (30 meV) below the heavy hole exciton transition, in agreement with the exciton–LO phonon lasing model [3.120]. In contrast, lasing in the sample with the SML superlattice was found to be very close to the exciton resonance energy ($\Delta E \approx 1.4$–9 meV at 80 K). Thus, lasing is resonant in the energy range of the exciton-induced enhancement of the refractive index.

While lasing was achieved at 80 K in [3.92], Krestnikov et al. [3.121] grew a slightly different structure, where a submonolayer superlattice was confined on both sides by 50 nm ZnMgSSe (instead of ZnSSe). Higher ZnMgSSe barriers hinder evaporation of carriers from the quantum dots and result in an enhanced temperature stability of the laser, allowing lasing up to room temperature.

Figure 3.64 illustrates the effect of resonant waveguiding and lasing. Figure 3.64a shows the absorption peak caused by multiple SML CdSe/ZnSe insertions evaluated from the optical reflectance spectra and the corresponding refractive index modulation derived from the Kramers–Kronig equation. In the case of carrier injection, the absorption peak is partly suppressed due to the partial population of QD states by electrons and holes, and a gain peak appears on its low energy side as larger islands provide higher capture efficiency. At finite temperatures, there also exists hopping transport of nonequilibrium carriers towards such low energy states. In the case of excitation densities far below those corresponding to the gain saturation level, the appearance of the gain peak increases the maximum n value (Fig. 3.64b) due to summing of positive contributions to n originating on the low energy side of the absorption peak and on the high energy side of the gain peak.

These facts allow waveguiding in a narrow spectral window. Contrary to the case of a conventional waveguide formed by two layers with lower refractive index than the core layer, an enhancement of the refractive index occurs in the active region in the waveguide for resonant emission processes to fit in the spectral window mentioned above. Thus the resonant character of the lasing mechanism in QDs is an important condition for such waveguiding. Sufficiently high absorption and/or gain values are needed to provide significant waveguiding. Absorption measurements demonstrated maximum absorption

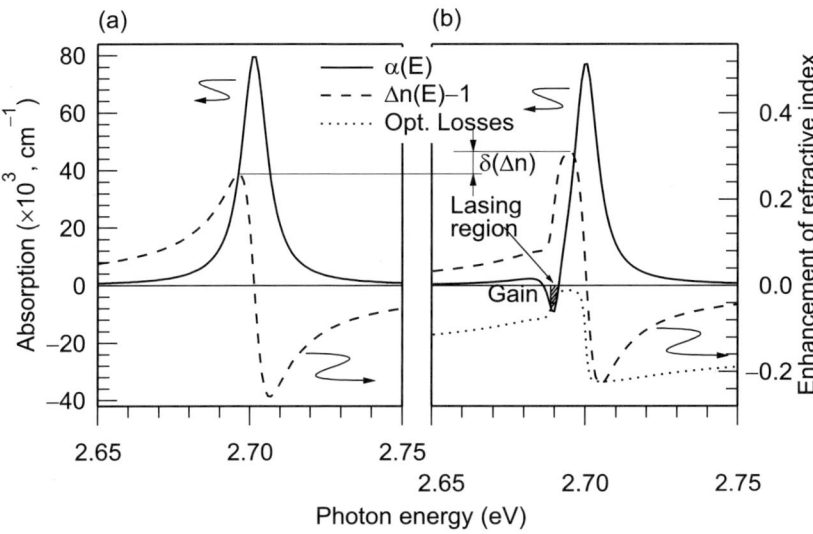

Fig. 3.64. Exciton-induced modulation of absorption and refractive index (**a**) where no excitation is applied and (**b**) for the case of carrier injection. The estimated optical losses (with negative sign) due to GaAs substrate absorption are shown by a *dotted line*. The region where losses cross the gain curve corresponds to the expected lasing energy

coefficients of the order of 10^5 cm^{-1} in the case of stacked SML QDs. This agrees with the estimate of the exciton oscillator strength from optical reflectance spectra [3.91], which corresponds to a refractive index enhancement of 0.2–0.3. The efficiency of exciton-induced waveguiding is sufficient to realize lasing both for II–VI [3.92] and III–V [3.122] SML structures.

Surface Lasing Without Bragg Reflectors. The high exciton absorption coefficients in QDs convert to ultrahigh exciton/biexciton gain coefficients at high excitation densities, since no screening of excitons occurs. This means that lasing can be achieved for very short cavity lengths in edge geometry, or surface lasing in vertical geometry, even if no highly reflecting Bragg mirrors are used. For example, ZnMgSSe/GaAs and ZnMgSSe/air interfaces could lead to 30% reflectivity, which allows surface lasing in structures with 20 sheets of CdSe SML insertions in a ZnMgSSe matrix [3.123].

The Eve of the Ge/Si Laser. Si/Ge-based nanostructures are attracting great interest and have already resulted in successful applications in transistors, photodetectors, and light-emitting diodes based on intersubband transitions in quantum wells. However, numerous attempts to fabricate efficient light-emitting devices based on interband transitions in quantum wells in Si/Ge material systems have failed. Efficient light-emitting devices like lasers or light-emitting diodes, if fabricated, would potentially provide the most direct integration of silicon technology with optoelectronic data transmission systems, both within silicon integrated circuits, and in telecommunications applications.

It was shown earlier that Si/SiGe quantum wells do not lead to any significant reduction in the radiative recombination time [3.124]. Furthermore, due to the particular alignment of conduction and valence bands, and due to the strain profile in coherent Si–Ge quantum wells, Si–Ge forms a heterojunction of the second type [3.125], and the overlap between electron and hole wave functions is significantly reduced, not only in k space, but also in real space. The spatial separation of electrons and holes at the heterojunction shifts the photoluminescence line towards shorter wavelengths with increasing excitation density, which is typical for quantum wells of the second type [3.124, 3.125].

Extensive studies of 3-dimensional quantum dots obtained via Stranski–Krastanow growth in SiGe/SiGeC/Si systems have aimed specifically at increasing the photoluminescence efficiency [3.126]. The large size of 3D QDs, along with a high Ge content, lead to even stronger spatial separation of the hole wave function, localized in the Ge-rich QD, and of the electron wave function, localized in the Si matrix. In the PL spectrum of these structures, a shift towards shorter wavelengths was observed with increasing excitation density. This is typical for quantum wells of the second type [3.127]. The relatively large size of the QDs (> 10 nm) makes it necessary to use rather thick (~ 10 nm) Si spacers in the growth of multilayered systems. The area density of 3-dimensional QDs is at the level of 10^9–10^{10} cm^{-2}, and the maximum volume density is also rather small (10^{15}–10^{16} cm^{-3}). Such a low density is a serious problem in achieving lasing, even for InAs–GaAs QDs based on direct bandgap materials [3.128]. Furthermore, the structure of electronic bands in the k space of Si changes very weakly with respect to that in the bulk material, since the typical extension of a localized hole state significantly exceeds the Bohr radius of the hole.

As emphasized above, another class of QDs based on submonolayer insertions of a narrow bandgap material in a wide bandgap matrix provides major advantages. The typical lateral size of these QDs is significantly smaller, and the density is considerably higher than in the case of 3D QDs [3.129]. The possibility of stacking such QDs, to be discussed in detail in the next chapter, allows us to achieve ultrahigh modal gain (at a level of 10^4–10^5 cm^{-1}) in wide bandgap materials with small exciton Bohr radius.

The realization of submonolayer QDs in an Si–Ge material system, if successful, will resolve all major problems for optoelectronic applications. Firstly, the small lateral size of the QDs (3–5 nm) effectively lifts momentum selection rules for radiative recombination with electrons from the indirect minimum of the conduction band. Secondly, the strength of the repulsive potential in the conduction band is small, allowing localization of both an electron and a hole in the same region of real space [3.130].

It was already shown in [3.130] that, in the case of ultrathin layers of the second type, efficient localization of an electron by a hole can be achieved even using type-II heterostructures with a high potential barrier in the conduction band [3.130], since the Coulomb attraction of an electron by a hole in the case of a thin barrier overcomes repulsion from the barrier. Indeed, for ultrathin insertions of Ge in an Si matrix, there is no short wavelength shift of the PL line with excitation density [3.124]. Using ultrasmall QDs further simplifies the task of localizing an electron, compared to the case of quantum wells, because the strength of the barrier in the lateral direction is weaker for QDs than for quantum wells.

Makarov et al. [3.131] suggested using ultrasmall QDs obtained via deposition of an amount of Ge with thickness below the critical thickness, whereupon 3-dimensional islands form. Two types of superlattice were grown by molecular beam epitaxy. One group of superlattices consisted of 20 layers of submonolayer insertions of Ge of various nominal thicknesses, separated by Si spacers of thickness 4–5 nm. The effective thickness of Ge varies from 0.07 nm to 0.14 nm. Another group of superlattices consists of 10 periods, each of which includes 0.5–0.7 nm Ge insertions, separated by 11 nm spacers of Si. Silicon spacers were doped by Sb with concentration 5×10^{16} cm^{-3} in the middle of the Si spacers. The growth temperature was 750°C for undoped superlattices and 700°C for doped superlattices. To prevent segregation of Sb, spacers were grown at 600°C. The growth velocity was 0.05 nm/s for Si and 0.005 nm/s for Ge. Total vapor pressure during growth was not below 5×10^{-9} torr. Surface morphology was controlled in situ by reflection high energy electron diffraction (RHEED). The formation of 3-dimensional islands was not observed even in the topmost layers.

Figure 3.65a shows cross-sectional high resolution transmission electron microscopy (HRTEM) images of a structure containing submonolayer insertions of Ge with nominal thickness 0.07 nm, grown at 650°C. The thickness

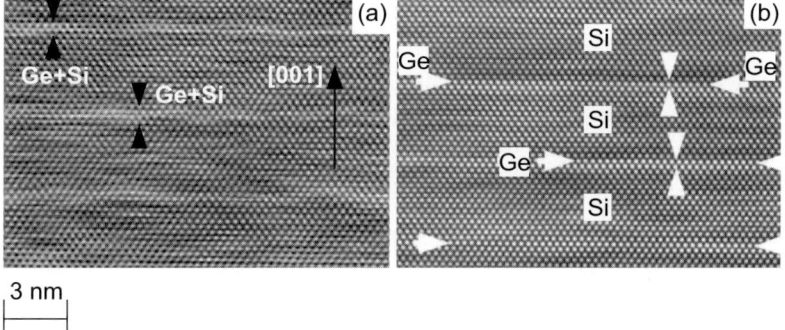

Fig. 3.65. Cross-sectional HRTEM images of multilayered Ge/Si structures. (**a**) Structure with submonolayer (0.07 nm) insertions of Ge. (**b**) Structure with monolayer (0.136 nm) insertions of Ge

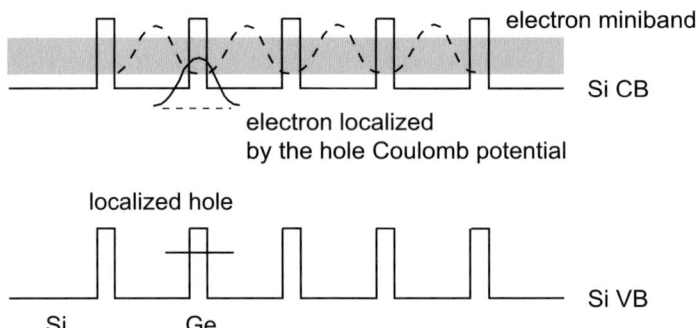

Fig. 3.66. Schematic zone diagram of a multilayered structure with Ge insertions in the Si matrix

of Si spacers is 4.4 nm. Ge insertions do not form a continuous layer, but are arranged in nanometer-scale domains with typical lateral size 3–5 nm and density ∼ 5 × 10^{11} cm^{-2} [3.132, 3.133]. In the case of Ge insertions of thickness about 1 ML and more (Fig. 3.65b), the typical lateral size of nanodomains is 7–10 nm.

Figure 3.66 shows a schematic zone diagram of the studied multilayer structures. Insertions of Ge in the Si–Ge system form potential wells in the valence band and potential barriers in the conduction band. In the case of multilayered structures, mini-zones form in the Si conduction band. Then the wave function of an electron has minima at Ge insertions. If nonequilibrium holes are inserted in the Si matrix, they are trapped by Ge potential wells, and an additional Coulomb potential occurs that attracts an electron to a hole. Since the Coulomb energy in Si is rather large (14.7 meV), and the barrier in the conduction band is rather small (< 100 meV [3.125]), electrons may be efficiently localized in Ge regions due to the Coulomb potential of holes, as was shown in a general case for ultrathin quantum wells of type II.

Figure 3.67 depicts typical photoluminescence (PL) spectra of a sample with a submonolayer (0.1 nm) Ge insertion in an Si matrix. Features of the PL spectra are due to acoustic and optical phonons of the Si matrix as well as PL from Ge insertions (Ge^{NP}, Ge^{TO}, Ge^{TO-O}) with maxima at 1.121 eV, 1.064 eV, and 1.004 eV, respectively. An interesting property of submonolayer insertions of Ge is a long wavelength shift of PL lines from Ge QDs with increasing excitation density. At the same time, the spectral position of the no-phonon PL line at low excitation density is close to the expected position, if one considers the dependence of the PL peak on the thickness of the insertion [3.134].

In structures with Ge insertions thicker than 1 ML, no shift was observed in the PL peak with changing excitation density, according to earlier observations [3.124]. The absence of any shift in the PL peak confirms the absence of any spatial separation between electrons and holes, as it should be for type-I quantum dots. A shift in the PL peak towards longer wavelengths with in-

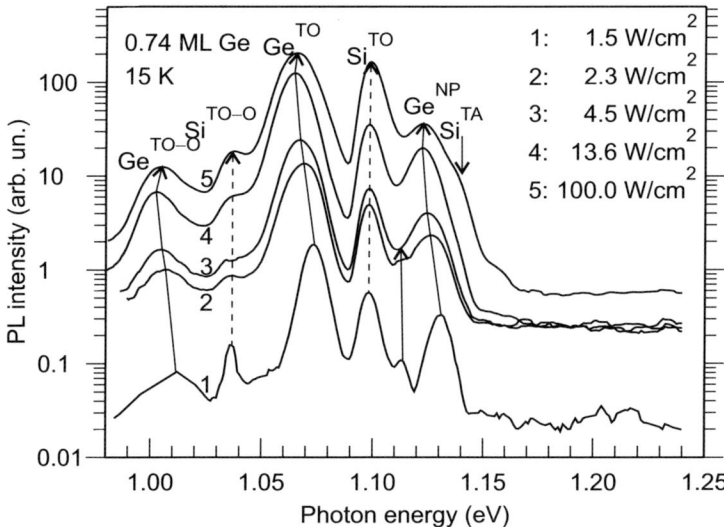

Fig. 3.67. Dependence of the photoluminescence (PL) spectrum on the excitation density for Si structures with submonolayer insertions of Ge. With increasing excitation density, PL peaks shift towards lower photon energies (longer wavelengths)

creasing excitation density in structures with submonolayer insertions may be related to the formation of multi-exciton complexes bounded by quantum dots. This emphasizes once again the increasing role of Coulomb attraction between electrons and holes in the case of relatively weak repulsion from Ge potential barriers in the Si conduction band.

A characteristic feature of PL spectra in Si–Ge structures is the rapid decay of PL intensity with temperature. This may be related to thermal escape of weakly localized electrons and their non-radiative recombination at the surface and in the bulk of the Si substrate. Even a relatively weak doping of the active region of the structure by a donor impurity (average concentration $\sim 10^{16}$ cm^{-3}), creating a moderate concentration of equilibrium electrons, can strongly enhance the PL intensity and make it observable up to room temperature. Figure 3.68 shows the dependence of the PL spectrum on temperature. A slower shift with temperature of the PL peak from Ge insertions relative to the PL line from Si is clearly visible in the figure. This was observed in all samples, i.e., in structures with both submonolayer and monolayer insertions, in both undoped and doped structures (see also Fig. 3.69). This fact together with the absence of any short wavelength shift in the PL spectrum with excitation density very likely attests to thermal population of the mini-band in Si by electrons at elevated temperatures. To overcome this effect, a significant increase in the doping level is needed, up to the formation of a degenerate electron gas.

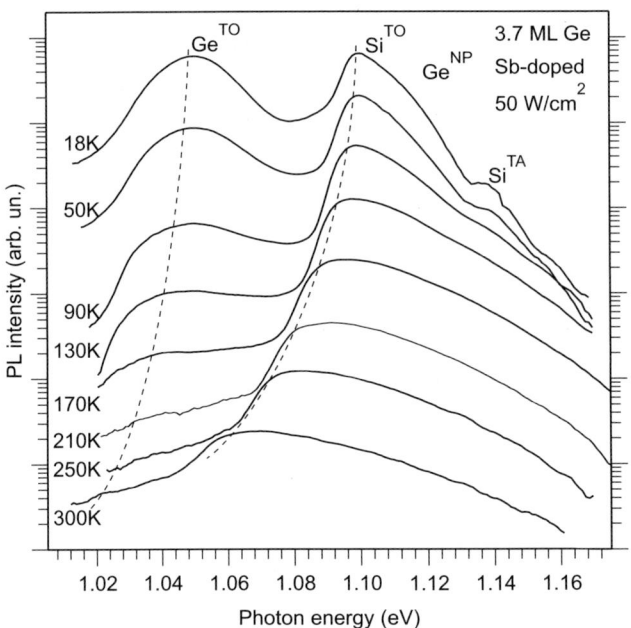

Fig. 3.68. Photoluminescence (PL) spectrum of a structure with submonolayer Ge insertions in Si as a function of observation temperature

The high intensity and temperature stability of the PL in doped samples with Ge–Si QDs made it possible to observe a narrowing of the PL line in both the following scenarios:

- with decreasing temperature, at a fixed high excitation density,
- at a fixed temperature, with increasing excitation density.

The narrowing of the PL line is accompanied by a tremendous increase in the integrated PL intensity. The effect is observed in the vertical direction and only in samples with polished back surface. It may attest to stimulated radiation in the vertical Si cavity with the active region based on a multi-sheet system of dense arrays of doped Ge QDs.

Hence, studies of structural and photoluminescence properties of dense arrays of Ge QDs in an Si matrix by Makarov et al. [3.131] have shown that these QDs are type-I QDs. The doping of QDs makes it possible to obtain a high PL intensity at elevated temperatures and to achieve a superlinear increase in PL intensity with excitation density. This may be a manifestation of stimulated radiation in Si–Ge heterostructures. Employing ultra-dense arrays of Ge–Si QDs of small size and high density, and heavily n-doped, it may be possible in the very near future to achieve lasing in room temperature Si–Ge structures.

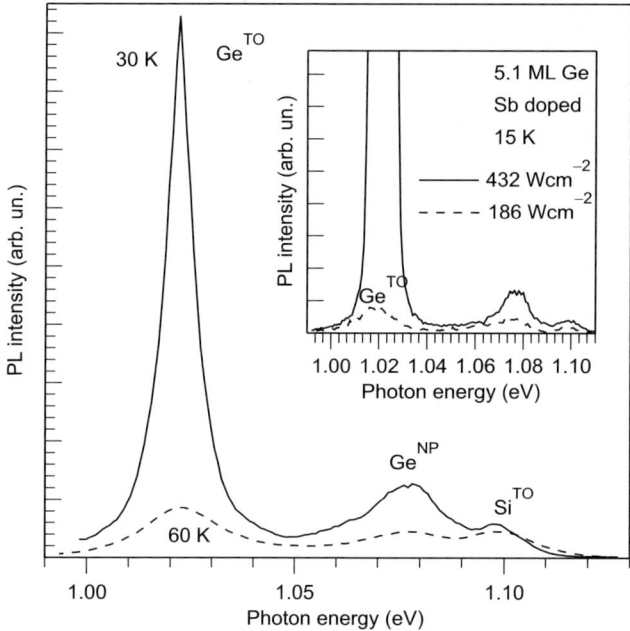

Fig. 3.69. Dependence of the photoluminescence (PL) spectra of Si–Ge structures on observation temperature and excitation density

Further Prospects. The use of dense arrays of very small quantum dots achieved by submonolayer (or, more generally, ultrathin) insertions, first developed for II–VI materials, can be successfully extended to other semiconductor compounds. The latter has been demonstrated for InAs SML insertions in an AlGaAs matrix and has resulted in resonant waveguiding and lasing. Wavefunction engineering has made it possible to realize SML QDs as active regions in high power lasers emitting at 940 nm [3.135].

Group III nitrides have become the most attractive candidates for application of ultrathin insertions. Such insertions have already achieved gain in the green spectral range in InGaN/GaN structures [3.136]. Adding arsenic to this system [3.137] may help to shift the photoluminescence wavelength towards the red spectral range. In this case, optoelectronic devices for the whole visible range will be possible using the same material system.

The application of resonant waveguiding and lasing concepts seems to be particularly important in systems where no convenient lattice-matched heterostructure exists with high contrast between refractive indices in the two materials. The concept may also lead to the development of very thin waveguides in existing heteroepitaxial systems, thereby achieving ultralow threshold current densities.

In view of the possibility of achieving ultrahigh modal gain, SML insertions are particularly promising in vertical cavity surface-emitting lasers, especially in cases where the creation of highly reflecting distributed Bragg reflectors is difficult due to the low contrast between the refractive indices of the materials used. This is the case for GaN-based VCSELs and for 1.3 µm and 1.55 µm emitting structures on InP substrates.

First observations of bright photoluminescence in structures with SML Ge insertions in Si would suggest realistic prospects for silicon-based optoelectronics.

3.3 Arrays of 3-Dimensional Coherently Strained Islands

If a heteroepitaxial system grows according to the Stranski–Krastanow (SK) mode (see Fig. 3.33) and the amount of deposited material exceeds a certain critical thickness, a transition from the 2-dimensional to the 3-dimensional morphology occurs and islands form with 3-dimensional shape.

3.3.1 The In(Ga)As/GaAs System: From 3-Dimensional Islands to Quantum Dots

For lattice-mismatched heteroepitaxial systems, two kinds of surface morphology have long been known. For example, in 1983, Schaffer et al. studied the growth of InAs films on GaAs(001) surfaces by molecular beam epitaxy (MBE) combined with in situ reflection high energy electron diffraction (RHEED) and X-ray diffraction [3.138]. A drastic difference in surface morphology was observed depending upon growth conditions. Indium-stabilized conditions result in a smooth growth surface while arsenic-stabilized conditions lead to a spotty RHEED pattern, indicating a rough surface due to 3D islands. In addition, it was noticed that "islands of InAs are not coherently constrained to the underlying GaAs substrate lattice constant" [3.138].

Similar results for MBE growth of $In_xGa_{1-x}As$/GaAs(001) were observed in 1984 by Lewis et al. [3.139]. It was reported that the initial growth is planar, but "after an amount of deposition depending upon InAs content and growth conditions, a sudden change from a streaked reflection pattern to a spotty transmission pattern is observed indicating formation of 3D islands" [3.139].

As the fabrication and study of quantum wells and quantum well superlattices constituted the mainstream of semiconductor physics and technology in the 1980s, the formation of 3D islands was considered to be a highly undesirable effect. Thus, the identification of growth conditions for InAs/GaAs multiple quantum well structures involving 4 monolayers of InAs alternating with 8 monolayers of GaAs by Grunthaner et al. [3.140] and Yen et al. [3.141], where cross-sectional transmission electron microscopy (TEM) revealed well formed interfaces and low defect density, was considered a big success in this direction. Only a small waviness was observed in the image contrast at the interfaces, revealing no 3D islands. Under other conditions used later

3.3 Arrays of Three-Dimensional Coherently Strained Islands 157

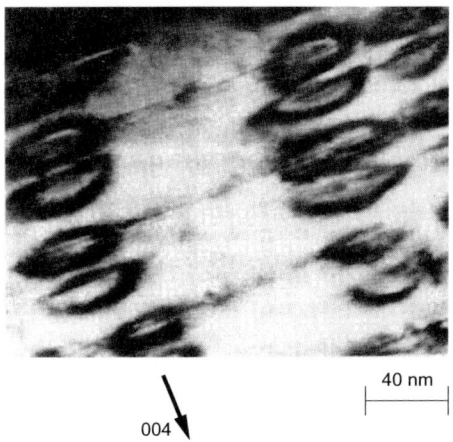

Fig. 3.70. Cross-sectional scanning transmission electron microscopy (STEM) images of InAs islands in GaAs, showing vertical columns of correlated coherent 3-dimensional islands. With kind permission of L. Goldstein et al. [3.143]

to grow InAs/GaAs superlattice structures in [3.141], cross-sectional transmission electron microscopy (XTEM) revealed a different behavior. It was claimed in [3.141] that: "One notes the presence of intensity contrast 'bands' over distance scales much larger than the 3 ML InAs/6 ML GaAs layers constituting the superlattice. These bands are associated with internal variations of strain in the TEM specimen and are not necessarily a reflection of structural defects," such as dislocations and twins.

In addition, following the work by Gibson et al. [3.142] on MBE-grown Ge_xSi_{1-x} superlattices, Yen et al. [3.141] noted that, although large variations in the local lattice parameter and modulation thickness can occur in thin samples for XTEM due to elastic relaxation of strain, the strain bands observed in their samples were dominantly intrinsic. We infer that the contrast bands observed in cross-sectional TEM images of InAs/GaAs superlattices in [3.141] are due to 3D island-like features.

Although it had long been recognized that 3D islands, e.g., InAs islands in a GaAs matrix, could potentially work as quantum dots, dislocations seemed to be unavoidable in such a system. An important step in this direction occurred when experimental studies of the InAs/GaAs heteroepitaxial system by Goldstein et al. [3.143] hinted at the formation of coherent, i.e., dislocation-free islands. InAs/GaAs superlattices were grown by molecular beam epitaxy, where the thickness of the InAs layer was varied from 1–4 ML, and GaAs spacers were relatively thick (200–300 Å).

- In situ reflection high energy electron diffraction (RHEED) measurements revealed an abrupt transition from a diffuse to a spotty pattern at InAs thicknesses > 2 ML, which corresponds to the formation of 3D islands.

- Figure 3.70 shows a cross-sectional scanning tunneling microscopy (STM) image of a superlattice in which insertions of InAs with a nominal thickness of 2.5 ML are separated by 280-Å-thick GaAs spacers. The image depicts stress centers whose strain fields extend into neighboring layers.
- X-ray microanalysis showed that these centers are richer in indium than the region surrounding them.
- These In-rich islands with sizes less than 100 Å tend to align along the growth direction between successive layers, although they are spaced by 280-Å-thick GaAs layers. This memory effect confirms that the whole thickness of the GaAs layer is strained.
- Photoluminescence (PL) spectroscopy performed at 77 K revealed that formation of 3D islands is always accompanied by the onset of a broad PL spectrum shifted to lower photon energies with respect to the spectrum of a system with flat InAs layers, the peak maximum being around 1.14 eV. Peak broadening is about 50–100 meV.

Thus, the occurrence of bright photoluminescence in the results of Goldstein et al. [3.143] obtained in 1985 suggests the formation of coherently strained 3D islands. It was claimed that "these kinds of structures are thus proved to be of interest to study low-dimensional (< 2) objects showing good optical properties". It should be noted that cross-sectional TEM images in a single vertical plane were presented in [3.143], and no statement about the shape of 3D InAs islands was made, particularly as to whether the islands have a compact dot-like or elongated wire-like shape.

The next important step was made in 1987 by Glas et al. [3.144], who performed plan-view TEM and scanning TEM (STEM) studies of similar InAs/GaAs MBE-grown structures. Specimens with deposits of nominal thickness ranging from 1 to 100 monolayers were imaged using various techniques to produce information about the density of islands at the onset of 3-dimensional growth, the preferred shape of islands, the island-induced strain in the substrate, and the balance between elastic and plastic deformation. Figure 3.71 shows STEM images of the uncovered samples with two distinct nominal thicknesses of InAs, viz., 2 and 7 ML.

In the sample with 2 ML of InAs, elements of black/white contrast have a compact, dot-like shape in the lateral plane and an area density of about 10^{11} cm^{-2}. Cross-sectional TEM images discussed in [3.144] showed that the islands are of very uniform size (about 8 nm along [110] and 2 nm along [001] for the main part), and no direct evidence of dislocations was found. It was reported that: "The segments joining nearest neighbor islands lie preferentially along the [100] type directions (with a smallest length of about 15 nm), and less frequently along [110] type directions; furthermore, alignments of several islands along the [100] directions are common" (Fig. 3.71a).

In the sample with 7 ML of InAs (Fig. 3.71b), the size of islands is much larger, and a high density of misfit dislocations was reported in [3.144], visible in plan-view images with a [400]-type g (not shown in Fig. 3.71). Thus,

3.3 Arrays of Three-Dimensional Coherently Strained Islands

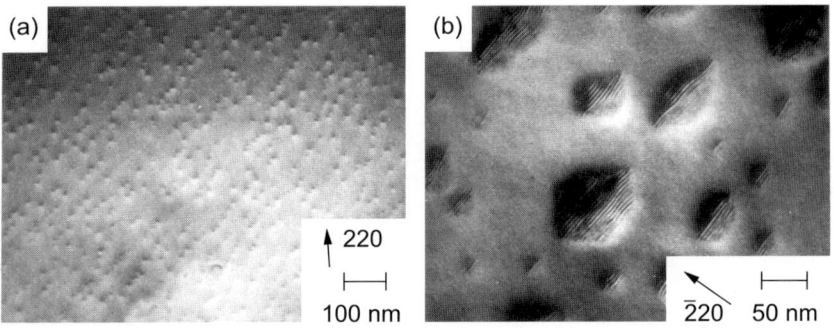

Fig. 3.71. Plan-view scanning transmission electron microscopy (STEM) images of InAs islands in GaAs. Two-beam dark field images ($g = 220$) of specimens with a nominal InAs thickness of 2 ML (**a**) and 7 ML (**b**). Note Moiré fringes in (**b**) pointing to plastic relaxation in InAs-rich regions. With kind permission of F. Glas et al. [3.144]

the study of InAs/GaAs islands in [3.144] revealed a compact dot-like island shape in the lateral plane, the importance of the elastic interaction between the islands, the alignment of islands along $\langle 100 \rangle$-type directions, the formation of coherent islands at a nominal InAs thickness of 2 ML, and the onset of a high density of misfit dislocations for thicker InAs layers.

While the occurrence of photoluminescence reported in [3.143, 3.144] was suggestive of the formation of coherently strained InAs islands, no high resolution transmission electron microscopy (HRTEM) studies were performed to exclude the possibility that local defects were forming inside InAs islands. An important step in identifying coherent vs. dislocated islands was made by Guha et al. [3.145] in 1990. A structure with 7 ML of $In_{0.5}Ga_{0.5}As$ was grown by MBE on a GaAs(100) substrate and capped by epoxy to simulate free-standing islands of InAs on the GaAs surface. Figure 3.72 shows cross-sectional HRTEM images of the structure. Micrographs were taken with the electron beam parallel to the [0$\bar{1}$1] and [011] directions. According to HRTEM images, islands below a certain critical size were found to be coherent, and those exceeding the critical size were defective.

From HRTEM images of coherent islands, Guha et al. [3.145] measured the spacing $d_{0\bar{1}1}$ of the $\{011\}$ planes that are normal to the interface as a function of distance from the interface on both sides (see Fig. 3.73). The data clearly show elastic strain relaxation in the islands when the interplanar distance close to the island apex approaches the bulk lattice constant of InAs.

An important step towards elucidating the nature of photoluminescence from InAs/GaAs structures was made by Tabuchi et al. [3.146] in 1992. Figure 3.74 shows TEM images of InAs/GaAs structures with different nominal thicknesses of deposited InAs. In the image of the sample with a 2 ML InAs layer, some characteristic features are observed. The strong oval contrast observed in this image (marked A in Fig. 3.74a) is likely to indicate a misfit

160   3. Self-Organization Phenomena at Crystal Surfaces

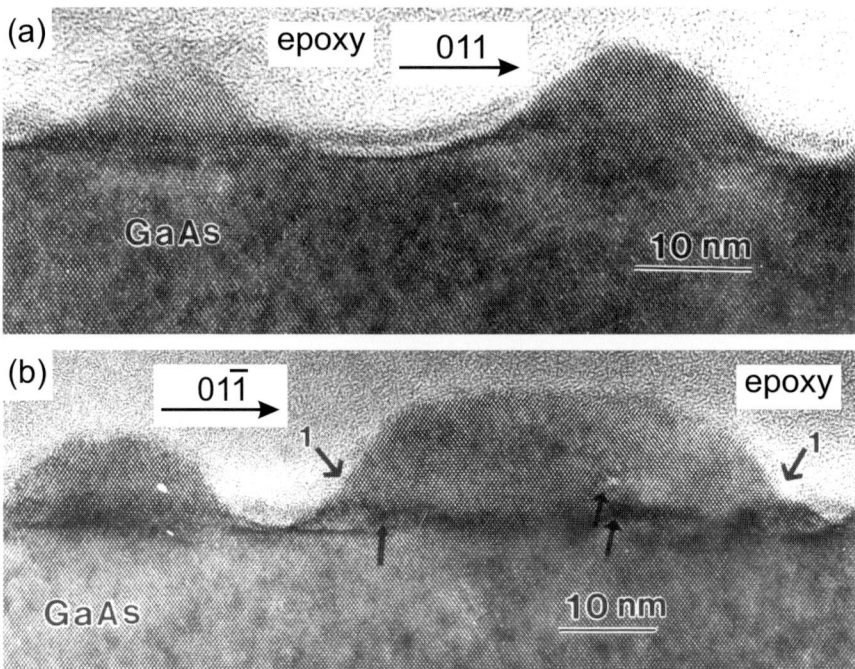

Fig. 3.72. Cross-sectional high resolution transmission electron microscopy (HRTEM) images showing (**a**) two coherent islands with the electron beam parallel to the [0$\bar{1}$1] direction and (**b**) a coherent (*left*) and a defective (*right*) island with the electron beam parallel to the [011] direction. *Arrows* point to defects. Those marked (1) point to stacking faults near the island edges. Note the coherent nature of small islands. With kind permission of S. Guha et al. [3.145]

dislocation. The weak dotted image (marked B in Fig. 3.74a) is thought to be the image of a 3D-grown island observed by RHEED. The sample with the 4 ML InAs layer shows both contrasts: the 3D-grown islands and the misfit dislocations (Fig. 3.74b). Images of dislocations observed in the 4 ML InAs samples were longer and of higher concentration than those in the 2 ML InAs sample. In the sample with the 10 ML InAs layer (Fig. 3.74), only the image which indicates dislocations was observed, while the image of the InAs islands was negligible.

The formation of 3D islands also agreed well with RHEED observations during growth. Photoluminescence studies of the sample with 1 ML of InAs showed a narrow line at 77 K with a full width at half maximum of 20 meV, which is likely to be from the 2D-grown quantum well structure. Samples with 2 ML and 4 ML InAs show broad PL spectra (about 100 meV wide) with peak intensity as strong as that from the 1-ML-grown InAs. However, no emission was observed in the wavelength range 800–1 600 nm from samples

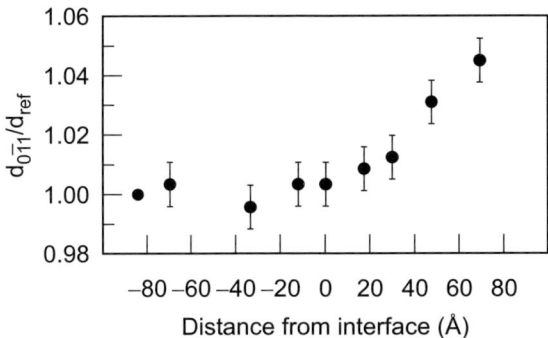

Fig. 3.73. Plot of measured $(0\bar{1}1)$ interplanar spacing vs. distance from the interface for a coherent island. The interplanar spacing is normalized to the lattice parameter value measured 84 Å below the interface in the substrate (d_{ref}). With kind permission of S. Guha et al. [3.145]

with a 10 ML InAs layer. It was thus concluded in [3.146] that a bright PL with a broad spectrum in InAs/GaAs samples does not come from misfit dislocations, but rather from 3D islands. In [3.146], the formation of 3D InAs islands was also studied on the vicinal GaAs(001) surface.

The next step forward in the study of 3D islands arose from a quantitative analysis of the distribution of island sizes. Moison et al. [3.147] (the paper was submitted April 27, 1993) grew InAs islands on GaAs by MBE and measured island base width, height, and density by atomic force microscopy (AFM). Island statistics revealed a remarkably low dispersion in sizes, with an average half-base of about 150 Å and dispersion down to ±10%, and an average height of about 40 Å and dispersion ±20%.

Leonard et al. [3.148] (the paper was submitted July 2, 1993) grew $\text{In}_{0.5}\text{Ga}_{0.5}\text{As}$ islands on GaAs(001) and examined capped structures by plan-view transmission electron microscopy (TEM). TEM studies revealed, as in [3.147], a remarkably narrow size distribution, within 10% of the average lateral size of 300 Å.

The next advance in the study of 3D coherently strained InAs and InGaAs islands on GaAs was connected with experimental proof of the formation of quantum dots exhibiting a δ-function-type electronic spectrum. In 1994, Ledentsov et al. [3.149] reported a spatially resolved cathodoluminescence (CL) study of an ensemble of InAs islands in a GaAs matrix. While the relatively broad photoluminescence spectrum represents the average over the ensemble of islands, CL with high spatial resolution allows selective excitation of a small part of the system. The observed spectrum consisted of a series of ultrasharp lines with full width at half maximum < 0.3 meV, limited by the spectral resolution, each of the lines originating from a single quantum dot. This directly visualizes the absence of inhomogeneous broadening and a δ-function-type density of electronic states.

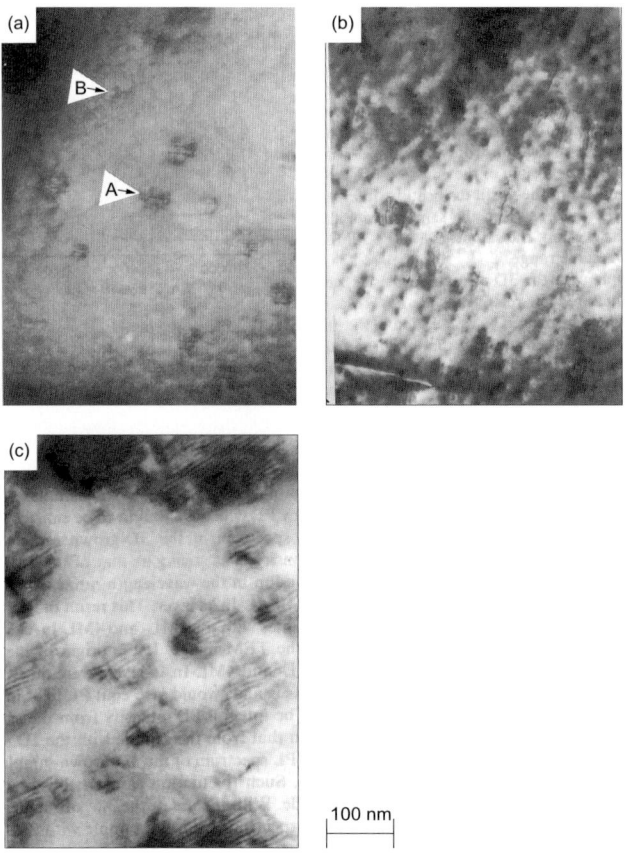

Fig. 3.74. Plan view transmission electron microscopy (TEM) images of InAs on GaAs, where the thickness of the InAs layer is (**a**) 2 ML, (**b**) 4 ML, and (**c**) 10 ML. Note strong oval contrasts A and weak dotted image B in (**b**). With kind permission of M. Tabuchi et al. [3.146]

Marzin et al. [3.150] also measured photoluminescence (PL) spectra from InAs/GaAs structures on small etched mesa structures with sizes down to 200 nm. The reported PL spectra measured at 10 K consisted of sharp individual lines. Both results, [3.149] and [3.150], were made public in August 1994.

The CL studies of [3.149] were extended to elevated temperatures by Grundmann et al. [3.151] (submitted May 23, 1994). Up to 60 K, the width of CL lines from individual dots remained < 0.15 meV, a value determined by the experimental resolution limit.

Thus, in 1994–95 it was experimentally established that in InAs/GaAs and InGaAs/GaAs systems:

- 3-dimensional islands form,

3.3 Arrays of Three-Dimensional Coherently Strained Islands

- these islands may be both coherently strained and defective,
- islands may exhibit a narrow distribution in sizes,
- coherent islands localize electrons and holes and reveal bright photoluminescence,
- coherent islands are indeed quantum dots exhibiting δ-function-like spectra and showing no inhomogeneous broadening of PL.

However, most work on fabrication and properties of In(Ga)As/GaAs quantum dots carried out before 1995 was limited to very specific cases, and hardly any systematic study, particularly the investigation of growth regimes, had been undertaken. Ledentsov et al. [3.152] carried out a detailed investigation of the formation and properties of In(Ga)As/GaAs islands in a wide variety of MBE growth regimes. It was established that:

- the formation of 3-dimensional islands with a narrow size distribution and bright photoluminescence occurs in a certain window of the substrate temperature, arsenic vapor pressure, and amount of deposited InAs;
- under these growth conditions, if growth interruption is introduced after deposition of InAs, the size of the islands reaches some characteristic values and does not change during further growth interruption, thus showing size-limited island growth;
- changes in the arsenic pressure from optimum to low lead to reversible changes in the surface morphology, unless large dislocated islands form on the surface;
- at high arsenic pressure, the onset of dislocations occurs in large islands, and large well separated dislocated islands form within a few seconds, whereupon changes in the surface morphology become irreversible.

The phenomenon of size-limited island growth for a certain range of growth conditions, where 3D islands reach a certain size and apparently show no Ostwald ripening, turns out to be the crucial one in the formation of quantum dots with a narrow distribution in size. In the rest of this section, we focus on the mechanisms underlying size-limited island growth. In Sect. 3.3.2, the formation of coherent vs. dislocated 3D islands is discussed, following Vanderbilt and Wickham [3.153] and Ratsch and Zangwill [3.154].

In the following sections, we consider two basic classes of theoretical model for size-limited island growth. The thermodynamic theory developed by Shchukin et al. [3.155] and extended by Daruka and Barabási [3.156] and Shchukin and Bimberg [3.157] shows the following. Surface energies in a strained system are a function of the local strain. In a certain range of material parameters, the formation of a 3D island from an initially flat surface, accompanied by a decrease in the elastic strain energy, may also be accompanied by a decrease in the surface energy, despite the increase in overall surface area. In this case, there is no thermodynamic driving force to ripening and an array of islands of optimum size corresponds to a stable or metastable state of the system.

Kinetic theories focus basically on various mechanisms for the slowing down of the ripening process which may eventually lead to size-limited growth, even if there is a global thermodynamic driving force to ripening. Priester and Lannoo [3.158] considered the formation of 2-dimensional platelets of optimum size. These transform to 3D islands, thereby ensuring a narrow size distribution for the latter. Madhukar et al. [3.159] and Kobayashi et al. [3.160] focused on the dependence of the barriers for attachment and detachment of adatoms to islands on the island-induced strain. This eventually results in size-limited growth. Moreover, they hypothesized that the strain field between islands gives rise to preferred migration of adatoms from larger islands to smaller islands, thus favoring equalization of island sizes [3.159, 3.160]. Chen et al. [3.161] and Jesson et al. [3.162] showed that the growth of every new atomic layer on flat side facets of a 3D strained faceted island occurs via nucleation. They also showed that the corresponding barriers become progressively higher with increasing island size and that this may eventually result in size-limited island growth. Wang et al. [3.112] considered the situation where the islands first nucleate from an initially 2D film, this process governing their density. A local equilibrium is then established between each island and the surrounding wetting layer, which determines the island volume.

Experimental data on 3D strained islands in various material systems are discussed with a view to comparing these models. The majority of detailed experimental studies specifically aimed at distinguishing between different formation mechanisms have been performed for the InAs/GaAs system [3.152, 3.163, 3.164]:

- Clear evidence was obtained of a phase-diagram-like range of growth parameters in which size-limited island growth occurs.
- Very fast (on a time scale of a few seconds) Ostwald ripening of the islands was observed beyond this region.
- A thorough examination of the volume vs. temperature dependence of InAs islands revealed an increase in the lateral size, a decrease in the height, and an overall decrease in the island volume vs. increase in the substrate temperature. This favors a thermodynamic rather than a kinetic picture of island formation.
- A high degree of reversibility was established in the island density, volume, and shape upon variations in arsenic pressure and/or substrate temperature.
- Experiments revealed evidence for a high density of indium adatoms and their condensation on islands during cooling [3.165].

The above data indicate that arrays of 3D InAs/GaAs islands are to a large extent thermodynamically dominated.

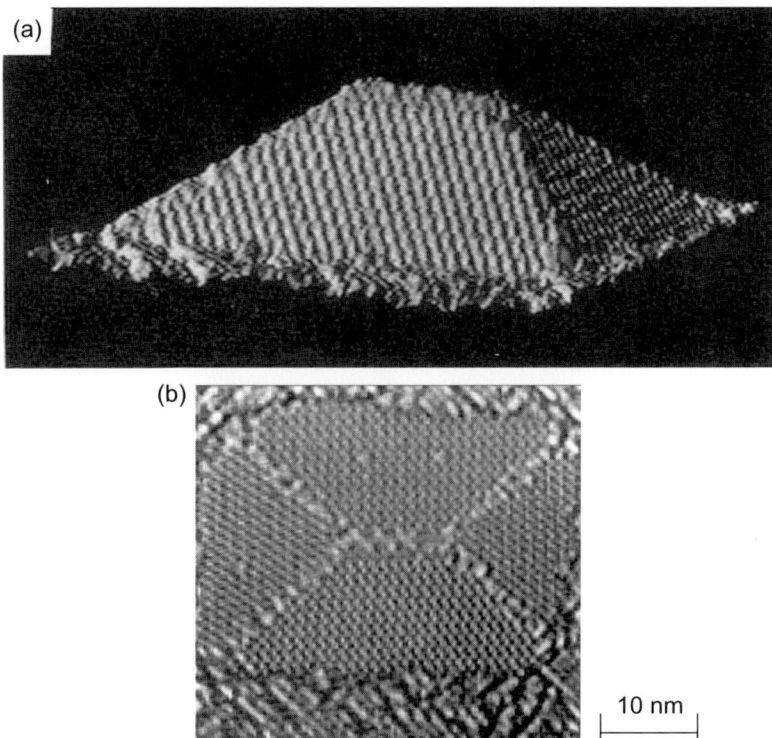

Fig. 3.75. Scanning tunneling microscopy (STM) images of a single 'hut' cluster. (**a**) Perspective plot. Scan area 400 Å × 400 Å. The height of the hut is 28 Å. (**b**) Curvature-mode grey-scale plot. The crystal structure on all four facets and the dimer rows in the 2D Ge layer around the cluster are visible. The 2D layer dimer rows are at 45° to the axis of the cluster. With kind permission of Y.-W. Mo et al. [3.167]

3.3.2 Coherent vs. Dislocated Islands in Lattice-Mismatched Systems

The InAs/GaAs material system is certainly not a unique one. Another classical system, namely Ge/Si, has revealed similar behavior, where 3D islands form when the nominal thickness of Ge exceeds a critical value of about 3 ML ≈ 10 Å (see [3.166] and references therein). Eaglesham and Cerullo [3.166] grew Ge/Si structures by MBE and examined them using plan-view and cross-sectional TEM. They concluded that dislocation-free Ge islands form on Si to a thickness of ≈ 500 Å, 50 times higher than the critical thickness of the 2D to 3D transition in Ge/Si epitaxy.

Mo et al. [3.167] investigated the transition from 2D to 3D growth using scanning tunneling microscopy (STM). Deposition beyond 3 ML of Ge leads to the formation of 3D islands. In addition to widely separated macroscopic

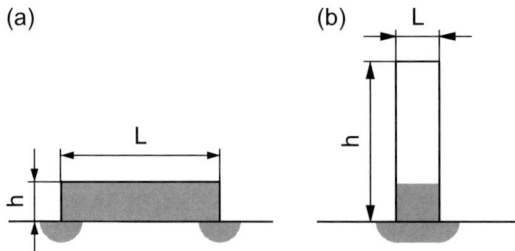

Fig. 3.76. Effect of island shape on volume elastic relaxation of a coherently strained island. The *grey area* is the region with a large elastic strain energy density. (**a**) The island with height-to-width ratio $h/L \ll 1$ is weakly relaxed. (**b**) The island with height-to-width ratio $h/L \gg 1$ is almost completely relaxed

clusters, a large concentration of generally much smaller islands was found. These islands have well defined shape and crystal structure. They are predominantly prism-shaped with canted ends, in some cases a four-sided pyramid, with the same atomic structure as shown in Fig. 3.75. They look like small huts and were termed hut clusters. By measuring the relevant length and angle parameters, all four of their facets were determined to be {105} planes.

The formation of 3D coherently strained islands in lattice-mismatched systems is one of several possibilities for relieving elastic strain energy due to the onset of a non-flat surface. This is related to the Asaro–Tiller–Grinfel'd instability of a strained layer against long-wavelength corrugation of the surface [3.168–3.171]. To illustrate the physical mechanism underlying elastic relaxation, it is convenient to consider a strongly pronounced corrugation, e.g., islands, troughs [3.153], surface cusps [3.172], and cracks [3.173]. The formation of troughs, cusps, or cracks can occur in a strained epitaxial film of a certain macroscopic thickness upon annealing. At the same time, for the first stages of heteroepitaxial growth on a substrate, island formation seems to be the only coherent mechanism for elastic relaxation.

Figure 3.76 illustrates two islands of different shape. A flat island with a small height-to-width ratio is practically unrelaxed, whereas a hypothetical island having the shape of a vertical bar with a large height-to-width ratio is almost completely relaxed. Hence, elastic relaxation depends strongly on island shape. For a given shape, the elastic relaxation energy is proportional to the volume of the island. In order to distinguish this type of elastic relaxation from that due to capillarity effects (see, for example, Sects. 3.1 and 3.2), we will call it volume elastic relaxation.

Consequently, the volume elastic relaxation of coherently strained islands is a relaxation mechanism which competes with the formation of dislocations. The theory developed by Vanderbilt and Wickham [3.153] compares the two mechanisms for elastic relaxation and yields the phase diagram of the lattice-mismatched system, where all possible morphologies are present, i.e., uniform films, dislocated islands, and coherent islands (Fig. 3.77). The formation of an island from a uniform film is accompanied, first of all, by a relaxation of the elastic energy, $\Delta E^{V}_{\text{elast}} < 0$ and, secondly, by a change in the surface area, $\Delta A > 0$. The corresponding change in the surface energy is then caused by the formation of side facets on the islands and by the disappearance of certain areas of planar surface. It is generally held that the change in the surface energy is positive, $\Delta E_{\text{surf}} > 0$. It was shown by Vanderbilt and Wickham [3.153] that the morphology of a mismatched system is determined by the relation between ΔE_{surf} and the energy of the dislocated interface $E^{\text{disl}}_{\text{interface}}$. The ratio of these two energies, denoted $\Lambda = E^{\text{disl}}_{\text{interface}}/\Delta E_{\text{surf}}$, is the control parameter that governs the morphological phase diagram of Fig. 3.78.

If ΔE_{surf} is positive and large, or the energy of the dislocated interface is relatively small, the corresponding value Λ on the phase diagram of Fig. 3.78 is smaller than Λ_0. Then the formation of coherently strained islands is not favorable. With increasing amounts of deposited material, a transition occurs from a uniform film to dislocated islands, and coherently strained islands are not formed.

If ΔE_{surf} is positive and small, or the energy of the dislocated interface is relatively large, the corresponding value Λ on the phase diagram of Fig. 3.78 is larger than Λ_0. As the amount of deposited material increases, a transition occurs from a uniform film to coherent islands. Further deposition may

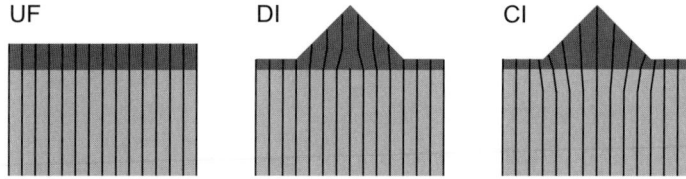

Fig. 3.77. Elastic strain relaxation during Stranski–Krastanow growth (schematic). *Light grey* denotes the substrate and *dark grey* the lattice-mismatched epilayer. *Lines* symbolize lattice planes. It is assumed that surface and interface energies are such that the formation of a wetting layer is energetically preferred. UF: uniform strained film, $E_{\text{UF}} = \lambda \varepsilon_0^2 V$. DI: dislocated relaxed islands, $E_{\text{DI}} = \gamma_2(\Delta A) + \gamma_{12} A_0$. CI: coherently strained islands, $E_{\text{CI}} = \gamma_2(\Delta A) + \lambda \varepsilon_0^2 V - |\Delta E_{\text{elast}}|$. With kind permission of E. Pehlke et al. [3.174]

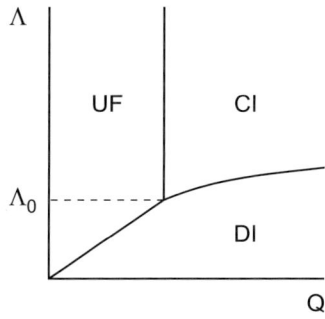

Fig. 3.78. Phase diagram showing the preferred morphology as a function of the amount of deposited material Q (*horizontal axis*) and of the quantity $\Lambda = \Delta E_{\text{surf}}/E^{\text{disl}}_{\text{interface}}$, where ΔE_{surf} is the change in the surface energy due to island formation, and $E^{\text{disl}}_{\text{interface}}$ is the energy of a dislocated interface. Labels UF, CI, and DI refer to uniform film, coherent island, and dislocated island, respectively. With kind permission of D. Vanderbilt and L.K. Wickham [3.153]

168   3. Self-Organization Phenomena at Crystal Surfaces

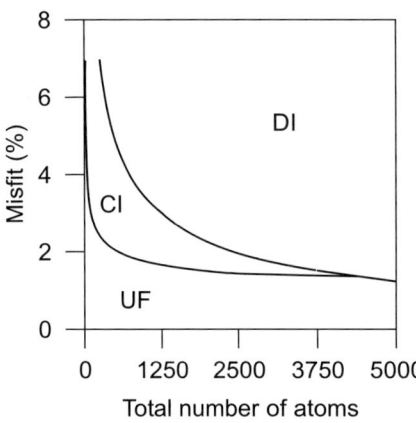

Fig. 3.79. Equilibrium morphological phase diagram as a function of lattice misfit and total particle number, determined by minimizing the total energy. Phase fields are found for uniform films (UF), coherent islands (CI), and dislocated islands (DI). With kind permission of C. Ratsch and A. Zangwill [3.154]

cause the onset of dislocations in the islands. The theory of Vanderbilt and Wickham [3.153] deals with islands having the shape of elongated prisms (or 'ridges'). The theory of Ratsch and Zangwill [3.154] implies the existence of the same morphologies, i.e., uniform films, coherent islands, and dislocated islands, in the case of pyramid-shaped islands (Fig. 3.79).

3.3.3 Size-Limited Island Growth: Are Islands Stable Against Ripening?

The classical picture of island formation refers to a general theory of first-order phase transitions. Consider, for definiteness, the case of a closed system with a constant number of atoms, realized during growth interruption or annealing. The description of island formation usually deals with a supersaturated dilute phase, e.g., a 2D gas of surface adatoms. The processes may be regarded as consisting of three stages. At the nucleation stage, islands form and dissolve due to thermal fluctuations in the system. The change ΔF in the total Helmholtz free energy of a system due to the formation of an island can be written as a sum of surface and volume contributions,

$$\Delta F = \Delta F_{\text{surf}} + \Delta F_{\text{V}} . \tag{3.134}$$

The surface contribution is usually positive, $\Delta F_{\text{surf}} > 0$, due to the energy cost of a new surface, while the volume contribution (in the case of a supersaturated gas of adatoms) is negative, $\Delta F_{\text{V}} < 0$. As a function of the island volume, ΔF first increases while the volume is below a certain value, $V < V_{\text{c}}$, and then decreases once $V > V_{\text{c}}$. V_{c} is the volume of critical nuclei. Islands form due to thermal fluctuations. An island with volume $V < V_{\text{c}}$ will dissolve, with the highest probability. If an island has volume exceeding that of the critical nuclei, $V > V_{\text{c}}$, it will grow further in volume, with the highest probability.

3.3 Arrays of Three-Dimensional Coherently Strained Islands 169

At the growth stage, a sufficient density of islands is present, exceeding the critical volume. Islands grow due to attachment of adatoms, the gas of adatoms depletes, and no new islands nucleate.

At the last stage, coarsening, or Ostwald ripening, the gas of adatoms is depleted to a large extent and supersaturation decreases to a very low value. Correspondingly, the volume of critical nuclei V_c increases to a high value, and islands which originally had volumes above the critical value and were growing, turn out to have volumes below the new value of V_c so that they start to dissolve. At any given instant of time, larger islands grow at the expense of dissolving smaller islands. This process is called Ostwald ripening. The general theory of Ostwald ripening in first-order phase transitions was first developed by Lifshits and Slyozov [3.97, 3.98] and Wagner [3.99]. For the particular case of 3D island growth on a surface, it was extended by Chakraverty [3.175]. The detailed kinetics depends on the slowest process, either attachment/detachment or diffusion of adatoms between islands. In both cases, ripening yields a rather broad size distribution of islands which evolves with time in a self-similar way, leading to an increase in the average volume of islands and a corresponding decrease in island density.

Although some slowing down has been predicted for the system of strained islands [3.176], this mechanism does not yield substantial narrowing of the size distribution and a halt in island growth. In a strained system, both coherent and defective islands may exist. Once a given island reaches a certain critical volume for the onset of dislocations, plastic relaxation occurs, and this accelerates the growth rate of the island.

Surprisingly, experimental studies on coherent islands of InAs/GaAs(001) [3.147, 3.149, 3.177, 3.178] and InGaAs/GaAs(001) [3.148] have revealed a narrow size distribution of islands which does not follow from the Stranski–Krastanow growth mode itself. Such a narrow size distribution has been observed in a wide range of heteroepitaxial systems, including GaSb/GaAs [3.179], InAs/InP(001) [3.180], AlInAs/GaAlAs(001) [3.181], GeSi/Si(001) [3.182–3.184], CdSe/ZnSe(001) [3.185], InAs/InAlAs(001) and InAs/InGaAs(001) [3.186], InAs/Si(001) [3.187].

The small size of these islands allows them to function as quantum dots and realize electron confinement in all three dimensions. The narrow size distribution and the absence of misfit dislocations make these quantum dots suitable for laser applications. Lasing from quantum dots was first reported by Ledentsov et al. [3.188]. Further progress (see, for example, [3.189, 3.190]) is discussed in detail in Chap. 5.

The nature of the narrow island size distribution and apparent absence of Ostwald ripening is a subject of intense debate. One can emphasize two major concepts for describing the effect. Thermodynamic theory states that, under certain conditions, the equilibrium state of a lattice-mismatched heteroepitaxial system is an ordered array of 3-dimensional coherently strained islands. No Ostwald ripening occurs in this case. Kinetic theories state that, once 3-dimensional islands have formed, there exists a thermodynamic tendency towards ripening, but that the growth of islands above a certain size can be kinetically slowed down due to strain-induced barriers.

Within this general debate, one can focus on two different questions. The first question is rather general: do equilibrium arrays of 3D islands really exist? The second question arises for any particular system: is an observed array of 3D islands thermodynamically or kinetically controlled?

There are a number of papers in which 3D islands, once formed, are immediately capped by the substrate material (e.g., [3.178, 3.181]) or immediately cooled down (e.g., [3.147]). This approach allows one to obtain rather small dislocation-free islands that may work as quantum dots. However, this does not help much in understanding the relative role of thermodynamics and kinetics in the formation of arrays of 3D strained islands.

In order to resolve the question experimentally, i.e., to decide whether a given array of islands is an equilibrium array, one must stop deposition of the material and introduce growth interruption before overgrowth of the islands, or before cooling down, or before any other change in the system. Then the heteroepitaxial system may be regarded as a closed system with the fixed amount of deposited material evolving towards equilibrium.

Ledentsov et al. specifically focused on the behavior of a system upon growth interruption [3.152] for MBE growth of InAs on GaAs(001). This will be discussed in detail in Sect. 3.3.13. Briefly, these results indicate that InAs islands reach a narrow size distribution around a certain size upon growth interruption, and that the distribution of island sizes does not change upon further growth interruption. Furthermore, these islands exhibit a strong correlation in the nearest-neighbor arrangement. This correlation is typical of a periodic square superlattice of islands [3.191–3.193]. These results indicate the existence of a new class of spontaneously ordered nanostructure, namely, ordered arrays of 3-dimensional coherently strained islands.

Motivated by these observations, Shchukin et al. [3.155] developed a thermodynamic theory of spontaneous formation of arrays of 3D coherently strained islands. Extending an earlier theoretical approach to describing heteroepitaxial growth on a periodically faceted surface [3.40] and the effects of externally applied stress on a faceted surface [3.42], the theory in [3.155] emphasizes the existence of two sources for the strain field in a system with 3D strained islands. One is the lattice mismatch between deposit and substrate, and the other is the discontinuity in the surface stress tensor at island edges. In addition, it is shown that the dependence of the surface energy on island facets, substrate and wetting layer may be of crucial importance. The analysis by Shchukin et al. [3.155] shows that, for a certain parameter region, the equilibrium in a system of 3D coherently strained islands corresponds to a periodically ordered array of identical islands, and that Ostwald ripening does not occur.

3.3 Arrays of Three-Dimensional Coherently Strained Islands

There are two driving forces ordering island sizes. The first is elastic relaxation due to the discontinuity in the intrinsic surface stress tensor at the island edges, similar to the one described in Sects. 3.1 and 3.2. The second driving force is the dependence of the surface energies on strain, which may be called 'strain-induced renormalization' of surface energies. This renormalization may result in the fact that, despite an increase in the total surface area, the creation of a 3D island leads to a decrease in total surface energy. Indeed, the energy cost of creating additional surface area may be dramatically reduced by elastic relaxation in the islands [3.194]. In this case, there is no energy benefit in Ostwald ripening, and the latter does not occur.

In a subsequent paper, Shchukin et al. [3.195] also emphasized that the wetting layer formed in Stranski–Krastanow growth has a microscopic thickness and is a surface distinct from the surface of the deposited material. Therefore the wetting layer has a different surface energy to the (001) surface of a bulk crystal of the deposited material. For example, the surface energy of the InAs wetting layer on GaAs(001) differs from the surface energy of the (001) surface of bulk InAs. The appearance of tilted facets via the formation of a 3D island competes with the disappearance of a certain area of the wetting layer. This fact yields an additional possibility: the formation of a 3D island in a heteroepitaxial system may be accompanied, despite an increase in the surface area, by a decrease in surface energy, even in the case where the renormalization of the surface energy is not sufficient to produce the effect and to render the array of islands stable.

Following the approach by Shchukin et al. [3.155], Daruka and Barabási [3.156] constructed an equilibrium phase diagram of a lattice-mismatched heteroepitaxial system which reproduces all possible morphologies observed experimentally, i.e., a flat film, an ordered array of islands over a wetting layer or a bare substrate, ripened islands over a wetting layer or a bare substrate, and a bimodal size distribution of islands, including both ordered and ripened islands.

Another branch of thermodynamic research into lattice-mismatched systems has focused on the equilibrium shape of a single isolated island. Kaminski and Suris [3.196], Chen et al. [3.197], and Duport et al. [3.198], making different assumptions about the facet energies, deduced that the shape of a strained island changes with its volume, larger islands having steeper side facets. The same results for a particular system were obtained from ab initio calculations by Pehlke et al. [3.174, 3.199]. Duport et al. [3.198] suggested that, as the volume increases, the ultimate shape of an island corresponds to an overhanging island, nearly detached from the substrate, and that the strain energy in such islands is completely relaxed. Spencer and Tersoff [3.200] developed a model in which they showed that the asymptotic shape of a large coherent island is a ball sitting atop the wetting layer. This way of achieving total relaxation is an alternative to the onset of dislocations, which is unavoidable in large islands if the elastic relaxation is only partial [3.153]. Based on this effect of the total elastic relaxation in overhanging coherent islands, it was concluded in [3.198, 3.200] that the equilibrium state of a lattice-mismatched system is in every case a single island formed via Ostwald ripening. Shchukin and Bimberg [3.157] showed that, despite the possibility of the formation of totally relaxed overhanging islands, there nevertheless exists a parameter region where the equilibrium state of a lattice-mismatched heteroepitaxial system is an ordered array of 3D islands, and ripening does not occur.

Later in the present section, we present the thermodynamic theory of the formation of stable arrays of 3D coherently strained islands. In Sect. 3.3.4, the general equation is derived for the total energy of an array of 3D coherently strained islands, provided that the total amount of deposited material is fixed, i.e., growth is interrupted, evaporation is neglected, and the system is regarded as a closed system which evolves towards equilibrium. As the deposited material is generally distributed between 3D islands and the wetting layer, the total energy is a function of the amount of material in the islands, the shape, the volume, and the relative arrangement of the islands. First, we fix the amount of material in the islands (and hence the thickness of the wetting layer) and minimize the total energy under this constraint.

In Sect. 3.3.5, we focus on a dilute array of islands, where the island–island elastic interaction is negligible, and obtain the equation for the energy of a single island. In Sect. 3.3.6, we describe how minimizing the energy of a single isolated island of fixed volume can give the shape of the island. The main objective in the theoretical part of this section is to demonstrate the theoretical possibility of equilibrium arrays of 3D islands for which there is no thermodynamic driving force to Ostwald ripening. In this connection, in Sects. 3.3.7–3.3.9, we assume a constant island shape and seek the optimum size of the island, following papers by Shchukin et al. [3.155, 3.194, 3.195]. We demonstrate that, in a certain parameter region, equilibrium corresponds to an ordered array of 3D coherently strained islands, and ripening is not energetically favorable.

In Sect. 3.3.10, we follow the paper by Daruka and Barabási [3.156] and show how the thermodynamic theory reveals the phase diagram which reproduces all possible experimentally observed morphologies. In Sect. 3.3.11, we refer to papers by Duport et al. [3.198] and Spencer and Tersoff [3.200], who pointed out the possibility that totally relaxed overhanging coherent islands might form, which should favor ripening in all systems. In contrast to [3.198, 3.200], we demonstrate that when the shape-vs.-volume dependence for 3D strained islands is taken into account, there still exists a parameter region where ripening is not favorable. In Sect. 3.3.12, we give a brief description of kinetic theories of island formation focusing on particular mechanisms which can eventually slow down Ostwald ripening and result in a narrow size distribution of islands.

In Sects. 3.3.13 and 3.3.14, we describe experimental data on 3D island (quantum dot) formation, mainly in InAs/GaAs and GaInAs/GaAs systems.

We focus particularly on key experiments which allow us to distinguish thermodynamically controlled arrays of islands from kinetically controlled arrays. Thorough and comprehensive experimental studies have revealed the following:

- The evolution of the dot size up to a limiting value upon growth interruption for MBE-grown InAs QDs, observed only in a certain range of growth conditions.
- A reversible phase transition in the InAs/GaAs system, from 3D to 2D morphology, driven by a reduction in As pressure.
- A reversible phase transition from 3D to 2D morphology in the GaInAs(P)/GaAs system, driven by off/on switching of AsH_3 and on/off switching of PH_3.
- An irreversible phase transition from coherently strained islands to dislocated islands, i.e., the 'switching on' of Ostwald ripening, driven by an increase in As pressure.
- The formation of coherent InAs QDs in MOCVD, which is possible only in MBE-like conditions, at very low AsH_3 pressures.
- The general tendency towards Ostwald ripening of InAs islands upon an increase in arsenic pressure in both MBE and MOCVD.
- The preferred alignment of nearest neighbor dots lies along elastically soft directions $\langle 100 \rangle$.
- A decrease in the average volume of InAs islands upon an increase in the substrate temperature.
- Reversible changes in island shape, volume, and density upon cyclic variations of the substrate temperature, observed for InAs/GaAs islands.

These results provide strong evidence for the close-to-equilibrium nature of the formation of 3D coherently strained islands of InAs on a GaAs(001) substrate.

Experimental data on 3D strained island formation in other material systems are briefly discussed in Sect. 3.3.16.

3.3.4 Energetics of a Lattice-Mismatched Heteroepitaxial System

We focus on the equilibrium structure of a heteroepitaxial lattice-mismatched system which may be reached upon growth interruption, for example. Let Q monolayers of material 2 be deposited on the (001) substrate of material 1. We treat both the substrate and the deposited material as elastically anisotropic cubic media with equal elastic moduli λ_{ijlm} and lattice mismatch $\varepsilon_0 = \Delta a/a$ between the two materials, where a is the lattice parameter. The total energy of the uniform planar film per unit surface area may be written

$$E_{\text{planar}}(Q) = \lambda \varepsilon_0^2 Q a + W(Q) \,. \tag{3.135}$$

Here the first term is the strain energy of the uniform film, and the elastic modulus λ equals $(c_{11} + 2c_{12})(c_{11} - c_{12})c_{11}^{-1}$, where c_{11} and c_{12} are elastic moduli in the Voigt notation. If $Q \to 0$, the quantity $W(Q)$ is the surface energy of material 1, $W(0) = \gamma_1$. If the film thickness is macroscopic, $Q \gg 1$, the quantity $W(Q)$ is the sum of the surface and interface energies, $W(Q) = \gamma_2 + \gamma_{12}^{\text{interface}}$. For a film of arbitrary microscopic thickness, contributions of surface, interface, and strain energy cannot be separated, and only the total energy $E_{\text{planar}}(Q)$ has a physical meaning. However, it is convenient to use (3.135) for arbitrary Q, and to consider this equation as the definition of the quantity $W(Q)$ for arbitrary Q. The energy $W(Q)$ defined in this way takes into account wetting or non-wetting effects as well as possible surface reconstruction effects.

Since we focus on the equilibrium array of islands in a closed system, the formation of islands obeys the conditions of mass conservation. The latter means that the initially deposited volume of material 2 equals the sum of the volume of the wetting layer and the volumes of all islands. If we denote the thickness of the wetting layer by Q' monolayers, the other $(Q-Q')$ monolayers of deposited material are assembled in islands.

To obtain the structure of the equilibrium array of islands, we assume that all islands have the same shape and volume and form a 2-dimensional periodic superlattice on the surface. If we denote by q the fraction of the surface covered by islands, the total energy of the array of islands per unit surface area may be written

$$E = \lambda \varepsilon_0^2 Q' a + (1-q) W(Q') + \frac{1}{A_0}\left(\widetilde{E}_{\text{island}} + \frac{\widetilde{E}_{\text{inter}}}{2}\right). \tag{3.136}$$

where the first two terms give the energy of the uniform planar film of thickness $Q'a$, the third term is the energy of a single island, and the fourth term is the interaction energy of a single island with all other islands. A_0 denotes the unit cell area of the superlattice comprised of islands.

Subtracting (3.135) from (3.136), one obtains the change in energy due to island formation,

$$\Delta E = W(Q') - W(Q) + E_{\text{array}}(Q', Q), \tag{3.137}$$

where the energy $E_{\text{array}}(Q', Q)$ is given by

$$E_{\text{array}}(Q', Q) = (Q - Q')a \left[-\lambda \varepsilon_0^2 + \frac{\widetilde{E}_{\text{island}} - W(Q')\widetilde{A}_{\text{island}}}{V} + \frac{\widetilde{E}_{\text{inter}}}{2V}\right], \tag{3.138}$$

with V the volume of a single island and $\widetilde{A}$ the area of the island base. The change in energy ΔE in (3.137) is a function of the total amount of material $(Q - Q')$ assembled in all islands, the shape of a single island, the volume V of a single island, and the lateral arrangement of islands, i.e., the type of lateral superlattice that the islands make up.

Equation (3.137) indicates that the structure of the equilibrium array of islands and the thickness of the wetting layer should be determined in a self-consistent way by minimizing ΔE. We will first fix the total amount of material $(Q - Q')$ assembled in all islands (thus fixing also the wetting layer thickness Q') and seek the minimum of E_{array} with respect to the island shape, volume, and arrangement, following the paper by Shchukin et al. [3.155]. Afterwards, we will substitute the minimum value of E_{array}, which will be a function of Q', into (3.137) and minimize ΔE with respect to Q'.

3.3.5 Dilute Array of 3D Strained Islands

The energy of a single island $\widetilde{E}_{\text{island}}$ which enters into (3.138) may be written as a sum of three contributions,

$$\widetilde{E}_{\text{island}} = \widetilde{E}_{\text{elast}} + \widetilde{E}_{\text{surf}} + \widetilde{E}_{\text{edges}} . \tag{3.139}$$

To write down the elastic energy of the heterophase system, we apply the concept of stress-free strain $\varepsilon_{ij}^{(0)}(\boldsymbol{r})$ developed by Khachaturyan [3.201, 3.202]. Since different phases, with different values of the equilibrium lattice parameter, are coherently conjugated, the elastic field is characterized by the strain tensor $\varepsilon_{ij}(\boldsymbol{r})$ defined throughout the entire heterophase system. If the strain $\varepsilon_{ij}(\boldsymbol{r})$ coincides locally with the stress-free strain $\varepsilon_{ij}^{(0)}(\boldsymbol{r})$ for a given material, both the elastic stress and the elastic energy density vanish. A deviation of the strain from the stress-free strain leads to a non-zero stress and a non-zero value of the elastic energy density. For the latter we use the definition by Roitburd [3.203],

$$f_{\text{elast}} = (1/2)\lambda_{ijlm}(\boldsymbol{r}) \left[\varepsilon_{ij}(\boldsymbol{r}) - \varepsilon_{ij}^{(0)}(\boldsymbol{r})\right] \left[\varepsilon_{lm}(\boldsymbol{r}) - \varepsilon_{lm}^{(0)}(\boldsymbol{r})\right] .$$

The total elastic energy is then given by the integral

$$\widetilde{E}_{\text{elast}} = \frac{1}{2} \int dV \lambda_{ijlm}(\boldsymbol{r}) \left[\varepsilon_{ij}(\boldsymbol{r}) - \varepsilon_{ij}^{(0)}(\boldsymbol{r})\right] \left[\varepsilon_{lm}(\boldsymbol{r}) - \varepsilon_{lm}^{(0)}(\boldsymbol{r})\right] . \tag{3.140}$$

For the heteroepitaxial system in question, it is natural to use the non-stressed substrate as the reference frame, so that the stress-free strain vanishes in the substrate, i.e., $\varepsilon_{ij}^{(0)}(\boldsymbol{r}) = \varepsilon_0 \delta_{ij} \vartheta(\boldsymbol{r})$, where $\vartheta(\boldsymbol{r}) = 1$ in the deposited material, $\vartheta(\boldsymbol{r}) = 0$ in the substrate, $\delta_{ij} = 1$ if $i = j$, and $\delta_{ij} = 0$ otherwise. It should be noted that $\varepsilon_{ij}^{(0)}(\boldsymbol{r}) = 0$ does not imply a rigid substrate, because the strain field $\varepsilon_{ij}(\boldsymbol{r})$ generally penetrates into the substrate.

The surface energy per unit area γ is renormalized in the strain field [3.21, 3.28] [see also (3.10)]:

$$\gamma(\varepsilon_{\alpha\beta}) = \gamma_0 + \tau_{\alpha\beta}(\varepsilon_{\alpha\beta} - \varepsilon_0 \delta_{\alpha\beta}) \tag{3.141}$$
$$+ \frac{1}{2} S_{\alpha\beta\mu\nu}(\varepsilon_{\alpha\beta} - \varepsilon_0 \delta_{\alpha\beta})(\varepsilon_{\mu\nu} - \varepsilon_0 \delta_{\mu\nu}) + \cdots ,$$

where $\tau_{\alpha\beta}$ is the intrinsic surface stress tensor, $S_{\alpha\beta\mu\nu}$ is the tensor of 'surface excess elastic moduli' [3.204], and α, β, μ, ν are 2D indices in the local facet plane. The total renormalized surface energy of the heterophase system is

$$\widetilde{E}_{\text{surf}} = \int dA \Big[\gamma_0(\hat{\boldsymbol{m}}) + \tau_{\alpha\beta}(\hat{\boldsymbol{m}}) \big[\varepsilon_{\alpha\beta}(\boldsymbol{r}) - \varepsilon_0 \delta_{\alpha\beta} \Theta(\boldsymbol{r}) \big] \tag{3.142}$$

$$+ \frac{1}{2} S_{\alpha\beta\mu\nu}(\hat{\boldsymbol{m}}) \big[\varepsilon_{\alpha\beta}(\boldsymbol{r}) - \varepsilon_0 \delta_{\alpha\beta} \Theta(\boldsymbol{r}) \big] \big[\varepsilon_{\mu\nu}(\boldsymbol{r}) - \varepsilon_0 \delta_{\mu\nu} \Theta(\boldsymbol{r}) \big] \Big] (\hat{\boldsymbol{m}} \cdot \hat{\boldsymbol{n}})^{-1} \,,$$

where $\hat{\boldsymbol{m}} = \hat{\boldsymbol{m}}(\boldsymbol{r})$ is the local normal to the facet, $\hat{\boldsymbol{n}} = (0,0,1)$ is the normal to the flat surface, and the integration in (3.142) is carried out over the reference flat surface. $\Theta(\boldsymbol{r}) = 1$ if the surface point belongs to the island facet, and $\Theta(\boldsymbol{r}) = 0$ on the surface of the wetting layer, since the wetting layer is considered as a complex surface of the substrate material with thickness Q' and surface energy $W(Q')$.

The third term in (3.139) is the short-range energy of edges. For lattice-mismatched systems with edges, the total strain field is the sum of two contributions, one due to the lattice mismatch, and the other due to the discontinuity in the intrinsic surface stress tensor τ_{ij} at the edges. Since excess elastic moduli $S_{\alpha\beta\mu\nu}(\hat{\boldsymbol{m}})$ exist only on the surface, their contribution to the energy is smaller than the elastic energy by a factor of $\sim a/L$, where L is the characteristic island size, and may be treated by perturbation theory.

In the zero approximation in $S_{\alpha\beta\mu\nu}(\hat{\boldsymbol{m}})$, the elastic energy, including both bulk and surface contributions, is given in terms of the sources of the strain field by [3.42]

$$\widetilde{E}_{\text{elast}} = \lambda \varepsilon_0^2 V - \frac{1}{2}(c_{11} + 2c_{12})\varepsilon_0 \oint dA \int dA' m_i(\boldsymbol{r}) G_{i\alpha}(\boldsymbol{r},\boldsymbol{r}') \widetilde{\sigma_{\alpha\beta}} m_\beta(\boldsymbol{r}')$$

$$- \int dl \int dA' F_i(\boldsymbol{r}) G_{i\alpha}(\boldsymbol{r},\boldsymbol{r}') \widetilde{\sigma_{\alpha\beta}} n_\beta(\boldsymbol{r}')$$

$$- \frac{1}{2} \int dl \int dl' F_i(\boldsymbol{r}) G_{ij}(\boldsymbol{r},\boldsymbol{r}') F_j(\boldsymbol{r}') \,. \tag{3.143}$$

Here the first term is the elastic energy of the volume V in a planar, uniformly strained film. The second term is the energy of the volume elastic relaxation. The contribution to this energy comes from tilted facets of the island where $m_\beta \neq 0$. The third term is the energy of the interaction between two strain fields, one due to the lattice mismatch, and the other due to the intrinsic surface stress discontinuity at the edges. The fourth term is the energy of elastic relaxation due to the surface stress discontinuity at the edges. $G_{ij}(\boldsymbol{r},\boldsymbol{r}')$ is the static Green tensor of elasticity theory defined for a semi-infinite crystal with a stress-free surface of given profile. The quantity

$$\widetilde{\sigma_{\alpha\beta}} = -(c_{11} + 2c_{12})(c_{11} - c_{12})c_{11}^{-1}\varepsilon_0 \delta_{\alpha\beta}$$

is the stress tensor in the uniformly stressed flat film and the forces $F_i(\boldsymbol{r})$ are due to the discontinuity in the surface stress tensor at the edges. The integration $\oint dA$ is defined over both the side facets of the island and the interface between the island and the wetting layer, whilst the integration $\int dA'$ contains non-vanishing contributions from the side facets of the island only, and the integrations $\int dl$ and $\int dl'$ are carried out over the edges of the island.

The scaling properties of the Green tensor of elasticity theory for the 3-dimensional elastic field can be used to obtain the scaling behavior of each of the contributions to the elastic energy in (3.143). For an infinite homogeneous medium, the Green tensor behaves as $G_{ij}(\boldsymbol{r},\boldsymbol{r}') \sim |\boldsymbol{r}-\boldsymbol{r}'|^{-1}$ [3.31], and for an arbitrary geometry of the system, it can be shown that $G_{ij}(\boldsymbol{r},\boldsymbol{r}')$ scales as L^{-1}. By substituting it into (3.143), one obtains the following scaling behavior for the various contributions to the elastic energy of the island $\widetilde{E}_{\text{elast}}$ [3.42, 3.155]:

$$E_{\varepsilon_0-\varepsilon_0} \sim L^3, \quad E_{\varepsilon_0-\tau} \sim L^2, \quad E_{\tau-\tau} \sim -L\ln L.$$

Surface excess elastic moduli give the correction to the energy of the volume elastic relaxation, which is proportional to L^2, and corrections to all other terms are of the order of L or smaller. Thus, summing up the scaling analysis of the energy of a single island, one can write it in the following schematic form [3.155]:

$$\widetilde{E}(V) = -f_1\lambda\varepsilon_0^2 V + (\Delta\varGamma)V^{2/3} - \frac{f_2\tau^2}{\lambda}V^{1/3}\ln\left(\frac{V^{1/3}}{2\pi a}\right) + f_3\eta V^{1/3}.$$

(3.144)

The first term in (3.144) is the energy of the volume elastic relaxation $\widetilde{\Delta E}_{\text{elast}}^V$, which is always negative. The second term is the change in the renormalized surface energy of the system due to island formation. For concreteness, we will write $(\Delta\varGamma)$ for a pyramidal island with $L\times L$ square base and tilt angle of the side facets equal to θ_0. Then $\widetilde{\Delta E}_{\text{surf}}^{\text{renorm}} = (\Delta\varGamma)(1/6\tan\theta_0)^{2/3}L^2$, and

$$(\Delta\varGamma) = (6\cot\theta_0)^{2/3}\Big[\gamma_2(\theta_0)\sec\theta_0 + \gamma_{12}^{\text{interface}} \qquad (3.145)$$
$$-W(Q') - g_1(\theta_0)\tau\varepsilon_0 - g_2(\theta_0)S\varepsilon_0^2\Big].$$

Here the change in the surface energy includes contributions due to:

- the appearance of the tilted facets of the island,
- the appearance of the interface between the deposited material and the substrate underneath the island,
- the disappearance of the planar surface area,
- renormalization terms, both linear and quadratic in ε_0.

The key point in what follows is that the quantity $(\Delta\varGamma)$ can be of either sign.

The third term in (3.144) is the contribution of the edges of the island to the elastic relaxation energy, $\widetilde{\Delta E}_{\text{elast}}^{\text{edges}} \sim -V^{1/3}\ln V^{1/3}$, which is always negative. The fourth term in (3.144) is the short range energy of edges, where η is a characteristic energy per unit length of the edge. Coefficients f_1, f_2, f_3 are geometric factors depending on the island shape. We note that

178   3. Self-Organization Phenomena at Crystal Surfaces

the parametrization of the energy (3.145) differs from that in [3.155, 3.195]. Unlike those papers, where we focused only on pyramid-shaped islands, here in Sect. 3.3.11, we consider islands of different shapes. That is why we have written the island energy (3.144) in terms of the island volume rather than in terms of the island lateral size.

3.3.6 Ordering of Islands in Terms of Shape

For a dilute array of islands, where the average distance between islands is large compared to the island size L, the equilibration of the island shape by atomic migration on island facets is faster than material exchange between islands. Then for any given volume of an island, there exists an equilibrium shape. For sufficiently large islands, the first two terms $\widetilde{\Delta E}_{\text{elast}}^V$ and $\widetilde{\Delta E}_{\text{surf}}^{\text{renorm}}$ dominate the island energy (3.144).

Several theoretical studies have been carried out in which the equilibrium shape of a single 3-dimensional coherently strained island has been calculated. Tersoff and Tromp [3.109] obtained the equilibrium shape of an island under the assumption that the height of the island is kinetically limited to a certain value h. The global geometry of such an island is rather 2D than 3D. This is why the corresponding discussion was included in Sect. 3.2.

The equilibrium shape of a single 3-dimensional coherently strained island was calculated by Kaminski and Suris [3.196] and by Pehlke et al. [3.199] by minimizing the total energy of an island. The total energy was approximated by the sum of the elastic energy and the surface energy, i.e., by the first two terms in (3.144).

Kaminski and Suris [3.196] showed that rather general assumptions about surface energies of the island facets yield a phase diagram containing regimes of 2D growth and 3D growth, and a regime where the 2D–3D transition occurs. Pehlke et al. [3.199] focused on InAs islands on a GaAs(001) substrate. They obtained surface energies for the (100), (110), (111), and $(\overline{1}\overline{1}\overline{1})$ surfaces of InAs from ab initio calculations, and applied Wulff's construction to these energies. This gave an equilibrium shape for InAs in As-rich conditions [3.174] which agrees with the observed shape of large, and thus presumably fully relaxed InAs islands grown by MOCVD on a GaAs(001) substrate [3.205].

To derive the equilibrium shape of a coherently strained InAs island, Pehlke et al. [3.199] considered a variety of possible configurations, which were restricted to surface orientations present on the equilibrium crystal shape of InAs. The elastic energy for each configuration was calculated within the framework of the continuum theory of elasticity using the finite-element method. Their results are displayed in Fig. 3.80. The optimum island shape for a given volume of the island $V_0 = 2.88 \times 10^5$ Å^3 is determined by that point where the line of constant total energy $E_{\text{total}}/V_0 = E_{\text{elast}}/V_0 + E_{\text{surf}}/V_0$ touches the manifold of island energies from below. Even when the volume is different, it is possible to find the optimum shape from the same graph. Using the scaling relations $E_{\text{elast}} \sim V$ and $E_{\text{surf}} \sim V^{2/3}$, one obtains

3.3 Arrays of Three-Dimensional Coherently Strained Islands   179

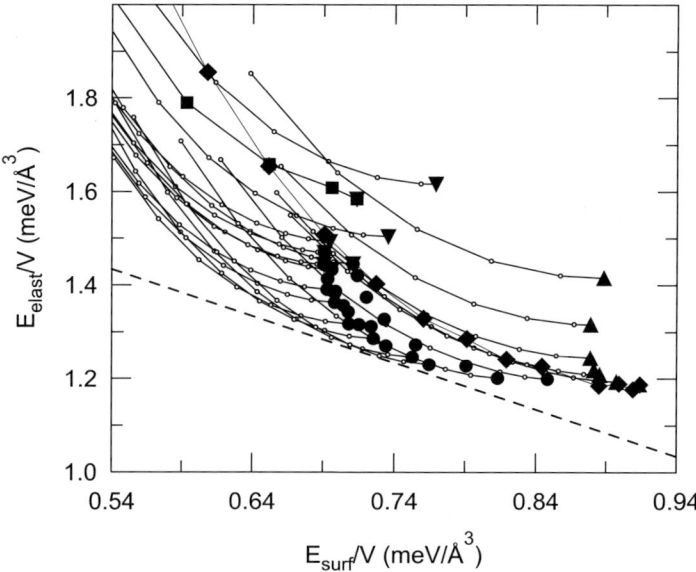

Fig. 3.80. Elastic energy per unit volume E_{elast}/V vs. surface energy per unit volume E_{surf}/V_0 for InAs islands with volume $V = 2.88 \times 10^5$ Å^3. ■ Square-based pyramids with $\{101\}$ faces and (001)-truncated $\{101\}$-pyramids. ♦ Square-based pyramids with $\{111\}$ and $\{\overline{1}\overline{1}\overline{1}\}$ faces and (001)-truncated pyramids. ▲ 'Huts' with $\{111\}$ and $\{\overline{1}\overline{1}\overline{1}\}$ faces. ▼ Square-based $\{101\}$ pyramids with $\{\overline{1}\overline{1}\overline{1}\}$-truncated edges. ● Islands with $\{101\}$, $\{111\}$, and $\{\overline{1}\overline{1}\overline{1}\}$ faces. *Filled symbols* denote numerical results, while *open circles* correspond to a simple analytical approximation for (001)-truncated 'mesa-shaped' islands. It is assumed that the elastic energy does not change when the (almost fully relaxed) top of an island is cut off. *Continuous lines* connect islands that are created in this way, varying the height of the (001) surface plane. The *dashed line* is the curve of constant total energy $E_{\text{elast}} + E_{\text{surf}}$ that selects the equilibrium shape. With kind permission of E. Pehlke et al. [3.174]

$$\frac{E_{\text{total}}(V)}{V} = \frac{E_{\text{elast}}(V_0)}{V_0} + \left(\frac{V_0}{V}\right)^{2/3} \frac{E_{\text{surf}}(V_0)}{V_0} . \qquad (3.146)$$

This says that only the slope of the total energy line changes when the volume changes, and the whole evolution of the island shape can be extracted from Fig. 3.80. For an island with volume $V_0 = 2.88 \times 10^5$ Å^3, the optimum shape deduced from Fig. 3.80 is a hill bounded by $\{101\}$, $\{111\}$, and $\{\overline{1}\overline{1}\overline{1}\}$ facets and a (001) surface on the top. This shape is similar to the shape of InP islands on GaInP, observed by Georgsson et al. [3.206]. However, since the surface energies of InP are different from those of InAs, a direct comparison of the calculated shape of InAs islands with the observed shape for InP islands is not possible. Various shapes have been observed for InAs islands grown on a GaAs substrate. Moison et al. [3.147] reported rather flat islands having $\{104\}$ facets. Leonard et al. [3.178] described their islands as planoconvex lenses with a radius-to-height aspect ratio of about 2. Ruvimov et al. [3.207, 3.208] reported a pyramid shape with {101} side facets. All observed shapes differ from the equilibrium shape predicted theoretically.

There exist several possible reasons for this disagreement, discussed by Pehlke et al. [3.174], for example. Firstly, the diversity of experimental results in itself indicates that islands are not equilibrium islands in most cases, but kinetically controlled islands. We will argue below that the pyramids observed by Ruvimov et al. [3.207] are the most likely equilibrium islands. However, this does not remove the existing disagreement with theory (Fig. 3.80).

Secondly, ab initio calculations of surface energies by Pehlke et al. [3.199] refer to $T = 0$ K, and finite-temperature corrections to the surface energies, which might affect the equilibrium shape of the islands, are not taken into account in Fig. 3.80.

We argue that different measured shapes of InAs islands are due firstly to different growth conditions, and secondly, to different measurement techniques. We will point out that carefully performed high resolution transmission electron microscopy (HRTEM) measurements combined with HRTEM simulations [3.208] reveal a pyramid bounded by {101} facets to be the true shape of InAs islands grown at temperatures 450–480°C. Moreover, we will emphasize that a set of experiments specifically aimed at distinguishing equilibrium islands from kinetically controlled islands indicate that the InAs pyramids with {101} facets are the most likely equilibrium islands.

3.3.7 Size Ordering of Islands vs. Ostwald Ripening

In this section, we underline the driving forces governing the narrow size distribution of islands and consider conditions in which Ostwald ripening does not occur. To focus on the essential physics, we use a constant shape approximation for islands, namely, a pyramid with $L \times L$ square base and a tilt angle of side facets θ_0.

The equilibrium state of an array of islands can be reached by material exchange between islands, which occurs via surface migration. For a dilute array of islands, the elastic interaction between islands via the strained substrate may be neglected. Then the energy of the array is the sum of contributions from single islands (3.144).

Equilibrium satisfies the condition that the total energy is minimum under the constraint of a fixed amount of material in all islands. Then, instead of minimizing the total energy, we may use an equivalent procedure and minimize the energy per atom in the island $E(L)$. Dividing $\widetilde{E}(L)$ from (3.144) by the volume $(1/6)\tan\theta_0 L^3$ of a single island and multiplying by the atomic volume Ω, one obtains [3.155]

3.3 Arrays of Three-Dimensional Coherently Strained Islands

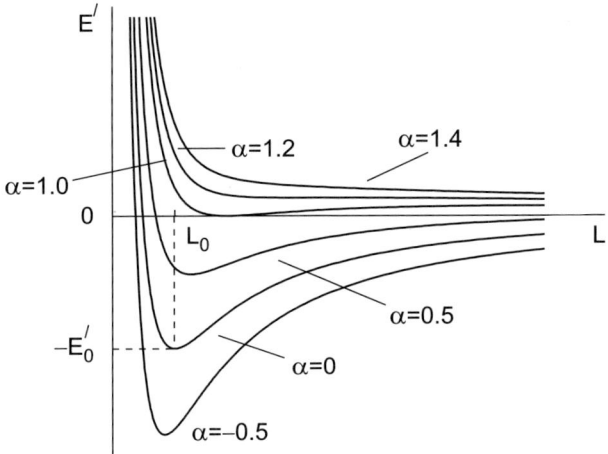

Fig. 3.81. Energy of a dilute array of 3D coherently strained islands per atom vs. island size. The parameter α is the ratio of the change in the surface energy due to island formation, $\Delta E_{\text{surf}}^{\text{renorm}}$, and the contribution of the edges to the elastic relaxation energy, $\left|\Delta E_{\text{elast}}^{\text{edges}}\right|$

$$E(L) = \Omega \left[-f_1(\theta_0)\lambda\varepsilon_0^2 + (6\cot\theta_0)^{1/3} \frac{(\Delta\Gamma)}{L} \right. \tag{3.147}$$

$$\left. -(6\cot\theta_0)^{2/3}\frac{f_2(\theta_0)\tau^2}{\lambda L^2}\ln\left(\frac{L}{2\pi a}\right) + (6\cot\theta_0)^{2/3}\frac{f_3(\theta_0)\eta}{L^2} \right].$$

It is worth noting that the volume elastic relaxation energy $\Delta E_{\text{elast}}^{\text{V}}$ [the first term in (3.147)] does not depend on the island size L. To seek the minima of $E(L)$ from (3.147), we introduce the characteristic length

$$L_0 = 2\pi a \exp\left[\frac{f_3(\theta_0)\eta\lambda}{f_2(\theta_0)\tau^2} + \frac{1}{2}\right], \tag{3.148}$$

and the characteristic energy per atom

$$E_0 = \frac{\Omega f_2(\theta_0)(6\cot\theta_0)^{2/3}\tau^2}{2\lambda L_0^2}. \tag{3.149}$$

Then we may write the sum of all L-dependent terms in $E(L)$ as [3.194]

$$E'(L) = E_0 \left[-2\left(\frac{L_0}{L}\right)^2 \ln\left(\frac{e^{1/2}L}{L_0}\right) + \frac{2\alpha}{e^{1/2}}\left(\frac{L_0}{L}\right) \right]. \tag{3.150}$$

The function $E'(L)$ is governed by the control parameter

$$\alpha = \frac{e^{1/2}\lambda L_0}{f_2(\theta_0)(6\cot\theta_0)^{1/3}\tau^2}(\Delta\Gamma), \qquad (3.151)$$

which is the ratio of the change in the surface energy due to island formation and the contribution from the edges to the elastic relaxation energy, $|\Delta E_{\text{elast}}^{\text{edges}}|$. The energy of the dilute array of islands per atom vs. the island size L is displayed in Fig. 3.81 for different values of α. If $\alpha \leq 1$, there exists an optimum island size L_{opt}, corresponding to the absolute minimum of the energy, $\min E'(L) \equiv E(L_{\text{opt}}) < 0$. On the other hand, the ripening of islands would correspond to $L \to \infty$, where the energy $E'(L) \to 0$. This means that the array of identical islands of optimum size L_{opt} is a stable array. Islands will show size-limited growth up to this value, and will not undergo further ripening. If $1 < \alpha < 2\,e^{-1/2} \approx 1.2$, there exists only a local energy minimum, corresponding to a metastable array in which $E'(L') > 0$. If $\alpha \geq 2\,e^{-1/2}$, the local minimum in the energy $E'(L)$ disappears. For both the latter cases where $\alpha > 1$, there exists a thermodynamic tendency to ripening. The energy minimum then corresponds to a single huge island where all deposited material is collected.

If $(\Delta\Gamma) < 0$ (and $\alpha < 0$), besides a decrease in the strain energy due to elastic relaxation, the formation of a 3D island also leads to a decrease in the renormalized surface energy.

For an InAs pyramid with {101}-type side facets over the InAs wetting layer deposited on the GaAs(001) surface, evaluation of $(\Delta\Gamma)$ yields [3.195]

$$(\Delta\Gamma) = 6^{2/3}\Big\{ 1.41\,\gamma_{\text{InAs}}^{(101)} + \gamma_{\text{interface}} - \gamma_{\text{WL}}^{(001)} \qquad (3.152\text{a})$$

$$- \Big[0.72\,\tau_{\mu\mu}^{(101)} + 0.40\,\tau_{\nu\nu}^{(101)} + 0.15\left(\tau_{\zeta\zeta}^{(001)} + \tau_{\eta\eta}^{(001)}\right) \Big]\varepsilon_0 \qquad (3.152\text{b})$$

$$+ \Big[0.22\,S_{\mu\mu\mu\mu}^{(101)} + 0.08\,S_{\nu\nu\nu\nu}^{(101)} + 0.25\,S_{\mu\mu\nu\nu}^{(101)} + 0.10\,S_{\mu\nu\mu\nu}^{(101)} \qquad (3.152\text{c})$$

$$+ 0.01\left(S_{\zeta\zeta\zeta\zeta}^{(001)} + S_{\eta\eta\eta\eta}^{(001)}\right) + 0.03\,S_{\zeta\zeta\eta\eta}^{(001)} \qquad (3.152\text{d})$$

$$+ 0.01\,S_{\zeta\eta\zeta\eta}^{(001)} \Big]\varepsilon_0^2 \Big\}, \qquad (3.152\text{e})$$

where the axes μ, ν, ζ, η are defined in Fig. 3.82. The change in the surface energy due to formation of a pyramid contains contributions due to:

- the appearance of tilted {101} facets of InAs [first term in (3.152a)],
- the appearance of the InAs/GaAs interface underneath the pyramid [second term in (3.152a)],
- the disappearance of the L^2 area of the wetting layer [third term in (3.152a)],
- linear renormalization terms (3.152b),
- quadratic renormalization terms (3.152c), (3.152d), (3.152e).

It should be noted that a decrease in surface energy in the InAs/GaAs(001) system, $(\Delta\Gamma) < 0$, may occur despite the fact that the (001) surface of bulk

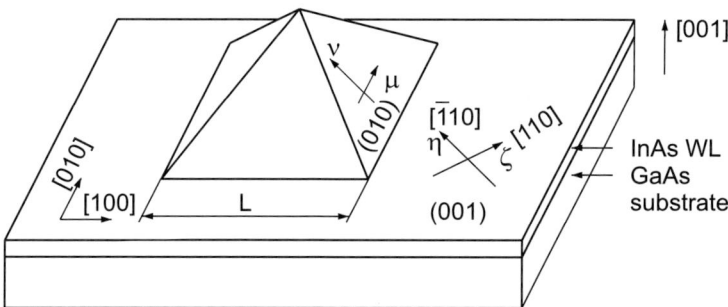

Fig. 3.82. Geometry of an InAs pyramid over the InAs wetting layer deposited on the GaAs(001) surface

InAs is stable against faceting. The reason is that the appearance of the tilted {101} facets of InAs competes with the disappearance of a certain area of the wetting layer. The wetting layer of InAs has a microscopic thickness of 1–2 ML and should be regarded as a surface whose surface energy is not necessarily equal to the surface energy of InAs(001).

The quantity ($\Delta \Gamma$) could be evaluated if all quantities entering (3.152a)–(3.152d) were known from ab initio calculations. However, one should be very careful when applying ab initio surface energies to evaluate crucial quantities entering the macroscopic theory. Firstly, preferred surface reconstructions and corresponding values of surface energies obtained so far by ab initio calculations refer mainly to temperature $T = 0$, while typical growth temperatures are 450°C and higher. It is known, however, that the surface reconstruction can and does change with temperature, as does the surface energy. Secondly, in order to apply the macroscopic theory to finite temperatures, one has to take into account the entropy contribution to the free energy, which has been neglected so far. It was shown in Sect. 3.2 that entropy effects play a major role in the temperature dependence of the average island size in the case of islands with a fixed height. Similar effects are also expected to govern the temperature behavior in the general case of 3D islands of variable height. Therefore, substantial efforts are required from both ab initio and macroscopic theories in order to get the ultimate theoretical answer concerning the nature of ordering in a given particular system. Below we will present experimental evidence for the close-to-equilibrium nature of 3D strained islands in the InAs/GaAs system.

3.3.8 Lateral Arrangement of Islands

For a dense system of islands, elastic interactions between islands via the substrate are essential. The system of interacting islands is then a system of elastic domains where the energy minimum corresponds to a periodic domain structure (see also [3.84, 3.86, 3.202, 3.209, 3.210]). To obtain the elastic energy

184   3. Self-Organization Phenomena at Crystal Surfaces

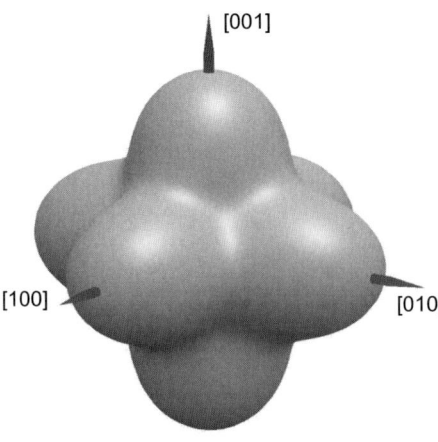

Fig. 3.83. Reciprocal Young's modulus of an elastically anisotropic cubic crystal. Typical behavior is shown for Si, Ge, and zinc-blende III–V and II–VI semiconductors. $1/Y$ reveals well pronounced maxima, and Y has minima in the elastically soft directions $\langle 100 \rangle$

of interacting islands, we calculate the second term on the right-hand side of (3.143), which is the major contribution to the interaction energy. We assume a small tilt angle of the island facets, where the elastic relaxation energy given by the second term of (3.143) reduces to [3.42]

$$\widetilde{\Delta E}_{\text{elastic}} = -C_0 \int dA \int dA' n_\alpha(\boldsymbol{r}_\|) G_{\alpha\beta}(\boldsymbol{r}_\| - \boldsymbol{r}'_\|; z, z') \bigg|_{\substack{z=0 \\ z'=0}} n_\beta(\boldsymbol{r}'_\|) .$$
(3.153)

Here $C_0 = (1/2)(c_{11} + 2c_{12})^2 (c_{11} - c_{12})^2 c_{11}^{-2} \varepsilon_0^2$, the integration is carried out over the planar substrate, and the integrand is only non-vanishing on the projections of the tilted facets of the islands, where $\boldsymbol{n}_\alpha \neq 0$, $\alpha = 1, 2$. $G_{\alpha\beta}(\boldsymbol{r}_\| - \boldsymbol{r}'_\|; z, z')$ is the static Green tensor of the semi-infinite elastic medium bounded by a planar, stress-free surface $z = 0$.

To get a hint of what will be the preferred arrangement of interacting islands, we evaluate the energy (3.153) for a pair of separated islands. We take into account the fact that Si, Ge, III–V, and II–VI semiconductors are cubic crystals with pronounced elastic anisotropy. The anisotropy is illustrated in Fig. 3.83, where the inverse Young's modulus of a cubic crystal is plotted as a function of crystallographic direction. The general equation for $1/Y$ is given in [3.31], for example (see the problem in Sect. 10), and particular calculations for GaAs have been carried out in [3.211]. To evaluate (3.153), we use the Fourier transform for the static Green tensor $\widetilde{G}_{\alpha\beta}(\boldsymbol{k}_\|; z, z')$ obtained by Portz and Maradudin [3.212] for elastically anisotropic cubic crystals bounded by a planar, stress-free (001) surface. We interpolate the angular dependence of the exact Green tensor by the lowest-order angular polynomial with cubic symmetry, $B_1 + B_2(8k_x^2 k_y^2 / k_\|^4 - 1)$. The accuracy of interpolation is less than 2%. Then we carry out an inverse Fourier transformation, and obtain the Green tensor in $\boldsymbol{r}$ space. By integrating (3.153) by parts, subtracting the energy of volume elastic relaxation of isolated islands, and retaining only

3.3 Arrays of Three-Dimensional Coherently Strained Islands   185

lowest-order terms, i.e., the dipole–dipole interaction between islands, we obtain [3.157, 3.213]

$$\widetilde{E}_{\text{inter}} = \frac{C_0}{2\pi} V^2 \frac{B_1 + 15 B_2 (1 - 8 m_x^2 m_y^2)}{R^3} \ , \tag{3.154}$$

where V is the island volume, R the distance between the two islands, and $\boldsymbol{m} = (m_x, m_y)$ the unit vector in the surface plane, parallel to the direction between the islands. For an elastically isotropic medium $B_2 = 0$, and the interaction between widely separated islands is the isotropic dipole–dipole repulsion. In systems with pronounced elastic anisotropy, like most III–V and II–VI semiconductors, the dipole–dipole interaction changes sign as a function of $\boldsymbol{m}$ and this results in attraction between islands within a certain range of directions $\boldsymbol{m}$ close to the elastically soft directions [100] or [010]. Therefore, a very dilute array of islands will be arranged in weakly coupled chains parallel to the [100] or [010] directions, where the dipole–dipole attraction is balanced by a higher-order (e.g., dipole–octopole) repulsion. Such an arrangement also manifests itself for a moderate area coverage.

To reveal the optimum arrangement of islands, one has to evaluate the energy from (3.153) for a number of periodic superlattices of islands on the surface. By expressing the normal vector via the surface profile gradient, $n_\alpha(\boldsymbol{r}_\parallel) \approx -\nabla_\alpha \zeta(\boldsymbol{r}_\parallel)$, one obtains the elastic energy of the array of interacting islands per atom as a sum over the vectors of the 2-dimensional reciprocal lattice corresponding to a given periodic array of islands,

$$\widetilde{\Delta E}_{\text{elast}} = -C_0 A_0 \frac{\Omega}{V} \sum_{\boldsymbol{K}_\parallel} \left| \widetilde{\zeta}(\boldsymbol{K}_\parallel) \right|^2 K_\alpha K_\beta \widetilde{G_{\alpha\beta}}(\boldsymbol{K}_\parallel; z, z') \bigg|_{\substack{z=0 \\ z'=0}} \ , \tag{3.155}$$

where $\widetilde{\zeta}(\boldsymbol{K}_\parallel)$ is the Fourier transform of the surface profile $\zeta(\boldsymbol{r}_\parallel)$ describing the shape of the islands, A_0 is the unit cell area of the superlattice made up of islands, V is the volume of the island, and Ω is the atomic volume.

Shchukin et al. [3.155] compared interaction energies for three arrays of islands, namely the square, hexagonal and checkerboard arrays. The energies for the three arrays are compared in Fig. 3.84. This reveals that the square array is the most favorable array of the three under investigation.

Later these calculations were extended to cover all possible superlattices [3.213] with primitive translation vectors $(\boldsymbol{e}_1, \boldsymbol{e}_2)$, where $\boldsymbol{e}_1$ is parallel to the elastically soft direction [100], and $\boldsymbol{e}_2$ has an arbitrary absolute value and orientation.

Equation (3.155) yields the sum of the energy of the volume elastic relaxation of the islands and the interaction energy between islands. Comparison of energies for different arrays reveals the following. If the fraction q of the surface covered by islands is less than 0.06, islands form a chain-like arrangement along the [100] (or [010]) direction, with weak coupling between chains. For $0.06 \leq q \leq 0.17$, the favored arrangement is a base-centered rectangular superlattice. The unit cell of such a superlattice is a parallelogram with the

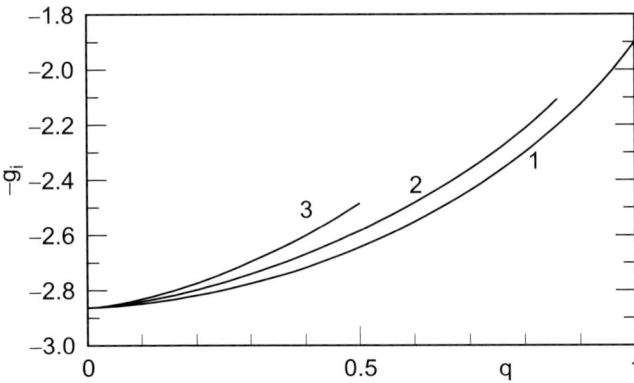

Fig. 3.84. The energy per unit area $\Delta E_{\text{elast}}^V + E_{\text{inter}} = (Q - Q')a\tilde{\lambda}\varepsilon_0^2\theta_0 \times [-g_i(q)]$ for different arrays of coherently strained islands vs. area coverage q. Curve 1: 2D square lattice of pyramids with primitive lattice vectors $(1, 0, 0)$ and $(0, 1, 0)$. Curve 2: 2D hexagonal lattice of pyramids with primitive lattice vectors $(-1/2, -\sqrt{3}/2, 0)$ and $(1, 0, 0)$. Curve 3: checkerboard square lattice of pyramids with primitive lattice vectors $(1, -1, 0)$ and $(1, 1, 0)$. Curves 2 and 3 terminate at the maximum possible coverage for the given arrays

angle between primitive lattice vectors changing from 78° for $q = 0.06$ to 73° for $q = 0.17$. Such an arrangement with the corresponding angle $\approx 75°$ has been observed by Ledentsov et al. [3.214]. For $0.17 \leq q \leq 0.33$, the preferred arrangement is rectangular. The aspect ratio of the rectangular unit cell decreases from 1.7 for $q = 0.17$ to 1 for $q = 0.33$, where the rectangle transforms into a square. For $0.33 \leq q \leq 1.0$, the square superlattice is the most favorable one. A perfect hexagonal superlattice with angle 60° is never favorable. Energy calculations for optimum arrays performed in [3.213] revealed that the difference in energies between the square array and the optimum array, if it is different from the square one, is considerably smaller than the difference between the checkerboard array and the square array. Therefore, one may approximate the energy of the square array to be the minimum energy of the array of interacting islands for any area coverage.

Figure 3.85a shows a plan-view TEM micrograph of a single-sheet array of MBE-grown InAs quantum dots [3.191]. The preferential alignment of dots in rows parallel to elastically soft $\langle 100 \rangle$ directions is clearly visible. Figure 3.85b displays a histogram of the direction between a given dot and the nearest neighboring dot. This histogram reveals a well-pronounced maximum for the $\langle 100 \rangle$ directions, in agreement with the above theoretical results. Since the interaction energy itself, and furthermore, the difference in energies between different arrangements of islands is rather small, it should be noted that the system of islands exhibits strong correlations only in the nearest neighbor arrangement.

3.3 Arrays of Three-Dimensional Coherently Strained Islands   187

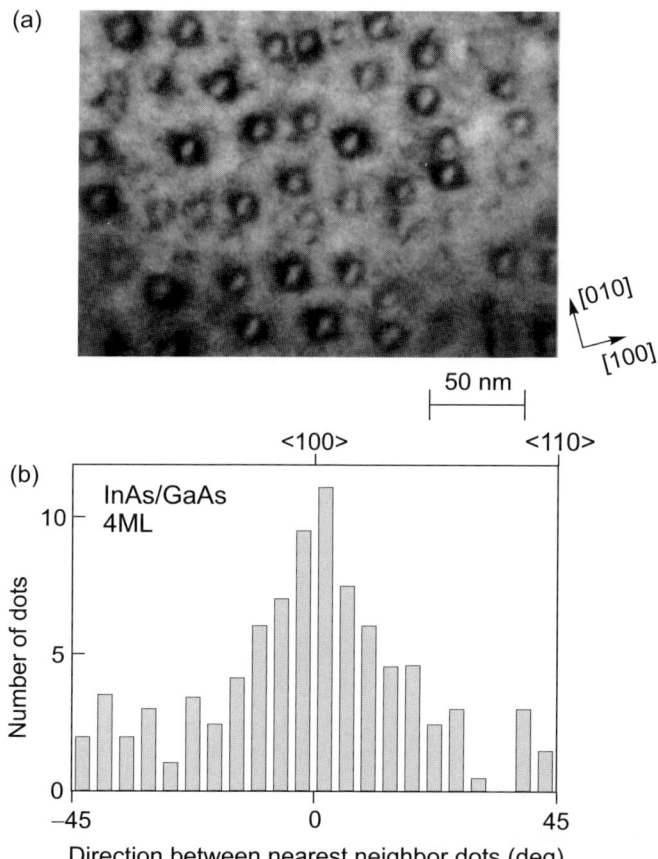

Fig. 3.85. (a) Plan-view transmission electron microscopy (TEM) micrograph of a single sheet of InAs dots grown in molecular beam epitaxy by 4 ML deposition of InAs. Dots are preferentially aligned in rows parallel to $\langle 100 \rangle$. (b) Histogram of the direction of nearest-neighboring dots

There exist two factors which favor the square lattice. The first is the cubic anisotropy of the elastic moduli of the medium, and the second is the square shape of the base of a single island.

The approximation of small tilt angles used in the present section does not have a significant impact on results. Firstly, this approximation works well for the volume elastic relaxation of isolated islands, even if the tilt angle of facets is 45°. The approximation of small tilt angles yields an energy of volume elastic relaxation equal to -64% of the strain energy for an equal volume in a uniformly strained flat film, whereas the exact calculations by the finite element method give the elastic relaxation energy of a pyramid as -60% of the strain energy of a flat film [3.194].

The close similarity between the approximation of small tilt angles and the exact numerical solution also holds for arrays of interacting islands. Evaluating the elastic energy for arrays of interacting islands by the finite element method for tilt angle $\theta_0 = 45°$ has shown that the cubic anisotropy of elastic moduli once again favors a 2D square lattice of islands with primitive lattice vectors parallel to the elastically 'soft' directions [100] and [010].

3.3.9 Phase Diagram of Arrays of Interacting Strained Islands

The main part of the elastic interaction energy in a system of strained islands is the energy of the dipole–dipole elastic interaction. The energy per atom is proportional to $\lambda\varepsilon_0^2\Omega(L/D)^3$, where D is the period of the lateral superlattice made up by the islands. For the square superlattice, the filling factor of the surface equals $q = L^2/D^2$, and the interaction energy per atom is

$$E_{\text{inter}} = \lambda\varepsilon_0^2 \Omega f_4(\theta_0) q^{3/2} , \tag{3.156}$$

where $f_4(\theta_0)$ is a geometrical factor. Due to mass conservation, the size L of the islands and the period D of the lateral superlattice are not independent quantities. To obtain the relationship between L and D, we note that $(Q-Q')$ monolayers of the deposited material are assembled in the islands, i.e., the volume $D^2(Q-Q')a$ of deposited material equals the volume of the island,

$$D^2(Q-Q')a = V . \tag{3.157}$$

For pyramid-shaped islands with an $L \times L$ square base and a tilt angle of the side facets equal to θ_0, we express the volume V in terms of L as $V = (1/6)\tan\theta_0 L^3$. The interaction energy per atom therefore takes the form

$$E_{\text{inter}} = \frac{\lambda\varepsilon_0^2 \Omega f_4(\theta_0) (6\cot\theta_0)^{3/2} [(Q-Q')a]^{3/2}}{L^{3/2}} . \tag{3.158}$$

In order to consider in detail the L-dependent terms in the energy per atom $E'(L) = E_{\text{dilute}}(L) + E_{\text{inter}}(L)$, it is convenient to introduce the characteristic length L_0 from (3.148) and the characteristic energy E_0 from (3.149). Then the sum of all L-dependent terms in $E(L)$ may be written as a function of the dimensionless length L/L_0:

$$E'(L) = E_0\left[-2\left(\frac{L_0}{L}\right)^2 \ln\left(\frac{e^{1/2}L}{L_0}\right) + \frac{4\beta}{e^{3/4}}\left(\frac{L_0}{L}\right)^{3/2} + \frac{2\alpha}{e^{1/2}}\left(\frac{L_0}{L}\right)\right] . \tag{3.159}$$

This function is governed by two control parameters, α defined in (3.151), and

$$\beta = [(Q-Q')a]^{3/2} \frac{e^{3/4} f_4(\theta_0)(6\cot\theta_0)^{5/6}(\lambda\varepsilon_0)^2 L_0^{1/2}}{2 f_3(\theta_0)\tau^2} . \tag{3.160}$$

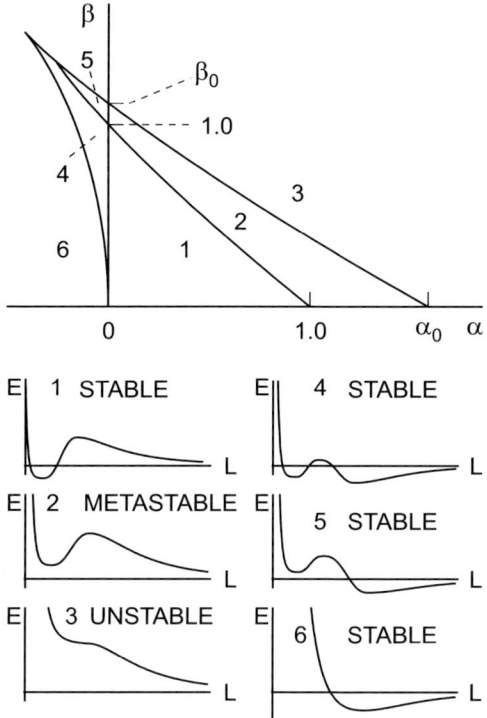

Fig. 3.86. *Upper*: schematic phase diagram of the stability of a square lattice of coherently strained islands in the plane of the control parameters α and β. Here α and β are defined in (3.151), (3.160), respectively. *Lower*: energy of the interacting array of islands vs. size of islands. Plots correspond to different regions of the phase diagram. In regions 1, 4, 5, and 6, there exist stable arrays of islands which do not undergo ripening. In regions 2 and 3, all arrays of islands undergo ripening. $\alpha_0 = 2\mathrm{e}^{-1/2} \approx 1.213$, $\beta_0 = (4/3)\mathrm{e}^{-1/4} \approx 1.038$

The parameter β is the ratio $E_{\mathrm{inter}}/\left|\Delta E^{\mathrm{edges}}_{\mathrm{elast}}\right|$. It increases as $(Q - Q')^{3/2}$ with the amount of material $(Q - Q')$ assembled in all islands.

By seeking the minima of the energy $E'(L)$ from (3.159) for different α and β, we obtain the phase diagram of Fig. 3.86. For region 1 of the phase diagram, there exists an optimum size of islands L_{opt}, corresponding to the absolute minimum of the energy, $\min E'(L) \equiv E'(L_{\mathrm{opt}}) < 0$. On the other hand, the ripening of islands would correspond to $L \to \infty$, where the energy $E'(L) \to 0$. This means that the 2D periodic square lattice of islands of optimum size L_{opt} is a stable array, and islands do not undergo ripening. For region 2 of the phase diagram, there exists only a local minimum of the energy, corresponding to a metastable array where $E'(L') > 0$. For region 3, the local minimum in the energy $E'(L)$ disappears. In both regions 2 and 3, there exists a thermodynamic tendency to ripening. If the system initially corresponds to region 1, and the amount of deposited material Q increases, then the point in the phase diagram moves to regions 2 and 3, and islands undergo ripening.

If $\alpha \leq 0$, there exists an absolute minimum of the energy $E'(L)$ for an arbitrary value of β, and $\min E' \equiv E'(L_{\text{opt}}) < 0$. Besides the absolute minimum of $E'(L)$, there may also exist a local minimum at $L = L'$, where the energy of a corresponding metastable state $E'(L') < 0$ in region 4 and $E'(L') > 0$ in region 5. No metastable state exists in region 6.

To estimate characteristic values of α and β, we substitute $L_0 \approx 100$ Å, $\lambda \approx 500$ meV/Å^3, $\tau \approx 100$ meV/Å^2, $a = 3$ Å into (3.160). This gives $\beta \approx 1$ for $|\varepsilon_0| \approx 1\%$ and $(Q-Q') = 0.5$ ML. Therefore, if $\alpha > 0$, the array of islands may correspond to region 1 of the phase diagram only for small values of $(Q-Q')$. It has been argued in Sect. 3.3.7 that the parameter α for the InAs/GaAs(001) system is very likely to be negative. If $\alpha < 0$, then an increase in $|\varepsilon_0|$, e.g., via an increase in x for the heteroepitaxial system In$_x$Ga$_{1-x}$As/GaAs(001), leads to a decrease in L_{opt}. This agrees with numerous experimental data on this material system (see, for example, [3.149]).

The key difference between the arrays of 3-dimensional coherently strained islands, on the one hand, and periodically faceted surfaces and periodic structures of surface domains, on the other hand, is the existence of both the ordering regime and the ripening regime for 3D islands, where a possible transition between these two regimes is governed by the surface energy of island facets.

For III–V semiconductor systems, there exists a possibility of tuning surface energies experimentally by varying the vapor pressure of the group V element(s). It is known (see, for example, [3.215, 3.216]) that the stoichiometry of the (001) surface of GaAs, and hence also the surface energy, depend strongly on the arsenic vapor pressure. This tendency should be rather general for all III–V systems, including the InAs wetting layer on GaAs(001). Therefore, one can expect the variation of arsenic pressure to drive the system from the regime in which islands are ordered in size to the regime of Ostwald ripening. A remarkable fact is that such a phase transition driven by the As pressure has been observed experimentally [3.152]. This transition will be discussed in Sect. 3.3.13.

3.3.10 Equilibrium Thickness of the Wetting Layer

The analysis in Sects. 3.3.7–3.3.9 reveals the equilibrium structure of an array of 3D coherently strained islands under the constraint of a fixed amount of material assembled in the islands. The treatment is based only on the scaling behavior of various contributions to the total energy, and reveals a criterion for the formation of coherent islands of optimum size vs. Ostwald ripening.

A more profound understanding of the equilibrium morphology of the system can be achieved if one takes into account the existence of the wetting layer. A common experiment on growth interruption implies a fixed amount

3.3 Arrays of Three-Dimensional Coherently Strained Islands

of the deposited material which is distributed between the wetting layer and 3D islands. Therefore neither the thickness of the wetting layer nor the total volume of all islands are fixed separately. To determine each of these quantities, one has to assume a certain microscopic model for the dependence of the energy of the wetting layer W on its thickness Q'.

Such a model has been introduced by Daruka and Barabási [3.156]. The energy per atom of a thick epitaxial film coherent to the substrate is given by $\lambda \varepsilon_0^2 \Omega - \Phi_{22}$, where $-\Phi_{22}$ is the energy of chemical bonds in the bulk of the deposited material 2, defined per atom. At the wetting layer–substrate interface, chemical bonds between substrate atoms and film atoms have energy $-\Phi_{12}$ such that $\Delta = \Phi_{22} - \Phi_{12} < 0$ (wetting condition). Due to the finite range of intermolecular interactions, the binding energies of the atoms in the film in the second and in successive monolayers are also modified: as we move away from the substrate, the binding energy density increases from $-\Phi_{12}$ in the first monolayer to its asymptotic value $-\Phi_{22}$. These intermolecular forces are responsible for a critical layer thickness larger than one monolayer [3.217, 3.218]. A model energy of the wetting layer proposed by Daruka and Barabási [3.156] takes this effect into account. The explicit form of the energy is

$$W(Q') = \int_0^{Q'} d\tilde{q} \left\{ -\Phi_{22} + \Delta \left[\vartheta(1-\tilde{q}) + \vartheta(\tilde{q}-1) \exp\left(-\frac{\tilde{q}-1}{\tilde{a}}\right) \right] \right\}. \tag{3.161}$$

Here $\vartheta(x) = 1$ if $x \geq 0$, $\vartheta(x) = 0$ if $x < 0$, and $\tilde{a}$ is the characteristic attenuation length of interatomic interaction.

By substituting $W(Q')$ from (3.161) into (3.137), it is possible in principle to find the optimum thickness of the wetting layer as a function of the total amount Q of deposited material. It follows from (3.137) that the formation of coherently strained islands leads to a decrease in the total energy, the decrease being a linear function of $(Q - Q')$. At the same time, the elastic repulsion between islands defined per unit surface area is proportional to $(Q - Q')^{5/2}$ [the last term in (3.137)], which exhibits a steep increase with Q at sufficiently large Q and hinders the formation of a dense array of islands. To take this effect into account, Daruka and Barabási [3.156] considered the possible coexistence of small islands of optimum size L_{opt} and 'ripened' islands having sizes considerably larger than L_{opt}. The total energy per unit cell of the substrate is

$$E = E_{\text{WL}}(Q_1) + Q_2 E_{\text{island}}(Q_2) + (Q - Q_1 - Q_2) E_{\text{rip}}. \tag{3.162}$$

Here the energy of the wetting layer equals $E_{\text{WL}}(Q_1) = \lambda \varepsilon_0^2 Q_1 + W(Q_1)$, where $W(Q_1)$ is given by (3.161). Equation (3.162) implies that Q monolayers of material 2 are deposited, Q_1 monolayers form the wetting layer, Q_2 monolayers are assembled in 3D coherently strained islands of a given pyramid-like shape and volume, and that the rest of material 2, namely $(Q - Q_1 - Q_2)$ monolayers, are assembled in ripened islands. The energy of 3D pyramids per atom equals $E_{\text{island}} = [1 - f_1(\theta_0)] \lambda \varepsilon_0^2 \Omega - \Phi_{22} + E'(L)$, where $f_1(\theta_0)$ is the geometrical factor describing the volume elastic relaxation [see (3.144)], and $E'(L)$ is defined in (3.159). Hence, the energy of 'ripened' islands can be obtained if one takes the limit $L \to \infty$, $E_{\text{rip}} = [1 - f_1(\theta_0)] \lambda \varepsilon_0^2 \Omega - \Phi_{22}$.

Equation (3.162) defines the total energy of the wetting layer and 3D pyramidal islands, where the latter may exhibit bimodal behavior, i.e., both small islands of size L_{opt} and large islands of a size considerably larger than L_{opt} may be present in the system. By minimizing the energy from (3.162) with respect to Q_1 and Q_2, Daruka and Barabási [3.156] obtained the equilibrium phase diagram of a lattice-mismatched heteroepitaxial system as a function of the lattice mismatch ε_0 and the total amount of deposited material Q. The domains of the phase diagram in Fig. 3.87 correspond to the following physical situations:

FM Phase. Deposited material contributes to the pseudomorphic growth of the wetting layer and 3D islands are absent, reminiscent of the Frank–van der Merwe (FM) growth mode. The total energy has its minima at $Q_2 = 0$ and $Q_1 = Q$, indicating that the thickness of the wetting layer coincides with the nominal thickness Q of deposited material.

R$_1$ Phase. Above a certain value of $Q_{c1}(\varepsilon_0)$, the total energy has new minima at $Q_2 = 0$ and $0 < Q_1 < Q$. This implies that, after formation of a wetting layer, the excess material contributes to formation of ripened islands. These ripened islands, being infinitely large, have zero area density.

SK$_1$ Phase. The total energy develops new minima at non-zero Q_1 and Q_2, such that $Q_1 + Q_2 = Q$, i.e., such that the deposited material (Q monolayers) is distributed between Q_1 monolayers of the wetting layer, and finite islands accumulating Q_2 monolayers of the deposited material, similar to the Stranski–Krastanow growth mode. It should be noted that, with an increase in the total amount of deposited material Q, the thickness of the wetting layer Q_1 continues to grow sub-linearly. This is a consequence of island–island repulsive interaction: in the dilute system limit, the wetting layer thickness is constant.

R$_2$ Phase. In this phase, the total energy has minima at $0 < Q_1 < Q$ and $0 < Q_2 < Q$, indicating that the deposited material is distributed between a wetting layer, finite islands, and ripened islands. The finite islands formed in the SK$_1$ phase will be preserved, being stable against ripening. Thus finite and ripened islands coexist in the R$_2$ phase.

VW Phase. For large mismatch and small coverage, the total energy has its minima at $Q_2 = Q$ and $Q_1 = 0$, indicating that all deposited material is accumulated in finite islands. Due to large mismatch, the wetting layer is absent and islands are formed directly on the substrate, similarly to what happens in the Volmer–Weber (VW) growth mode.

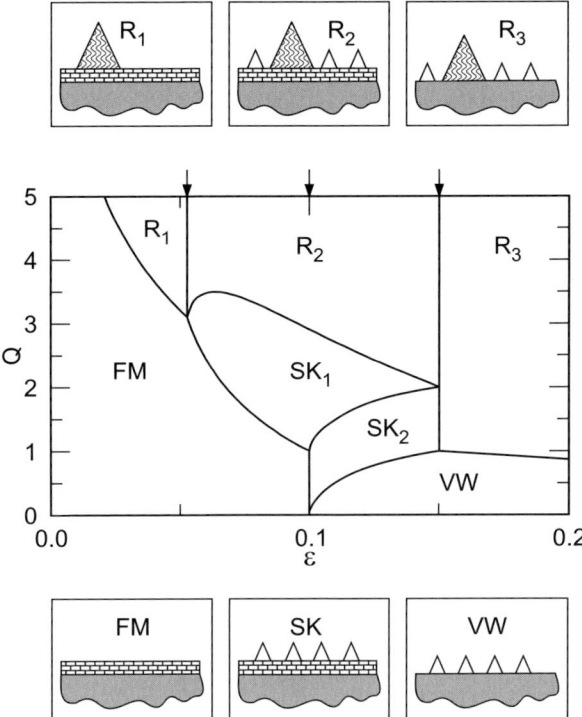

Fig. 3.87. Equilibrium phase diagram of a lattice-mismatched heteroepitaxial system as a function of the total amount of deposited material Q and the lattice mismatch ε_0. The *small panels* at top and bottom illustrate the morphology of the surface in the six growth modes. *Small empty* islands indicate the presence of stable islands, while *large shaded* ones refer to ripened islands. With kind permission of I. Daruka and A.-L. Barabási [3.156]

SK$_2$ Phase. By increasing Q, we reach the SK$_2$ phase. The behavior of the system is not the same as in the SK$_1$ growth mode. For a given ε_0, islands are already formed in VW phase. In the SK$_2$ phase, the island density and island size remain unchanged, and a wetting layer starts to form, until its thickness reaches 1 ML, at which point we enter the SK$_1$ phase.

R$_3$ Phase. The total energy has its minima at $Q_1 = 0$ and $0 < Q_2 < Q$, indicating the formation of ripened islands. Finite islands formed in the VW mode are preserved and coexist with ripened islands. However, in contrast to the R$_2$ phase, the wetting layer is absent.

Thus, the papers by Shchukin et al. [3.155] and Daruka and Barabási [3.156] solve the theoretical problem of the equilibrium state of a lattice-mismatched heteroepitaxial system.

3.3.11 Two Exact Theorems on the Shape vs. Volume Dependence of 3D Islands

The above thermodynamic theories developed by Shchukin et al. [3.155,3.194, 3.195] and Daruka and Barabási [3.156], and presented in Sects. 3.3.7–3.3.10, are based on the approximation that individual islands have fixed shape. These theories reveal that, in a certain parameter range, the equilibrium state of a lattice-mismatched system is an ordered array of 3D coherently strained islands, and that Ostwald ripening does not occur. On the other hand, Duport et al. [3.198] and Spencer and Tersoff [3.200] pointed out that the equilibrium shape of a sufficiently large coherent island is an overhanging ball nearly detached from the substrate or from the wetting layer. Since such islands are totally relaxed, it was concluded in [3.198,3.200] that equilibrium always corresponds to a single overhanging island formed as a result of ripening.

In the present section we show that, despite the possible existence of totally relaxed overhanging islands (or totally relaxed dislocated islands), there is, nevertheless, a parameter region where the equilibrium corresponds to an ordered array of 3D coherently strained islands, and ripening is not energetically favorable. The transition from a dense array of 3D coherent, partially relaxed islands over a wetting layer to a single overhanging, totally relaxed island is accompanied first by a reduction in the strain energy, and second by a change in the surface energy. In the case, where the surface energy of the wetting layer is higher than the surface energies of island facets, the disappearance of islands implies an increase in the total surface energy. If the amount of deposited material is not too large, the above-mentioned increase in the surface energy can outweigh the reduction in the strain energy, and the formation of a single island via ripening will not be favorable energetically.

In order to show this, we take into account the dependence of the island shape on the island volume and discuss how this can affect the conclusions of Sects. 3.3.7–3.3.10. The dependence of the equilibrium shape of a 3D coherently strained island on its volume has been obtained under certain model assumptions about facet energies [3.196–3.198, 3.200] or ab initio [3.199]. In contrast to the equilibrium shape of unstrained crystals, where exact theorems have already been formulated and proved by Wulff [3.11] and Herring [3.12], there has until recently been a lack of exact theorems concerning the shapes of strained islands. Shchukin and Bimberg [3.157] have formulated and proven two exact theorems. Since the main concern of this section is the existence of an optimum volume of the island vs. ripening, we consider the energy per atom of material in islands.

If we keep only the elastic strain energy and the surface energy (including the strain-induced renormalization terms) in the energy of a single island, i.e., the first and the second terms in (3.144), and if we express the energy in terms of the island volume and island shape, we obtain the energy per atom as

3.3 Arrays of Three-Dimensional Coherently Strained Islands

$$E = \Omega \left[\lambda \varepsilon_0^2 R + \frac{(\Delta \Gamma)}{V^{1/3}} \right] . \tag{3.163}$$

Unlike (3.144), which gives the change in the energy due to formation of a 3D island from the uniformly strained flat film, (3.163) gives the energy of the island itself. Here the shape-dependent coefficient R ($0 \leq R \leq 1$), determined by the volume elastic relaxation, is related to f_1 from (3.144) by $R = 1 - f_1$. The shape-dependent coefficient $(\Delta \Gamma)$ includes effects due to the appearance of side facets and the top facet, disappearance of a certain area of the wetting layer underneath the island, and strain-induced renormalization of the surface energies. Since the optimum shape of the island depends on the island volume, coefficients $R = R_{\text{opt}}$ and $(\Delta \Gamma) = (\Delta \Gamma)_{\text{opt}}$ also depend on the volume.

Theorem 1. *For the optimum shape of a 3D coherently strained island, the coefficient $R = R_{\text{opt}}$ in (3.163) is a non-increasing function of the island volume V, and the coefficient $(\Delta \Gamma) = (\Delta \Gamma)_{\text{opt}}$ is a non-decreasing function of the island volume.*

This theorem is, in fact, clear from an intuitive point of view. The validity of this statement has been demonstrated for several particular examples [3.196–3.200]. However, to the best of our knowledge, the first rigorous proof was given by Shchukin and Bimberg [3.157].

Proof. Let an island of volume V_1 have equilibrium shape 1, with corresponding coefficients $R = R_1$ and $(\Delta \Gamma) = (\Delta \Gamma)_1$. Let the corresponding coefficients for an island of volume V_2 be R_2 and $(\Delta \Gamma)_2$. For the island of volume V_1, the minimum energy is achieved by shape 1. Islands of any other shape, in particular, those of shape 2, have larger energy. Hence,

$$\lambda \varepsilon_0^2 R_2 + \frac{(\Delta \Gamma)_2}{V_1^{1/3}} \geq \lambda \varepsilon_0^2 R_1 + \frac{(\Delta \Gamma)_1}{V_1^{1/3}} . \tag{3.164}$$

Similar consideration of islands of volume V_2 yields

$$\lambda \varepsilon_0^2 R_1 + \frac{(\Delta \Gamma)_1}{V_2^{1/3}} \geq \lambda \varepsilon_0^2 R_2 + \frac{(\Delta \Gamma)_2}{V_2^{1/3}} . \tag{3.165}$$

Now, in each of inequalities (3.164) and (3.165), let us put the combination $(R_1 - R_2)$ on one side of the inequality, and the combination $[(\Delta \Gamma)_2 - (\Delta \Gamma)_1]$ on the other side. This procedure yields a double inequality,

$$\frac{(\Delta \Gamma)_2 - (\Delta \Gamma)_1}{V_2^{1/3}} \leq \lambda \varepsilon_0^2 (R_1 - R_2) \leq \frac{(\Delta \Gamma)_2 - (\Delta \Gamma)_1}{V_1^{1/3}} . \tag{3.166}$$

For concreteness, let $(R_1 - R_2) \geq 0$. Then, from the left inequality of (3.166), it follows that $(\Delta \Gamma)_2 - (\Delta \Gamma)_1 \geq 0$. Comparison of the left- and right-hand sides of the double inequality (3.166) yields $V_2 \geq V_1$. Similar considerations show that, in the case $(R_1 - R_2) < 0$, one obtains $(\Delta \Gamma)_2 - (\Delta \Gamma)_1 < 0$ and $V_2 < V_1$. QED

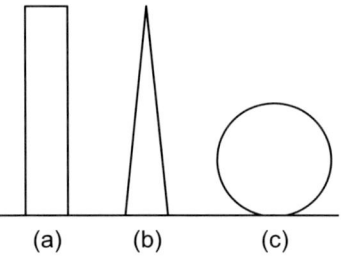

Fig. 3.88. Sample shapes of coherent islands with asymptotically small elastic energy, i.e., $R \to 0$. (**a**) vertical prism, (**b**) vertical pyramid, (**c**) overhanging island

One should keep in mind that the proof of the above theorem is based on the assumption that, for a given volume of the material, the island chooses its optimum shape among all possible shapes. This assumption is no longer valid if discrete atomistic effects become important. For example, the height of the islands cannot be less than one monolayer. The latter is important for sufficiently small island volume V, where islands are 2-dimensional.

As the volume increases, the equilibrium shape becomes 3D, and the above theorem applies. In particular, the coefficient R decreases. As the island volume tends to infinity, the coefficient R tends asymptotically to zero. Such behavior is realized, for example, for the island shapes given in Fig. 3.88. For all the islands in Fig. 3.88, the elastic energy is concentrated in a small volume of the island and in the substrate in the vicinity of the interface. In (3.145), where the quantity $(\Delta \Gamma)$ is defined, the first term, i.e., the positive surface energy of the island surface, dominates due to the large side surface of the island. Therefore $(\Delta \Gamma)$ is always positive for sufficiently large islands.

To discuss the possible dependence of the energy per atom vs. island volume, we differentiate (3.163) over V. One has to take into account the fact that there are two contributions to the change in E with island volume. The first contribution comes from the fact that the larger the volume, the smaller the fraction of atoms that are surface atoms. This corresponds to the factor $V^{-1/3}$ in the second term of (3.163). The second contribution comes from the change in island shape vs. volume. This corresponds to the dependence of R and $(\Delta \Gamma)$ on volume in (3.163). In the derivative dE/dV, these two contributions enter as follows:

$$\frac{dE}{dV} = \frac{\partial E}{\partial V}\bigg|_{\mathcal{G}=\text{const}} + \frac{d\mathcal{G}_{\text{opt}}}{dV} \frac{\partial E}{\partial \mathcal{G}}\bigg|_{V=\text{const}} . \tag{3.167}$$

Here $\mathcal{G}$ denotes the set of variables which define the island shape. For example, for truncated square-based pyramids, these are the facet tilt angle and the level of truncation. The definition of the optimum shape reads that $\partial E/\partial \mathcal{G} = 0$. Hence,

$$\frac{dE}{dV} = \frac{\partial E}{\partial V}\bigg|_{\mathcal{G}=\text{const}} = -\frac{1}{3}\Omega \frac{(\Delta \Gamma)_{\text{opt}}}{V^{4/3}} . \tag{3.168}$$

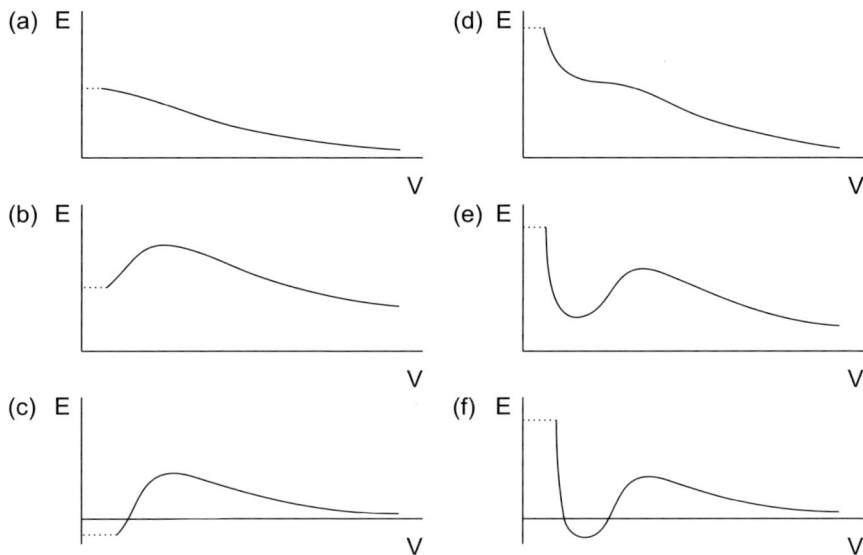

Fig. 3.89. The energy per atom for an array of 3D islands vs. the volume of the islands, $E(V)$. If the amount of material in all islands is less than 1 ML, islands of small volumes are 2D. If the amount of material in all islands exceeds 1 ML, islands of a certain volume cover 100% of the surface, and islands of smaller volumes cannot exist. In all graphs, a *dashed line* plotted in the region of small volumes implies that the real dependence $E(V)$ is not considered for such volumes. (**a**)–(**c**). Approximation of a dilute array of non-interacting islands. (**a**) $(\Delta \Gamma)_{\text{opt}} > 0$ for all volumes. The stable state corresponds to a single, huge island formed via Ostwald ripening. (**b**) $(\Delta \Gamma)_{\text{opt}} < 0$ for some volumes. The stable state corresponds to a single, huge island formed via Ostwald ripening. A metastable state exists, corresponding to an array of finite islands of optimum volume. (**c**) $(\Delta \Gamma)_{\text{opt}} < 0$ for some volumes. The stable state corresponds to an array of finite islands. Ostwald ripening is not favorable energetically and will not occur. (**d**)–(**f**). Approximation of an array of interacting 3D islands. (**d**) The stable state corresponds to a single, huge island formed via Ostwald ripening. (**e**) The stable state corresponds to a single, huge island formed via Ostwald ripening. A metastable state exists, corresponding to an array of finite 3D islands of optimum volume. (**f**) The stable state corresponds to an array of finite 3D islands of optimum volume. Ostwald ripening is not favorable energetically and will not occur

It follows from (3.168) that if for a given volume V of the island $(\Delta \Gamma)_{\text{opt}} > 0$, then the energy per atom is a locally decreasing function of the volume. And otherwise, if for a given volume V, $(\Delta \Gamma)_{\text{opt}} < 0$, then the energy per atom is a locally increasing function of the volume. Since $(\Delta \Gamma)_{\text{opt}}$ is an increasing function of the island volume (see Theorem 1), it is either positive for all volumes or changes sign once. From the behavior of $(\Delta \Gamma)_{\text{opt}}$, one can reveal the possible dependence of the energy per atom vs. volume. Figures 3.89a, b, and c demonstrate three possibilities for the energy per atom vs. island volume dependence for a dilute array of islands.

For small volumes, islands are either 2-dimensional or form a dense array of 3D interacting islands, and the above approximation of a dilute array of 3D islands does not apply. Therefore the behavior of the energy at small volumes is not shown.

If $(\Delta\Gamma)_{\text{opt}} > 0$ for all volumes (Fig. 3.89a), then the energy per atom decreases monotonically with island volume. The minimum energy corresponds to $V \to \infty$, i.e., to Ostwald ripening. If for some volume $(\Delta\Gamma)_{\text{opt}} < 0$, then there are two possibilities. Figure 3.89b relates to the situation where the absolute minimum still corresponds to $V \to \infty$, i.e., to Ostwald ripening. However, there exists a local minimum for small volumes. This local minimum corresponds to either an array of 2D islands or a dense array of 3D islands. In Fig. 3.89c, the absolute minimum of the energy per atom is attained by either an array of 2D islands or a dense array of 3D islands.

Analysis of the behavior of the energy per atom vs. island volume gives us a proof of the following result.

Theorem 2. *If for some island volume, $(\Delta\Gamma)_{\text{opt}} < 0$, then there exists a stable or a metastable array of islands of finite optimum volume.*

For a dense array of islands, we must take into account the island–island elastic interaction. As has been shown in Sect. 3.3.8, this interaction can be attractive only at a very small surface coverage ($< 1\%$), and is otherwise repulsive. The interaction energy per atom in the islands depends on the island volume as $E_{\text{inter}} \sim C(\mathcal{G})(Q - Q')^{3/2} V^{-1/2}$, where the coefficient $C(\mathcal{G})$ depends on the island shape. Therefore, for a given volume of the island, the optimum island shape depends on the total amount of material in all islands $(Q - Q')$. Such a dependence manifests itself in experiment [3.219]. Thus, for GeSi/Si(001) islands, the shape transition from a pyramid bounded by {105} facets to a dome bounded by steeper {113} facets occurs in dense arrays at a smaller volume of the islands than in dilute arrays.

The effect of the island–island interaction on the behavior of the energy per atom vs. island volume $E(V)$ is governed by two factors. Firstly, this energy is repulsive. Secondly, it decreases with the island volume. A quantitative analysis, based on a particular model of islands, will be published elsewhere. Qualitatively, the effect of the island–island interaction can be understood if one adds a positive decreasing function to the curves of Fig. 3.89a–c. Then, if $(\Delta\Gamma)_{\text{opt}} > 0$ for all island volumes, the tendency to Ostwald ripening remains unchanged (Fig. 3.89d). If $(\Delta\Gamma)_{\text{opt}} < 0$ for some volumes, then the local minimum in Fig. 3.89b is shifted to larger volumes (Fig. 3.89e). The same is valid for the absolute energy minimum in Fig. 3.89c, which is shifted to larger volumes (Fig. 3.89f). This shift means that the stable (Fig. 3.89f) or metastable (Fig. 3.89e) state of the system corresponds not to an array of 2D islands, but to an array of 3D islands. For sufficiently large amounts of material assembled in all islands, the local minimum in Fig. 3.89e or the absolute minimum in Fig. 3.89f will disappear, and islands will undergo ripening.

We note that the elastic relaxation due to island edges, which has not been considered in the present section, also contributes to the creation of a local or absolute minimum on the curve $E(V)$, and hinders the tendency to ripening.

Summing up the discussion of this section, we emphasize the main effect of the shape vs. volume dependence for 3D coherently strained islands. The parameter region where the absolute minimum of the total energy corresponds to an ordered array of 3D islands is narrower than the one obtained in Sect. 3.3.9. Otherwise, an ordered array of 3D islands is no longer stable, but metastable. It is worth noting that this metastable state has nothing to do with self-limiting kinetics of island growth. This metastable state is determined by energetic arguments alone, i.e., by a local minimum in the dependence of the energy per atom vs. island volume. On a rather long time scale, no difference between the situations of Fig. 3.89e and Fig. 3.89f can be observed in experiment. Such metastable arrays behave like stable ones, including the reversibility of the 2D to 3D morphology transition, the control of island volume by the vapor pressure of the group V element, etc. In this sense, both the above-described metastable array of islands and the stable array are thermodynamically controlled. In Sect. 3.3.13, we will focus on key experiments which allow one to distinguish thermodynamically controlled arrays from kinetically controlled arrays.

3.3.12 Kinetic Theories of Size-Limited Island Growth

The kinetic approach is a basic one in the theoretical description of crystal growth. The importance of strain effects in growth kinetics has been recognized for a number of growth-related phenomena. Thus, the step-flow growth of a material lattice-mismatched to the substrate results in step bunching [3.220]. The step-flow growth of an alloy can be unstable against either vertical modulation of composition due to stress induced by step bunching [3.221], or lateral modulation of composition due to stress induced by 'frozen' compositional fluctuations [3.222–3.225].

In this connection, two types of kinetics must be distinguished. The most common situation corresponds to the kinetics in an open crystal growth system, where the deposition of atoms and the evolution of a surface nanostructure are two simultaneous processes. A steady-state structure, if it exists, does not correspond to equilibrium, but is determined by the particular growth kinetics.

Another type of kinetics occurs in a closed system. This is realized by growth interruption or annealing, where the total amount of deposited material is preserved, provided that evaporation is negligibly small. The kinetic pathway for nanostructure evolution is a pathway towards thermodynamic equilibrium.

One should bear in mind the importance of kinetic effects, even in closed systems. Firstly, thermodynamic equilibrium relies on elementary kinetic processes in detailed balance. Secondly, it is possible to bring out kinetic effects in the formation of 3D coherently strained islands by rapidly freezing a certain configuration, e.g., by cooling down or burying the structure underneath a cap layer before it can reach thermodynamic equilibrium.

All debates on whether a narrow size distribution of 3D islands and an apparent absence of Ostwald ripening have a thermodynamic or a kinetic origin are only relevant if a closed system is considered and the system is subject to annealing or growth interruption, i.e., if it is allowed to equilibrate.

Kinetic models generally consider the situation where an array of 3D strained islands exhibits a global thermodynamic driving force to Ostwald ripening. However, various basically strain-induced barriers may occur, thereby hindering island ripening beyond a certain volume. This will eventually explain size-limited island growth.

3D Island Formation via a Precursor State of 2D Platelets. Priester and Lannoo [3.158] proposed a model addressing the 2D → 3D transition at a certain critical coverage and a narrow size distribution of 3D islands. Below critical coverage, deposited material forms 2D islands with monolayer height. In the case of Stranski–Krastanow growth, if the critical thickness of the wetting layer lies between 1 and 2 ML, these are platelets of the second incomplete monolayer over the first complete one. The energetics of such platelets is similar to that of submonolayer islands, discussed in detail in Sect. 3.2. Platelets have an equilibrium size which increases with increasing surface coverage q. The energy per atom in the optimum platelets also increases with q and, at a certain critical coverage, becomes larger than in 3D islands. At this coverage, 3D islands become more stable than platelets. According to [3.158], at the critical coverage $q = q_c$, the 2D platelets transform to 3D islands at an almost constant number of atoms in each island, and the resulting size distribution of these 3D islands thus reflects the platelet size distribution. A rather narrow Gaussian curve is then obtained, as observed experimentally.

Detailed experimental studies of formation mechanisms for InAs/GaAs 3D islands by Ledenstov et al. [3.152] clearly show a completely different behavior in this material system. In the initial stages, after 3D islands have been formed and before growth interruption is introduced, 3D islands have a broad distribution of sizes and a small average size. A few seconds or tens of seconds of growth interruption are needed before 3D islands grow to a limiting size and the size distribution narrows. Thus, 2D platelets, if they exist at all, are not related to the final state of the system in this case.

Strain-Induced Barriers at Island Edges. Kinetic considerations of the elementary processes occurring in epitaxial growth, i.e., attachment of surface adatoms to islands, detachment of atoms from islands into an 'adatom sea', and migration of adatoms on the surface between islands has revealed two important kinetic processes providing mechanisms consistent with the narrowing of the island size distribution.

3.3 Arrays of Three-Dimensional Coherently Strained Islands   201

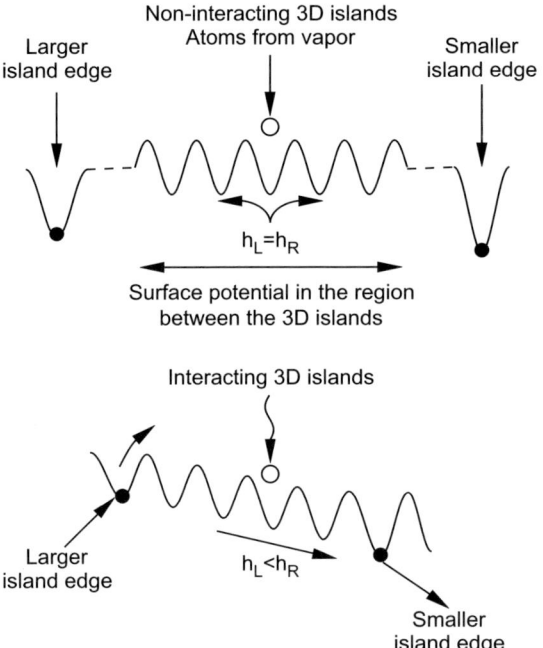

Fig. 3.90. Schematic diagram showing the surface potentials experienced by migrating adatoms in the regions between 3D islands in the absence (*upper panel*) and presence (*lower panel*) of overlapping substrate strain fields. The binding energies at the edges of two islands and the behavior of the surface diffusion potential in-between change with decreasing island separation, thus providing a potential mechanism for island size equalization via (**a**) a break in the left–right symmetry ($h_L = h_R$) of the migration rate in the regions between the islands and (**b**) a destabilization of the initially formed largest islands. With kind permission of N. Kobayashi et al. [3.160]

Carrying out kinetic Monte Carlo simulations that mimic highly strained growth of tetrahedrally-bonded compound semiconductor combinations such as InAs/GaAs, Ghaisas and Madhukar [3.226, 3.227] made a suggestion, based upon energy considerations, that the growth rate of 3D islands would diminish when an island grows to a certain size, due to the accumulation of elastic strain energy. This was modeled as an attendant activated rate for atom incorporation at island edges with an energy barrier that increases with island size. As a consequence, at a later stage, smaller islands will grow more rapidly, and new islands may form.

The second contribution was put forward by Madhukar et al. [3.159] and Kobayashi et al. [3.160]. It was suggested that, with increasing island density, the island-induced strain fields in the substrate would begin to interact. At this stage, a driving force for preferential migration of atoms towards smaller islands can arise due to a downward tilt in the surface potential towards smaller islands (Fig. 3.90). Equally importantly, the interacting strain fields in the substrate and the wetting layer also provide a means to destabilize the initially largest islands by not accommodating as much strain relief (in the substrate). This would decrease the barrier for detachment of atoms from larger islands and contribute to island size equalization.

A similar model was proposed by Chen and Washburn [3.228], who pointed out that: "after an island forms, the strain relaxation in the island causes a strain concentration at the island edge," from which they concluded that: "the adatoms deposited on the wetting layer surface will have to overcome an energy barrier $\Delta\mu_s$ before they can attach to the island". As each new adatom attached to the island tends to increase the strain concentration at the island edge, the energy barrier $\Delta\mu_s$ increases monotonically with increasing island radius R, and this eventually slows down the island growth rate. The larger islands grow more slowly than the smaller islands, leading to a homogeneous distribution of island size, as observed experimentally.

We note that it is not clear from the above-discussed models why a moderate increase in arsenic pressure may cause ultrafast ripening of the islands, whereas a moderate decrease in arsenic pressure tends to flatten the surface.

Barabási [3.229] carried out kinetic Monte Carlo simulations of island formation during heteroepitaxial growth using a 1-dimensional model. Two strain-related effects were discussed:

- Strain created by islands in the substrate makes the chemical potentials of migrating adatoms position-dependent and generates a net current of adatoms away from the island.
- For large islands, the strain energy at the edge becomes comparable with the bonding energy of the edge atom, enhancing its detachment, and leading to a gradual dissolution of the island.

The simulated action of these two effects leads to a kinetic mechanism stabilizing island size: as an island grows, a strain field develops, which helps to 'dissolve' edge atoms and push them away from the island. The newly deposited atoms also diffuse away from the larger islands. These combined effects slow the growth rate of larger islands and increase the nucleation of new islands, eventually leading to a narrow size distribution in the system for sufficiently large lattice mismatch ($> 5\%$). Results from the simulations of [3.229] thus confirm the qualitative model of [3.159, 3.160].

The above kinetic models address one of the key properties of lattice-mismatched heteroepitaxial systems, namely, the possibility of forming 3D islands with a narrow size distribution. From these models one may infer that, with increasing temperature, barriers of the same height will become less important. Then island growth will be limited only by islands of a larger volume generating stronger barriers. This implies an increase in the island volume with temperature. We will show below that InAs/GaAs 3-dimensional islands show the opposite behavior: a decrease in island volume with temperature [3.164]. Moreover, reversible changes in the substrate temperature or

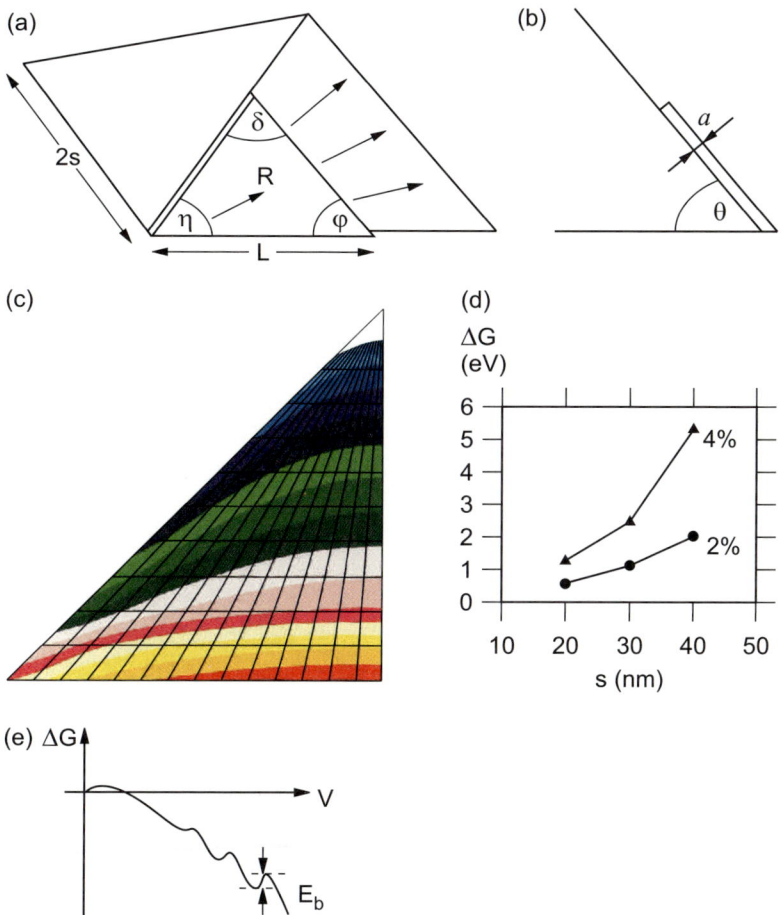

Fig. 3.91. Schematic representation of strained facet growth. (**a**) An embryo emerges from the bottom left-hand corner of the facet and expands across the facet with the geometry shown. (**b**) The embryo of height a increments the basal dimension of the island, as shown in cross section. (**c**) Elastic energy density associated with half the triangular facet of a pyramidal island $s = 40$ nm at 2% misfit. The elastic energy density is represented by a *color scale* which decreases linearly from 96 MJ/m^3 (*red level*) at the base to 18 MJ/m^3 (*white level*) close to the peak. (**d**) Energy barrier ΔG to complete the facet as a function of island size s and misfit strain. (**e**) Schematic diagram showing a typical energy curve during layer-by-layer growth of island facets. Undulations in the energy curve induce kinetic barriers for island growth. (**a**)–(**d**) With kind permission of D.E. Jesson et al. [3.162], (**e**) with kind permission of K.M. Chen et al. [3.161]

arsenic pressure lead to reversible changes in the surface morphology, showing that the system is close to equilibrium.

Strain-Induced Barriers in the Growth of Faceted Islands. In the above kinetic models, attachment of adatoms to islands, their detachment from the islands, and migration of the surface between islands are assumed to be the only relevant processes involved in island formation. However, for faceted islands, atomistic features of island shape may be of key importance, and strain-induced barriers may produce an even more severe effect. Chen et al. [3.161] and Jesson et al. [3.162] emphasized the importance of nucleation for every new atomic layer in the growth of faceted islands. If an island has a shape with smooth facets which do not contain any steps or kinks, a new layer starts to nucleate at the edge between the island facet and the substrate. The embryo of a new facet is shown schematically in Fig. 3.91a and b. This is a highly strained area, as demonstrated in the elastic energy density plot of Fig. 3.91c. Strain provides a barrier for such nucleation, and the height of the barrier is larger for larger islands (Fig. 3.91d). The energy of a single island vs. the island volume will thus contain local minima at volumes where the island facets are complete (Fig. 3.91e). Because the barriers are higher for larger islands, this leads to self-limited growth and may result in a narrow size distribution.

In the growth of 3D strained faceted islands, the strain-induced decrease in the attachment rate and the strain-induced barrier for nucleation of new atomic layers on facets both occur during growth and eventually lead to size-limited island growth with a narrow distribution of island sizes. The relative importance of the two effects depends drastically on the numerical values of the barrier heights, which may be obtained only via first-principles calculations. One may infer qualitatively that an increase in the substrate temperature should lead to an increase in the limited island volume. This is likely to be consistent with the observations for Ge/Si islands.

However, we emphasize below that certain experimental data for arrays of InAs/GaAs and InGaAs/GaAs islands can hardly be explained via kinetic concepts. These are the reversibility of the 3D–2D phase transition driven by changes in As pressure for MBE-grown InAs/GaAs islands, or by the replacement of AsH_3 by PH_3 for MOCVD-grown InGaAs/GaAs islands, an As pressure-driven irreversible transition from an ordered array of 3D islands to ripening, and reversible changes in the surface morphology upon variation of the substrate temperature. At the same time, these observations agree well with the thermodynamic theory of island formation.

Local Equilibrium between 3D Islands and the Wetting Layer. Wang et al. [3.112] proposed a view of the growth process divided into three phases:

- an early nucleation phase which largely determines the island density,
- a second phase where islands grow mostly at the expense of the wetting layer,
- a third phase characterized by Ostwald ripening.

3.3 Arrays of Three-Dimensional Coherently Strained Islands

As long as the wetting layer acts as a source of material, existing islands grow rapidly and the island density n remains constant. The island density is treated in [3.112] as an input to the model, n being determined by the growth kinetics. The island size is discussed in terms of a constrained thermodynamic equilibrium between the islands and the wetting layer, for a fixed island density. The particular example considered refers to strained dislocation-free InAs islands with a fixed shape, viz., square-based pyramids with {110} side facets on the GaAs(001) surface (with a wetting layer). Figure 3.92 schematically illustrates island formation on the substrate surface. Q_0 and Q are the nominal coverage and wetting layer thickness, respectively, after island formation.

The total energy gain per unit volume of a single island can be expressed as

$$\frac{E_{\text{total}}}{V} = \varepsilon_{\text{isl}}^{\text{elast}} - \varepsilon_{\text{film}}^{\text{elast}} + \frac{S\gamma_{\text{facet}} - L^2 \gamma_{\text{WL}}(Q_0)}{V} \quad (3.169)$$
$$+ \left(\frac{1}{n} - L^2\right) \frac{\gamma_{\text{WL}}(Q) - \gamma_{\text{WL}}(Q_0)}{V},$$

where $\varepsilon_{\text{isl}}^{\text{elast}}$ and $\varepsilon_{\text{film}}^{\text{elast}}$ are the elastic energy densities of the island and uniformly strained film. The third term describes the change in surface energy due to the island, with γ_{facet} being the surface energy of the island facets and S their area. The fourth term accounts for the thinning of that part of the wetting layer which feeds the island. $\gamma_{\text{WL}}(Q_0)$ and $\gamma_{\text{WL}}(Q)$ are the formation energies of the wetting layer as a function of its thickness Q, measured relative to the InAs bulk kept at the GaAs lattice constant. From mass conservation, the volume of an island V is given by $V = (1/6)a^3 \tan\alpha = (1/n)(Q_0 - Q)L$, where α and a are the tilt angle of the island facets and the monolayer thickness, respectively.

The surface energies of the island facets and wetting layer were calculated ab initio [3.112], and the elastic strain energy was calculated using continuum elasticity theory. The size of an optimum island was obtained as a function of the island density and nominal coverage Q. The dependence of the lateral

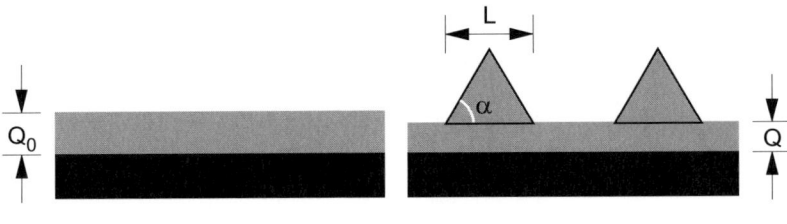

Fig. 3.92. Schematic illustration of the formation of coherent islands on the substrate surface. Q_0 and Q are the nominal coverage and wetting layer thickness, respectively. α is the tilt angle of island facets and L is the island base length. With kind permission of L.G. Wang et al. [3.112]

size of islands on the amount of deposited material obtained in this way reproduces the experimental data of [3.147, 3.230], if suitable island densities are used as input. In addition, the theoretical results of Wang et al. [3.112] show an increase in the critical thickness of InAs at which the 2D → 3D transition occurs, with increasing island density. The critical thickness varies from 1.20 ML to 1.79 ML when the island density increases from 10^9 cm^{-2} to 3×10^{11} cm^{-2}, which matches the experimentally observed range 1.2–2 ML [3.230–3.232].

The model of [3.112] is, however, limited to an intermediate stage of island formation, and does not address the final state of a heteroepitaxial system attained upon a long growth interruption or annealing. We will show below that experimental data on the InAs/GaAs system also show dramatic changes in island density if the substrate is subject to heating or cooling after the formation of 3D islands.

3.3.13 Experimental Studies of 3D Island Formation in the In(Ga)As/GaAs System

As discussed above, one of the key issues of size-limited island growth is the relative role of thermodynamic and kinetic effects. Specifically focusing on this debate, criteria have been formulated [3.157] to test, for any particular system, whether an observed array of islands is controlled predominantly by thermodynamics or by kinetics. To resolve this question, it is important to verify experimentally whether the following properties of equilibrium systems are present in a given system.

- Upon growth interruption or annealing, the system evolves towards equilibrium. When equilibrium is reached, no further changes occur.
- The equilibrium state of the system depends only on thermodynamic parameters, and not on prehistory. For the heteroepitaxial systems in question, these are the amount of deposited material and temperature. For III–V and II–VI compound semiconductors, there exists one more thermodynamic parameter. Growth interruption is provided by switching off the supply of cations, whatever the anion (e.g., arsenic) vapor pressure. The anion vapor pressure is the third thermodynamic parameter which can affect the morphology of the system.
- For the system in equilibrium, we may cause reversible changes in the morphology by varying the thermodynamic parameters of the system.

When analysing experimental data, it is important to note the following. The thermodynamic theory of size-limited island growth described in detail above only deals with coherently strained islands. At equilibrium, islands are distributed in volumes around an average value. In this distribution, islands are present, even at an extremely low density, with volumes exceeding the critical volume V_c against the onset of defects. The free energy per atom in such defective islands is considerably lower than in coherent islands. Once defective islands form even at a very low density, they become efficient sinks for the deposited material, accumulating the latter from coherent islands. In this connection, an array of coherent islands should be regarded as a metastable state of the system, whereas the stable state corresponds to defective islands. For an extremely long growth interruption time the onset of defects in large islands is expected for any strained heteroepitaxial system. This imposes certain restrictions on possible experiments. Testing the evolution of the system upon growth interruption, and reversibility upon changes in substrate temperature and/or arsenic pressure, should be carried out on a time scale where defective islands have not yet formed.

These are the main items in our discussion of experimental data on quantum dot arrays in the present section. Despite a large amount of experimental work on the formation of QDs in various semiconductor systems, there are only a few papers which specifically focus on whether dots are in equilibrium or kinetically controlled. In our discussion, we will focus on these experimental data, referring mainly to In(Ga)As/GaAs islands.

Effect of Growth Interruption on the Morphology of InAs/GaAs Systems. Specific studies of the behavior of InAs/GaAs heteroepitaxial systems upon growth interruption were carried out by Ledentsov et al. [3.152]. Systems were grown by MBE. Growth interruption was then introduced, and systems were subsequently overgrown by GaAs. The capped structure was studied by transmission electron microscopy (TEM) and photoluminescence (PL) spectroscopy. Experiments were carried out at the same growth temperature 480°C, with different amounts of deposited material and different growth interruption times.

These studies revealed the following. When the critical average thickness of InAs deposition is reached (1.6–1.7 ML), the morphological transition to 3-dimensional InAs islands occurs. Since the process of dot formation is very fast at this stage, it can hardly be frozen for STM and AFM studies. Therefore, TEM investigation of immediately covered dots is the only reliable way of characterizing dots.

TEM and PL studies of dots formed after 2 ML deposition of InAs have demonstrated that dots are small, do not show well-resolved crystalline shape, and exhibit wide dispersion in size (Fig. 3.93a).

If 4 ML of InAs are deposited, a dense array of well-developed dots is formed. In plan-view, the bases of dots are square-shaped, with sides parallel to the [100] and [010] directions. The average lateral side of the dots is ~ 140 Å (Fig. 3.93c).

The introduction of growth interruption results in dramatic changes in the heterophase morphology with respect to as-grown samples. If 3 ML of InAs are deposited and a growth interruption of 10 s is introduced, the dots reach the same lateral size ~ 140 Å. If 2.5 ML of InAs are deposited and a growth interruption of 40 s is introduced, the dots reach the same lateral size ~ 140 Å.

208 3. Self-Organization Phenomena at Crystal Surfaces

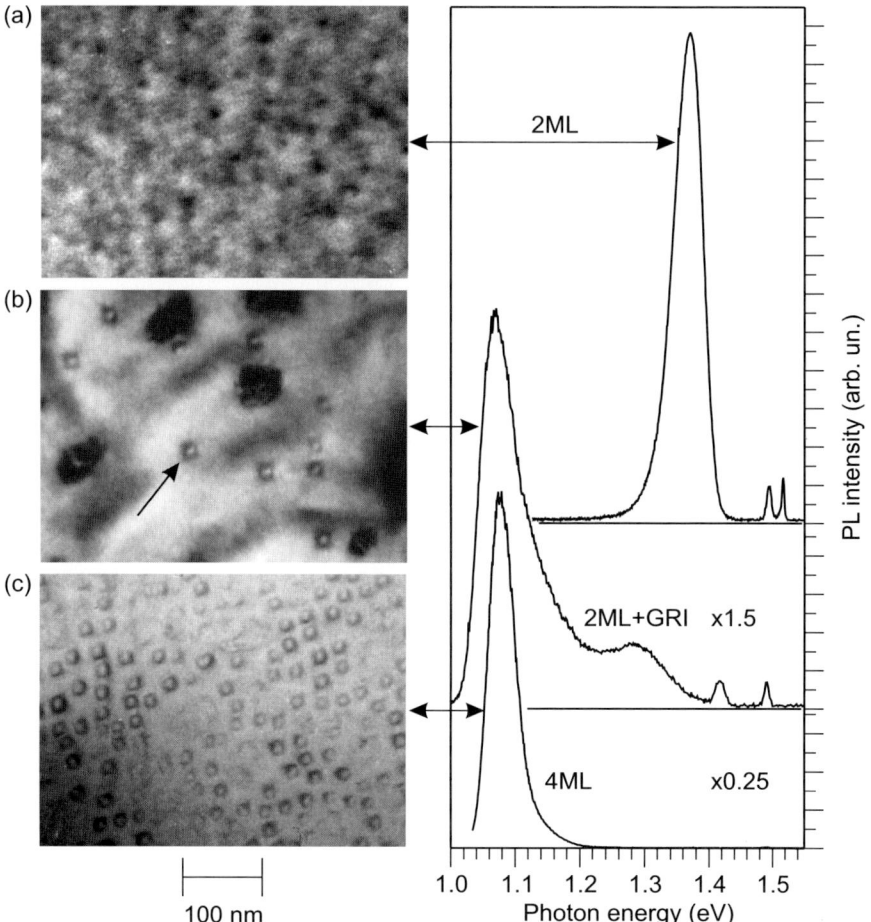

Fig. 3.93. Effect of growth interruption on the morphology of the InAs/GaAs system. Plan-view transmission electron microscopy (TEM) images and photoluminescence (PL) spectra. (**a**) 2 ML of InAs, no growth interruption. (**b**) 2 ML of InAs deposited with submonolayer cycles (0.3 ML InAs) separated by 100 s growth interruptions. (**c**) 4 ML of InAs, no growth interruption

If only 2 ML of InAs are deposited, a very long (100 s) growth interruption forces initially small dots to reach the same lateral size (Fig. 3.93b). When the dots have reached this lateral size, further growth interruption does not result in any change in the shape, size, or density of dots.

The data cited above allow us to conclude that the InAs dot size ~ 140 Å is the equilibrium size, which can be attained by growth interruption. However, one should keep in mind, that the allowed time for growth interruption has its natural upper limit. For growth temperature $\sim 480°$C, a growth interruption much longer than 100 s can lead to evaporation of In, alloying of quantum

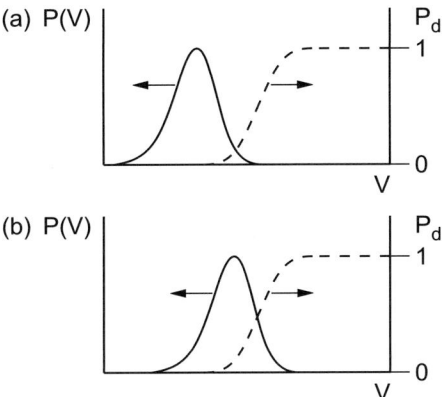

Fig. 3.94. Distribution function of island volumes $P(V)$ (*solid line*) and the probability that an island of given volume contains defects P_d (*dashed line*). (**a**) If the average volume of islands is moderate, a small fraction of islands contains defects. (**b**) With an increase in the average volume of the islands, a larger fraction of islands is defective

dots or the wetting layer with the substrate, or migration of intrinsic defects from the substrate to the surface. Any of these processes make both the energetics and the kinetics in the system much more complex.

Conventional treatment of QD array energetics or the kinetics of dot formation is based on the assumption that there is no evaporation of deposited material and no intermixing with the substrate, and that the properties of the substrate, e.g., its lattice parameter, do not change. If any of these assumptions is not valid, growth interruption does not give an answer concerning the nature of the dot array.

In the TEM image from the sample with 2 ML InAs deposited and 100 s growth interruption, large defective islands are seen along with coherent islands. The following note should be made. Figure 3.94 shows schematically the equilibrium distribution function of island volumes $P(V)$ and the probability $P_d(V)$ that an island with volume V contains dislocations or other defects. At finite temperatures, the equilibrium distribution of finite volumes has finite width. The same is valid for $P_d(V)$, because the probability of defect formation also has a statistical nature. The transition from coherent to incoherent islands is therefore smooth at finite T.

Thus, if the average equilibrium volume of coherent islands is far below the onset of defects, there still exists a small finite probability that a very small fraction of islands will be defective. As the onset of dislocations is an activation process, a certain time is needed before a critical fluctuation occurs. Thus, the probability that defective islands are present in the system increases with time. This explains the observation of defective islands after a long growth interruption (Fig. 3.93b).

Reversible 3D-to-2D Transitions in Heteroepitaxial Systems. For III–V compound semiconductors, the vapor pressure of the group V element is a control parameter which can affect the morphology of the heteroepitaxial system. The dependence of the morphology of InAs/GaAs heteroepitaxial systems on the arsenic pressure has been studied in detail by Ledentsov et al. [3.152]. MBE growth of InAs at temperature $T = 480°C$ and standard MBE arsenic pressure ($P_{As}^0 \approx 2.0 \times 10^{-6}$ torr) results in an array of dots of high density (5×10^{10} cm^{-2}). The dot array is stable and islands do not undergo Ostwald ripening upon growth interruption.

Growth at an arsenic pressure of $(1/6)P_{As}^0$ does not lead to the formation of 3D islands, and only 2-dimensional InAs islands ($\sim 1\,000$ Å) appear. These growth conditions are close to 'virtual surfactant epitaxy' [3.46] and the corresponding reflection high energy electron diffraction (RHEED) pattern is streaky, indicating a planar surface morphology. The photoluminescence (PL) peak shifts towards higher energies than in the case of 3D dots. The broadening of the PL peak corresponds to a highly non-uniform corrugated 2D layer of InAs.

The same state of the system can be obtained in quite a different way. If 3D islands are formed at the standard arsenic pressure P_{As}^0, the As flux is interrupted, and 0.15 ML In are deposited, then 3D islands disappear and the RHEED pattern converts from spotty to streaky within a few seconds. The similarity of the final morphology in the two above-mentioned cases indicates the reversibility of the phase transition between a corrugated 2D layer of InAs and an array of 3D coherently strained islands of InAs.

A similar reversible 3D-to-2D transition was observed independently by Ozasa et al. [3.233] for GaInAs QDs grown by chemical beam epitaxy (CBE). The experiments were carried out on GaAs(001) substrates using triethylgallium (TEGa), trimethylindium (TMIn), and AsH$_3$ as sources. QDs were grown at temperature $480°C$ and the indium content in GaInAs was 0.50. The onset of 3D island formation was detected in situ by the change in the RHEED pattern from streaky to spotty. The density, height, and diameter of QDs were 7×10^{10} cm^{-2}, approximately 13 nm, and 24 ± 3 nm, respectively. These values were obtained by atomic force microscopy (AFM) observations and by high resolution scanning electron microscopy (HRSEM) observations.

After cutting the supply of TEGa and TMIn, the AsH$_3$ flux was switched off, and the PH$_3$ flux was switched on. The spotty RHEED pattern gradually changed into a streaky pattern. This change indicates that the 3D dot structure became a flatter layer of InGaAsP through the replacement of arsenic in the dots by phosphorus. This transition can be attributed to a smaller lattice mismatch in the InGaAsP/GaAs system than in the InGaAs/GaAs system. The RHEED pattern obtained 12 s after switching from AsH$_3$ to PH$_3$ still reveals spots, but a shift was observed in the spots. This shift was explained by the partial replacement of As by P in the dots, and the composition of the dots was In$_{0.52}$Ga$_{0.48}$As$_{0.68}$P$_{0.32}$. The shifted spots became thinner after

12 s, and finally fused into streaks. The shift of spots/streaks stopped after around 30 s, but the diffusion of spots into streaks proceeded more slowly thereafter. The observed RHEED changes revealed that the dot structure became a flatter surface upon phosphorus exposure via two steps:

- first, some arsenic in the dots was replaced by phosphorus with preservation of dot shapes,
- then the dots started to flatten with further arsenic/phosphorus replacement.

The RHEED pattern obtained 85 s after switching from AsH_3 to PH_3 was similar to that of the InGaAs layer observed before the onset of dot formation. The composition of $In_{0.52}Ga_{0.48}As_{0.23}P_{0.77}$ was estimated for the resulting 2-dimensional layer.

When the AsH_3 beam was reapplied instead of PH_3, a reverse transition took place from the InGaAsP flat surface to the InGaAs dot structure. The RHEED pattern at 120 s was identical to that obtained for the initial 3D dot formation with InGaAs deposition. Photoluminescence studies of capped reproduced dots, on the one hand, and of capped as-grown dots, on the other hand, indicated a minor difference in peak positions which can be attributed to some residual phosphorus in the dots.

The reversible 3D-to-2D transition is driven by the reduction in lattice mismatch when arsenic is replaced by phosphorus, in agreement with the equilibrium phase diagram of Fig. 3.87. These data indicate that the morphology of the heteroepitaxial system is determined by the vapor pressure of arsenic and phosphorus. The reversibility of such a transition clearly indicates the equilibrium nature of the quantum dot arrays under consideration.

Transition from an Ordered Array of InAs Quantum Dots to Ostwald Ripening. An increase in arsenic pressure affects the morphology of the InAs/GaAs system in a different way. The effect of arsenic pressure on MBE-grown InAs QDs has been studied in detail by Ledentsov et al. [3.152]. As mentioned earlier in the present section, the growth of InAs at temperature $T = 480°C$ and standard MBE arsenic pressure ($P_{As}^0 \approx 2.0 \times 10^{-6}$ torr) results in an array of dots of high density (5×10^{10} cm^{-2}) (see Fig. 3.95). No qualitative change occurs in the morphology if the arsenic pressure is changed by $\approx 50\%$ around P_{As}^0. An increase in As pressure by a factor of 3 ($3P_{As}^0$) leads to a dramatic change in morphology. The size of dots reduces and a high density of large (~ 500–$1\,000$ Å) dislocated InAs islands appears (see also [3.177]). A further increase in arsenic pressure ($5P_{As}^0$) almost completely suppresses the formation of small dots, and only dislocated InAs islands can be resolved on the InAs wetting layer. PL emission is dominated by the wetting layer peak at low temperatures and no emission from dots can be resolved at 300 K. The integral intensity of PL strongly degrades, in agreement with the formation of large dislocated islands.

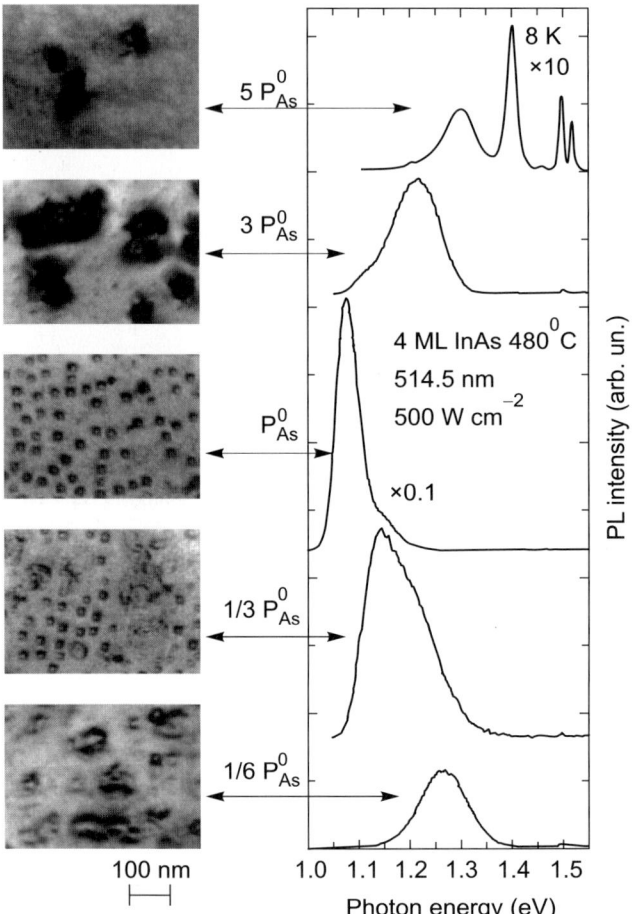

Fig. 3.95. Effect of As pressure on InAs/GaAs quantum dot arrays. Plan-view transmission electron microscopy (TEM) images and photoluminescence (PL) spectra are given for 5 values of the arsenic pressure, with $P_{As}^0 = 2 \times 10^{-6}$ torr. The dense array of 3D coherent dots existing for $P = P_{As}^0$ undergoes a reversible transition to a planar morphology when the pressure is reduced to $P = (1/6)P_{As}^0$. An increase in As pressure results in Ostwald ripening, the formation of dislocated islands at $P = 3P_{As}^0$, and complete disappearance of coherent dots at $P = 5P_{As}^0$

We emphasize here that the kinetics of surface migration is very fast in the above experiments. At the arsenic pressure P_{As}^0, 3D coherent islands of InAs have base length 140 Å and typical separation 250–350 Å. At high As pressure, dislocated islands of InAs are separated by 0.2–1 µm. This means that migration kinetics is sufficient to rearrange InAs and deliver material over a distance of 0.2–1 µm.

Such fast kinetics confirms that the array of 3D coherent InAs islands of approximately the same size is not formed as a result of 'frozen' kinetics, but

3.3 Arrays of Three-Dimensional Coherently Strained Islands

rather due to the fact that this array corresponds to a thermodynamically preferred state.

The change in morphology of the InAs/GaAs system upon an increase in arsenic pressure can be explained by considering surface energies. The stoichiometry of (001) surfaces of III–V semiconductors which are in equilibrium with a vapor is known to depend on the pressure of the group V element in the vapor. Thus, a change in As pressure leads to a change in the surface energy of the GaAs(001) surface and even to a change in the surface reconstruction [3.215, 3.216]. The wetting layer of InAs on a GaAs(001) surface may be considered as a complex semiconductor surface which is expected to exhibit similar behavior with changing arsenic pressure. An increase in the arsenic pressure leads to a decrease in the surface energy of GaAs(001) [3.215, 3.216] or InAs(001) [3.174]. It is natural to assume that the same tendency will occur for the InAs wetting layer on GaAs(001). A decrease in the surface energy of the (001) surface will lead to an increase in the control parameter α in (3.151) and favor transition from the ordering regime to the ripening regime. This is likely to be the case when the arsenic pressure increases from P_{As}^0 to $3P_{As}^0$.

The effect of arsenic pressure on the formation of InAs quantum dots is even more pronounced for QDs grown by metalorganic chemical vapor deposition (MOCVD). It is important to compare QD growth by MOCVD and by MBE, above all because it allows us to extract and distinguish the effects of both energetics and kinetics on the formation of quantum dots.

The 'standard' MOCVD procedure for growing arsenides of group III elements requires the AsH_3 flux to be switched on during the whole epitaxy process, i.e., during growth and growth interruptions. This 'standard' procedure is mandatory if we are to guarantee group V stabilization of the crystal surface at elevated temperatures and prevent formation of group III droplets. However, an absence of AsH_3 for just a few seconds is not critical. Heinrichsdorff et al. [3.234] found quite the opposite: for the growth of InAs/GaAs QDs, the presence of AsH_3 is critical and degrades the structure quality.

A set of experiments was carried out in which a growth interruption $t_{GI}=14$ s was introduced after deposition of 1.65 ML of InAs. Over the same $t_{GI}=14$ s, the AsH_3 flux was switched off during a first time interval t_{GI}^{off}, and switched on during the rest $t_{GI} - t_{GI}^{off}$. If AsH_3 is switched off for 12 s, a high-quality array of QDs is formed. If AsH_3 is switched off for only 3 s, the photoluminescence peak decreases by a factor of ≈ 2. The presence of AsH_3 ($P = 0.7$ mbar) during the entire GRI reduces the QD luminescence intensity by more than two orders of magnitude. Dislocated islands with low area density are formed, giving rise to macroscopic surface roughness. To sum up, during growth interruption, the AsH_3 pressure is the key parameter for the growth of defect-free InAs QDs in MOCVD, and this for two reasons:

- QD nucleation seems to be hindered by AsH_3. Fewer and larger objects are formed in the presence of AsH_3.

- Even after nucleation, exposure to AsH$_3$ leads to material redistribution and the formation of larger dots and dislocated islands.

The simplest possible explanation for the preferred formation of dislocated islands under AsH$_3$ pressure is based on the fact that typical MOCVD growth conditions can be regarded as As-rich compared to the MBE case. For As-rich conditions in MBE, the QD density reduces, and dislocated islands are formed (see above).

A major difference between MOCVD and MBE growth concerns the reactants used for epitaxy. In particular, atomic hydrogen stemming from AsH$_3$ decomposition in MOCVD is expected to affect both the energetics and kinetics of QD formation, since hydrogen radicals are known to interact with the surface by breaking bonds, due to their high reactivity, whereby they become incorporated into the bulk crystal [3.235]. It is nevertheless clear that switching off the AsH$_3$ flux during growth interruption makes QD nucleation and development more 'MBE-like', since it reduces both the As and the H pressure.

The effect of AsH$_3$ pressure on the formation of QDs during growth is similar to the effect of AsH$_3$ pressure during growth interruption. High AsH$_3$ pressure results in a redshift of the PL spectrum and a decrease in the integral PL intensity. This corresponds to an increase in QD size, broadening of the QD size distribution, and the onset of misfit dislocations. This indicates that an increase in the AsH$_3$ pressure promotes Ostwald ripening.

Consequently, the effect of arsenic pressure on the formation of InAs QDs on a GaAs(001) substrate is similar for both MBE and MOCVD growth. Low arsenic pressure results in the formation of a dense array of QDs, whereas high arsenic pressure leads to Ostwald ripening and the formation of dislocated islands. These results indicate that the morphology of the InAs/GaAs system depends on energetics rather than kinetics, which is quite different for MBE and MOCVD growth.

3.3.14 Temperature Ramping and Cooling in InAs/GaAs Systems: Evidence of Close-to-Equilibrium Behavior

It was demonstrated above that the behavior of an array of strained islands with respect to variations in the substrate temperature can be used as an experimental tool for distinguishing thermodynamically controlled arrays of islands from kinetically controlled arrays. So far, the detailed theory has been developed for islands with fixed height and variable lateral dimensions, e.g., submonolayer islands. It was shown in Sect. 3.2 that the average island volume decreases if the substrate temperature increases for quasi-equilibrium arrays of islands. In contrast, when we consider kinetically controlled arrays, the average island volume increases when the temperature increases. We argue that this statement is quite general and can be extended qualitatively to arrays of 3D strained islands.

The behavior of a system with respect to reversible changes in thermodynamic parameters, such as vapor pressure or substrate temperature, may be viewed as the ultimate criterion for the physical nature of a system. The structure of an equilibrium system, e.g., a heteroepitaxial system, depends only on thermodynamic parameters, such as the amount of deposited material, substrate temperature, vapor pressure, etc. It should not depend on the prehistory of the system, and the structure should be restored after reversible changes in thermodynamic parameters. On the other hand, the structure of a kinetically controlled system can depend on the prehistory and should not generally be restored after reversible changes in thermodynamic parameters.

Adatom Condensation upon Substrate Cooling. The importance of adatoms in arrays of strained islands at elevated temperatures has been emphasized by Shchukin et al. [3.157] for the particular case of submonolayer islands. In a heteroepitaxial system, islands and adatoms coexist, the adatom density increases with temperature, and at typical growth temperatures, a substantial fraction of the deposited material is present in the form of adatoms.

The role of adatoms was independently emphasized by Leon et al. [3.165], who considered the relationship between the size of capped and uncapped InGaAs/GaAs quantum dots. InGaAs/GaAs QDs were grown by metalorganic chemical vapor deposition (MOCVD). $(CH_3)_3Ga$, $(CH_3)_3In$, and AsH_3 were used as precursors. Nanometer-sized InGaAs islands were grown by depositing 5 ML with nominal composition $In_{0.6}Ga_{0.4}As$. For a set of capped samples, the GaAs capping layer thickness was 30 nm. After QD formation, the flow of group III sources was interrupted, whilst the arsine flow was maintained and the structures cooled to below 400°C. All precursor gases were removed as the samples cooled from 400°C to room temperature. Island sizes, shapes, and densities were measured using atomic force microscopy (AFM) and plan-view transmission electron microscopy (TEM). Comparing the two sets of images, it was observed that:

- uncapped dots are larger than capped ones,
- this size discrepancy increases with decreasing QD surface density.

Two different types of growth were used, one with constant growth rate, and another with a graded growth rate.

Leon et al. [3.165] reported differences between capped and uncapped islands observed from different growth experiments. Surface island size estimates also agree across both types of structural technique. Therefore, one can believe these differences to be real, despite potential errors from technique-related artifacts. Uncapped surface islands, mainly for low density, are significantly larger than their capped counterparts. Good agreement was observed between values of the density determined from capped and uncapped islands [3.165]. The latter observation eliminates Ostwald ripening during sample cooling as a major contributor to this size discrepancy.

Leon et al. [3.165] suggested a mechanism to explain the effect: "The island size difference can be explained by group III thermal adatom condensation on existing islands during sample cooling." Islands act as 'sinks' for condensing adatoms, which migrate from the surface area surrounding each island. Since cooling is slow, adatom diffusion still permits migration to existing islands for small and intermediate inter-island separations. For widely spaced islands, migration lengths become a limitation, and this may explain the saturation in surface island dimensions.

Impact of Growth Temperature on InAs/GaAs Systems. However, comparing lateral sizes of islands in different structures (capped and uncapped), measured using two different techniques, does not yield sufficiently accurate information on island shape. Moreover, the reaction of the system to cyclic temperature changes was not addressed in [3.165]. A detailed understanding of the temperature behavior in InAs/GaAs systems was achieved by Shchukin et al. [3.163] and Ledentsov et al. [3.164].

With a view to settling the long dispute over the relative role of thermodynamic and kinetic effects in InAs/GaAs(001) heteroepitaxial systems, the influence of substrate temperature, growth interruptions, and tuning of substrate temperature after InAs deposition on the resulting shape, volume, and density of the islands was studied in detail in [3.163, 3.164].

Samples were grown on semi-insulating GaAs substrates using conventional molecular beam epitaxy (MBE). The sample geometry, growth rates and arsenic pressure were similar to those used by Ledentsov et al. [3.152]. Island formation starts at about 1.7 ML InAs on GaAs(001). Islands formed by 3 ML InAs deposition have sizes close to the limiting size possible for the particular substrate temperature chosen for InAs deposition. A 10 s growth interruption (GI) was introduced after InAs deposition to ensure that this limiting size was completely reached. InAs islands were overgrown by 10 nm of GaAs at the same substrate temperature. For some of the samples, the substrate temperature was tuned after InAs deposition and InAs islands were covered by GaAs at the final temperature. Transmission electron microscopy (TEM) measurements were performed using a Philips EM 420 microscope with acceleration voltage 100 kV. Photoluminescence (PL) was excited by the 514.5 nm line of an Ar^+ laser and detected at room temperature. The excitation density was ≈ 200 W cm^{-2}.

To begin with, we consider the effect of temperature on 3D island shape, volume and density. Figure 3.96 shows the dependence of the localization energy of the exciton in a quantum dot with respect to the GaAs bandgap energy (a), average lateral size (b) and density (c) of the islands revealed in plan-view TEM images (see also Figs. 3.97a and b). As can be seen from Fig. 3.96 and Figs. 3.97a, b, d and e, the lateral size and height of the islands exhibits the opposite behavior, i.e., an increase in the substrate temperature leads to an increase in lateral size and a decrease in island height. We note that the lateral size and height of the islands revealed in TEM images are

3.3 Arrays of Three-Dimensional Coherently Strained Islands   217

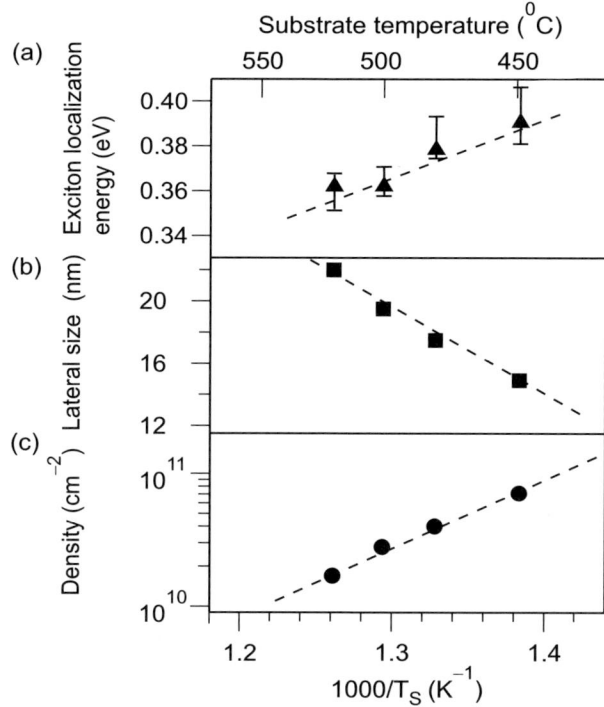

Fig. 3.96. Effect of formation temperature on parameters of an array of 3D coherently strained islands of InAs on GaAs(001). (**a**) Position of the photoluminescence peak from quantum dots. (**b**) Average lateral size of islands. (**c**) Island density

very sensitive to imaging conditions and the TEM foil thickness. Thus, the exact numbers and shapes given in Figs. 3.96 and 3.97 serve only to indicate a general trend for comparable imaging conditions.

In order to independently evaluate the dependence of the volume on temperature, we focus on the localization energy of electron–hole pairs confined by QDs. Figure 3.96a shows a decrease in localization energy with increasing temperature, which generally corresponds to a decrease in the QD volume. However, the opposite changes in the lateral size and height of the QDs mean a drastic change in the QD shape. The effect of the QD shape on the optical transition energy was calculated in [3.163]. It was shown that, for an InAs QD of fixed volume in a GaAs matrix, a flattening of the QD shape leads to a redshift of about 30 meV (Fig. 3.98). Hence, the observed behavior of the PL peak in Fig. 3.96a can only be explained by a decrease in the QD volume with increasing temperature, where the blueshift due to the volume decrease overcomes the redshift due to flattening of the islands. We note that the change in the height-to-lateral size ratio for islands with the volume change agrees with the predicted behavior of strained islands (flatter

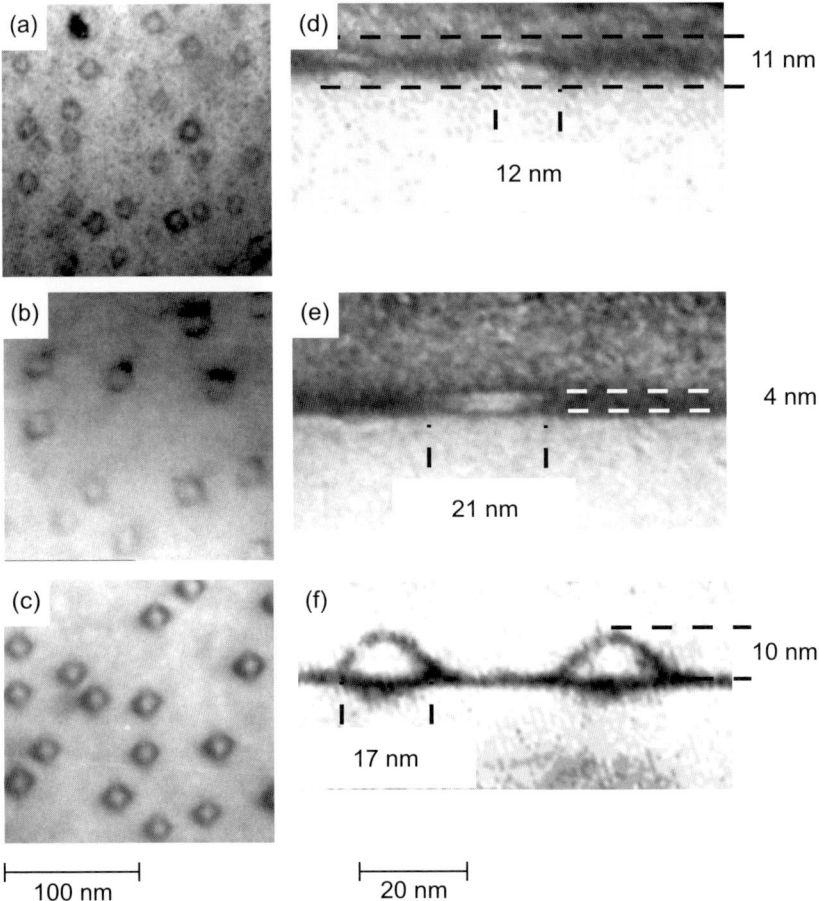

Fig. 3.97. Effects of temperature and temperature ramping to lower values on the lateral size and height of InAs/GaAs(001) islands. (**a**)–(**c**) Plan-view symmetrical [001] zone-axis TEM images of arrays of InAs/GaAs islands. (**a**) Islands grown at 450°C, growth interruption (GI) 10 s. The density of the islands is 7.1×10^{10} cm^{-2} and the average lateral size 14.9 ± 0.5 nm. (**b**) Islands grown at 500°C, growth interruption 10 s. The island density is 2.8×10^{10} cm^{-2} and the average lateral size 19.5 ± 0.5 nm. (**c**) Islands grown at 500°C and cooled down to 450°C over 120 s. The island density is 4.5×10^{10} cm^{-2} and the average lateral size 17.5 ± 0.5 nm. (**d**)–(**f**) Cross-sectional (010) dark-field TEM images with $g = 002$ of the arrays of InAs/GaAs islands. Growth conditions are as in (**a**), (**b**), and (**c**), respectively

islands at smaller volumes) (see Sect. 3.3.11). We also note that, at lower temperatures, the actual volume of size-limited island growth (SLIG) is not yet completely reached, since longer growth interruption times are required to reach equilibrium, due to the slower kinetics of exchange reactions on the surface. Furthermore, going to substrate temperatures of about 420°C and less also results in the local formation of much larger nano-islands (up to 30–

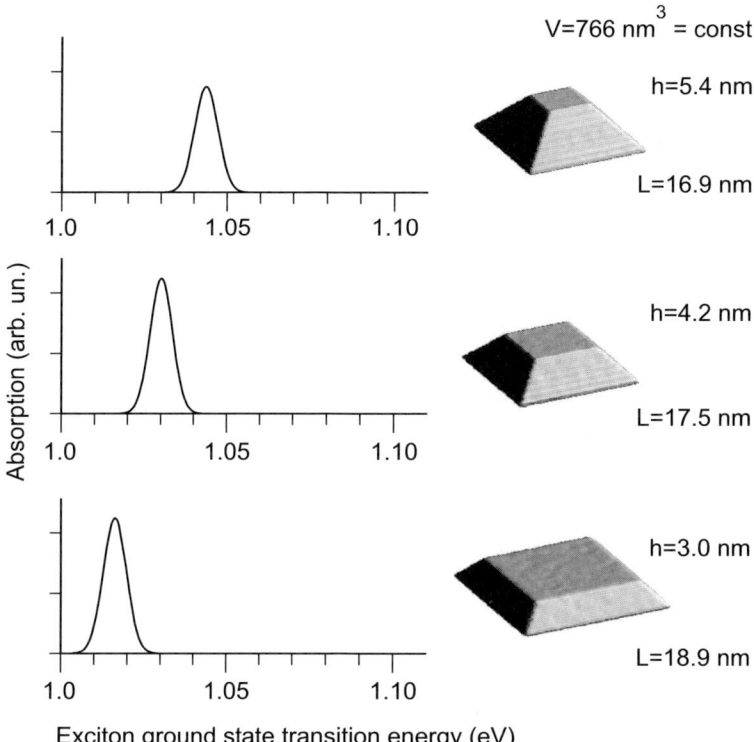

Fig. 3.98. Calculated linear absorption spectra of an InAs/GaAs quantum dot with volume 766 nm^3 and the shape of a truncated pyramid at different levels of truncation

100 nm in lateral size), with complex internal structure. An increase in island volume with temperature decrease clearly disagrees with the kinetic models of SLIG discussed in Sect. 3.3.12. The kinetic models predict an increase in island volume with temperature due to the increased adatom diffusion coefficient and weaker influence of diffusion barriers at higher temperatures. On the other hand, the thermodynamic model of the equilibrium distribution of island volumes relates only to the equilibrium properties of the surface and the bulk.

In Sect. 3.2, it was demonstrated that entropy effects involved in island formation favor smaller island volumes at higher temperatures. This particular model was developed for islands with 2D-like shape, fixed height, and variable lateral dimensions. The model implies that the characteristic temperature at which entropy effects dominate the temperature dependence is of the order of several hundred kelvin. An observed decrease in the average island volume with temperature emphasizes entropy effects and thus strongly supports the thermodynamic picture of island formation. Although the actual system of 3D islands with variable shape and height requires a much more complicated theoretical treatment than the one given in Sect. 3.2, general trends remain the same, since entropy effects in equilibrium arrays of islands favor islands of smaller volumes at higher temperatures.

A decrease in island volume occurring simultaneously with a decrease in island density also indicates that an excess of indium atoms is trapped by the surface of the wetting layer in the form of adatoms. After overgrowth, these adatoms contribute to the effective thickness of the wetting layer. The effects of indium segregation during the overgrowth of In-containing nanostructures on the broadening of the compositional profile were estimated for InGaAs/GaAs quantum well structures by Muraki et al. [3.236] and for InGaAs/GaAs QDs by Woggon et al. [3.237] and Rosenauer et al. [3.238]. These data revealed that the extension of the In profile in samples grown at 500°C only exceeds that in samples grown at 450°C by about 1 nm. Such a change can only weakly affect the PL spectrum of the QDs and is at the resolution limit of the above-mentioned TEM studies. We do not therefore expect the capping procedure at temperatures at and below 500°C to strongly affect the volume and shape of QDs for the experimental conditions discussed in the present section.

Impact of Growth Interruption. Growth interruption is a widely used experimental tool allowing a system to approach equilibrium. Ledentsov et al. [3.164] focused particularly on the behavior of the InAs/GaAs system under very long growth interruptions and temperature ramping. A growth interruption of 10 s leads to an array of coherent islands formed in size-limited island growth (SLIG) mode. A further increase in the growth interruption time to ≈ 120 s at the same temperature does not lead to a change in island size and density, but causes local formation of random large islands, which become dislocated and attract material from coherent islands in their vicinity. As the growth interruption time increases and with it the density of these dislocated islands, the density of coherent islands decreases, and the size and shape are also weakly affected (see also [3.152]).

The formation of large dislocated islands is traditionally attributed to Ostwald ripening. However, since no noticeable density of 'intermediate' sized islands was found, an alternative explanation can be sought within the framework of thermodynamics. An equilibrium distribution function of island volumes at finite temperatures contains a tail extended towards large volumes. Very large islands should appear locally, although with very low probability. If an island exceeds a certain critical volume, it can become dislocated and create a strong trap for indium atoms. Since the equilibrium distribution of island volumes refers only to coherent islands, its validity is limited by the appearance of even a small density of dislocated islands. Then coherent islands will gradually vanish.

Impact of Temperature Ramping on QD Shape, Volume and Defect Density. Let us now consider the effect of ramping the substrate tempera-

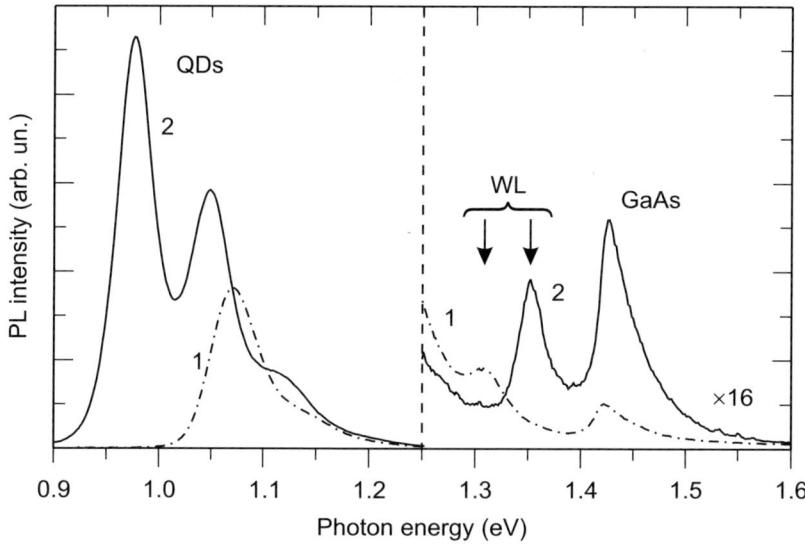

Fig. 3.99. Photoluminescence (PL) spectra (at 300 K) of an array of InAs/GaAs islands formed and capped at 500°C (curve 1) and InAs islands formed at 500°C and cooled down to 450°C, before capping at the final temperature (curve 2)

ture on density, volume, shape and PL spectra of QD arrays. Comparing the array of islands after 120 s cooling (Figs. 3.97b and e) with the array formed and capped at 500°C (Figs. 3.97c and f), we observe that cooling causes a decrease in the lateral size of islands. Hence, the lateral size of islands after cooling is intermediate between its value for the array deposited and capped at 450°C (Figs. 3.97a and d), and the value for the array deposited and capped at 500°C (Fig. 3.97b and e). The same goes for the island density and height, which are significantly increased after cooling (Figs. 3.97c and f) from 2.8×10^{10} cm^{-2} to 4.5×10^{10} cm^{-2}. An even more dramatic increase in QD density was manifested for fast cooling of QDs formed at 520°C (from 1.7×10^{10} cm^{-2} to 5.0×10^{10} cm^{-2}) accompanied by a strong reduction in lateral size and an increase in height. This indicates a partial reversibility of changes in arrays of InAs strained islands under temperature variations.

Figure 3.99 shows the photoluminescence (PL) spectra of the two arrays of islands. Curve 1 refers to the array formed at 500°C, subjected to 10 s growth interruption, and capped by GaAs at the same temperature. Curve 2 shows the PL spectrum of the array formed at 500°C, cooled down to 450°C over 120 s, and capped by GaAs at 450°C. The comparison shows two effects. Firstly, cooling leads to a redshift of the QD PL peak to 1 270 nm (with only 30% drop in intensity at 1 300 nm) indicating an increase in the QD volume. This again shows a reversibility of the volume change upon change in temperature. Secondly, cooling leads to a blueshift of the PL peak from the wetting layer. This is related to a strong reduction in adatom density as

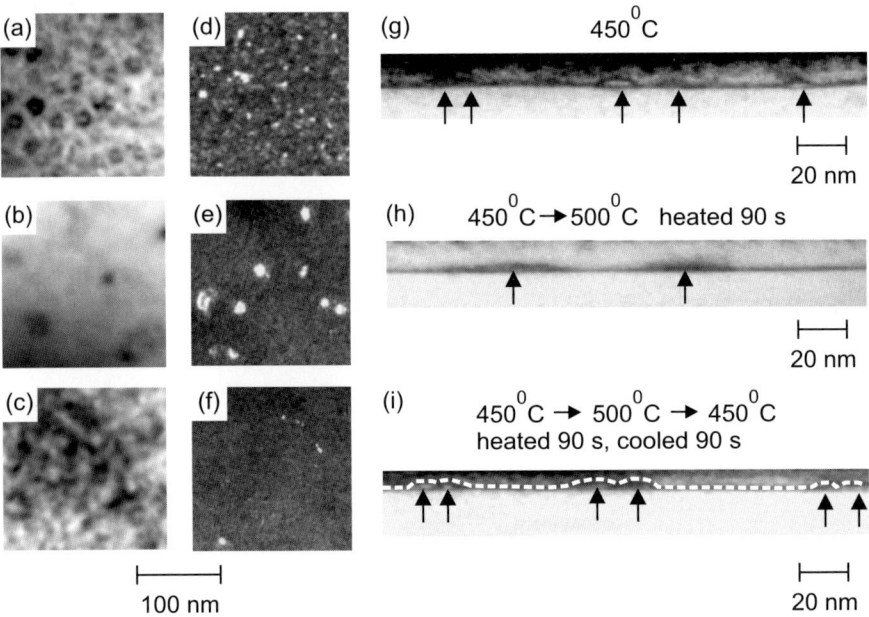

Fig. 3.100. Effects of temperature and temperature ramping to higher values on the lateral size and height of InAs/GaAs(001) islands and the size and density of defects. (**a**)–(**c**) Plan-view symmetrical [001] zone-axis TEM images of the arrays of InAs/GaAs islands. (**d**)–(**f**) Weak-beam (g, $2g$) (001) plan-view image with g=(220). (**g**)–(**i**) Cross-sectional (010) dark-field TEM images with g=(002) of the arrays of InAs/GaAs islands. Images (**d**)–(**f**) were taken under defect-sensitive conditions

adatoms condense, increasing the volume of existing 3D islands and forming new islands. This shows that the effect of a higher overgrowth temperature leading to an asymmetric quantum well profile in the wetting layer (WL), which should blueshift the WL emission for the same amount of material deposited, is a much weaker effect than the thermodynamically-driven growth phenomenon, even for the case of ultrathin insertions. Reversibility is partial because it is difficult for the system to create or eliminate enough islands in a limited time. The improved PL efficiency in the cooled sample is likely to be caused by the decomposition of dislocated islands and will be explained below.

We now consider the impact of heating on the size, shape and density of QDs. Figures 3.100a, d, and g are TEM images of dots formed at 450°C by 3 ML InAs deposition and immediately covered by GaAs. The average lateral size of the QDs is comparable to that in Fig. 3.97a, while the height is slightly smaller (7 nm) and the density is higher (10^{11} cm^{-2}) due to the absence of the 10 s growth interruption phase in the latter case. The plan-view TEM image presented in Fig. 3.100d was taken under defect-sensitive conditions and shows that about 0.5×10^{11} cm^{-2} islands contain defects (small dislocation loops). Figures 3.100b, e, and h show TEM images of the sample with QDs formed in the same mode as in the previous case, but with a subsequent increase in substrate temperature to 500°C within 90 s, followed by deposition of a GaAs cap layer. It is clear from the TEM images, that small InAs islands disappear and large flat islands with height 3–4 nm and lateral size 30–70 nm are formed in this case. TEM images taken under defect-sensitive conditions show that the large flat islands contain defects (dislocation loops) placed at their boundaries.

While the PL spectrum of the sample with QDs formed and covered at 450°C is dominated by QD emission, the PL spectrum of the sample with heated QDs recorded at room temperature (RT) exhibits very weak integrated PL intensity, dominated by the wetting layer luminescence. This is in agreement with a strongly reduced concentration of small coherent islands and a high concentration of defects, acting as efficient centers of non-radiative recombination. Figures 3.100c, f, and i show TEM images of the sample with QDs formed at 450°C, ramped to 500°C within 90 s, cooled to 450°C within 90 s, and capped by a GaAs layer at the final temperature. Both plan-view (Fig. 3.100c) and cross-sectional (Fig. 3.100i) TEM images demonstrate the reappearance of small islands, together with locally formed agglomerates of small islands. The RT PL spectrum of this sample is dominated by the quantum dot emission, which is about 4 times brighter than the QD emission in the sample with QDs formed and capped at 450°C. The reason for this effect is clear from Fig. 3.100f, taken in defect-sensitive conditions. It can be seen from the image that the concentration of dislocation loops is much smaller in this sample, and the average size of these defects is also smaller. Thus, during the segmentation of large dislocated islands into small coherent nanodomains, most of the defects annihilate. The reappearance of small islands after cooling the sample, ramped to higher temperature, shows the reversibility of the QD size under reversible tuning of the substrate temperature.

A similar reversible phenomenon was described in Sect. 3.3.13. Studies of the effect of arsenic pressure on the InAs/GaAs system showed the reversibility of the transition between the 2D InAs layer and SK InAs islands, with periodic appearance (disappearance) of InAs islands upon periodic increase (decrease) in the arsenic pressure, respectively. The observed cleaning of the QD structure from defects via temperature cycling opens up new possibilities for improving QD device characteristics.

The present discussion has been concerned with the effects of formation temperature and temperature ramping on the volume, shape, density and optical spectra of 3D InAs islands on a GaAs(001) substrate. Let us conclude with the following note. Plan-view and cross-sectional TEM and PL spectroscopy indicate a decrease in island volume with an increase in formation temperature, accompanied by a flattening of the island shape. By cooling and heating the array of InAs islands after deposition and before capping, we can show that the parameters of the array of islands are mainly determined by the final temperature, and demonstrate reversible changes upon temperature variations. Our observations reveal the importance of thermodynamics in the formation of arrays of 3D strained islands of InAs on GaAs(001), obtained in the size-limited island growth (SLIG) regime.

3.3.15 Formation of InAs/GaAs Islands at Ultra-Low Temperatures

InAs islands also form on the GaAs surface at temperatures far below 400°C. Maximov et al. [3.239] carried out structural and optical studies of InAs/GaAs systems grown by molecular beam epitaxy (MBE) at 325–350°C. Figure 3.101a shows a plan-view transmission electron microscopy (TEM) image of a structure where 4 ML of InAs were deposited at 350°C. The image reveals several distinct features. Besides individual quantum dots, other objects with larger size and complex shape are brought out. These objects may be called lateral associations of quantum dots (LAQDs). The HRTEM images of Fig. 3.101b–d show distinct objects with a larger magnification. Each LAQD consists of a number of well-defined segments with lateral size 6–7 nm. The density of LAQDs is about 2×10^{10} cm^{-2}. Two different types of LAQD are visible in Fig. 3.101c and d. In the first case, segments forming an LAQD are attached along one $\langle 110 \rangle$ direction, so that the LAQD has a chain-like shape. In the second case, the segments form a rectangular array aligned along the [110] and [$\bar{1}$10] directions.

The formation of LAQDs manifests itself by a pronounced change in the PL spectra. The temperature dependence of the PL spectra for samples with 4 ML InAs deposited at 325°C and 350°C is shown in Figs. 3.102a and b, respectively. There are two PL lines in the spectra recorded at 10 K. The short wavelength line at about 1.05 μm is characteristic for samples with QDs formed by deposition of 4 ML InAs at 480–520°C, where LAQD formation is not observed. However, samples grown at 325°C and 350°C have an additional long wavelength PL line at 1.5–1.7 μm. This emission is not observed for samples in which LAQDs do not form, e.g., in the case where the average thickness of InAs deposited at low substrate temperatures is below the value needed for island formation, as controlled by reflection high energy electron diffraction (RHEED) studies. We therefore attribute this line to PL from LAQDs.

As both coherent and dislocated islands were present among LAQDs in the HRTEM studies, it was not clear from the beginning which islands determine photoluminescence in the region around 1.7 μm. Later studies using a special defect reduction technique (see Sect. 4.1) showed that the photoluminescence at 1.7 μm persists when defects are eliminated, and even becomes much stronger. This confirms that coherent LAQDs are the source of long wavelength PL.

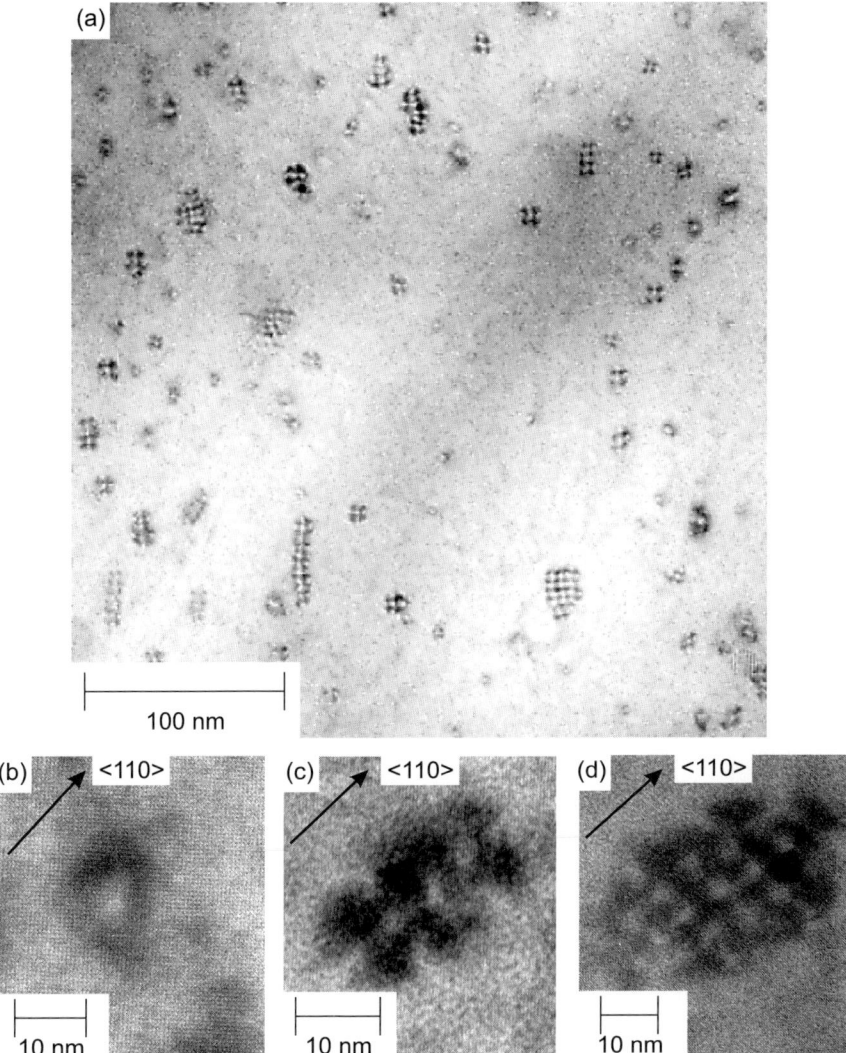

Fig. 3.101. InAs/GaAs islands formed at ultra-low temperatures. (**a**) Bright-field plan-view TEM image of InAs/GaAs islands formed by deposition of 4 ML of InAs at 350°C. (**b**) High resolution transmission electron microscopy (HRTEM) image of an individual InAs island (quantum dot). (**c**) HRTEM image of a lateral association of quantum dots (LAQD) with chain-like shape. (**d**) HRTEM image of an LAQD forming a rectangular array

The substrate temperature of 320–340°C seems to be optimum for LAQD formation. InAs deposition at substrate temperature 350°C results in an LAQD density decrease down to 1×10^{10} cm^{-2}, together with an increase in the density of individual QDs. PL spectra from the two samples (Fig. 3.102)

226   3. Self-Organization Phenomena at Crystal Surfaces

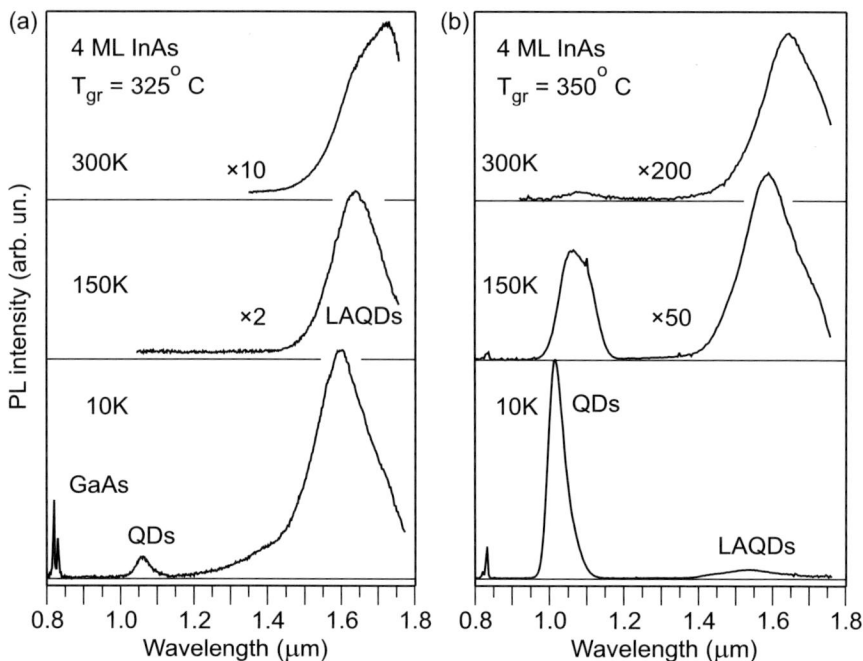

Fig. 3.102. Photoluminescence (PL) spectra recorded at different temperatures for samples with 4 ML InAs deposited at 325°C (**a**) and 350°C (**b**)

confirm this trend, showing a weaker PL in the 1.7 μm region from a sample grown at 350°C. On the other hand, a decrease in the substrate temperature down to 300°C suppresses the formation of any kind of QD, presumably due to kinetic limitations.

3.3.16 3D Islands in Other Material Systems

3D Coherent Islands in Ge(Si)/Si Systems. Among other material systems, Ge and SiGe islands on Si(001) are particularly interesting, given the prominence of Si technology for electronic device applications. The formation of Ge and SiGe islands upon deposition on Si(001) has been found in a number of works (e.g., [3.161, 3.166, 3.167, 3.183, 3.240], and for a review [3.7]). However, only a few works have attempted to distinguish thermodynamically dominated arrays of islands from kinetically controlled arrays.

Chen et al. [3.161] deposited $Si_{0.5}Ge_{0.5}$ on a Si(001) substrate at the low temperature of 400°C. Post-deposition annealing at 590°C for 6 minutes resulted in an array of islands bounded by {105} facet planes and exhibiting a narrow size distribution.

When the sample was annealed at a much higher temperature ($\geq 650°C$) for 10 minutes, a much broader distribution of island sizes was observed.

3.3 Arrays of Three-Dimensional Coherently Strained Islands

This indicates that islands were undergoing Ostwald ripening. Based on these observations, Chen et al. [3.161] concluded that the formation of an array of islands with a narrow size distribution at 590°C is kinetically dominated, rather than thermodynamically dominated. A model of self-limited growth kinetics was proposed [3.161], in which the evolution of the island towards Ostwald ripening is stopped due to strain-induced barriers associated with the nucleation of each new facet layer. (For a detailed discussion, see Sect. 3.3.12.) Since it is easier to overcome the barrier at a higher temperature, the average island size increases with temperature, and Ostwald ripening will occur only at sufficiently high temperatures.

In principle, the observations of [3.161] agree with the model of self-limiting growth kinetics. However, this does not give an unambiguous proof of the kinetic nature of the array of 3D islands. The equilibrium state of a heteroepitaxial system depends on temperature via the entropy contribution to the free energy. This entropy contribution should account for the finite area concentration of adatoms, as well as fluctuations in island shape, size, and relative arrangement. The entropy for an array of 3D islands has not yet been considered in the literature. However, in view of such effects, it is difficult to distinguish thermodynamically dominated arrays of islands from kinetically controlled ones if islands form at one temperature and further annealing is performed at another. To obtain an unambiguous answer, one should arrange for the island formation and further annealing (or growth interruption) to be carried out under identical conditions, in particular, at the same temperature.

Just such a careful study has been carried out by Kamins et al. [3.240] for Ge islands grown on Si by chemical vapor deposition. Layers deposited with 6 ML of Ge at approximately 580°C exhibited a narrow size distribution of Ge islands. The layers were annealed for times up to 20 min without cooling or exposing the wafers to air between deposition and annealing. The distribution of islands becomes less uniform as annealing continues, indicating some transfer of Ge from small to large islands (Ostwald ripening). This indicates that the array of Ge islands, when it has a narrow size distribution, is a kinetically controlled array rather than an equilibrium one.

In the next paper, Kamins et al. [3.241] revealed a reversible transition between the two shapes of Ge islands on Si(001), one being a square-based pyramid bounded by {105} facets, and the other an octagonal-based 'dome'. Ge islands are pyramids when they are small and transform to domes when they exceed a certain critical volume. Annealing at 650°C results in some alloying of Ge islands with the Si substrate, the lattice mismatch reduces, and the island shape can change from a dome back to a pyramid, even though the island size increases substantially. However, this result does not mean real reversibility, since the overall morphology of the heteroepitaxial system, including the island size, density, and chemical composition, is not restored.

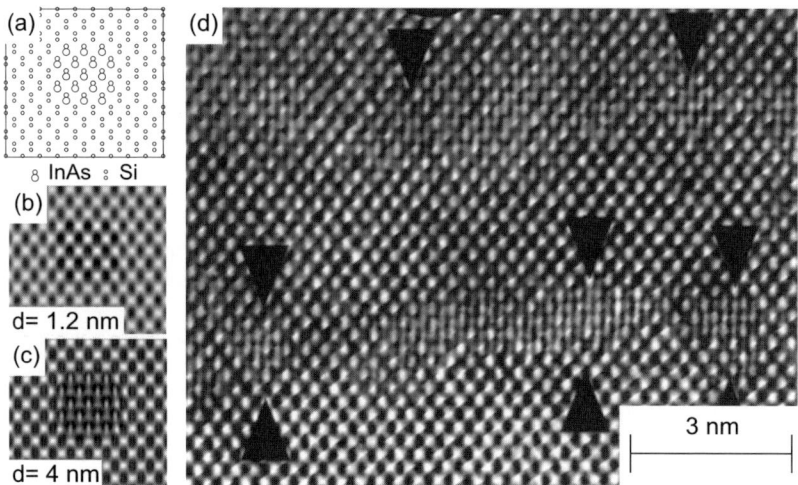

Fig. 3.103. Three-dimensional coherently strained InAs islands in an Si matrix. (**a**) Modeled positions of atoms in a coherent inclusion of InAs in an Si matrix. (**b**) Modeled cross-sectional HRTEM image in the case of a 1.2-nm-thick sample. (**c**) Modeled cross-sectional HRTEM image in the case of a 0.4-nm-thick sample. (**d**) Experimental cross-sectional HRTEM image of InAs inclusions in an Si matrix. *Arrows* indicate coherent InAs inclusions

Other aspects of the formation of strained islands in Ge/Si and SiGe/Si systems are discussed in the review article [3.7]. (See also a recent theoretical paper on island formation mechanisms in (Si)Ge/Si systems [3.242].)

InAs Coherently Strained Islands in an Si Matrix. The potential benefits of combining the advantageous optical properties and flexibility of III–V semiconductors with the kind of silicon technology widely used in microelectronics has attracted great interest for decades. The large misfit between Si and, for example, InAs ($\varepsilon = 10.6\%$) renders the growth of electronic or optical quality material a practically insoluble problem, if one tends to grow continuous layers. Ledentsov [3.243] pointed out the possibility of exploiting the formation of narrow-gap III–V islands on Si substrates. Indeed, such small InAs islands have been observed on the Si(001) surface by scanning tunneling microscopy and high-resolution transmission elctron microscopy (HRTEM) [3.244, 3.245]. HRTEM investigations of capped InAs/Si structures revealed a high density of coherent InAs inclusions with typical dimensions in the 3 nm region at the InAs/Si interface, for optimized growth conditions [3.246, 3.247]. Detailed photoluminescence (PL) studies of the samples by Heitz et al. [3.248] revealed a broad PL spectrum in the 1.3 μm region at 10 K. The excitation density dependence and time-resolved PL data [3.248] support the assignment of PL observed in the 1.3 μm spectral region to nanometer-scale InAs QDs.

3.3 Arrays of Three-Dimensional Coherently Strained Islands

By optimizing the growth conditions, Petrov et al. [3.249] were able to grow basically defect-free structures containing only coherent InAs islands in an Si matrix. The procedure involved growing 3 nm InAs at 450°C, 2 nm Si at 450°C, 4 nm Si at 620°C, 4 nm InAs at 460°C, 2 nm Si at 460°C, and 18 nm Si at 620°C. Capping by Si at an elevated temperature of 620°C efficiently evaporates excess InAs, preventing the formation of large dislocated islands. Figure 3.103 shows modeled HRTEM images together with the experimentally obtained image. In the case of a 4-nm-thick sample, the model corresponds well to the observed image. Such agreement confirms that inclusions in an Si matrix are indeed coherent and contain pure InAs. HRTEM images of a series of samples show that the observed PL spectrum arises only from samples where coherent inclusions are formed, and no PL comes from samples with dislocated islands.

The results suggest inserting direct bandgap inclusions of InAs into Si as a potential way of improving the optical properties of Si. This approach might lead to efficient Si-based light-emitting diodes.

Quantum Dots in III–N Systems. During the past few years, the group III nitrides have attracted a lot of attention as materials for optoelectronic devices. The bandgap of nitride-based semiconductors varies from 1.9 eV for InN up to 6.2 eV for AlN. This variation of the bandgap extends over the whole visible spectrum of wavelengths and into the ultraviolet spectral range. High-brightness light-emitting diodes [3.250] and laser diodes [3.251–3.253] have been fabricated. These were based on heteroepitaxial technology which allows tuning of emitted wavelengths. The best device structures were mainly produced by metalorganic chemical vapor deposition (MOCVD) or similar methods. Their optical, electrical, and structural properties have been extensively studied [3.254, 3.255]. However, despite the successful fabrication of optoelectronic devices, the processes responsible for light emission in these devices are still under discussion. It has been shown that local fluctuations in indium content play an important role in the optical properties of In-GaN/GaN heterostructures [3.256], but the actual size and density of these localized states have not been systematically studied [3.257–3.259]. The transition energy and shape of photoluminescence (PL) spectra may be governed completely by the shape and density of In-rich nano-domains, rather than by the average indium composition. InGaN layers have been shown to be inhomogeneous [3.129, 3.259] and to contain In-rich regions. Musikhin et al. [3.260] performed a detailed study of the influence of growth conditions on the structural properties of InGaN/GaN heteroepitaxial structures.

Samples were grown by MOCVD on sapphire substrates. The structures consisted of a thick (2 µm) buffer GaN layer, multiple quantum well (MQW) InGaN/GaN heterostructures, and a 100-nm-thick GaN cap layer. The MQW consisted of five or ten InGaN quantum wells. The GaN buffer and cap layers were grown at 1050°C. Temperature cycling was performed during deposition of the MQW region. This controls the indium composition in the InGaN

230   3. Self-Organization Phenomena at Crystal Surfaces

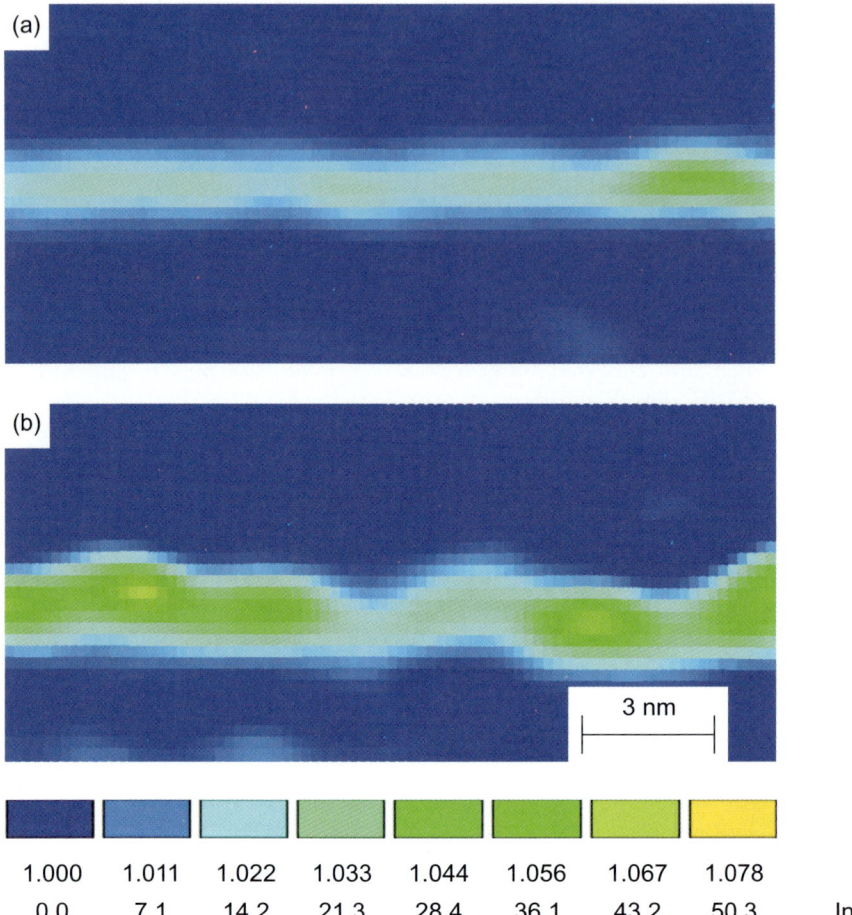

Fig. 3.104. Cross-sectional HRTEM image of InGaN/GaN heterostructures processed by the DALI evaluation program. The *color-coded map* depicts the local lattice parameter in the vertical direction relative to the vertical lattice parameter of GaN. Corresponding values of the local In content are given. (**a**) Sample grown at higher substrate temperature. (**b**) Sample grown at lower substrate temperature

quantum wells. Figure 3.104 shows cross-sectional high resolution transmission electron microscopy (HRTEM) images of the two samples processed by the digital analysis of lattice images (DALI) technique [3.68,3.261]. All growth conditions were identical for the two samples, apart from the substrate temperature. For the sample in Fig. 3.104a, the growth temperatures of the GaN barrier and InGaN QW were 900°C and 785°C, respectively, while for the sample in Fig. 3.104b, the temperatures were 30°C lower (870°C and 755°C, respectively). The total amount of InN incorporated into the QW was calculated on the basis of the maximum displacement at the top InGaN layer, yielding 1.5 ML and 2 ML for the samples grown at higher and lower temperatures, respectively. Photoluminescence (PL) spectra of the sample grown at lower temperature show a 200 meV shift in the emission peak to the long wavelength side.

The maps in Fig. 3.104 are the characteristic parts of several HRTEM images (50 × 50 nm). It is clear from the images that the InGaN quantum well is inhomogeneous. It contains nano-islands with an increased In content which exceeds the average values by a factor of approximately two. The lateral size of these nano-islands is about 3 nm for the lower growth temperature (Fig. 3.104b). Increasing the InGaN growth temperature from 755°C to 785°C results in:

- a slight decrease in nano-island size,
- a decrease in the maximum In content in the nano-islands from 50% to 35%,
- a decrease in the nano-island density from 3×10^{12} cm^{-2} to 0.5×10^{12} cm^{-2}.

Thus a decrease in the substrate temperature leads to an increase in nano-island lateral size and a blueshift in the PL spectrum, which again confirms a decrease in island volume. These results do not agree with kinetic models of size-limited island growth, discussed in Sect. 3.3.12, but rather favor the thermodynamic model. The latter predicts a decrease in island volume due to entropy effects. In alloy-based nanostructures, the entropy of mixing of alloy components will contribute, along with the configuration entropy.

3.3.17 What Have we Learned about 3D Coherently Strained Islands?

In this section, we have presented an overview of the key experimental results and theoretical concepts relevant to the formation of arrays of 3-dimensional coherently strained islands in lattice-mismatched heteroepitaxial systems. It is understood that both coherent and defective, e.g., dislocated, islands can form in lattice-mismatched systems. Generally, islands below a critical volume are coherent, whereas islands exceeding the critical volume contain defects.

A striking feature of arrays of 3D coherently strained islands that allows us to use them as quantum dots is the narrow distribution in island sizes and the stability of the array. This stability may be regarded as an apparent absence of Ostwald ripening.

Two major classes of theoretical models have been proposed for explaining this feature of size-limited island growth. Thermodynamically-based models bring out the importance of the strain dependence of surface energies associated with island facets, the wetting layer, and the substrate. These models show that, under a certain relationship between these surface energies and surface stresses, an array of equally-shaped and equally-sized 3D coherently strained islands is thermodynamically stable against Ostwald ripening. Kinetically-based models emphasize various mechanisms for slowing down the evolution in an array of 3D islands. This slowing down may be due to strain-induced barriers for adatoms to attach to islands or form a new atomic layer on island facets. As islands increase in size, the barriers become higher, eventually resulting in size-limited island growth. The validity of each model depends strongly on a particular material system.

Detailed investigations of the formation mechanisms in In(Ga)As/GaAs systems have revealed the following. An array of coherently strained islands with narrow size distribution and stable against ripening will only form in a certain parameter region, where the parameters include substrate temperature, arsenic vapor pressure, or the amount of deposited material.

The above overview of experimental results shows that many observations of InAs/GaAs(001) or InGaAs/GaAs(001) islands favor the thermodynamically dominated nature of arrays of islands. The following results corroborate this:

- the evolution of dot size up to a limiting value upon growth interruption for MBE-grown InAs QDs, observed only in a certain range of growth conditions;
- a reversible phase transition in the InAs/GaAs system from 3D to 2D morphology, driven by a reduction in As pressure;
- a reversible phase transition from 3D to 2D morphology in the GaInAs(P)/GaAs system, driven by switching AsH_3 off/on and switching PH_3 on/off;
- an irreversible phase transition from coherently strained islands to dislocated islands, i.e., the 'switching on' of Ostwald ripening driven by an increase in As pressure;
- the formation of coherent InAs QDs in MOCVD, which is possible only in MBE-like conditions, at very low AsH_3 pressures;
- the general tendency of InAs islands towards Ostwald ripening with increasing arsenic pressure, for both MBE and MOCVD;
- the preferred alignment of nearest-neighbor dots along elastically soft directions $\langle 100 \rangle$;
- a decrease in the average volume of InAs islands with increasing substrate temperature;
- reversible changes in island shape, volume, and density with cyclic variation of the substrate temperature, observed for InAs/GaAs islands.

Concerning the comparison of experimental results obtained by different research groups, the following two points should be noted. Firstly, growth conditions must be precisely defined, including the substrate temperature, the vapor pressure of arsenic, the amount of deposited material, the nominal alloy composition of the deposit, the degree of substrate misorientation (if used), etc. Secondly, any surface investigation using scanning tunneling microscopy (STM) or atomic force microscopy (AFM) inevitably involves cooling the sample from the growth temperature down to room temperature. It has been unambiguously demonstrated for the InAs/GaAs system that cooling, even from 480°C down to 420°C, leads to dramatic changes in island shape, volume, and density. Furthermore, InAs islands form even at low substrate temperatures, in the range 320–350°C, revealing the coexistence of individual dots and lateral associations of QDs and showing that there is sufficiently fast adatom diffusion at these temperatures. If STM and AFM measurements obtained after 'fast quench' are indeed to be relevant to the surface morphology at a high island formation temperature, it is absolutely necessary to carry out a thorough experimental study of the way STM and AFM observations depend on the cooling rate. To the best of our knowledge, no such measurements have been reported in the literature so far.

To conclude the chapter, we note that self-organization effects on crystal surfaces offer us a wide variety of nanostructures in a large class of material systems. In the next chapter, we will show that the combined effects of self-organization and further engineering of complex nanostructures can provide a practically infinite choice of nanoworlds, with easily controllable parameters. These complex nanostructures reveal even richer physics and a higher potential for applications.

4. Engineering of Complex Nanostructures: Working Together with Nature

> Природа не храм, а мастерская, и человек в ней работник.
>
> Иван Сергеевич Тургенев.
> *Отцы и дети*

> Nature is not a temple but a workshop, and man is the workman in it.
>
> Ivan Sergeevich Turgenev.
> *Fathers and Sons.*
> Translated from the Russian
> by Richard Hare

The spontaneous formation of nanostructures on crystal surfaces discussed in detail in Chap. 3 makes it possible, among other things, to create quantum wire and quantum dot (QD) structures with exciting physical properties and great potential for applications. It turns out, however, that merely exploiting natural phenomena is far from sufficient either for the physics or for the applications. We may mention just a few limitations:

- Although geometrical and electronic parameters of quantum wire and quantum dot structures can be tuned by varying the amount of deposited material, the substrate temperature and the vapor pressure (in the case of compound semiconductors, e.g., III–V, II–VI, or III–N), a much higher degree of flexibility and tunability is actually required.
- The volume density of QDs is rather low so that, for laser applications, the overlap of the active medium with the optical mode is rather small, and this suppresses major device characteristics.
- The maximum obtained wavelength of the photoluminescence peak from three-dimensional (3D) InAs/GaAs QDs is 1.24 µm, which does not reach the practically important spectral region 1.3–1.55 µm.

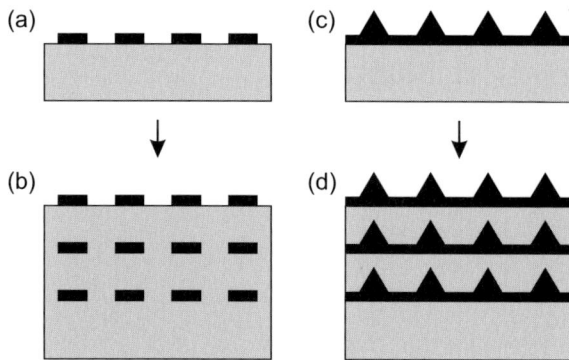

Fig. 4.1. Schematics of the growth of multisheet arrays of quantum dots. (**a**) Single-sheet array of submonolayer islands. (**b**) Multisheet array of submonolayer islands. (**c**) Single-sheet array of 3-dimensional (3D) coherently strained islands. (**d**) Multisheet array of 3D coherently strained islands

- The InAs/GaAs structure with PL maximum at 1.24 μm contains, along with coherent islands, a significant density of dislocated islands, which make the structure unsuitable for laser applications. Thus the lasing wavelength is limited to even smaller values.
- Localization of electrons and holes by 3D InAs/GaAs QDs is not sufficiently strong to maintain the performance of QD structures up to room temperature.
- Submonolayer InAs/GaAs islands provide even weaker localization of carriers. Hence, lasers based on these structures would not operate at room temperature.

One way to overcome these problems has been generally recognized. The idea is to work together with Nature to combine the phenomena of spontaneous nanostructure formation with the engineering of complex systems. Figure 4.1 shows a basic approach to this kind of engineering. This is the growth of multilayered structures, where layers of a narrow bandgap material forming quantum wires or quantum dots alternate with layers of a wide bandgap material forming barriers for electrons and holes. Thus multisheet arrays of islands form in a matrix. For the right combination of materials, these arrays may be multisheet arrays of quantum wires or dots. Such complex nanostructures exhibit significant advantages over single-sheet arrays of wires or dots:

- Multisheet arrays of islands have a higher degree of tunability. The material and thickness of each sheet of wires or dots, as well as the material and thickness of each spacer layer, can be adjusted independently.
- Multisheet arrays of QDs have a larger overlap with the optical wave in lasers, thus improving laser characteristics.

- In multisheet arrays the dots of neighboring sheets can be electronically coupled. This coupling will shift electron and hole levels to lower energies, and the photoluminescence (PL) spectra should shift towards lower photon energies or longer wavelengths (redshift). For InAs/GaAs QDs this will mean a shift towards the highly prized spectral region of 1.3–1.55 µm.
- A shift of electronic levels to lower energies will mean higher electron and hole localization energies and higher thermal stability of the lasers.

In the present chapter, we demonstrate how the above advantages, which have motivated the construction and study of multisheet structures, work in reality. In addition, the engineering of complex nanostructures has revealed unique properties that were hardly expected before:

- the possibility of independently controlling the density and volume of QDs using a 'seeding' concept,
- the possibility of controlling the PL polarization via exciton wave function engineering,
- the possibility of tuning in different types of vertical correlation between the islands,
- a defect-reduction technique for growing structures with large quantum dots and almost no defects,
- the overgrowth of initial strained islands by an alloy material that is accompanied by activated alloy phase separation in the cap layer.

4.1 Multisheet Arrays of Strained Islands

The first arrays of 3-dimensional coherently strained islands observed experimentally by Goldstein et al. [4.1] were in fact multisheet arrays of InAs islands in GaAs. An important feature is that islands form vertical columns. Islands in the next sheet form directly above islands in the previous sheet (Fig. 3.70). Similar effects in the formation of columns were observed in Ge/Si [4.2], InGaAs/GaAs [4.3], and InAs/GaAs [4.1, 4.4–4.10] systems. Typically, the multisheet growth regime involves:

- deposition of material 2 on substrate 1, resulting in the formation of an array of islands of material 2,
- overgrowth of islands by the substrate material 1,
- deposition of a second sheet of material 2 resulting in the formation of a second sheet of islands,
- overgrowth of islands by the substrate material 1, etc.

Certainly, more complicated structures can be grown, using layers of more than two materials.

A key feature of such growth regimes is the dramatic difference between surface and bulk diffusivities of atoms. For a typical growth temperature of a few hundred degrees C, the bulk diffusivity of atoms in semiconductors is a

238   4. Engineering of Complex Nanostructures

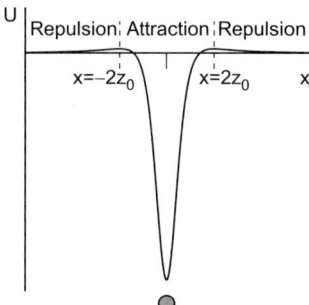

Fig. 4.2. Interaction energy of a surface point elastic defect with a buried point elastic defect. The energy profile shows a region of strong attraction to a potential energy minimum directly over the buried defect, and regions of weak repulsion far from the buried defect

few orders of magnitude lower than the surface diffusivity. Thus, once islands are overgrown, their volume, shape, and position remain unchanged during further growth. Through the long-range strain field, 'buried' islands affect surface migration of adatoms. The latter is then a combination of drift in the elastic field created by buried islands, and diffusion.

If deposition of each sheet in a multisheet structure is followed by growth interruption, the array of islands in the first sheet evolves towards equilibrium and the final structure may be described by thermodynamics. In every subsequent sheet, an array of islands evolves upon growth interruption to a certain partial equilibrium, i.e., to an equilibrium in the strain field created by all the sheets of buried islands. The structure of each subsequent sheet may be addressed by constraint thermodynamics.

4.1.1 Vertical Correlation of Strained Islands

The strain field created at the surface by the buried islands can be described by continuum elasticity theory. Maradudin and Wallis [4.11] considered the elastic interaction between a point elastic defect buried at depth z_0 [coordinates $(0,0,-z_0)$], with a point elastic defect at the surface $(x_0, y_0, 0)$. The strain (the hydrostatic component) equals

$$\frac{1}{3}\varepsilon_{ii}(x,y,0) = \frac{C}{(x^2+y^2+z_0^2)^{3/2}}\left[1 - \frac{3z_0^2}{x^2+y^2+z_0^2}\right], \quad (4.1)$$

where C is the strength of the point defect. In the case of strained islands the 'strength' of the point defect C is proportional to the island volume and the lattice mismatch between the island material and the matrix. If the island material has a larger lattice parameter than the matrix (which is the case for InAs/GaAs and Ge/Si systems), then $C > 0$. The interaction energy U of an adatom and the buried point defect is proportional to (4.1). This is plotted in Fig. 4.2, showing a region of strong attraction to the position exactly over the buried defect and regions of weak repulsion away from the buried defect.

The modulated strain field on the surface leads to modulation of the chemical potential of surface adatoms $\mu(x,y) = \text{Const.} + U(x,y)$ [4.12]. In its turn this modulation of the chemical potential means that the migration of adatoms on the surface consists of diffusion and drift [4.13]. (The term 'directional migration' is sometimes used for the drift of adatoms.) Drift is the driving force behind kinetic self-organization in this complex growth mode. The same kinetic mechanism is responsible for the instability of epitaxial growth of a homogeneous alloy and may result in the formation of compositionally modulated structures. (A detailed discussion is given in Sect. 4.3.)

Since for InAs islands in a GaAs matrix, the coefficient C in (4.1) is positive, the position above the buried island, i.e., at $x = y = 0$ corresponds to the maximum of the tensile strain ε_{ii}. The migration of In atoms on the surface of GaAs in the strain field due to a buried island of InAs was considered by Xie et al. [4.4] and by Xie et al. [4.5]. It was argued that In atoms with larger radius than Ga atoms migrate preferentially to surface regions with maximum tensile strain ε_{ii}, i.e., to a position directly above a buried island. Therefore, islands in the next sheet form with a higher probability above buried islands. Xie et al. [4.5] calculated this vertical pair correlation probability as a function of the spacer layer thickness and island volume and found a good fit to experiment.

4.1.2 Order Enhancement in Multisheet Arrays

Tersoff et al. [4.14] pointed out that the strain profile (and hence preferred sites for nucleation of new islands) depends strongly on island arrangement in the buried sheet. If two islands are close together, like the left two islands in Fig. 4.3a, their strain fields overlap significantly and there is only a single deep minimum in the potential energy of adatoms on the surface. If islands have a medium spacing, a single minimum occurs over each of the two islands (like the two islands on the right of Fig. 4.3a). Finally, if two islands have a large spacing (like the second and third islands in Fig. 4.3a), an additional shallow minimum occurs on the surface. Thus, two very close islands in the buried sheet create a single nucleation site on the surface, while two widely spaced islands create an additional nucleation site in-between. Figures 4.3b and c show the results of modeling multisheet growth. This reveals that, irrespective of volumes and spacings in the first sheet, islands in subsequent sheets become more uniform in volume and exhibit a more uniform spacing, providing a specific strain-mediated mechanism for self-organization of islands in multisheet structures.

Liu et al. [4.15] extended the model by Tersoff et al. [4.14] by taking into account the finite volume of the islands and different island shapes. The growth of a multilayer of islands was simulated in [4.15] by the following model. Compressively strained islands buried in the embedding layer produce a non-uniform strain field on the surface. The regions right above the islands become expanded (tensile strain) while the regions between islands become compressed. An ensemble of islands randomly distributed in positions and sizes was assumed for the first layer. In all subsequent layers, it was assumed

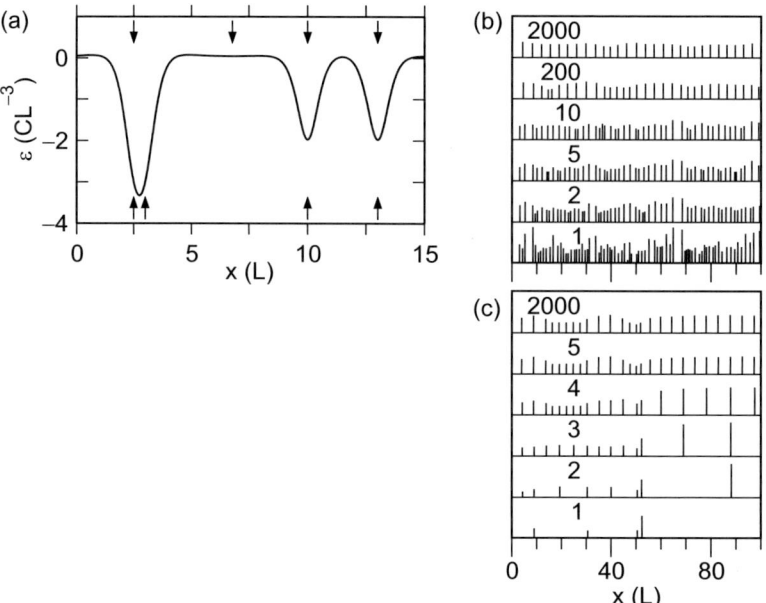

Fig. 4.3. Self-organization mechanism in the growth of multisheet arrays of strained islands. (**a**) Surface strain ε vs. lateral displacement x (in units of spacer thickness L), for four islands buried at depth L, from (4.1). *Arrows at the bottom* indicate lateral positions of buried islands. *Arrows at the top* indicate minima in ε, i.e., favored positions for subsequent nucleation. These positions include both deep minima over buried islands and shallow minima midway between widely separated islands. (**b**) and (**c**) Calculated island positions and sizes in successive selected layers. Layer numbers are indicated. Heights of vertical lines represent island volumes relative to the average for that layer. Sequences begin with (**b**) closely spaced islands and (**c**) widely spaced islands. The periodic cell size is $300L$. For clarity, only a third of the cell is shown. With kind permission of J. Tersoff et al. [4.14]

that islands nucleate only in tensile regions and that they will grow larger when they are sitting in a larger tensile region because doing so will lower the strain energy. Particular simulations were carried out in [4.15] for a 2-dimensional case, i.e., for wire-shaped islands elongated in the y direction. The crucial quantity that determines the growth regime is the ratio B_i/x_{0i}, where $2B_i$ is the base size of the ith nucleated islands, and $2x_{0i}$ is the extent of the ith tensile region. In the simulations, an island first forms with triangular shape (a 2D pyramid), with base size $2B_i$ and a fixed angle. As it gets buried, it transforms into a rectangular shape with conserved volume and a fixed height-to-base ratio.

It was shown by Liu et al. [4.15] that, in a certain range of the parameter B_i/x_{0i}, the growth regime includes vertical self-organization of islands, leading to a narrower size distribution and a higher degree of spatial ordering.

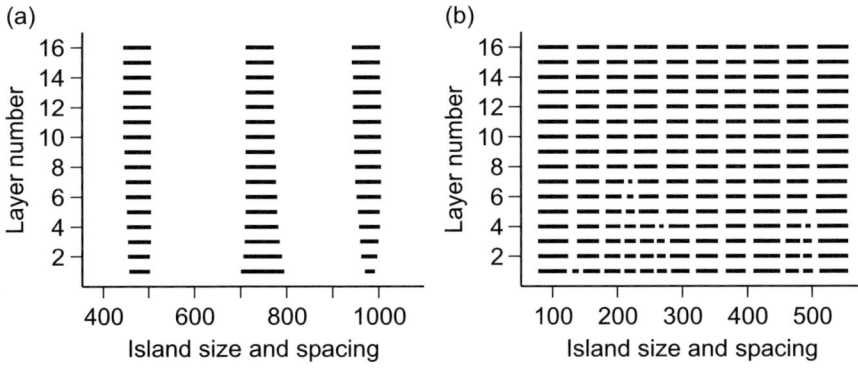

Fig. 4.4. Evolution of island size in a multilayer film of 16 bilayers simulated for island arrays with embedding layer thickness $L = 12$ and ratio B_i/x_{0i} of island size to the extent of the tensile region equal to 0.75. The *abscissa* shows island position and size, whilst the *ordinate* shows the bilayer number of the film. Lengths of horizontal lines represent island base sizes. Island heights (not shown) scale with base sizes. (a) A sparse island array. Three typical cases are selected for illustration. Islands all reach the same size. (b) A dense island array. Islands achieve uniform spacing, in addition to converging in size. With kind permission of F. Liu et al. [4.15]

Figure 4.4a shows typical examples of the growth of a multilayer film for an initially sparse island array. The spacer layer thickness was chosen as $L = 12$ and the ratio of base area to tensile strain was kept constant at $B_i/x_{0i} = 0.75$. The three islands in the first layer have starting base sizes of $4.5L$, $10L$, and $2.3L$, respectively. For such a sparse island array, the strain distribution produced at the embedding-layer surface consists of a deep negative minimum at each island position, so that a new island nucleates on top of each buried island in the next layer, leading to the formation of vertical columns of islands as more bilayers are added. Even more remarkably, the island sizes in each column either increase or decrease and gradually converge toward the same final size and then stay fixed. The stable island size is defined by the spacer layer thickness and the ratio of base area to tensile strain area. With the parameters used in Fig. 4.4a, it is $6.4L$.

The formation of island columns and the convergence of island sizes in different columns have been observed in both III–V [4.1, 4.5, 4.9, 4.10, 4.16–4.19] and group IV multilayer films [4.14, 4.20–4.22]. This can be explained as follows. Islands nucleate and grow with a random size distribution on the initial strain-free surface. New islands in subsequent layers adopt a size distribution in accordance with the relative lateral extents of the tensile regions they are sitting on. A small buried island produces a tensile region above itself, much larger in lateral extent than its own size. This region collects more atoms and the new islands above it grow rapidly in the first few subsequent layers (right-hand column in Fig. 4.4a). An initially larger island produces a tensile region whose extent is more like its own size. Islands above it in subsequent

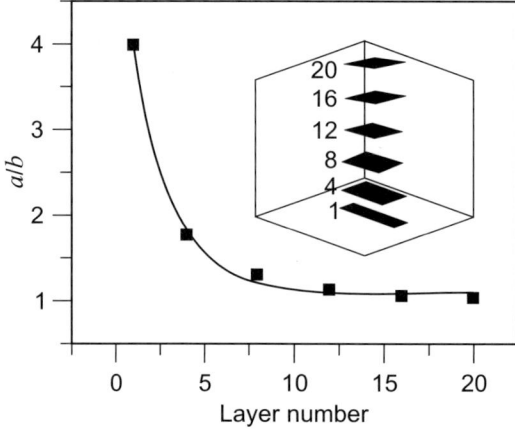

Fig. 4.5. Evolution of a 3D island shape in a multilayer film of 20 bilayers, simulated for a single island column with embedding layer thickness $L = 12$ and island-size-to-tensile-region ratio $B_i/x_{0i} = 0.75$. The initial island has a rectangular base with an aspect ratio $a{:}b$ of 4:1. The aspect ratio of the island decreases with increasing number of bilayers. *Squares* are simulated values and the *solid curve* is a spline fit. The *inset* shows a schematic view of the island shape evolution. With kind permission of F. Liu et al. [4.15]

layers grow more slowly or may even shrink (middle column in Fig. 4.4a) when $B_i/x_{0i} < 1$. As the distribution of the lateral extents of tensile regions becomes more and more uniform in each successive layer, islands evolve to a common stable size.

For an initially dense island array, lateral coalescence of close-lying islands within the same layer provides a second mechanism for improving island size uniformity (in addition to vertical replication of islands in successive layers). Coalescence and vertical size convergence lead to both uniform island size and uniform spacing. Figure 4.4b shows an example of the evolution of a dense island array in a multilayer film, for $L = 12$ and $B_i/x_{0i} = 0.75$. The starting islands are created randomly with a mean size of $3.0L$ and a standard deviation of $1.1L$, with their bases in contact. When two islands are sufficiently close, a point where the strain equals zero may not exist between them. The boundary of the tensile region created by each island is then defined by the local maximum in the negative strain curve, i.e., the point of least strain between them. The evolution of larger islands is dominated by vertical replication. These islands form columns and gradually converge in size, as for sparse islands. Smaller islands are eliminated and/or combined into larger islands by coalescence. The latter causes a decrease in area island density and an increase in average island volume, because islands maintain their shape (i.e., facet angle) upon coalescence. For example, the average island size in Fig. 4.4b increases from $3.0L$ to $4.3L$.

Simulations for 3D islands demonstrate that these mechanisms also transform the shapes of vertically replicated islands in a multilayer film, independently of the initial density of islands. As an island with anisotropic base shape is buried, it produces a tensile region that is anisotropic. However, the aspect ratio of the tensile area, and hence the aspect ratio of the nucleated island base, is smaller than the original aspect ratio of the buried island base, leading to a more isotropic shape in the next layer of islands. Figure 4.5 shows an example of the evolution of 3D island shape in a single island column. The initial island, a rectangular disk, has a base aspect ratio of 4:1, which decreases rapidly towards 1:1 with an increasing number of bilayers. The improvement in island shape uniformity in a multilayer film has been observed in both III–V [4.19] and group IV systems [4.20].

Consequently, the theory of multilayered growth of strained islands developed by Tersoff et al. [4.14] and Liu et al. [4.15] demonstrated the vertical self-organization of islands. Self-organization manifests itself through an increasing uniformity of islands in size and spacing, together with shape transformation from an elongated rectangle to a square. The results of the modeling agree with numerous observations of III–V and group IV multilayer films.

4.1.3 Electronically Coupled Multisheet Quantum Dots

Multisheet arrays of 3D coherently strained islands (quantum dots, QDs) form a new class of spontaneously formed nanostructures, in which the ordering occurs both in the lateral plane and in the vertical direction. However, as far as applications of QDs are concerned, an array of merely vertically stacked islands does not provide any significant improvement over a single-sheet array. The reason is that vertically stacked islands are essentially separated, i.e., the 3D islands of the first sheet are completely covered by a cap layer, and deposition of the second sheet of islands starts only afterwards. In this case both electron and hole wave functions are localized inside each quantum dot. Therefore the regularity of the above arrangement does not result in a modification of the basic electronic properties of the structures.

In order to obtain advantage from vertically correlated arrays of quantum dots, electronically coupled quantum dots were fabricated [4.7, 4.8, 4.10, 4.23]. Ledentsov et al. [4.8] performed a multi-cycle InAs–GaAs growth, in which deposition of 5.5 Å of InAs was alternated by deposition of 15 Å of GaAs. The deposition of the first sheet of InAs results in the formation of InAs pyramids over the wetting layer of InAs. Since the characteristic height of InAs pyramids is ≈ 60 Å, the cap layer of GaAs only partially covers the pyramids, whereupon the next layer of InAs is deposited.

Figures 4.6a and b show cross-section and plan-view transmission electron microscopy (TEM) images, respectively, of the structure containing vertically coupled InAs quantum dots formed by three-cycle InAs–GaAs deposition. Figure 4.6b shows the average lateral size of islands in the top layer, 170 ±

244   4. Engineering of Complex Nanostructures

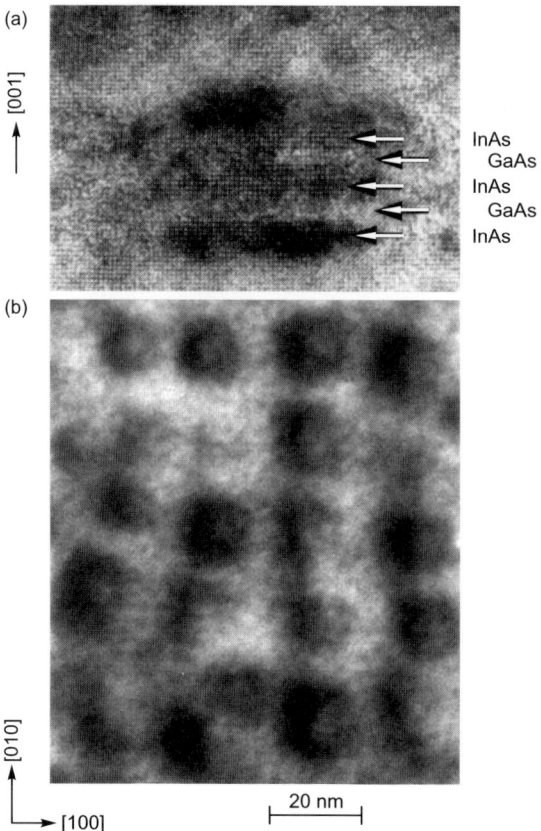

InAs
GaAs
InAs
GaAs
InAs

Fig. 4.6. Vertically coupled InAs quantum dots (QDs) in a GaAs matrix. (**a**) [010] cross-sectional HRTEM image formed by 9 beams with defocusing 60 nm. Note the different spot density in InAs and GaAs regions. (**b**) Bright-field plan-view TEM micrograph under [001] zone axis illumination

10 Å. Islands have a square base with main axes along the [100] and [010] directions.

The histogram of nearest-neighbor dot orientations (Figs. 4.7a and b) demonstrates the preferred orientation in the [100] and [010] directions, typical of the 2D square lattice with primitive lattice vectors along the same directions. Each vertically coupled quantum dot is composed of three vertically aligned parts separated by GaAs regions 2–4 ML thick (see Fig. 4.7a). The top parts have a larger lateral size (≈ 170 Å) than the lower part (≈ 110 Å).

The important properties of the resulting structure are:

- a larger lateral size of InAs islands in the top sheet than in the bottom sheet,

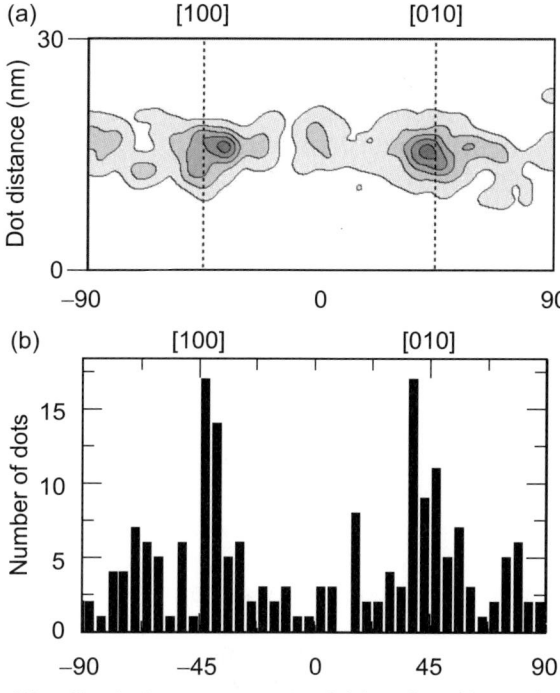

Fig. 4.7. Vertically coupled InAs quantum dots (QDs) in a GaAs matrix: (**a**) 2D histogram of nearest-neighbor center-to-center distances and directions for the vertically coupled QDs from Fig. 4.6b. (**b**) Projection of (**a**) onto the angular axis. Maxima in the [100] and [010] directions prove the correlation in the nearest-neighbor arrangement of the islands, which corresponds to the lateral square superlattice composed of vertically coupled QDs

- the splitting of InAs pyramids and appearance of thin layers of GaAs separating the parts of the coupled QDs,
- the lateral correlation in the nearest-neighbor dot orientation, which is typical for the 2D square lattice and is more pronounced than in a single-sheet structure.

An increase in the lateral size of InAs islands can be explained by the transfer during growth of In from buried parts of the structure, where the InAs is strained, to the top part, where the InAs is partially relaxed.

The enhancement of the lateral correlation is similar to that observed for vertically stacked Ge/Si(001) islands, where a similar effect was explained theoretically by Tersoff et al. [4.14]. However, since the InAs islands of each sheet are only partially covered by GaAs, the detailed growth kinetics differs from that described in [4.14]. A semi-qualitative kinetic model for the system in question was proposed in [4.8].

Since successive parts of vertically coupled InAs QDs are separated only by very thin barriers, they are electronically coupled. By varying the thicknesses of deposited GaAs and InAs in each cycle, and varying the number of cycles, one can significantly modify the electronic spectrum of vertically coupled QDs. Laser characteristics can then be optimized [4.8, 4.24].

4.1.4 Seeding of Quantum Dots

As discussed in detail in Sect. 2.3, volume, shape and density of quantum dots in a single-sheet are determined by material parameters such as surface energies, surface stress, and lattice mismatch between deposit and substrate. For a given material system, the QD array can be controlled by varying the amount of deposited material, substrate temperature, and, for III–V material systems, vapor pressure of the group V element. Then, if the QD density increases upon such variations, the average QD volume decreases, and vice versa.

It is of great importance to be able to fabricate QD arrays with high density and large volume, particularly for laser applications. QD lasers provide ultrahigh material gains of 10^5 cm^{-1} and differential gains of 10^{-13} cm^2 [4.25]. Theoretical studies of the threshold current density J_{th} in QD lasers [4.26] revealed that optimization of QD surface density and uniformity can significantly affect laser characteristics. The optimum surface density minimizing J_{th} was estimated to be of the order of 10^{11} cm^{-2} for realistic QD size dispersion and losses [4.26]. As the typical density of InGaAs QDs is usually below this value, one may ask if it is possible to increase the QD density without changing the major deposition parameters. Growth of vertically coupled QDs [4.8] allows one to increase the homogeneity of the QDs in the upper sheets and improve laser characteristics. This approach, however, does not solve the problem of poor QD uniformity in the first rows and does not increase the QD density.

A concept for separately controlling quantum dot density and volume was proposed independently by Maximov et al. [4.27] and Mukhametzhanov et al. [4.28]. Both approaches involve different deposition schemes in the first sheet and in subsequent sheets.

Maximov et al. [4.27] used the fact that InAlAs QDs have an area density of 2×10^{11} cm^{-2}, which is 4 times higher than that for InGaAs QDs deposited under the same growth conditions, as shown in [4.29] (see Fig. 4.8). InAlAs QDs have a much higher bandgap energy and hence a much smaller localization energy for the same matrix, and it is therefore difficult to use them to improve laser characteristics. On the other hand, the large bandgap of InAlAs QDs, which causes them to depopulate even at low temperatures, can be advantageous if these QDs are used as passive prelayers for seeding QDs with larger localization energy (InAs or InGaAs QDs).

Maximov et al. [4.27] proposed and carried out seeding of InGaAs islands on top of a sheet of InAlAs islands used as stressors, thereby exploiting the

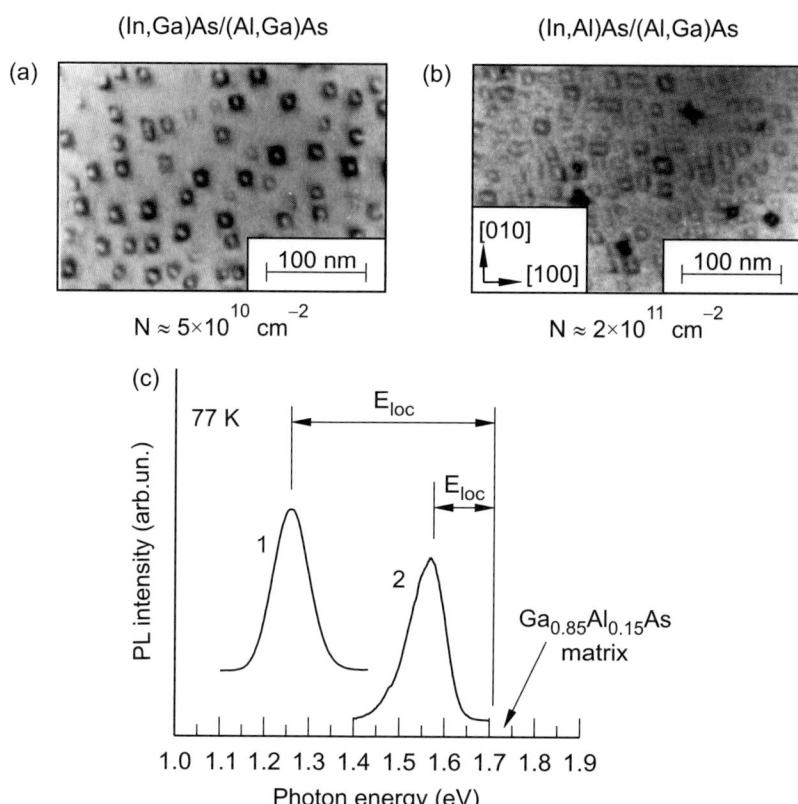

Fig. 4.8. Comparison of InGaAs and InAlAs quantum dots in a GaAlAs matrix. Plan-view TEM images showing (**a**) InGaAs QDs in a GaAlAs matrix with a relatively low area density and (**b**) InAlAs QDs in a GaAlAs matrix with a high area density. (**c**) Photoluminescence (PL) spectra of the two arrays of QDs. InAlAs QDs have a small localization energy and are easily depopulated at room temperature, while InGaAs QDs have a large localization energy and are more stable against carrier evaporation at room temperature

idea of vertically coupled quantum dots [4.8]. This is illustrated in Figs. 4.9a and b. In the first sheet, InAlAs is deposited on a GaAlAs substrate, resulting in the formation of small InAlAs islands. These islands serve as stressors, leading to the growth of columns of vertically correlated islands (Fig. 4.9a). Islands in each subsequent sheet have a larger volume than those in the first sheet. Figure 4.9b depicts energy bandgaps in all the sheets of QDs, showing a shallow localization potential for electrons and holes in the first sheet and a deeper potential in subsequent sheets.

The experimental data in Figs. 4.9c–e confirm the anticipated effect. Cross-sectional TEM (Fig. 4.9c) reveals smaller islands in the first sheet and larger islands in subsequent sheets. The plan-view TEM image in Fig. 4.9d

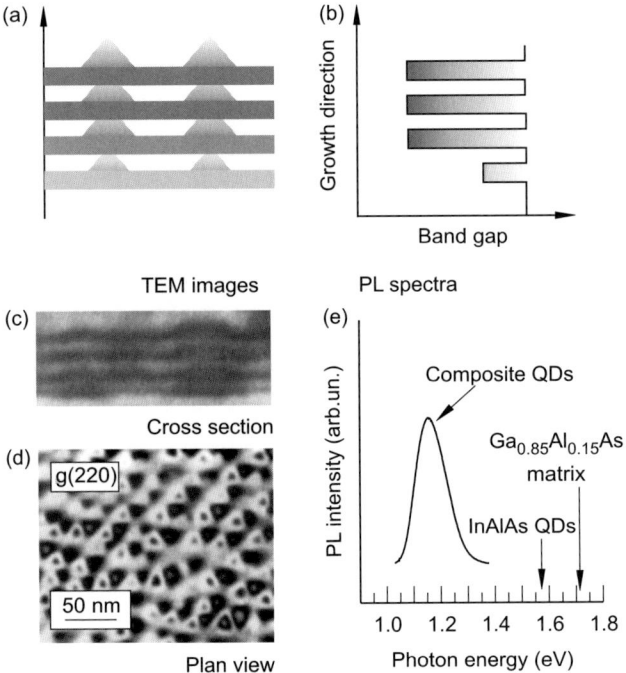

Fig. 4.9. Seeding of InGaAs quantum dots on a layer of InAlAs stressors. (**a**) Growth of columns of vertically correlated islands on small stressors. (**b**) Energy bandgaps of the multisheet structure showing a shallow localization potential for carriers in the first layer and a deeper one in subsequent layers. (**c**) Cross-sectional TEM image of the multisheet structure revealing smaller islands in the first sheet and larger islands in subsequent sheets. (**d**) Plan-view TEM image of the multi-sheet structure revealing a highly uniform array of QDs with a high density. (**e**) Photoluminescence (PL) spectrum of composite quantum dots

demonstrates a high-density of QDs with a high degree of uniformity in size. The photoluminescence (PL) spectrum of composite QDs (Fig. 4.9e) reveals a single maximum around 1.15 eV. It is shifted with respect to the PL maximum from a single sheet of InGaAs QDs (Fig. 4.8c) to the value 1.26 eV due to electronic coupling in a vertical column of QDs. At the same time, PL from composite QDs reveals no contribution from high-energy states in the InAlAs QDs of the first layer.

Thus the novel growth regime proposed by Maximov et al. [4.27], which involves seeding a high density of small QDs in the first sheet and growing vertical columns of large, vertically correlated and electronically coupled QDs, has indeed provided a high density of large QDs with a large localization energy.

Mukhametzhanov et al. [4.28] studied bilayer stacks of InAs/GaAs QDs with different InAs deposition amounts in the initial (seed) and second layers. They were able to enhance the average volume and improve the uniformity

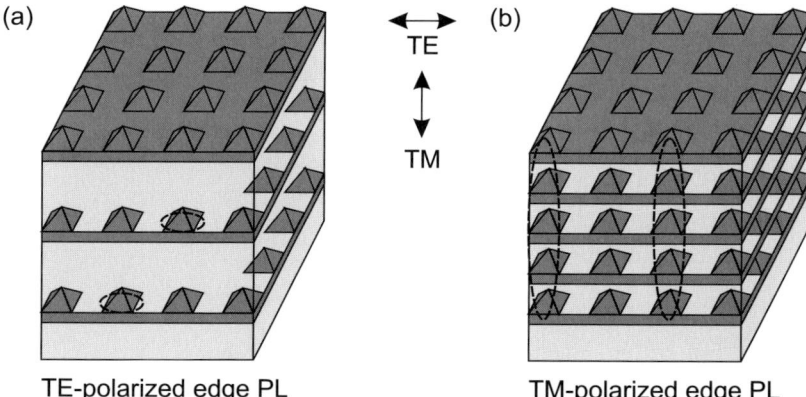

Fig. 4.10. Schematics of exciton wave function engineering by stacking quantum dots. (**a**) Sheets of quantum dots are separated by thick spacers, so that the dots are not coupled electronically. The wave function of an exciton localized by quantum dots roughly follows the shape of the QD and is extended in the lateral plane. The edge photoluminescence (PL) is TE polarized. (**b**) Sheets of QDs are separated by thin spacers, QDs are electronically coupled, and edge PL is TM polarized. *Dashed lines* show exciton wave functions schematically

of InAs QDs, resulting in room temperature photoluminescence at 0.955 eV ($\sim$ 1.3 µm) for 1.74 ML (seed)/3.00 ML InAs stacking. No data were reported on lasers fabricated via such structures.

On the other hand, lasers fabricated using seeded arrays of QDs by Maximov et al. [4.27] demonstrated substantial improvements in key laser characteristics, e.g., lower threshold current density, increased gain, reduced gain saturation effect, and higher characteristic temperature T_0 near room temperature. Further technological innovations aimed to improve QD laser parameters are discussed in more detail in Chap. 5.

4.1.5 Engineering the Exciton Wave Function by Stacking Quantum Dots

The growth of multisheet arrays of quantum dots, along with a shift in the energy of electronic states, allows one to control electron and hole wave functions, and hence also the wave function of excitons which determines such optical characteristics as the oscillator strength of the optical transition and polarization of absorption and photoluminescence. Figures 4.10a and b are schematic representations of multisheet arrays of 3-dimensional islands with thick and thin spacers, respectively. In the case of a thick spacer, quantum dots of neighboring sheets are electronically uncoupled and the wave functions of electrons and holes are localized in separate QDs.

For InAs QDs in a GaAs matrix having the shape of square-based pyramids with {101} side facets, Stier et al. [4.30] calculated electron and hole

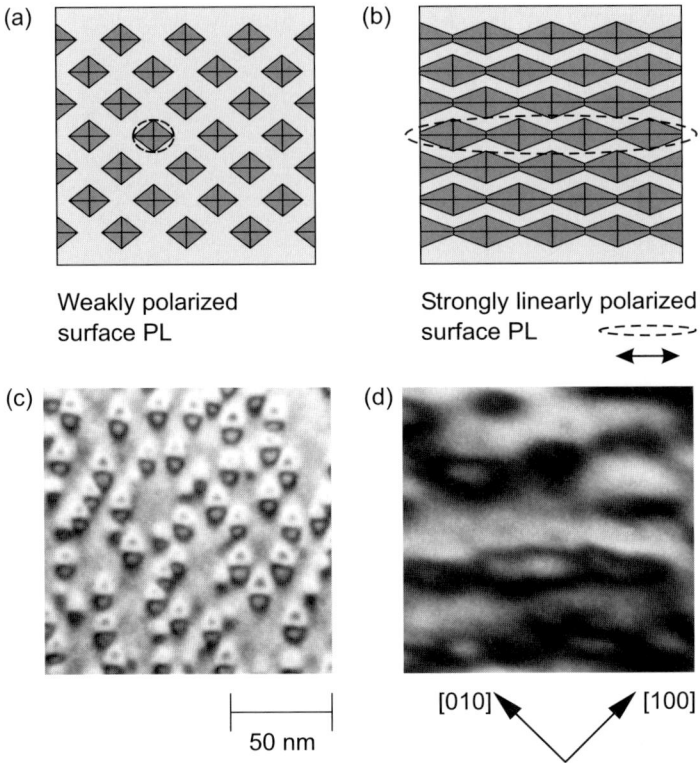

Fig. 4.11. Engineering of exciton wave functions in the lateral plane. (**a**) Plan view of a sheet of laterally uncoupled QDs with nearly symmetric shape. The surface PL is only weakly polarized. (**b**) Plan view of a sheet of laterally overlapping QDs forming quantum wires. The surface PL is strongly linearly polarized. (**c**) Plan-view TEM image of three-sheet structures with InGaAs/GaAs QDs. Islands have a compact shape. (**d**) Plan-view TEM of the 20-sheet InGaAs/GaAs structure. Islands have a strongly elongated shape, forming quantum wires in the lateral plane rather than quantum dots

states using an 8-band ($\mathbf{k} \cdot \mathbf{p}$) model and taking into account the inhomogeneous strain distribution and piezoelectric effects. It was shown that the hole wave functions are confined at the pyramid base and are therefore strongly flattened compared with the shape of the pyramid itself. As a consequence, both the absorption and emission of light will be polarized in the (001) plane. The edge photoluminescence (PL) will thus be TE polarized. For InGaAs QDs with a more flattened shape, this effect will be even stronger.

In contrast, in multisheet arrays of electronically coupled QDs, electron and hole wave functions extend over the entire vertical columns of QDs. Then PL will be TM polarized. The transition from TE polarized to TM polarized PL can be driven by varying the spacer thickness, or by increasing the number of QD sheets in a structure.

Yu et al. [4.31] studied PL from multisheet arrays of InGaAs/GaAs QDs. Structures were grown by molecular beam epitaxy (MBE) and contained 1, 3, 10, and 20 layers of $In_{0.5}Ga_{0.5}As$ QDs deposited at 485°C and separated by 5-nm-thick GaAs spacers. The QDs have pyramidal shape with a base length of about 18 nm, a height of about 5 nm, and an average lateral separation of 55 nm. The vertical alignment of the QDs was identified in cross-sectional TEM studies [4.32]. The edge PL from the structures with 1, 3, and 10 sheets exhibited TE polarization, while the edge PL from the 20-sheet structure was TM polarized, in accordance with Fig. 4.10.

Plan-view TEM studies of these structures by Yu et al. [4.31] revealed that the dots are slightly elongated along the [1$\bar{1}$0] direction in the sample with a small number of sheets (see Fig. 4.11a). Note that Fig. 4.11c shows a strain contrast rather than the real shape of the QDs. In the sample with a large number of sheets, the TEM image shows the formation of long chains of QDs which make the structure wire-like rather than dot-like (Fig. 4.11b). Surface PL investigations showed a dominating [1$\bar{1}$0] polarization for all samples and bands. The polarization anisotropy, defined as $P = (I_\| - I_\perp)/(I_\| + I_\perp)$, where $I_\|$ and $I_\perp$ are PL intensities along the [1$\bar{1}$0] and [110] directions, reaches its maximum value of 0.75 at the 10-sheet structure. Emission is then almost linearly polarized in the surface plane.

The possibility of controlling PL polarization is of great importance, for example, in semiconductor laser applications. Hence, the control of edge PL polarization and an existing transition from TE to TM polarization plays a significant role in edge-emitting lasers. The possibility of obtaining nearly linear polarization of the surface emission is of key importance for surface-emitting lasers.

In Sect. 4.2, we demonstrate the possibility of controlling PL polarization in a different system, namely, multisheet arrays of submonolayer CdSe/ZnSe QDs. This confirms once again that the concept of exciton wave function engineering is universal and applies to any QD system.

4.1.6 Surface Evolution During Overgrowth of Strained Islands

The stacking of strained islands in vertical columns and the detailed behavior of the system during overgrowth of strained islands are key issues that provide new tools for nanostructure engineering. Xie et al. [4.4] focused on the evolution of the growth front during the overgrowth of InAs islands by GaAs. Four monolayers (ML) of InAs were deposited on GaAs at a substrate temperature of 480°C and an As_4 beam equivalent pressure of 6×10^{-6} torr. For the cap layer, five periods of $(Al_{0.25}Ga_{0.75}As)_3/(GaAs)_{15}$ were grown at 480°C. Figure 4.12a shows a cross-sectional transmission electron microscopy image taken along the [011] direction (perpendicular to the As dimer on an As-stabilized (2×4) GaAs(100) surface. Visible in these images are the InAs islands, the InAs wetting layer between the islands, the GaAs cap layer with AlGaAs marker layers, and the strained region (dark) in the substrate below

252   4. Engineering of Complex Nanostructures

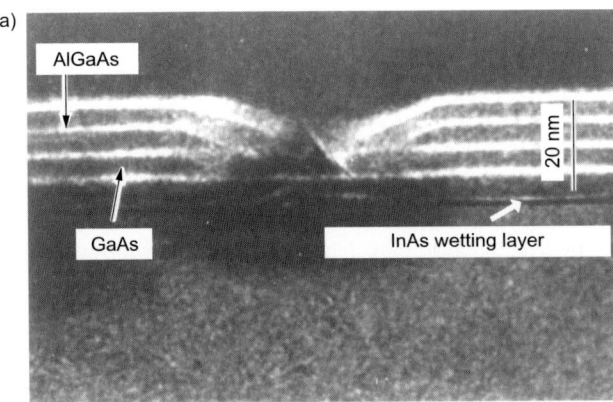

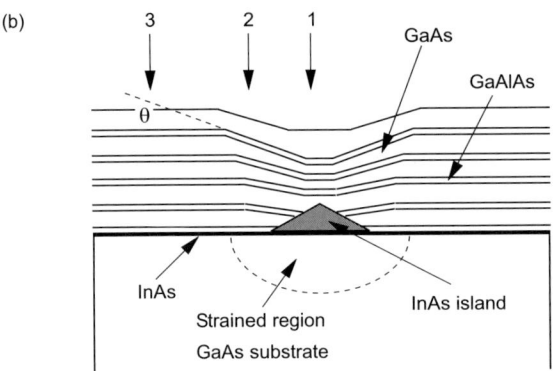

Fig. 4.12. Evolution of the growth front during overgrowth of strained InAs islands. (**a**) Dark-field cross-sectional TEM image showing the tilted nature of AlGaAs marker layers viewed along the [011] azimuth. (**b**) Schematics of GaAs growth front evolution identifying the distinct regions. With kind permission of Q. Xie et al. [4.4]

the island. An important feature are depressions with lateral dimensions of roughly 450 Å, visible in the GaAs cap layer just above the InAs islands. These regions can be classified according to the AlGaAs marker layer images in Fig. 4.12a:

- a central region, where the contrast of the AlGaAs marker layers is lost,
- the tilted region, which represents the transition from the central region to a flat region,
- the flat region, away from the InAs islands, where marker layers are approximately parallel to the interface plane.

A simplified schematic of the marker layer profile evolution around an InAs island is shown in Fig. 4.12b. A cross-sectional TEM image in a perpendicular ([01$\bar{1}$]) direction reveals similar features.

Xie et al. [4.4] explained the observed features as being caused by mass transport away from the island surface during GaAs cap layer growth. This is a consequence of the inhomogeneity of the strain field due to the partially strain-relaxed InAs islands and the wetting layer in-between having a local lattice parameter close to that of bulk GaAs. Kinetic modeling of the overgrowth of GaAs islands [4.4] showed qualitative agreement with the observed values of growth rate reduction around InAs islands. In addition, it was argued that strain-driven migration away from islands was the dominant cause underlying growth reduction (rather than surface curvature) at the initiation of cap layer growth.

The substrate temperature was found to play an important role, because overgrowth at the lower temperature 420°C leads to fairly flat marker layers around InAs islands [4.4]. In contrast, overgrowth at higher temperatures reveals dramatic changes in the shape of the InAs islands themselves. Garcia et al. [4.33] grew InAs islands by MBE on GaAs(100) substrates, annealing for 40 s under 1×10^5 torr arsenic pressure at 530°C to narrow the island size distribution. Islands were capped with GaAs from (20 to 75 Å) at the same temperature and cooled down for atomic force microscopy (AFM) measurements. AFM data from partially covered islands present a crater-like depression in the middle for the 20 Å capping layer, which becomes a clear hole for 40 Å and, for thicker GaAs (50 Å), changes into a depression (≈ 5 Å). This feature disappears for thicker capping layers, where initial InAs islands are likely to be completely capped. The formation of craters inside initial InAs islands may be employed to fabricate nanoscopic semiconductor quantum rings [4.34] with unique magnetic properties.

4.1.7 Defect-Reduction Techniques

The problem of lattice matching between the constituent materials in semiconductor heterostructures is critically important. The first double heterostructure laser, realized using lattice-mismatched GaAs–GaAsP materials, did not perform sufficiently well to be used in practical applications. Progress in this area only began once lattice-matched heterostructures had been developed. Lasers were obtained with low threshold current density at room temperature and continuous wave operation was achieved at room temperature. Only a relatively small lattice mismatch can be tolerated in high-performance devices, for example, in GaAs–AlGaAs heterostructures. In the case of a small lattice mismatch, layer growth occurs pseudomorphically, and the layer accumulates significant strain energy. For example, Tsang [4.35] described a GaAs/AlGaAs heterostructure laser with InGaAs active layer. Indium was incorporated into the active layer in order to increase the output wavelength to 0.94 µm. However, above a certain thickness, or composition, the strain energy becomes very high and dislocations start to form, thus ruining device performance. The critical thickness for dislocation formation rapidly decreases with increasing lattice mismatch. For InGaAs–GaAs layers, this results in a fast degradation of luminescence properties at InGaAs layer thicknesses corresponding to the practically important wavelength range 1.3–1.6 μm.

Propagating dislocations typically occupy a negligibly small surface area of the plastically relaxed layer. Regions between dislocations may remain structurally and optically perfect and their sizes may approach the micrometer scale, even with a high lattice mismatch and thick plastically-relaxed layers. The exact thickness for dislocation formation, the density of dislocations formed after plastic relaxation, and the degree of deterioration of optical properties may depend on the particular surface morphology and deposition conditions. Under certain growth sequences, bright luminescence in the range up to 1.35 μm may be realized at room temperature using highly strained InGaAs quantum wells, and up to 1.7 μm using thicker graded-composition InGaAs layers on GaAs substrates. Injection lasing has been demonstrated at 1.17 μm using an InGaAs quantum well grown on top of the GaAs-inserted strained ultrathin InGaAs buffer layer. Earlier attempts failed to move the lasing wavelength to larger values using conventional InGaAs quantum wells. Therefore, in order to use longer wavelengths with GaAs substrates, it was found necessary to use either different material systems, such as InGaAsN–GaAs or GaAsSb–GaAs, or to apply different growth approaches, such as those using an elastic strain relaxation effect in the Stranski–Krastanow growth mode. This process, accompanied by overgrowth of the islands, may result in the formation of strained coherent nanodomains, also called quantum dots, emitting up to and beyond 1.3 μm.

However, these approaches do not themselves yield a device with practically acceptable parameters, and cost-effective and reliable technology. In many cases, they result in the formation of propagating dislocations and other defects (e.g., dislocation loops, defect dipoles, dislocated islands). Numerous patents have tried to overcome the problem of lattice mismatch in fabricating semiconductor heterostructure devices. U.S. Patents [4.37–4.39], for example, used strain-compensation regions inside or near the active region of the device. Other patents, such as [4.40–4.43], employed dislocation filtering techniques to prevent dislocation propagation in the active layer or reduce their density. U.S. Patents [4.44–4.46] applied complicated growth methods on profiled substrates. All these approaches, however, had only a limited success or were not cost-effective.

There was thus a pressing need for a method which would eliminate local regions of the relaxed semiconductor structure in the vicinity of dislocations in situ, without any of the additional processing steps required by previous techniques.

Sample Techniques for Defect Reduction. The core of the defect-reduction technique proposed by Ledentsov [4.36] involves the application of a quite revolutionary approach, where dislocations are used as key elements to produce defect-free nanostructures or epilayers. The technique uses a com-

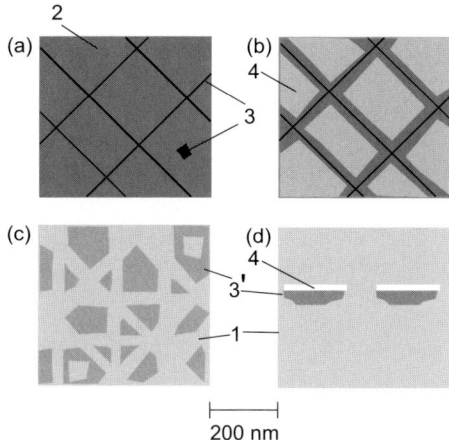

Fig. 4.13. Major steps in the defect-reduction process. (**a**) A lattice-mismatched epitaxial layer is deposited on top of the substrate (1), which contains dislocations and local defects (3), separating defect-free regions (2). (**b**) A temperature stable material (4) is deposited. It has a different lattice parameter from the epilayer, closer to that of the substrate. Due to elastic repulsion between the deposit (4) and plastically-relaxed areas of the epilayer (2) in the vicinity of defects (3), the regions near the defects remain uncapped with material (4). (**c**) Thermal etching eliminates dislocated regions and/or forms trenches in the epilayer. (**d**) Schematic cross-section of the structure after overgrowth (e.g., with matrix material)

bination of defect engineering and self-organized nanofabrication. It includes epitaxial overgrowth of the initially-dislocated substrate (or epitaxial layer) with a lattice-mismatched film, deposition of a thermally-stable cap material with a different lattice parameter, in such a way that dislocated areas become uncapped due to elastic interactions, and in situ thermal etching. The dislocated regions may be selectively evaporated leaving nanoscale domains on the surface. These domains can either be covered by matrix material or used as templates for the epitaxy of dislocation-free materials, in a similar way to epitaxy on porous substrates, even if no ex situ oxidation/etching is applied.

Numerous modifications of the technique are available to meet almost every demand of the modern semiconductor industry, from the growth of defect-free III–V epilayers on silicon and sapphire substrates to defect-free long-wavelength (1.3 μm) lasers on GaAs substrates using InGaAs layers. Transistors, lasers, semiconductor optical amplifiers, and photodetectors can all be produced with enhanced performance using this generally complex multistage nanotechnology.

Formation of Coherent Nanodomains from Initially Heavily Dislocated Films. An extension of the defect-reduction technique (Fig. 4.13) can be used for direct fabrication of nanostructures using dislocation networks as templates. The elastic interaction of the deposited material

256   4. Engineering of Complex Nanostructures

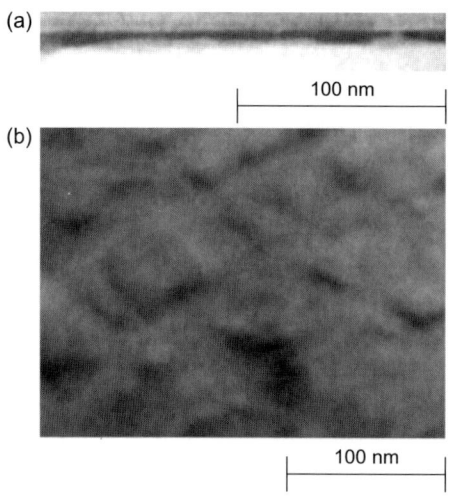

Fig. 4.14. TEM images of coherent InGaAs domains fabricated from a heavily-dislocated $In_{0.3}Ga_{0.7}As$ layer. (**a**) Cross-section. (**b**) Plan view

with existing defects and dislocations in the lattice-mismatched layer may result either in decoration of the dislocations with the deposit material or, on the contrary, in the repulsion of the deposit from the dislocation regions, depending on the lattice mismatch between the deposit, the substrate, and the dislocated layer. If the deposit is being repelled from the plastically-relaxed dislocated region (e.g., AlAs deposition on a plastically-relaxed InGaAs layer grown on top of a GaAs substrate), the defect-containing region remains uncovered by the deposit. If the deposit has a higher temperature stability than the defect-rich layer, an in situ evaporation technique can be applied. We note that structures with dense and ordered arrays of dislocations can be used for both nanostructure fabrication and production of coherent pedestals for further nanoepitaxy (epitaxial lateral overgrowth, etc.). InGaAs domains measuring 30–50 nm (see Fig. 4.14) were successfully fabricated from plastically relaxed 20-nm-thick $In_{0.3}Ga_{0.7}As$ layers by 20-nm-thick AlAs cap layer deposition and annealing at 750°C. After the annealing step, the structure was capped with AlGaAs material. The resulting PL efficiency at room temperature was increased by a factor of 400. Injection lasers were created using InGaAs domains emitting in the long-wavelength spectral range.

Elimination of Dislocated Islands. Spontaneous formation of mesoscopic dislocated islands is characteristic of the growth of highly-strained materials. The typical size of these defects is 5–20 nm. Such objects ruin device performance of the related structures in many applications. A dislocated island creates a local narrow gap region with ultrahigh nonradiative recombination rate and acts as an enormous leakage channel. It may also severely affect optical losses and cause short cuts in p–n junctions.

The traditional way of eliminating dislocated islands was based on the assumption that dislocated islands are larger than coherent ones and, in par-

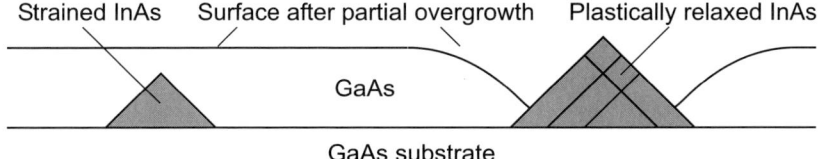

Fig. 4.15. Schematic view of the partial overgrowth of islands. It is possible to stop the overgrowth at a moment when islands below a certain height are covered, while islands exceeding that height remain uncovered. Subsequent annealing leads to evaporation of uncovered islands

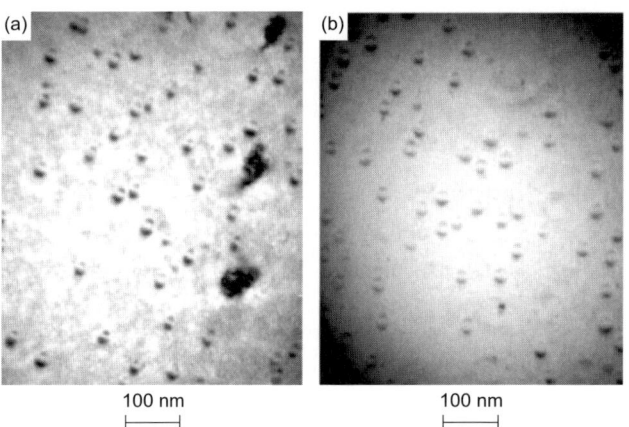

Fig. 4.16. Plan-view TEM images of the structures grown without (**a**) and with (**b**) an annealing step at 600°C after 8 nm GaAs deposition and prior to further GaAs overgrowth

ticular, exceed coherent islands in height. Figure 4.15 shows schematically the method for partial overgrowth of the islands. This method allows us to select islands by height, so that islands exceeding a certain height remain uncovered. Subsequent annealing leads to the evaporation of uncovered InGaAs islands. Holes forming at the positions of evaporated islands are filled by the subsequent overgrowth. Figure 4.16 shows experimental results on the overgrowth of small InGaAs islands.

However, this technique is based on the simplest possible assumption, namely that dislocated islands are higher than coherent ones. This is not always the case. Furthermore, even flat dislocated islands are eliminated. It has been shown by Xie et al. [4.4] that the overgrowth of coherently strained InAs islands by GaAs proceeds in such a way that the growth front is curved in the vicinity of the islands, and the overgrowth is faster in regions between islands than close to islands. It is argued in [4.4] that it is energetically more favorable for gallium atoms to migrate away from the strained regions close to the islands, where the local lattice parameter differs from that of the bulk

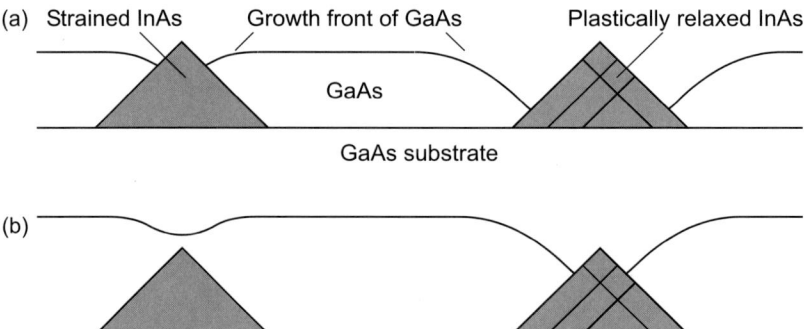

Fig. 4.17. Using the effect of the curved growth front to eliminate dislocated islands. (**a**) In the first stage, fully plastically-relaxed InAs islands, with local lattice parameter close to that of bulk InAs, repel gallium atoms much more strongly than coherently strained islands, which have a lattice parameter intermediate between those of GaAs and InAs. (**b**) In the next stage, coherent islands are overgrown, whereas dislocated islands remain uncapped

GaAs, to less strained regions between the islands, where the local lattice parameter is close to that of GaAs.

These arguments can be extended to the case where both coherent and dislocated islands are present in the system and subject to overgrowth by GaAs. Plastically-relaxed InAs islands have a local lattice parameter close to that of InAs, whereas coherently strained InAs islands have a lattice parameter with an intermediate value between a_{GaAs} and a_{InAs}. The growth front at dislocated islands is then more curved than at coherent islands (Fig. 4.17a), and the overgrowth of dislocated islands proceeds much more slowly than that of coherent islands. At some point, coherent islands are overgrown, whereas dislocated ones remain uncovered (Fig. 4.17b) and can be annealed.

Stacking of Nano-Islands. Even if the density of large defective islands is small, stacking of strained nano-islands represents a significant problem. The underlying small nano-inclusions, which contain local defects, create long-range strain fields in the matrix and affect the growth of nano-islands in the upper layers.

In this case small islands containing defects must also be selectively eliminated. In Fig. 4.18, we show plan-view (only the upper sheet of nano-islands is trapped by the TEM foil) and cross-sectional TEM images of 3-fold-stacked InGaAs–GaAs nano-islands formed by InGaAs deposition using MOCVD. In this example of a defect-reduction technique, a thin GaAs layer was deposited. It covered coherent nano-islands, but did not cover dislocated InAs islands, which impose a higher lattice mismatch with respect to the cap GaAs layer and remain open. A properly performed annealing step left the coherent islands, while the dislocated small islands were selectively evaporated (Figs. 4.18c and d). The structure grown without the annealing step (Fig. 4.18a) was completely dislocated. Coherent dots persisted only in the

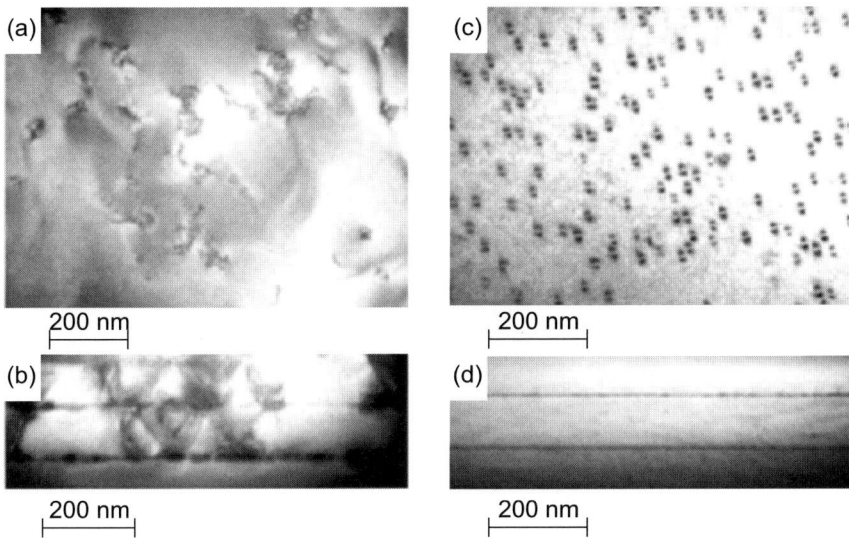

Fig. 4.18. Defect reduction in a three-layered array of MOCVD-grown InGaAs QDs. TEM images of the structure grown without [(**a**) and (**b**)] and with [(**c**) and (**d**)] overgrowth by a thin GaAs film and subsequent annealing. (**a**) Plan-view TEM image. (**b**) Cross-sectional TEM image. The structure is completely dislocated. Coherent islands persist only in the first sheet, whereas only dislocations are resolved in the upper sheets. (**c**) Plan-view TEM image. (**d**) Cross-sectional TEM image. Coherent islands persist in all layers

lower sheet in this case (Fig. 4.18b), while in the upper sheets, all InGaAs exceeding the thickness of the wetting layer was absorbed by the dislocations.

Double-Step Annealing of Defect-Rich Structures. A schematic diagram for double-step evaporation of defects is shown in Fig. 4.19. In this case, the first stage of the evaporation process occurs at 600°C. Most of the larger defect-containing nano-islands are eliminated. However, there exist small islands which contain defects, but the repulsion potential is not enough to prevent the overgrowth of these islands by GaAs. To eliminate these defective nano-inclusions, the structure is overgrown with an AlAs layer and annealed at higher temperature, sufficient for evaporation of the GaAs cap layer, which is followed by evaporation of InAs and redistribution of the material over the surface. Defect-containing islands, which are not covered by AlAs, become open once their GaAs capping has evaporated, and defect-containing InAs nano-islands are eliminated.

Defect Reduction in Structures Grown at Low Temperature. Low temperature (200–350°C) growth offers some advantages for the fabrication of structures on GaAs for long-wavelength light emission. Suppression of InGaAs redistribution allows the growth of thicker planar InGaAs layers, particularly at low arsenic overpressures in molecular beam epitaxy. For highly-

260   4. Engineering of Complex Nanostructures

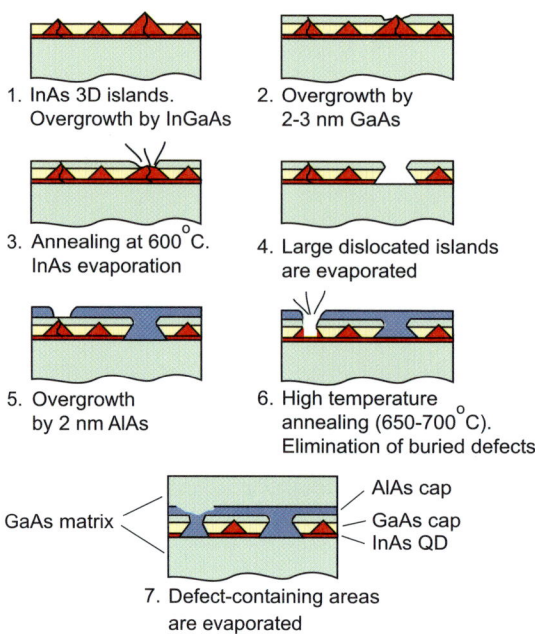

Fig. 4.19. Evaporation of defect-contaning islands using a double-step annealing procedure

strained InAs deposition on GaAs under the conventional arsenic pressure, chains or agglomerates of small islands can be formed locally, resulting in a second PL peak in the spectral range 1.5–1.6 µm (see Sect. 3.3.15). However, most of these chains contain defects (see Figs. 4.20a and b) and the overall material quality is poor.

The key points of the defect-reduction technique in this case are similar to the previous one:

- a thin GaAs cap layer is deposited so as to completely cover coherent InAs nano-objects, while plastically-relaxed objects with higher strain remain uncovered due to GaAs/AlAs repulsion from regions with high lattice mismatch with respect to GaAs;
- InAs QD structures covered by thin GaAs caps are annealed to evaporate defects;
- multi-stage annealing is applied, using AlAs sealing to evaporate the remaining defects trapped in GaAs but save coherent regions.

The impact of the annealing approach is summarized in Fig. 4.20. From TEM images taken under defect-sensitive conditions, capping of lateral associations of QDs with a thick (15 nm) GaAs cap followed by high-temperature annealing (HTA) at 700°C only slightly decreases the defect density (white spots in Fig. 4.20b). Thus, annealing itself is not a solution to the problem. On

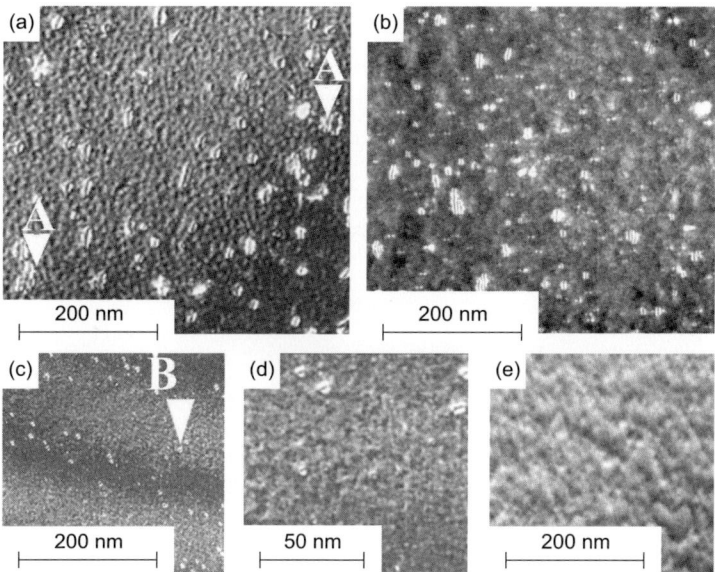

Fig. 4.20. (001) plan-view dark-field (220) TEM images of laterally associated QD structures, 3 ML InAs, 350°C. (**a**) As-grown and covered with 15 nm GaAs at the same temperature. The final GaAs cap layer is grown at 600°C. (**b**) With 2 nm AlAs sealing, 15 nm GaAs overgrowth and subsequent high-temperature annealing (HTA) at 700°C for 10 minutes. (**c**) With 4 nm GaAs overgrowth and medium-temperature annealing (MTA) at 600°C for 10 minutes. (**d**) Coherent InGaAs QDs look like fine-scale contrast modulations. (**e**) With the MTA procedure following 2 nm AlAs sealing and HTA. The label A denotes large dislocated $In_xGa_{1-x}As$ islands, whilst B refers to small dislocated $In_xGa_{1-x}As$ islands in the sample containing dislocation loops. Lateral associations of QDs look like merged chains of coherent QDs

the other hand, depositing a thin cap of 4 nm and performing growth interruption at the moderate temperature of 600°C (MTA) results in almost complete disappearance of large dislocated islands (Fig. 4.20c). However, one can still observe a high density ($N_d = 5.6 \times 10^{10}$ cm^{-2}) of QDs containing small edge-type dislocation loops (labeled B in Fig. 4.20c). TEM and high resolution TEM (HRTEM) images show that QDs subjected to MTA combined with sealing and HTA at 700°C with a thin AlAs cap exhibit a complete elimination of dislocation dipoles (Fig. 4.20e).

The integrated PL intensity was increased by about three orders of magnitude by proper adjustment of the defect-elimination approach in this case.

Defects in low temperature alloy materials (InGaAsN–GaAs, InGaN–GaN) can be eliminated using a similar approach.

Defect Reduction in QD Structures Overgrown by an Alloy. Initial InAs or GaInAs QDs can be overgrown by InGaAs or InGaAlAs alloy materials. This can eventually lead to alloy phase separation in the cap layer, activated by initial islands which act as stressors. (A detailed discussion is

(a) InAs/InGaAs + 2 nm GaAs + 600°C + 2 nm AlAs + GaAs
DF (002) along [010]

(b) InAs/InGaAs + 2 nm GaAs + 600°C + 2 nm AlAs + 700°C + GaAs
DF (002) along [010]

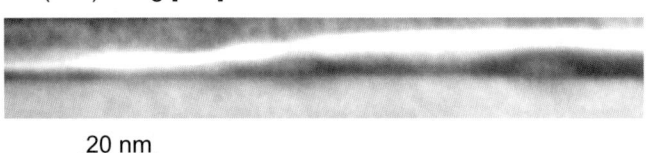

|20 nm|

Fig. 4.21. Cross-sectional TEM images. (**a**) 2.5 ML InAs islands are capped with 5 nm $In_{0.15}Ga_{0.85}As$ (activated alloy phase separation) and 2 nm GaAs layer overgrowth, followed by a temperature increase to 600°C. Larger defect-containing islands are eliminated at this stage. Then 2 nm AlAs is deposited, followed by GaAs overgrowth. Note that some QDs remain uncovered by AlAs. (**b**) A similar growth sequence, in which annealing at 700°C was introduced after AlAs deposition. Note that coherent islands remain, but those which were not covered by AlAs have evaporated

given in Sect. 4.3.) The defect-reduction technique can also be applied in this regime to eliminate dislocated islands (Fig. 4.21).

Survey of Novel Defect-Reduction Techniques. We believe that the proposed techniques for defect reduction have a significant advantage over all other reported approaches to the growth of long-wavelength (1.3–1.55 µm) structures on GaAs substrates. The technique may also be applied to:

- defect reduction in substrates,
- defect reduction in thick epilayers (GaAs on Si, InP on Si or GaAs, GaN on sapphire or Si, etc.),
- in situ formation of templates for epitaxial lateral overgrowth and pendoepitaxy,
- numerous other applications in modern solid-state technology.

To conclude this discussion, a method has been proposed for in situ fabrication of dislocation-free structures from plastically-relaxed layers grown on a semiconductor surface suitable for epitaxial growth. This method solves the problem of lattice-mismatched growth. It produces coherent dislocation-free regions from initially dislocated and/or defect-rich layers lattice-mismatched with respect to the underlying substrate, and does not contain any processing steps before or after the formation of the defect-free regions. The process uses in situ formation of a cap layer on top of a dislocated layer. The cap layer has

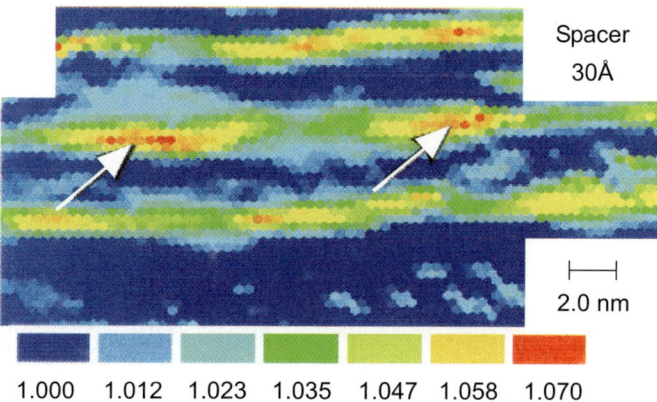

Fig. 4.22. Anticorrelation in multisheet arrays of 2-dimensional CdSe islands in a ZnSe matrix. Cross-sectional HRTEM image processed by the DALI evaluation program. A local map of the lattice parameter in the vertical direction is plotted. Larger values of the local lattice parameters (*yellow* and *red areas*) correspond to a high Cd content

a lattice parameter close to that in the underlying substrate, and different from that in the lattice-mismatched epilayer in the no-strain state. Under these conditions, the cap layer material is elastically repelled from regions in the vicinity of dislocations, where the lattice parameter differs most from that in the substrate. The cap layer is absent in these regions. When the cap layer has a lower thermal evaporation rate than the underlying dislocation layer, the regions of this dislocation layer in the vicinity of dislocations are selectively evaporated at sufficiently high temperatures, and only the coherent defect-free regions of the initially-dislocated epilayer remain on the surface.

4.2 Anticorrelation in Multisheet Arrays of Strained Islands

Vertical correlations and the formation of vertical columns of strained islands have been well understood, observed in various material systems, and widely employed to fabricate arrays of vertically coupled quantum dots (QDs). The concept of seeding has been proposed to manipulate the density and volume of QDs independently. Surprisingly, Strassburg et al. [4.47] have observed a different type of relative arrangement of islands in multi-layered arrays of CdSe islands in a ZnSe matrix. Figure 4.22 shows the anticorrelation between islands in neighboring sheets. Islands of each subsequent sheet form above spaces between islands in the previous sheet.

The observations of Strassburg et al. [4.47] were a real puzzle, contradicting all previous experimental and theoretical work. However, they were explained theoretically by Shchukin et al. [4.48].

4. Engineering of Complex Nanostructures

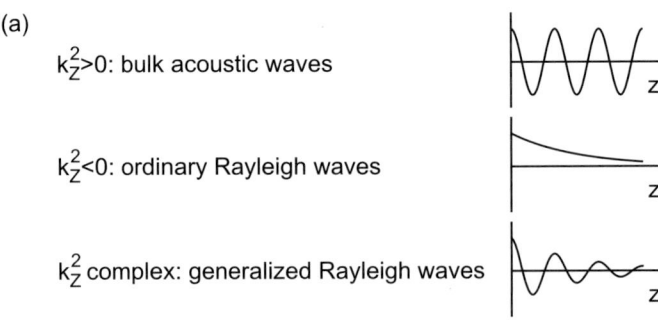

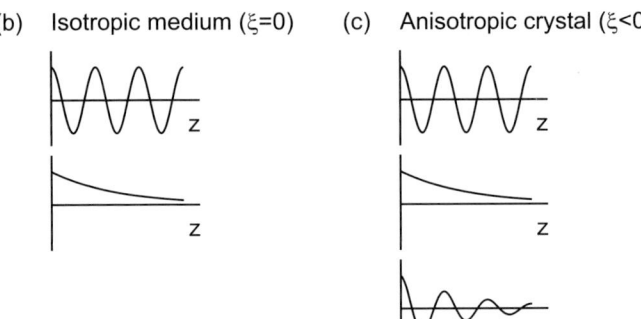

Fig. 4.23. Effect of elastic anisotropy on acoustic waves in a crystal. (**a**) Three possible types of acoustic wave. (**b**) Acoustic waves existing in elastically isotropic media. (**c**) Possible acoustic waves in elastically-anisotropic cubic crystals

4.2.1 Generalized Rayleigh Waves in Elastically Anisotropic Crystals

The key point is that typical semiconductors like Si and Ge with a diamond structure, or III–V and II–VI semiconductors with a zinc-blende structure, are elastically anisotropic cubic materials. The anisotropy of elastic properties is governed by the dimensionless parameter

$$\xi = \frac{c_{11} - c_{12} - 2c_{44}}{c_{44}}, \tag{4.2}$$

where c_{11}, c_{12}, and c_{44} are the elastic moduli in the Voigt notation. The approximation of an elastically isotropic medium corresponds to $\xi = 0$, and in typical semiconductors, first, $\xi < 0$, and second, $|\xi| \sim 1$. The elastic anisotropy turns out to be crucial for acoustic waves to propagate in the crystal [4.49]. Consider an acoustic wave with frequency ω and 2-dimensional wave vector $\boldsymbol{k} = (k_x, k_y)$ excited at a crystal surface $z = 0$. Then the propagating acoustic wave inside the crystal may be sought in the form

$$u_i(x, y, z; t) \propto \exp(-\mathrm{i}\omega t) \exp(\mathrm{i}k_x x) \exp(\mathrm{i}k_y y) \exp(\mathrm{i}k_z z), \tag{4.3}$$

where u_i is the elastic displacement vector. The wave vector k_z describing the propagation of the wave in the z direction normal to the surface is determined by the dispersion law for the acoustic waves. Figure 4.23a shows three possibilities:

- bulk acoustic waves propagating without attenuation,
- ordinary Rayleigh waves decaying exponentially away from the surface,
- generalized Rayleigh waves exhibiting oscillatory decay.

The particular realization depends critically on the symmetries of the elastic modulus tensor of the crystal. In elastically isotropic media, only two types of acoustic wave are possible (Fig. 4.23b). In elastically anisotropic cubic crystals with a negative elastic anisotropy parameter ($\xi < 0$) and bounded by a (001) surface, all three types of acoustic waves are possible (Fig. 4.23c) [4.49, 4.50].

The spatial distribution of static strain fields may be considered as a static limit ($\omega \to 0$) of acoustic waves. Correspondingly, the static Green tensor of the theory of elasticity for a semi-infinite elastically anisotropic cubic crystal bounded by a (001) stress-free surface $G_{ij}(k_x, k_y; z, z')$ obtained by Portz and Maradudin [4.51] exhibits an oscillatory decay in the z direction for a negative anisotropy parameter $\xi < 0$.

4.2.2 Formation of Multisheet Arrays in Elastically Anisotropic Crystals

To elucidate how such strain field behavior affects the formation of multisheet arrays, Shchukin et al. [4.48] examined a double-sheet array comprising one sheet of buried islands and one sheet of surface islands. As a simple model system, a double-sheet array of submonolayer islands was considered.

Let material 2 be deposited on the (001) surface of the cubic substrate 1. Upon submonolayer deposition, a periodic array of monolayer-high islands forms [4.52–4.56]. Such an array was discussed in detail in Sect. 3.2. Then the structure is capped by substrate material 1 and the second cycle of deposition of material 2 is introduced. The total energy of the surface array of islands in the strain field of the buried islands is

$$E_{\text{total}} = E_{\text{surf}} + E_{\text{bound}} + \Delta E_{\text{elast}}^{(\text{SS})} + E_{\text{elast}}^{(\text{SB})}, \tag{4.4}$$

where E_{surf} is the sum of the surface energy of surface islands and the surface energy of uncovered parts of material 1, E_{bound} is the energy of island boundaries, $\Delta E_{\text{elast}}^{(\text{SS})}$ is the elastic relaxation energy of surface islands (S) due to the discontinuity in the intrinsic surface stress tensor at island boundaries, and $E_{\text{elast}}^{(\text{SB})}$ is the elastic energy of the interaction of surface islands (S) and buried islands (B). Since we address the effects of the finite lateral size of islands and elastic anisotropy, and since we aim to avoid other complications, we focus on the typical experimental situation in which there is an equal amount of

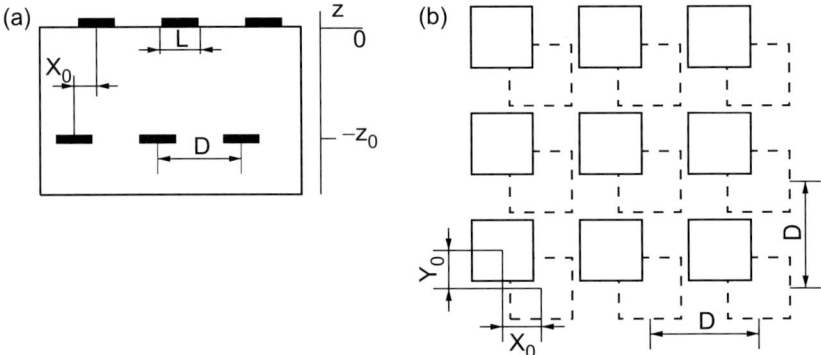

Fig. 4.24. Geometry of double-sheet arrays of 2-dimensional islands. The array of surface islands has the same structure as the array of buried islands but is shifted as a whole. (**a**) Each sheet of islands forms a 1-dimensional array of stripes. The cross-section of the double-sheet structure is shown. (**b**) Each sheet of islands forms a 2-dimensional array of square-shaped islands. The plan view of the double-sheet structure is plotted. Buried islands are depicted by *dashed lines*, and surface islands by *solid lines*

deposited material in each deposition cycle. Then each sheet of islands alone tends to form the same periodic structure, corresponding to the minimum of the sum of the first three terms on the right-hand side of (4.4). If the interaction between surface islands and buried islands is neglected, the surface array of islands as a whole can be subject to an arbitrary shift in the xy plane. The strain field created by buried islands has the same periodicity as the array of surface islands alone. Therefore the fourth term in (4.4) does not change the periodicity of the surface structure, and merely serves to define its position relative to the array of buried islands (Fig. 4.24).

Since the interaction energy $E_{\text{elast}}^{(\text{SB})}$ is the only term in (4.4) which depends on the shift of the array of surface islands as a whole with respect to the buried islands, we will focus on this energy term. Below, we consider the dependence of $E_{\text{elast}}^{(\text{SB})}$ on the shift X_0 for a 1D array of stripes (Fig. 4.24a) and on the shifts X_0 and Y_0 for a 2D array of compact islands. For simplicity, we focus on the extreme case of compact islands, as distinct from infinitely elongated stripes, i.e., on square-shaped islands (Fig. 4.24b).

To evaluate the strain due to buried islands, we first consider the strain due to point defects. Eshelby [4.57] proposed an approach whereby a point defect located at $\widetilde{r}$ is represented by the superposition of three mutually perpendicular double forces (by an elastic dipole), and the effective body force density is

$$f_i(r) = a_{ij} \nabla_j \delta(r - \widetilde{r}) . \qquad (4.5)$$

A monolayer-thick inclusion in the plane $z = \widetilde{z}$ with macroscopic lateral dimensions is a 2D array of point defects occupying every atomic site within a certain area. The island can be described by a 2D shape function $\Theta^{\mathrm{B}}(\boldsymbol{r}_\parallel)$ which equals 1 inside the inclusion and 0 otherwise. The body force density associated with the given inclusion can be obtained by adding the contributions of single point defects from (4.5). In the macroscopic approach, this summation can be replaced by integration to give

$$f_i(\boldsymbol{r}) = \frac{1}{A_0} \int d^2 \widetilde{\boldsymbol{r}}_\parallel a_{ij} \nabla_j \left[\delta(\boldsymbol{r}_\parallel - \widetilde{\boldsymbol{r}_\parallel})\delta(z - \widetilde{z})\right] \Theta^{\mathrm{B}}(\widetilde{\boldsymbol{r}_\parallel}) , \qquad (4.6)$$

where A_0 is the unit cell area in the xy plane. Equation (4.6) is derived under the assumption of no mutual influence between the point defects comprising the inclusion. Generally speaking, the tensor a_{ij} characterizing the double-force density is different for a single point defect and a monolayer-thick inclusion. A substitutional impurity atom in a zinc-blende crystal of a III–V or II–VI semiconductor has T_d site symmetry, and the corresponding double-force tensor a_{ij} has cubic symmetry. On the other hand, if an inclusion of equal substitutional impurity atoms is oriented in the (001) plane of the zinc-blende crystal and has monolayer thickness and infinite lateral dimensions, each atom of the inclusion has $D_{2\mathrm{d}}$ symmetry. Therefore, the tensor a_{ij} characterizing a monolayer-thick buried island has uniaxial symmetry, with $a_{xx} = a_{yy}$, and $a_{zz} = P^{\mathrm{B}} a_{xx}$, where P^{B} is the parameter of the uniaxial anisotropy of the double forces created by buried islands. For an inclusion having all three macroscopic dimensions, the double-force density is related to the lattice mismatch between the inclusion and the matrix and, for a cubic inclusion in a cubic matrix, has cubic symmetry once again:

$$a_{ij} = \frac{1}{v}(c_{11} + 2c_{12})\frac{\Delta a}{a}\delta_{ij} , \qquad (4.7)$$

where v is the unit cell volume.

The elastic properties of surface islands are described by the difference between the 2-dimensional intrinsic surface stress tensors of the two materials ($\Delta \tau_{\alpha\beta}$). The energy of the elastic interaction between a periodic array of buried islands and a similar periodic array of surface islands may be written in the form of a sum over the reciprocal lattice vectors [4.48], viz.,

$$E_{\mathrm{elast}}^{(\mathrm{SB})} = \frac{h^{\mathrm{B}}}{A_0} \sum_{\boldsymbol{k}_\parallel} \left|\widetilde{\Theta}(\boldsymbol{k}_\parallel)\right|^2 \exp\left(i\boldsymbol{k}_\parallel \cdot \boldsymbol{R}_0\right) (\Delta\tau_{\alpha\beta}) a_{lm} \qquad (4.8)$$

$$\times \nabla'_m \left[\nabla_\alpha \widetilde{G_{\beta l}}\left(\boldsymbol{k}_\parallel; z, z'\right) + \nabla_\beta \widetilde{G_{\alpha l}}\left(\boldsymbol{k}_\parallel; z, z'\right)\right]\bigg|_{\substack{z=0 \\ z'=-z_0}} ,$$

where $\boldsymbol{R}_0$ is the relative shift of the two arrays, h^{B} is the thickness of the buried islands, and $\nabla_x \equiv ik_x$, $\nabla'_x \equiv -ik_x$, $\nabla_y \equiv ik_y$, $\nabla'_y \equiv -ik_y$, $\alpha, \beta=1,2$, $l, m=1,2,3$. Since we focus on III–V and II–VI semiconductors with zinc-blende structure, the crystal is an elastically anisotropic cubic medium. The static elastic Green tensor $\widetilde{G}_{il}(\boldsymbol{k}_\parallel; z, z')$ was obtained by Portz and Maradudin [4.51].

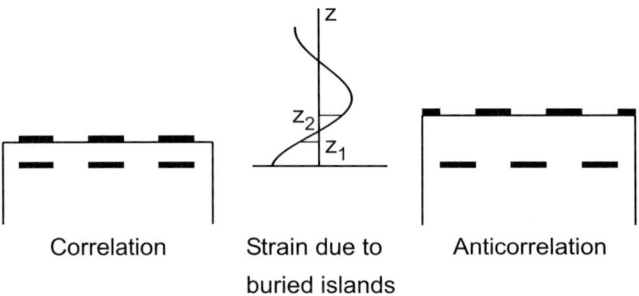

Fig. 4.25. Strain field at the surface created by a sheet of buried islands. The schematic plot shows an oscillatory decay of the strain with spacer thickness. Depending on the spacer thickness, the two sheets of islands exhibit a correlated (*left*) or anticorrelated (*right*) arrangement

The dependence of $E_{\text{elast}}^{(\text{SB})}$ on the separation between the two sheets is determined by the behavior of $\widetilde{G}_{il}(\mathbf{k}_\parallel; z, z')$ as a function of z_0. The Green tensor $\widetilde{G}_{il}(\mathbf{k}_\parallel; z, z')$ is a linear combination of three exponentials $\exp(-\alpha_s k_\parallel z_0)$, where the three attenuation coefficients α_s are functions of the direction of $\mathbf{k}_\parallel$ in the surface plane. The key point is that, in a cubic crystal with a negative elastic anisotropy parameter $\xi < 0$, which is the case for all III–V and II–VI cubic semiconductors, two of the three α_s are complex conjugate [4.51], corresponding to generalized Rayleigh waves.

As a consequence, the elastic strain created by buried islands exhibits an oscillatory decay with the distance of the islands from the sheet (Fig. 4.25). Therefore, the interaction between successive sheets of islands exhibits an oscillatory decay with separation between the sheets, i.e., with the spacer thickness.

The interaction energy (4.8) was evaluated in [4.48] for double-sheet arrays of stripes and double-sheet arrays of square islands. For a separation between the two sheets $z_0 \leq 0.5 D_0$, where D_0 is the lateral period, the difference between the two values of $E_{\text{elast}}^{(\text{SB})}$, the one for the most favorable relative arrangement, and the other for the most unfavorable arrangement, is of the order of 0.1 meV/Å^2. This is the same order of magnitude as the typical energy of a single sheet of surface islands. This comparison confirms that the elastic interaction between the two sheets of islands can indeed result in vertical correlation or anticorrelation between the two sheets.

Figure 4.26 shows the phase diagram of a double-sheet array of monolayer-high islands. In particular, a periodic array of stripes is treated at surface coverage $q = 0.5$. The relative arrangement of the two sheets of stripes is determined by two parameters. One parameter is the ratio z_0/D of the spacer thickness to the in-plane period. The other parameter is the anisotropy $P^{\text{B}} = F_z/F_x$ of the forces characterizing an elementary elastic dipole. The diagram shows alternating correlation and anticorrelation. At a small spacer

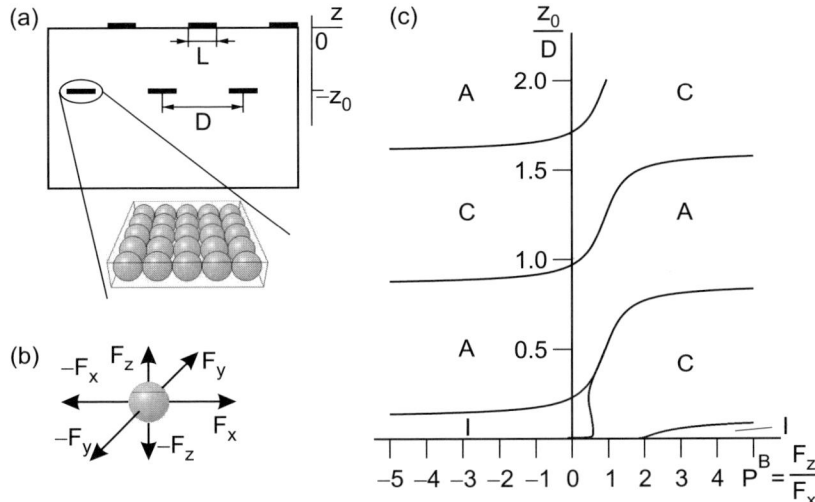

Fig. 4.26. Schematic structure and phase diagram of a double-sheet array of stripe-shaped monolayer-high islands. (**a**) Cross-section of the structure. One buried island is shown at a larger magnification to consist of elastic dipoles. (**b**) A single atom in the buried island is shown schematically as an elastic dipole with unequal pairs of forces in the vertical and lateral directions. The anisotropy of the elastic dipole is characterized by the parameter P^B. (**c**) Phase diagram of the double-sheet array of islands exhibiting parameter regions of correlation (C), anticorrelation (A), and an intermediate arrangement (I)

thickness, some intermediate arrangements occur, in which the two sheets are shifted in the lateral direction by less than half a period, $X_0 < D/2$. A similar phase diagram for a double sheet array of square-shaped islands has been constructed by Shchukin et al. [4.48] and shows a similar alternation of correlation and anticorrelation with the spacer thickness.

One should bear in mind that alternating correlation and anticorrelation is a fundamental property of elastically anisotropic crystals and should persist also for other types of islands, e.g., for multisheet arrays of 3-dimensional pyramid-shaped islands. For those islands, the dipole forces are isotropic and the parameter $P^B = 1$.

4.2.3 Multisheet Arrays of CdSe/ZnSe Submonolayer Islands

After anticorrelation had been observed experimentally by Strassburg et al. [4.47] and explained theoretically by Shchukin et al. [4.48], further experimental studies of a similar multisheet system of CdSe submonolayer islands in a ZnSe matrix were carried out by Strassburg et al. [4.58] and, in more detail, by Krestnikov et al. [4.59].

Structural Characterization. To investigate the structural properties of the samples, high resolution transmission electron microscopy (HRTEM)

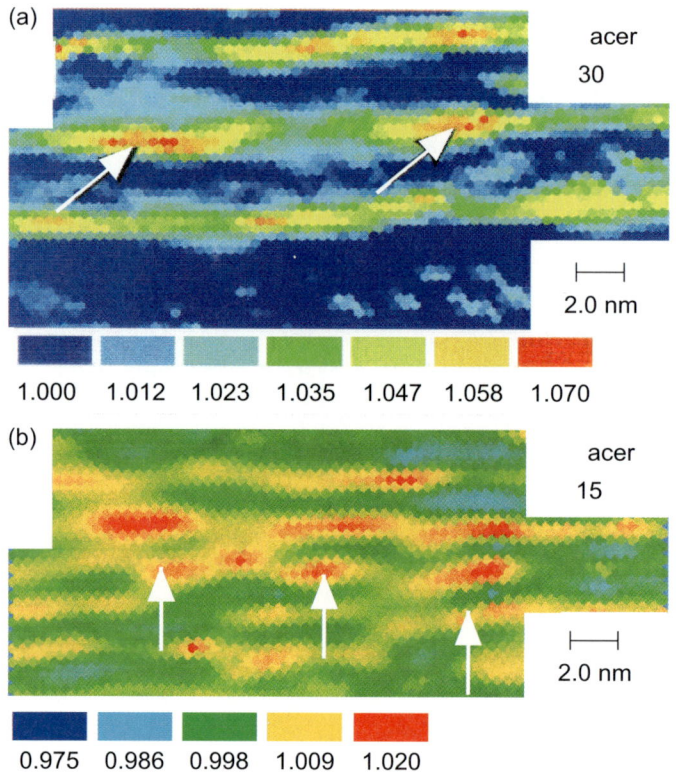

Fig. 4.27. Local lattice parameter map for multisheet arrays of CdSe islands in a ZnSe matrix. Images were obtained by HRTEM and processed by the DALI evaluation program. (**a**) Spacer thickness 30 Å. The map is normalized to the ZnSe local lattice parameter. (**b**) Spacer thickness 15 Å. The average local lattice parameter is used for normalization

measurements were carried out [4.58, 4.59]. DALI processing was performed to reveal the shape and size of islands. Figures 4.27a and b show color-coded maps of the local lattice parameter (LLP) in the growth direction for structures with 30 Å and 15 Å spacer layer thickness, respectively. The local lattice parameters are measured for every projected unit cell with dimensions $a_{[110]} \times a_{[001]}$. A reference lattice parameter is determined by which the LLPs are normalized. The reference lattice parameter for Fig. 4.27a was chosen to be the ZnSSe lattice constant resulting in LLPs between 1 and 1.07. An average lattice parameter was determined from the whole image as the reference value in Fig. 4.27b. Therefore, normalized LLPs smaller than 1 are observed in the ZnSe spacers. A shift in the color from blue to red corresponds to an increase in the lattice parameter in the vertical direction a_z. Thus, green, yellow and red areas indicate (Cd,Zn)Se layers with larger LLP and, consequently, with higher Cd content, due to the larger bulk lattice parameter of CdSe (6.081 Å) compared to that of ZnSe (5.6697 Å).

Figure 4.27 clearly shows the different behaviour of the two samples with different spacer thicknesses. Areas with increased Cd content have lateral sizes of 40–50 Å for 30 Å spacers and 25–30 Å for 15 Å spacers. It should be noted that the measured Cd composition of the islands may be reduced due to the averaging effect along the thickness of the foil, which ranges between 5 and 20 nm and may therefore exceed the lateral island size. The estimated height of Cd-rich insertions is about 2 ML and 4 ML. A broadening of the Cd distribution along the growth direction may be induced by steps along the electron beam or by segregation of cadmium.

The relative arrangement of CdSe islands demonstrates a remarkable difference in the two structures. Close inspection of Fig. 4.27a reveals that the islands in the top sheet tend to form at positions between islands in the bottom sheet, i.e., an anticorrelated arrangement occurs. On the other hand, in the structure with 15 Å spacers (Fig. 4.27b), islands form vertical columns locally, extended over 2–5 sheets, i.e., they exhibit a tendency to correlate vertically. These observations are in agreement with the theory of Shchukin et al. [4.48], where transitions were predicted from correlation to anticorrelation with changing spacer thickness.

It should be noted, however, that no continuous vertical columns were observed, like those in the structures with 3-dimensional islands. Furthermore, some of the columns are tilted. These effects may be attributed to the great complexity of the system. Moreover, the HRTEM images of Fig. 4.27 show only a small area of the cross-section. Information about the whole structure may be extracted from optical measurements.

Optical Characterization. The properties of the electronic states in quantum islands have been studied by optical investigations. Photoluminescence (PL) spectra of structures with different spacer layer thickness are shown in Fig. 4.28. For a large spacer thickness, only one line appears. The transition energies correspond to heavy-hole-like excitons, localized at 2-dimensional nanoscale islands with lateral extensions comparable to the exciton diameter. On the other hand, at a spacer layer thickness of 30 Å, a second lower energy luminescence line evolves. This may be attributed to the local formation of vertically correlated islands (while most islands are anticorrelated!), which results in an efficient coupling of electronic states between islands of the same column, since the main part of the wave function extends into the barrier. The low energy shift in luminescence is similar to what happens for vertically coupled QDs, as demonstrated first for 3D InAs–GaAs QDs [4.8]. A further decrease in the spacer thickness to 15 Å results in an increased intensity of the PL due to enhanced coupling of QDs and a complete suppression of luminescence from uncoupled islands.

The complex character of the structure with 30 Å spacer thickness required additional optical measurements to obtain further information about

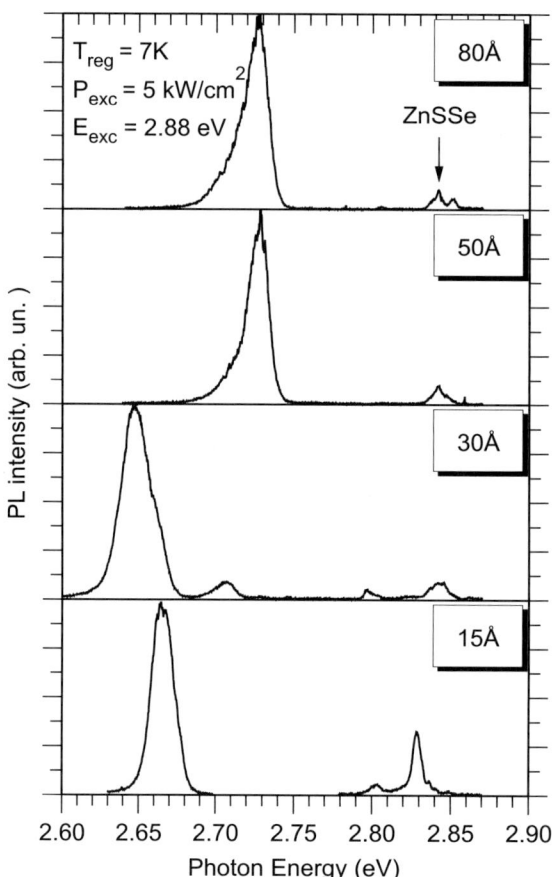

Fig. 4.28. Photoluminescence (PL) spectra of multisheet arrays of CdSe islands in a ZnSe matrix with different ZnSe spacer thicknesses. The registration temperature T_{reg} is 7 K, the excitation density P_{exc} is 5 kW/cm^2, and the energy E_{exc} of the exciting photon is 2.88 eV

electronic states. Figure 4.29 displays the photoluminescence (PL) spectrum and the second derivative of the optical reflectivity (OR) spectrum. The main PL line originates from coupled QDs, while the main feature in the OR spectrum appears at 2.71 eV and corresponds to uncoupled states. This can be explained by the fact that, even if the density of uncoupled QDs is still larger than that of coupled QDs, the efficient hopping and tunneling of carriers from uncoupled states with a higher energy results in predominant population of the coupled states and photoluminescence from these states.

Further important information about the electronic states may be extracted by analyzing the polarization of edge emission from the multisheet array of 2-dimensional islands (submonolayer quantum dots, SML QDs). The edge emission of uncorrelated or anticorrelated SML QDs, i.e., from electron-

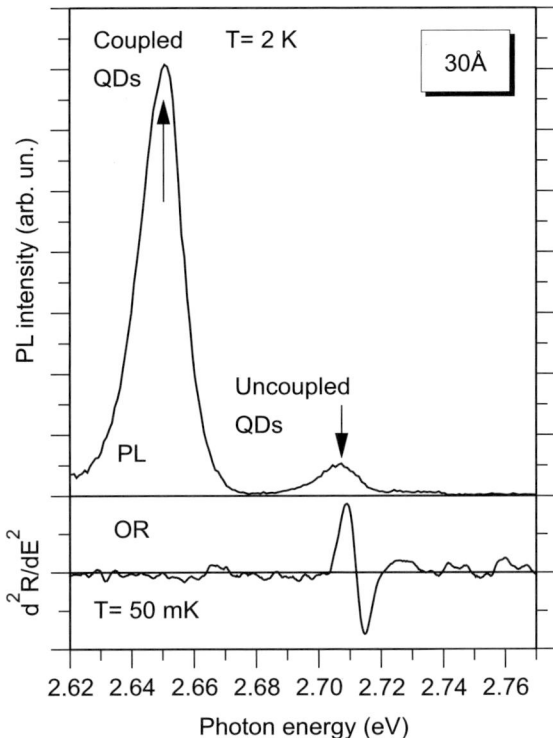

Fig. 4.29. Photoluminescence (PL) spectrum and the second derivative of the optical reflectance spectrum for the multisheet CdSe/ZnSe structure with 30 Å spacers

ically uncoupled states, is predominantly TE polarized [4.60]. This means that the electric field vector in the emitted light is directed perpendicularly to the plane of the islands, i.e., parallel to the growth direction. The reason is that a 2D island shape results in much stronger heavy-hole quantization in the growth direction than in the lateral direction. Thus, the heavy-hole-like exciton emission is predominantly TE polarized (50–60%), even if lateral quantization also enables a TM-like emission, where the electric field is directed in the plane of the islands. This depolarization is not possible in a quantum well, where the heavy-hole exciton emission is completely TE polarized. It should also be noted that for spherical or cubic QDs, no predominant polarization of the exciton luminescence should be observed. Since vertical coupling of islands results in an extension of the heavy-hole wave function in the growth direction, the quantization effect is reduced. Therefore, a different polarization of the low energy line is expected. If the extension of the wave function in the vertical direction exceeds the lateral extension, the PL spectrum is expected to be predominantly TM polarized.

274   4. Engineering of Complex Nanostructures

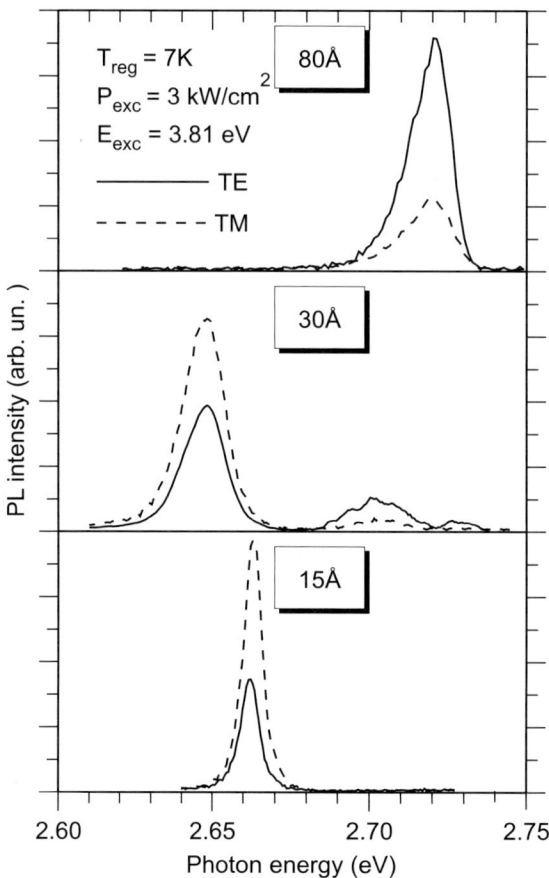

Fig. 4.30. Linearly polarized photoluminescence (PL) spectra of edge emission for structures with 80, 30, and 15 Å spacers. $T_{\mathrm{reg}} = 7$ K, $P_{\mathrm{exc}} = 3$ W/cm^2, and $E_{\mathrm{exc}} = 3.81$ eV. The PL energies of the low-energy transition of the 30 and 15 Å spacer samples differ due to the different lateral sizes of the islands

Figure 4.30 shows linearly polarized PL spectra of structures with 80 Å, 30 Å, and 15 Å spacers, measured in the edge geometry. Focusing on the structure with 30 Å spacers, one should note the following. Firstly, the polarization of the low-energy line which is related to coupled QDs is reversed with respect to the polarization of anticorrelated islands. Secondly, the high energy line shows the same polarization as the PL line from uncoupled islands for the sample with 80 Å spacers. This confirms that PL in the 30 Å sample comes from uncoupled electronic states, i.e., from anticorrelated islands. Furthermore, the low energy PL exhibits similar polarization to that from the sample with 15 Å spacers, indicating the formation of columns of vertically correlated islands and hence of coupled electronic states. This result clearly

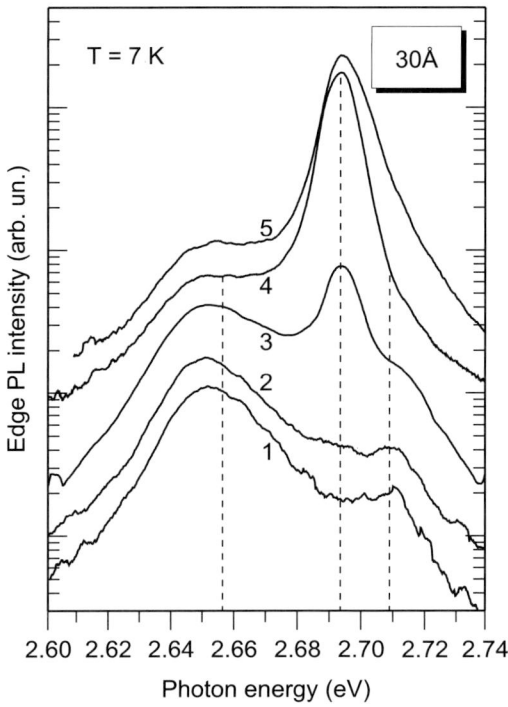

Fig. 4.31. Edge photoluminescence (PL) spectrum from the multisheet CdSe/ZnSe structure with 30 Å spacer at different excitation densities P_{exc}: (curve 1) 20 kW/cm^2, (curve 2) 100 kW/cm^2, (curve 3) 500 kW/cm^2, (curve 4) 1 MW/cm^2, (curve 5) 2 MW/cm^2

points to a delocalization of the heavy-hole-like QD exciton state along the growth axis and manifests the formation of coupled QD states. The effects are similar to those observed in multisheet arrays of 3-dimensional InGaAs/GaAs islands (see Sect. 4.1.5).

The change in PL polarization in multisheet arrays of CdSe/ZnSe submonolayer islands clearly demonstrates a new way of engineering QD exciton wave functions using the correlated QD growth and provides unique possibilities for polarization control of edge-emitting lasers. We note here that the degree of polarization of the edge emission is uniform within the shape of the corresponding line.

Figure 4.31 demonstrates the dependence of edge photoluminescence from the structure with 30 Å spacers on the level of excitation. At moderate excitation density, the PL spectrum reveals a maximum at 2.65 eV, corresponding to light emitted from coupled QDs. As the excitation density increases, the low energy levels saturate, and PL from higher electronic levels increases. At the excitation density 500 kW/cm^2 and higher, the maximum at $\approx$ 2.69 eV dominates in the PL spectrum. This maximum corresponds to PL from uncoupled QDs. At the excitation density 1 MW/cm² and higher, the PL from uncoupled QDs exceeds that from coupled QDs by more than an order of magnitude. These data confirm once again that most QDs in the structure with 30 Å spacers are electronically uncoupled.

Thus, comprehensive structural and optical characterization of multisheet arrays of submonolayer CdSe islands in a ZnSe matrix with different thicknesses of ZnSe spacers, carried out by Strassburg et al. [4.58] and Krestnikov et al. [4.59], confirmed the theory of Shchukin et al. [4.48], which had predicted transitions from vertically correlated to vertically anticorrelated growth. The complex structure of the sample with 30 Å spacer, where both anticorrelated and correlated islands exist, suggests that this structure is close to the transition point. From Fig. 4.27a, one may approximate the characteristic lateral period of the structure as 120 Å. For spacer thickness $z_0 = 30$ Å, the parameter $z_0/D = 0.25$. From the phase diagram of Fig. 4.26, it follows that a transition from correlation to anticorrelation occurs at $z_0/D = 0.25$, if the anisotropy P^B of the dipole forces characterizing buried islands equals 0.5–0.6. Comparison with another phase diagram given in [4.48] yields the same result.

To conclude, there is good agreement between the theory of multisheet array formation in elastically anisotropic cubic systems [4.48] and experiments with multisheet arrays of submonolayer CdSe islands in a ZnSe matrix [4.47, 4.58, 4.59]. The results demonstrate a novel approach to wave function engineering in arrays of submonolayer islands, where both the spectral position of the electronic states and polarization of photoluminescence can be controlled by varying growth parameters, e.g., the thickness of the spacer.

4.2.4 Highly-Ordered Quantum Dot Superlattices

After anticorrelation had been discovered experimentally in the CdSe/ZnSe system by Strassburg et al. [4.47] and explained theoretically by Shchukin et al. [4.48], similar anti-correlated arrangements of QDs were observed by Springholz et al. [4.61] in the PbSe/PbEuTe system.

Holý et al. [4.62] investigated theoretically the elastic energy density on the surface above the buried island for a wide variety of materials and for different orientations of the surface. The buried island was approximated as an isotropic point force dipole. Focusing on the formation of islands in the second sheet of the deposited material, the strain distribution on the wetting layer was considered. It is given by the superposition of the homogeneous lattice-mismatch strain $\varepsilon_{ij}^{(0)}$ and the contribution $\varepsilon_{ij}^{(1)}(\boldsymbol{r})$ from buried islands, viz.,

$$\varepsilon_{ij}(\boldsymbol{r}) = \varepsilon_{ij}^{(0)} + \frac{\Delta\varepsilon_{ij}(\boldsymbol{r})V_{\text{island}}}{\Delta V} \ . \tag{4.9}$$

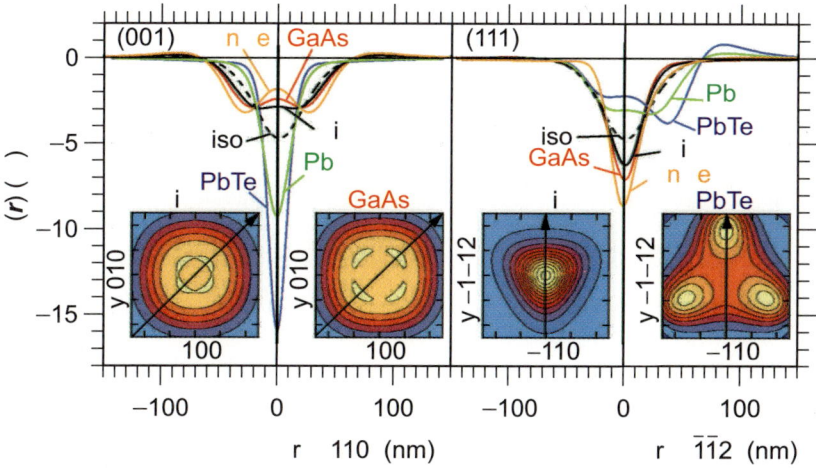

Fig. 4.32. Normalized elastic energy density $\rho(x,y)$ on the surface above a strained quantum dot at 50 nm below the surface for various matrix materials and for the (001) and (111) surface orientations. *Curves* represent cross-sections of ρ in the directions indicated by *arrows* in the *inserts*. *Inserts*: 2D contour plots with $-50 \leq (x,y) \leq +50$ nm and with a 0.5% step between contours. *Yellow areas* correspond to the minima of ρ. With kind permission of V. Holý et al. [4.62]

Here $\Delta\varepsilon_{ij}(\mathbf{r})$ is the strain introduced by a buried point stress source of elementary volume ΔV. The relative change in the strain energy on the surface above a buried island is given by

$$\rho(x,y) = \left.\frac{E_1(\mathbf{r}) - E_0}{E_0}\right|_{z=0}, \qquad (4.10)$$

where E_0 is the constant energy density in the wetting layer due to the misfit strain, and E_1 is the energy density with a buried island underneath. With this definition, ρ does not depend on the lattice mismatch between the island and the matrix material. To allow for a comparison of multisheet structures in different materials, the island volume was fixed at $7\,200$ nm^3 (a pyramid of height 20 nm and length 50 nm, with the spacer thickness fixed at 50 nm).

Elastic energy distributions are determined by two key parameters, namely, the elastic anisotropy of the matrix material, and the surface orientation. For cubic materials, the elastic anisotropy is characterized by the anisotropy parameter ξ defined in (4.2). The group IV, III–V and II–VI semiconductors with diamond or zinc-blende structure and with chemical bonds along the $\langle 111 \rangle$, the $\langle 111 \rangle$ directions are the elastically hard directions, and the $\langle 100 \rangle$ directions are the elastically soft directions. The anisotropy parameter ξ equals -0.721 for Si, -0.907 for GaAs, and -1.27 for ZnSe. It is generally bigger for the II–VI compounds. For rock salt materials, the nearest neighbors are along $\langle 100 \rangle$, the $\langle 100 \rangle$ are hard directions, and the $\langle 111 \rangle$ are

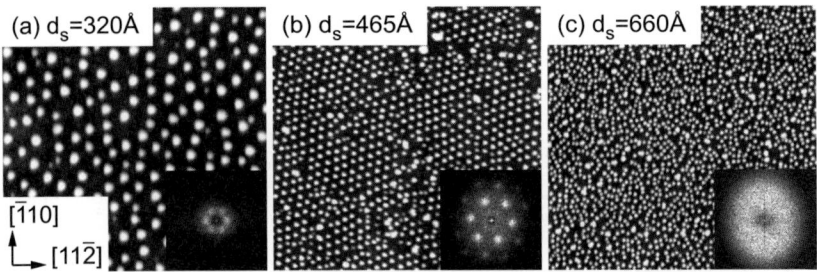

Fig. 4.33. Atomic force microscopy (AFM) image (2×2 µm^2) of PbSe quantum dots on the surface of PbSe/Pb$_{1-x}$Eu$_x$Te superlattices with a Pb$_{1-x}$Eu$_x$Te spacer layer thickness of (**a**) 320 Å, (**b**) 465 Å, and (**c**) 660 Å. *Insets*: 2-dimensional fast Fourier transformation power spectra of the AFM images. With kind permission of G. Springholz et al. [4.63]

soft directions. For IV–VI semiconductors, the elastic anisotropy parameter ξ is positive and particularly large, reaching 1.92 for PbS and 5.41 for PbTe.

Figure 4.32 displays the normalized elastic energy density $\rho(x,y)$ on the surface above the buried island for various matrix materials and various surface orientations. For elastically isotropic spacer layers (dashed lines), the energy minimum is always at the origin of $\rho(x,y)$, regardless of the surface orientation. However, the elastic anisotropy changes both the depth and the position of energy minima.

For growth along the hard direction ([111] for diamond and zinc blende and [001] for rock salt materials), the depth of the energy minimum significantly increases with increasing anisotropy, but its position above the buried dot remains unchanged. The opposite behavior is observed for growth along the soft direction ([001] for diamond and zinc blende and [111] for IV–VI materials), i.e., the depth of the minimum decreases as the anisotropy increases. Moreover, as shown by Holý et al. [4.62], when the anisotropy parameter exceeds a critical value, the central minimum is replaced by several side minima in the $\rho(x,y)$ distributions. For the (001) surface and $\xi < -0.67$, $\rho(x,y)$ exhibits four minima along the $\langle 110 \rangle$ directions. For Si and GaAs, these minima are quite shallow, but become more pronounced for higher anisotropies, e.g., for ZnSe. For the (111) surfaces and $\xi > 1.33$ (as in PbS, PbTe), three deep minima occur along the $\langle \overline{1}\overline{1}2 \rangle$ directions. At intermediate orientations of the surface, a single ρ minimum occurs, which is not necessarily located above the buried dot but can be displaced in a lateral direction, as shown in [4.62] for the (113) surface.

Springholz et al. [4.63] carried out experimental and theoretical studies of how finite-size effects change the vertical and lateral arrangement of islands. A series of multisheet superlattices comprising PbSe dots in a Pb$_{1-x}$Eu$_x$Te matrix were grown with varying spacer thickness but a fixed number of 30 sheets of QDs. Figure 4.33 shows atomic force microscopy (AFM) images of the final PbSe dot layer. For spacer thickness $d_s < 370$ Å (Fig. 4.33a), large and widely spaced dots form with a density of about 10^{10} cm^{-2}. This is much lower than for single dot layers grown under identical conditions. Correspondingly, the average dot height of 180 Å and dot base width of 700 Å are much larger than for single dot layers. Increasing the spacer thickness to 370–540 Å leads to a completely different dot structure. Figure 4.33b shows a nearly perfect hexagonal 2-dimensional lattice of dots with a substantial narrowing of the size dispersion. In addition, the dot density is significantly increased to about 2.5×10^{10} cm^{-2}, with a corresponding decrease in dot sizes. For a spacer thickness above 550 Å (Fig. 4.33c), the hexagonal dot structure is replaced by a highly disordered dot arrangement with a characteristic dot density of 5.5×10^{10} cm^{-2}.

The different lateral dot arrangements are clearly indicated by the fast Fourier transformation (FFT) power spectra of the AFM images shown as insets in Fig. 4.33. For samples with intermediate spacer layers (Fig. 4.33b), well-defined FFT satellite peaks appear. The clear sixfold symmetry and the large number of higher order peaks indicate the existence of large, perfectly ordered hexagonal dot domains. For large spacer thickness (Fig. 4.33c), only a diffuse ring is observed due to the lack of any preferred ordering direction and the broad dispersion of dot spacings. For small spacer thickness (Fig. 4.33a), although the dots seem to form at random positions, six broad satellite peaks are still visible in the FFT spectra, indicating a preferred hexagonal dot arrangement. However, the absence of high-order satellites shows that only a short-range order exists in this case.

The origin of the characteristic transitions in the dot arrangements was revealed by cross-sectional transmission electron microscopy (TEM) [4.63]. As shown in Fig. 4.34, for samples with different spacer thickness, qualitatively different dot correlations are formed. For smaller spacer thickness (Fig. 4.34a), the dots are vertically correlated and form columns, similar to an InAs/GaAs or SiGe/Si QD superlattice. In contrast, for spacers of intermediate thickness (Fig. 4.34b), the dots are aligned in directions inclined at 39° to the growth direction (dashed lines in Fig. 4.34b). Thus, well-ordered 3-dimensional superlattices of dots are formed with an $ABCABC\ldots$ stacking sequence. For even larger spacer thickness (Fig. 4.34c), the vertical interaction between the dots decreases below a critical value, so that correlations are no longer formed. Then the lateral ordering tendency disappears as well.

To explain the different dot correlations, Springholz et al. [4.63] considered the elastic interaction between the buried dots and the dots on the surface, where the finite dot size was taken into account. It was shown that, for spacer thickness D larger than the island base width L, the strain energy distribution closely resembles those obtained from a point source model [4.61], with three well-defined minima along the $\langle \overline{11}2 \rangle$ directions. The directions of these minima are inclined at $\approx 35°$ to the surface normal, which is close to the experimental correlation angle of 39°. However, when D/L decreases

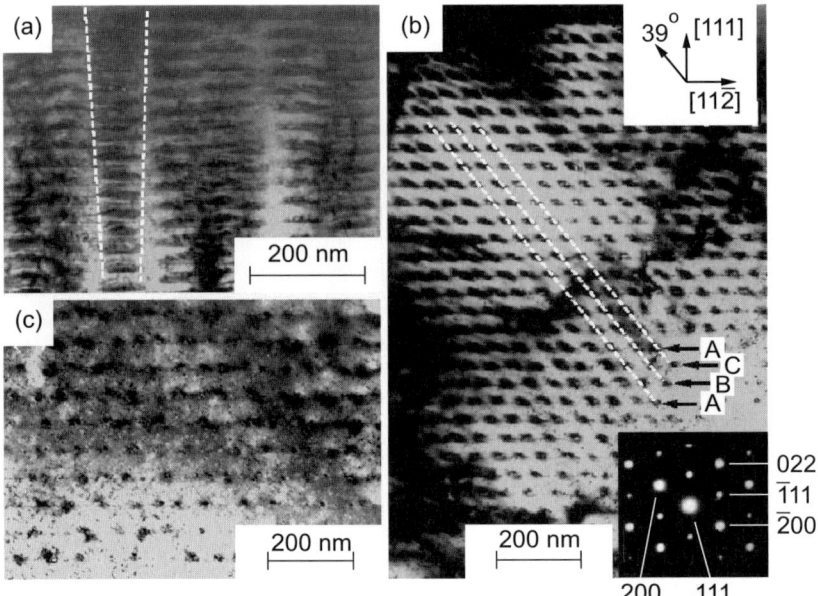

Fig. 4.34. ($\bar{1}10$) cross-sectional TEM image of self-organized PbSe/Pb$_{1-x}$Eu$_x$Te quantum dot superlattices with different Pb$_{1-x}$Eu$_x$Te spacer layer thicknesses: (**a**) 320 Å, (**b**) 465 Å, and (**c**) 680 Å. With kind permission of G. Springholz et al. [4.63]

below 1, the lateral separation between the energy minima becomes smaller than the lateral size of the island. Then the island size becomes incompatible with a trigonal dot arrangement, and the inclined dot correlation gives way to an alignment in vertical columns, similar to the experimental observation in Fig. 4.34a. For $D/L > 1$, although the correlation direction is almost constant, the depth of the energy minima decreases rapidly and the lateral ordering is finally smeared out by the formation of dots at random positions (Figs. 4.33c and 4.34c).

Thus, comprehensive experimental and theoretical investigations of the formation of self-organized multisheet PbSe/Pb$_{1-x}$Eu$_x$Te quantum dot superlattices [4.61–4.63] have clearly demonstrated various types of vertical correlation between the dots. The extremely high elastic anisotropy $\xi = 5.4$ of the Pb$_{1-x}$Eu$_x$Te matrix results in the formation of a 3-dimensional quantum dot superlattice with highly uniform dot sizes and arrangement.

4.2.5 Anticorrelated Multisheet Nanostructures in III–V Semiconductors

Although II–VI and IV–VI semiconductors are materials with the highest elastic anisotropy (for $\xi < 0$ and $\xi > 0$, respectively), anticorrelated arrangements have been observed in III–V semiconductors as well. Li et al. [4.64]

4.2 Anticorrelation in Multisheet Arrays of Strained Islands

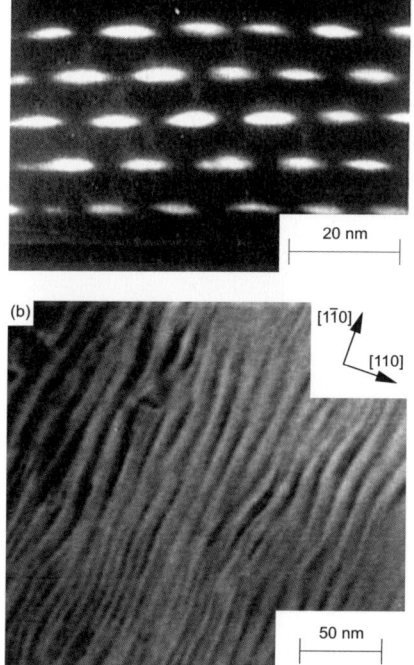

Fig. 4.35. InAs quantum wires in an $In_{0.52}Al_{0.48}As$ matrix. (a) Plan-view TEM image of stacked nanowire arrays with 6 period InAs (6 ML)/$In_{0.52}Al_{0.48}As$ (20 nm). (b) $[\bar{1}10]$ cross-sectional TEM image of stacked nanowire arrays with 6 period InAs (6 ML)/$In_{0.52}Al_{0.48}As$ (10 nm). With kind permission of H. Li et al. [4.65]

(see also [4.65, 4.66]) grew a six period InAs/$In_{0.52}Al_{0.48}As$ superlattice on an InP(001) substrate. The alloy $In_{0.52}Al_{0.48}As$ is nearly lattice-matched to InP, whereas lattice-mismatched InAs tends to form islands. Figure 4.35a shows a plan-view transmission electron microscopy (TEM) image of the structure with 20-nm-thick spacers. The image reveals the formation of islands elongated in the $[\bar{1}10]$ direction. The same morphology was obtained from the atomic force microscopy (AFT) image [4.64]. The cross-sectional TEM image (Fig. 4.35b) clearly reveals an anticorrelated arrangement of wires. Anticorrelation was observed for different spacer thicknesses $5 \leq D \leq 20$ nm, whereas no sample showed vertical correlation. Such behavior might be attributed to the effects of a particular cross-sectional shape of the wires. However, no theoretical model of this system has yet been made.

Anticorrelation has also been observed in a 'classical' quantum dot system, namely, in multisheet arrays of InGaAs QDs in a GaAs matrix. Liu et al. [4.67] and Wang et al. [4.68] reported a transition from vertical correlation to vertical anticorrelation with an increase in the spacer thickness, observed in cross-sectional scanning tunneling microscopy (STM).

To conclude, the growth of multisheet arrays of strained islands (quantum dots and quantum wires) may reveal both vertical correlation and vertical anticorrelation of the islands. Anticorrelation has been observed in various systems, including II–VI, IV–VI, and III–V semiconductors. The theory of the formation of multisheet arrays of strained islands demonstrates that the transition between correlated and anticorrelated arrangements is governed by elastic anisotropy of the matrix, and is therefore a universal phenomenon. These transitions may be controlled by the spacer thickness, which gives a powerful experimental tool for tuning the density, shape, and size of quantum dots or wires, as well as the electronic spectrum of the nanostructures.

4.3 Activated Alloy Phase Separation During Overgrowth of Quantum Dots

Most semiconductor materials are in fact alloys. $Si_{1-x}Ge_x$ is a well-known example of a binary alloy. Widely used compound semiconductors are pseudobinary or ternary alloys, e.g., $Ga_{1-x}Al_xAs$, $Ga_{1-x}In_xAs$, $GaAs_{1-y}N_y$, $Zn_{1-x}Cd_xSe$, etc. A great advantage of alloys is the possibility of continuously tuning the lattice parameter and/or the energy bandgap and/or the refractive index. In quaternary alloys, like $Ga_{1-x-y}Al_yIn_xAs$, $Ga_{1-x}In_xAs_{1-y}P_y$, $Ga_{1-x}In_xAs_{1-y}N_y$, etc., an independent variation of the composition parameters x and y allows us to tune the lattice parameter and the energy bandgap independently.

Another intrinsic property of alloy materials, namely, the possibility of spontaneous formation of structures with modulated composition, is much less frequently employed. It has long been known that most III–V semiconductor ternary and quaternary alloys are unstable against phase separation in a certain region of temperatures and compositions [4.69].

4.3.1 Basic Physics of Phase Separation in Alloys

Phase separation in alloys resulting in spontaneous formation of composition-modulated structures was first discovered in metal alloys. These were binary metal alloys $A_{1-c}B_c$, for which the theory of phase separation was developed. Cahn [4.70–4.72] introduced the concept of spinodal decomposition of alloys and developed the theory of thermodynamic instability of bulk alloys against small compositional fluctuations $\delta c(\mathbf{r})$. It was shown in [4.70, 4.71] that the most unstable modes of compositional fluctuations are those with the composition modulation in one of the elastically soft directions. Khachaturyan [4.73–4.75] developed the theory of the final state of alloys and showed that this final state is a structure with an alloy composition, periodically modulated in one or several elastically soft directions.

The theory of spinodal decomposition initially developed for metal alloys by Cahn [4.70–4.72] and Khachaturyan [4.73–4.75] was extended to ternary and quaternary alloys of III–V semiconductors in papers by De Cremoux et al. [4.76], De Cremoux [4.77], Stringfellow [4.78], and Onabe [4.79]. The theory

4.3 Activated Alloy Phase Separation During Overgrowth of Quantum Dots

of alloy decomposition of ternary alloys $A_{1-c}B_cC$ is, in fact, very similar to that for binary alloys $A_{1-c}B_c$. We therefore consider binary alloys in the following brief overview. The extension to ternaries is then straightforward.

Spinodal Decomposition in Bulk Alloys. This section contains a brief overview of the theory of spinodal decomposition of bulk binary alloys developed by Cahn [4.70–4.72] and Khachaturyan [4.73–4.75]. If the alloy composition is perturbed in a binary alloy $A_{1-c}B_c$,

$$c(\mathbf{r}) = \bar{c} + \delta c(\mathbf{r}) , \qquad (4.11)$$

this leads to a change in the Helmholtz free energy. Here $\bar{c}$ is the composition of a homogeneous alloy (e.g., $A_{1-c}B_c$), $\delta c(\mathbf{r})$ denotes fluctuation in composition, and $c(\mathbf{r})$ is the inhomogeneous composition profile. If the free energy of an inhomogeneous alloy with some profile $\delta c(\mathbf{r})$ of composition fluctuations is lower than that of a homogeneous alloy,

$$F_{\text{inh}} < F_{\text{hom}} , \qquad (4.12)$$

the homogeneous alloy is thermodynamically unstable.

The theory of thermodynamic instability of an alloy may be considered as consisting of two parts. The first problem is to obtain criteria determining a region of temperatures and average compositions in which a homogeneous alloy is unstable. The second problem is to find the final inhomogeneous composition-modulated structure resulting from decomposition of an unstable homogeneous alloy. The final equilibrium structure meets the conditions of the free energy minimum,

$$F_{\text{eq}} = \min\{F\} , \qquad (4.13)$$

under the constraint that the number of atoms of each kind is conserved. The latter may be written as the constraint that the average composition of the alloy is conserved, i.e.,

$$\frac{1}{V}\int d^3\mathbf{r}\, c(\mathbf{r}) = \bar{c} . \qquad (4.14)$$

The thermodynamic theory requires us to consider the Helmholtz free energy F. For an inhomogeneous alloy, F may be written as a sum of three contributions [4.70],

$$F_{\text{total}} = F_{\text{chem}} + F_{\text{grad}} + E_{\text{elast}} , \qquad (4.15)$$

where the first term is the chemical free energy (the term introduced by Cahn [4.70]). The density of the chemical free energy is a local function of the alloy composition,

$$F_{\text{chem}} = \int d^3\mathbf{r}\, f_{\text{chem}}\left[c(\mathbf{r})\right] . \qquad (4.16)$$

The second term in (4.15) is a non-local contribution to the chemical free energy,

$$F_{\text{grad}} = \frac{1}{2} \int \mathrm{d}^3 r \kappa(c) [\nabla c(\boldsymbol{r})]^2 , \qquad (4.17)$$

called the gradient energy. The third term in (4.15) is the elastic energy. The change in the total free energy due to fluctuations in composition is

$$\Delta F_{\text{total}} = \int \mathrm{d}^3 \boldsymbol{r} \left[f_{\text{chem}}\left(\bar{c} + \delta c(\boldsymbol{r})\right) - f_{\text{chem}}(\bar{c}) \right] + F_{\text{grad}} + E_{\text{elast}} . \qquad (4.18)$$

In (4.18), the gradient and the elastic energy are positive. Therefore an instability of the alloy, when $\Delta F_{\text{total}} < 0$, may only occur due to the chemical free energy.

The chemical free energy may be written as a sum of enthalpy and entropy contributions,

$$f_{\text{chem}}(c) = H(c) - TS(c) . \qquad (4.19)$$

At temperature $T = 0$ K, the thermodynamic driving force for the decomposition of an alloy is the positive enthalpy of formation,

$$\Delta H_{\text{formation}} (\mathrm{A}_{1-c}\mathrm{B}_c) \equiv H (\mathrm{A}_{1-c}\mathrm{B}_c) - (1 - c) H(A) - cH(B) > 0 . \qquad (4.20)$$

The positive enthalpy of formation implies that at $T = 0$ K, in order to form an alloy from pure constituents A and B, one needs to add heat to the system. Therefore, an inverse reaction of decomposition of an alloy into a mixture of pure A and pure B will be accompanied by a release of heat, as a mixture of two phases has a lower Helmholtz free energy than a homogeneous alloy. Then the alloy is unstable and will decompose. At high temperatures, the entropy contribution to the free energy tends to mix the alloy constituents and supports the random distribution of atoms A and B over crystal lattice sites. The entropy contribution thus hinders decomposition and stabilizes the homogeneous alloy.

One of the key properties of alloys with a major influence on alloy stability or instability is the dependence of the alloy lattice parameter on composition, according to Vegard's rule $a = a(c)$. Then an inhomogeneity in alloy composition is always accompanied by strain fields, and hence also by an elastic energy. Elastic energy associated with fluctuations in alloy composition hinders decomposition and stabilizes the homogeneous alloy [4.70]. Due to the long-range nature of the elastic interaction, the elastic energy depends non-locally on compositional fluctuations δc,

$$E_{\text{elast}} = \frac{1}{2} \int \frac{\mathrm{d}^3 \boldsymbol{k}}{(2\pi)^3} B\left(\frac{\boldsymbol{k}}{k}\right) |\widetilde{\delta c}(\boldsymbol{k})|^2 , \qquad (4.21)$$

where $\widetilde{\delta c}(\boldsymbol{k})$ is the Fourier transform of the compositional fluctuations $\delta c(\boldsymbol{r})$, and the coefficient $B(\boldsymbol{k}/k)$ in the elastic energy depends on the direction of $\boldsymbol{k}$ through the elastic anisotropy.

4.3 Activated Alloy Phase Separation During Overgrowth of Quantum Dots

Since equations are known for all the contributions to the free energy, one may obtain a criterion for the absolute instability of an alloy, i.e., for its instability against infinitesimal fluctuations. By substituting (4.16), (4.17), and (4.21) into (4.18) and expanding the free energy in powers of δc up to quadratic terms, one obtains the variation in the free energy as

$$\Delta F = \int \frac{d^3 k}{(2\pi)^3} \left[\frac{1}{2} \left(\frac{\partial^2 f_{\text{chem}}(c)}{\partial c^2} \right) + \kappa k^2 + \frac{1}{2} B\left(\frac{\boldsymbol{k}}{k}\right) \right] \left| \widetilde{\delta c}(\boldsymbol{k}) \right|^2 . \quad (4.22)$$

In III–V semiconductors with zinc-blende structure, $\langle 001 \rangle$ are elastically soft directions. Correspondingly, the coefficient $B(\boldsymbol{k}/k)$ reaches its minimum value just for these directions [4.70, 4.71, 4.74, 4.75]:

$$B\left(\frac{\boldsymbol{k}}{k}\right)_{\min} = B\left(\frac{\boldsymbol{k}}{k}\right)_{\boldsymbol{k} \| [001]} = B_0 \quad (4.23)$$

$$= \frac{2 (c_{11} + 2c_{12})(c_{11} - c_{12})}{c_{11}} \left(\frac{1}{\bar{a}} \frac{\partial a(c)}{\partial c} \right)^2 ,$$

where $\bar{a} \equiv a(\bar{c})$ is the lattice parameter of the homogeneous alloy with composition $\bar{c}$. The elastic anisotropy implies that the most unstable, so called 'soft' modes of compositional fluctuations are those in which the alloy composition is modulated in one or more elastically soft directions [4.75],

$$\delta c(x, y, z) = \delta c(x) , \quad (4.24\text{a})$$
$$\delta c(x, y, z) = \delta c_1(x) + \delta c_2(y) , \quad (4.24\text{b})$$
$$\delta c(x, y, z) = \delta c_1(x) + \delta c_2(y) + \delta c_3(z) , \quad (4.24\text{c})$$

and in addition the typical wavelength of the fluctuations is sufficiently large for the second term in the integrand of (4.22), i.e., the gradient energy, to be small compared to the other terms. Hence, the criterion for the absolute instability of an alloy may be written as

$$\left. \frac{\partial^2 f_{\text{chem}}(c; T)}{\partial c^2} \right|_{c=\bar{c}} + B_0 < 0 . \quad (4.25)$$

To obtain the critical temperature at which the alloy becomes unstable, one has to use some particular form of the chemical free energy f_{chem}. The simplest model for f_{chem} which demonstrates the interplay between the enthalpy and entropy terms is the regular solution approximation [4.80, 4.81],

$$f_{\text{chem}}(T; c) = \mu_{\text{A}}^{(0)}(T)(1-c) + \mu_{\text{B}}^{(0)}(T) c \quad (4.26)$$
$$+ \Omega c(1-c) + k_{\text{B}} T \left[(1-c) \ln(1-c) + c \ln c \right] .$$

Here $\mu_{\text{A}}^{(0)}(T)$ is the chemical potential of a pure A material, $\mu_{\text{B}}^{(0)}(T)$ is the chemical potential of a pure B material, and Ω is the interaction parameter. The third term in (4.26) is just the formation enthalpy,

$$\Delta H_{\text{formation}}(c) = \Omega c(1-c) . \quad (4.27)$$

The fourth and last term in (4.26) is the entropy contribution to the free energy due to the entropy of mixing,

$$S_{\text{mix}} = -k_B \left[(1-c)\ln(1-c) + c \ln c \right] . \tag{4.28}$$

The thermodynamic driving force to alloy instability occurs if the interaction parameter Ω is positive. By substituting the chemical free energy (4.26) into (4.25), we deduce that instability occurs for the first time at alloy composition $c = 0.5$ and temperature

$$T_c^{\text{bulk}} = \frac{1}{k_B}\left(\frac{\Omega}{2} - \frac{B_0}{4}\right) . \tag{4.29}$$

If decomposition of an alloy occurs in such a way that the two resulting phases are completely separated, and not coherently conjugated in a single crystal, then the critical temperature of the alloy instability will be

$$T_c^{(0)} = \frac{\Omega}{2k_B} . \tag{4.30}$$

Such a decomposition may occur, for example, during crystallization from the melt, where two solid phases and the melt are at equilibrium, but solid phases with different alloy composition are just disconnected pieces.

The critical temperature for the coherent instability of a bulk alloy, T_c^{bulk} from (4.29), is lower than $T_c^{(0)}$,

$$T_c^{\text{bulk}} = T_c^{(0)} - \frac{B_0}{4k_B} . \tag{4.31}$$

The lowering of the critical temperature is due to elastic strain energy associated with fluctuations in the alloy composition [4.70].

At temperatures $T < T_c^{\text{bulk}}$, the stability or instability of an alloy depends on the alloy composition. It was shown by Khachaturyan [4.75] that, if a homogeneous alloy is unstable, the final inhomogeneous state may be written as a combination of 'soft' mode fluctuations of the form (4.24a), (4.24b), or (4.24c). For these types of composition modulation, the elastic energy (4.21) may be greatly simplified. The coefficient $B(\mathbf{k}/k)$ is then constant and may be taken out of the integral. The elastic energy, which generally depends non-locally on fluctuations, reduces to a local function of $\delta c(\mathbf{r})$,

$$E_{\text{elast}} = B_0 \int d^3\mathbf{r} [\delta c(\mathbf{r})]^2 . \tag{4.32}$$

Then the total Helmholtz free energy of the inhomogeneous alloy may be written as a sum of the local chemical term and the effectively local elastic term,

$$F_{\text{total}} = \int d^3\mathbf{r} \left[f_{\text{chem}}\left[c(\mathbf{r})\right] + \frac{1}{2} B_0 \left[c(\mathbf{r}) - \bar{c}\right]^2 \right] . \tag{4.33}$$

The effectively local form of the free energy allows us to consider the effective free energy density,

4.3 Activated Alloy Phase Separation During Overgrowth of Quantum Dots

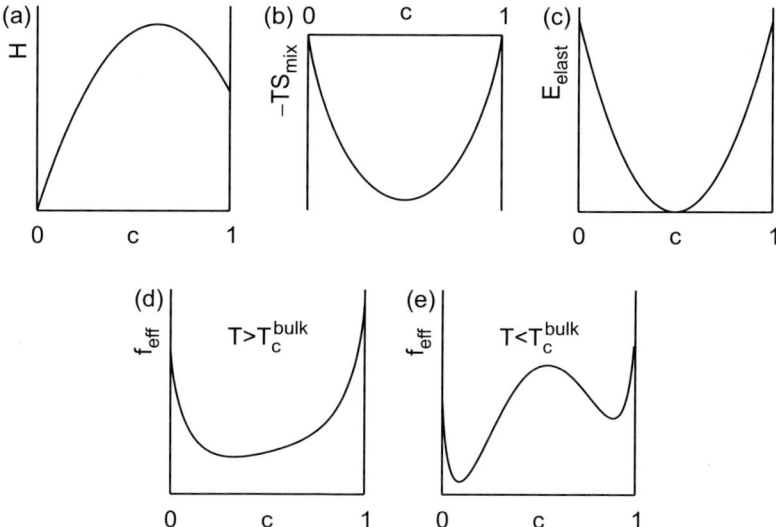

Fig. 4.36. Effective free energy density of a binary alloy $A_{1-c}B_c$ as a function of composition. (**a**) Enthalpy H of an alloy. (**b**) Entropy contribution to the free energy $-TS_{\text{mix}}$. (**c**) Effective elastic energy of a 'soft' mode. (**d**) Effective Helmholtz free energy density at high temperatures $T > T_c^{\text{bulk}}$. (**e**) Effective Helmholtz free energy density at low temperatures $T < T_c^{\text{bulk}}$

$$f_{\text{eff}}(c) = f_{\text{chem}}(c) + \frac{1}{2}B_0\left[c(\mathbf{r}) - \bar{c}\right]^2 . \tag{4.34}$$

Substituting (4.26) into (4.34), it is convenient to write the effective free energy density as a sum of three contributions,

$$f_{\text{eff}}(c) = H(c) - TS_{\text{mix}}(c) + E_{\text{elast}}(c) . \tag{4.35}$$

Figure 4.36 displays the three contributions to the effective density of the Helmholtz free energy. The enthalpy H (Fig. 4.36a) has a negative second derivative and thus destabilizes a homogeneous alloy, whereas the entropy contribution (Fig. 4.36b) and the elastic energy (Fig. 4.36c) play a stabilizing role. At high temperatures $T > T_c^{\text{bulk}}$ (Fig. 4.36d), the effective free energy density $f_{\text{eff}}(c)$ is everywhere convex, and the homogeneous alloy is stable. At low temperatures $T < T_c^{\text{bulk}}$, the plot $f_{\text{eff}}(c)$ has a concave part, where the alloy is unstable.

Figure 4.37 shows a plot of the effective free energy density $f_{\text{eff}}(c)$ at a temperature below T_c^{bulk}. Since the total free energy may be written as an effective local function of the alloy composition (4.33), the stability analysis can be performed in a standard way (see, for example, [4.70, 4.75]), by constructing the common tangent to the curve $f_{\text{eff}}(c)$. The common tangent construction (the straight line K_1K_2) determines the region

$$c_1^{(0)} < c < c_2^{(0)} , \tag{4.36}$$

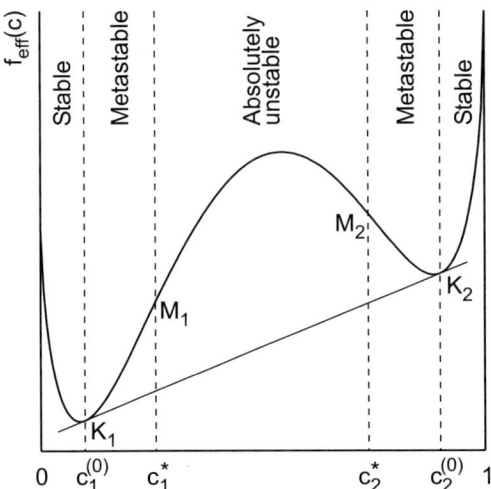

Fig. 4.37. Diagram showing the stability of a binary alloy $A_{1-c}B_c$ against fluctuations in composition

where a homogeneous alloy is unstable. Within this region, there is a narrower region bounded by two inflection points M_1 and M_2 where the absolute instability criterion (4.25) holds. Thus, in the region of composition

$$c_1^* \leq c \leq c_2^* , \tag{4.37}$$

a homogeneous alloy is absolutely unstable. Unstable alloys that are not absolutely unstable, i.e., at compositions

$$c_1^{(0)} < c < c_1^* \quad \text{and} \quad c_2^* < c < c_2^{(0)} , \tag{4.38}$$

are metastable. Finally, alloys outside the instability region, i.e., with

$$0 \leq c \leq c_1^{(0)} \quad \text{and} \quad c_2^{(0)} \leq c \leq 1 , \tag{4.39}$$

are stable.

Points K_1, M_1, M_2, and K_2 at different temperatures form the phase diagram of Fig. 4.38 in temperature–alloy composition variables. The curve separating the regions of absolutely unstable and metastable alloys is called the spinodal curve. The curve separating the regions of metastable and stable alloys is called the miscibility curve, or the solubility curve, or the binodal curve.

If an alloy is quenched to an unstable region of the phase diagram, it will undergo phase separation. Cahn [4.70, 4.72] introduced the concept of spinodal decomposition and showed that there exist two possible kinetic pathways for alloy phase separation. If the initial state of an alloy corresponds to the region of absolute instability, it decomposes according to the spinodal mechanism. In the initial, linear stage of decomposition, any small fluctuation in

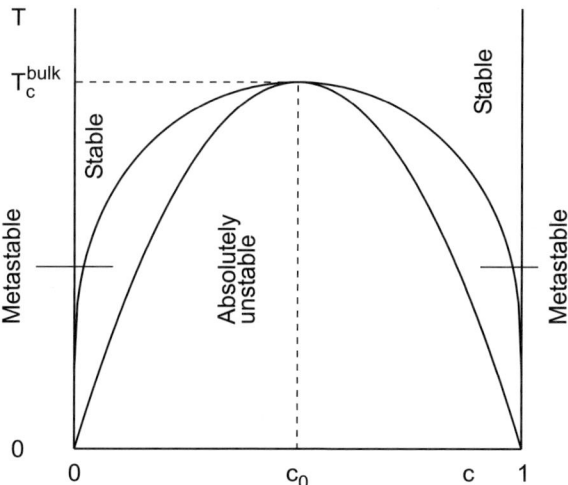

Fig. 4.38. Schematic phase diagram for the stability of a binary alloy $A_{1-c}B_c$ against compositional fluctuations in temperature–alloy composition variables

the alloy composition corresponding to unstable modes increases in amplitude with time. In the medium term, the increase in amplitude is saturated due to non-linear effects. The latter basically result in the formation of two phases with different alloy composition. At a later stage, domains of each phase undergo coarsening, and the average size of a domain grows with time.

If the initial state of an alloy corresponds to the metastable region of the phase diagram, it decomposes according to the nucleation and growth mechanism. Fluctuations with small amplitude do not lead to decomposition. A critical nucleus forms with a certain volume and a certain amplitude of compositional fluctuations. This then grows in volume to form a two-phase system, which subsequently coarsens.

Concerning the final state of the decomposing alloy, coherent phase separation of the whole sample will generally result in a complicated domain structure, determined by the particular shape of the sample. A different approach to the problem was proposed by Khachaturyan [4.73–4.75]. Phase separation typically starts in different parts of the crystal and proceeds within separate complexes surrounded by a matrix of non-decomposed alloy. The volume of such complexes may be bounded by different extended defects like stacking faults, dislocations, etc. The phase separation within a complex occurs much faster than the complex changes its volume and shape. The final state of the decomposed complex coherently conjugated to the surrounding matrix is then well defined. It has been shown [4.73–4.75] that the final equilibrium state of the complex is a domain structure periodically modulated in one of the elastically soft directions. Theoretical calculations of the final state take place in four stages.

- The structure comprises alternating layers of composition $c_1^{(0)}$ and $c_2^{(0)}$, corresponding to the common tangent points in the diagram of Fig. 4.37.
- The volume fraction of each phase is determined by the conservation of the average alloy composition,

$$c_1 V_1 + c_2 V_2 = \bar{c} V \ . \tag{4.40}$$

- The structure of the domain boundary and its energy is determined according to the theory by Khachaturyan and Suris [4.82].
- The period of the structure is determined by the balance between the part of the elastic energy associated with the penetration of the strain field from the complex into the surrounding matrix, which hinders structures with a large period, and the energy of domain boundaries, which hinders structures with a small period.

A detailed theory can be found in the monograph by Khachaturyan [4.74, 4.75]. It should be noted that a structure with 2-dimensional (4.24b) and 3-dimensional (4.24c) composition modulation can only occur as a metastable state, not as the stable state of the decomposing alloy.

A rigorous theoretical approach to the final inhomogeneous state of the decomposing alloy is possible, if an alloy decomposes within a film of a certain thickness H sandwiched between thicker layers of a matrix. A structure can be realized where:

- the alloy film is lattice-matched on average to the matrix,
- the elastic moduli of the alloy are approximately equal to those of the matrix,
- the diffusivity of atoms in the matrix is substantially lower than in the film.

Then decomposition occurs only in the film, and the matrix serves only as a medium into which strain fields may penetrate. The theory of the final state of the phase-separating alloy by Khachaturyan [4.74, 4.75] may be extended to the case of a film without additional restrictions on the size of the decomposing complex.

To explain the instability of III–V semiconductor alloys observed in numerous III–V ternary and quaternary systems (see, for example, [4.69]), De Cremoux [4.77], Stringfellow [4.78], and Onabe [4.79] extended the thermodynamic theory of Cahn [4.70–4.72] to ternary and quaternary III–V alloys. The critical temperatures $T_c^{(0)}$ of instability against bulk incoherent phase separation were calculated. The typical critical temperature was $\sim 1\,000$ K for ternary and $\sim 1\,500$ K for quaternary alloys. These results imply that practically all III–V semiconductor alloys should be unstable at typical growth temperatures, in disagreement with experimental data.

To overcome this discrepancy, De Cremoux [4.77] and Stringfellow [4.83, 4.84] took into account, not only the chemical energy, but also the elastic energy of the system. De Cremoux [4.77] used the elastic energy of a thin film on a substrate in the calculations. Stringfellow [4.83, 4.84] used the bulk elastic energy from [4.70] associated with compositional fluctuations in a bulk sample. The main result of [4.77, 4.83, 4.84] is that the elastic energy stabilizes a homogeneous alloy, vigorously suppressing the driving force to phase separation. The critical temperature of coherent alloy phase separation T_c^{bulk} lowers almost to 0 K for all ternary and most quaternary alloys, disagreeing once again with experimental data.

Spinodal Decomposition in Epitaxial Films. The next step in the theory of spinodal decomposition was made by Glas [4.85]. He considered the elastic energy of an epitaxial alloy film grown coherently on a substrate, where the alloy is lattice-matched to the substrate on average, and the alloy composition in the epilayer is modulated in a direction parallel to the substrate surface and is constant in the vertical direction. He found that the elastic energy (in the case where the period of composition modulation is of the same order of magnitude as the epilayer thickness) is reduced by a factor of 0.45 with respect to the elastic energy of (4.23). The corresponding critical temperatures of semiconductor alloy decomposition were found to be of the order of a few hundred kelvin, within the typical range of temperatures used in epitaxial crystal growth of semiconductors.

Ipatova et al. [4.86] extended the approach of Glas [4.85] in two directions. Firstly, abandoning the approximation of an elastically isotropic medium used in [4.85], Ipatova et al. [4.86] took into account the cubic elastic anisotropy of both the substrate and the epitaxial film. Secondly, in [4.86], arbitrary alloy composition profiles were examined, modulated in both the lateral and vertical directions.

The theoretical problem of the alloy phase separation in an epitaxial film, like that in the bulk, consists of two parts. The first problem is to find criteria according to which the alloy is absolutely unstable. The thermodynamic linear stability analysis of the problem was carried out in [4.86], and the 'soft' modes were obtained.

The variation of the total Helmholtz free energy of the system, caused by fluctuations δc in alloy composition in the epitaxial film, may be written up to quadratic terms as a sum of chemical and elastic contributions, viz.,

$$\delta F_{\text{total}} = \frac{1}{2v} \int \frac{d^2 \mathbf{k}}{(2\pi)^2} \int_0^h dz \int_0^h dz' \, \widetilde{\delta c}^*(\mathbf{k}, z) \qquad (4.41)$$
$$\left[\left(\frac{\partial^2 f_{\text{chem}}}{\partial c^2} \right) \delta(z - z') + B_{\text{elast}}(\mathbf{k}; z, z') \right].$$

Here $\mathbf{k}$ is a 2-dimensional wave vector, h is the film thickness, and v is the unit cell volume. The elastic energy operator $B_{\text{elast}}(\mathbf{k}; z, z')$ was found in [4.86] in terms of the static Green tensor $\widetilde{G_{ij}}(\mathbf{k}; z, z')$ from elasticity theory. The Green tensor $\widetilde{G_{ij}}(\mathbf{k}; z, z')$ for a semi-infinite elastically anisotropic cubic crystal bounded by a stress-free (001) surface was obtained by Portz and Maradudin [4.51]. The integral operator B_{elast} is

$$B_{\text{elast}}(\boldsymbol{k}; z, z') = B_0 \delta(z - z') + \sum_{s=1}^{3} C_s(\varphi) k \exp\left[-\alpha_s(\varphi) k |z - z'|\right] \quad (4.42)$$

$$+ \sum_{s=1}^{3} \sum_{s'=1}^{3} D_{ss'}(\varphi) k \exp\left[-\alpha_s(\varphi) k (h - z)\right] \exp\left[-\alpha_{s'}(\varphi) k (h - z')\right] ,$$

where the first two terms are the bulk contribution to the elastic interaction, and the third term is the surface contribution. The summation over s and s' takes into account the contributions from the three acoustic branches of the phonon spectrum of the crystal in the static limit. α_s are the dimensionless attenuation coefficients of static acoustic waves. Due to the elastic anisotropy, the coefficients C_s, $D_{ss'}$ and α_s depend on the angle φ between the wave vector $\boldsymbol{k}$ and the elastically soft direction [100].

To obtain the criterion of absolute instability for an alloy in the epitaxial film, it is useful to reduce δF_{total} in (4.41) to diagonal form. This quantity is quadratic in the functional δc. Since the contribution of the chemical energy is diagonal, the problem reduces to finding the eigenvalues and eigenfunctions of the elastic energy integral operator,

$$\frac{1}{h} \int_0^h dz' B_{\text{elast}}(\boldsymbol{k}; z, z') \chi_p(\boldsymbol{k}; z') = \lambda_p(\boldsymbol{k}) \chi_p(\boldsymbol{k}; z) . \quad (4.43)$$

Eigenfunctions $\chi_p(\boldsymbol{k}; z)$ determine the profile of alloy composition fluctuations, and eigenvalues $\lambda_p(\boldsymbol{k})$ characterize the elastic energy associated with the fluctuation in composition $\widetilde{\delta c}(\boldsymbol{k}; z) \sim \chi_p(\boldsymbol{k}; z)$. The absolute instability of the alloy occurs for the first time for the minimum eigenvalue.

In the paper by Ipatova et al. [4.86], the minimum eigenvalue $\lambda_0(\boldsymbol{k})$ was found for every wave vector. Figure 4.39 displays this value as a function of the wave vector for three directions of $\boldsymbol{k}$ in the surface plane. The minimum eigenvalue corresponds to wave vectors parallel to elastically soft directions [100] and [010]. It is reached asymptotically for short wavelength fluctuations, where $kh \to \infty$, and equals

$$\lambda_0^{\min} = \frac{c_{11}}{2(c_{11} + c_{12})} B_0 \approx \frac{1}{3} B_0 . \quad (4.44)$$

It was shown in [4.86] that the eigenfunction of the elastic energy operator, corresponding to the minimum eigenvalue, is localized near the stress-free surface $(z = h)$ and decays with depth in the epitaxial film, $\chi_0(k_x, 0; z) \propto \exp[-|k_x|(h - z)]$. The soft mode of alloy composition fluctuations has the form

$$\delta c(x, y, z) \sim \exp(ik_x x) \exp\left[-|k_x|(h - z)\right] , \quad (4.45\text{a})$$
$$\delta c(x, y, z) \sim \exp(ik_y y) \exp\left[-|k_y|(h - z)\right] . \quad (4.45\text{b})$$

4.3 Activated Alloy Phase Separation During Overgrowth of Quantum Dots

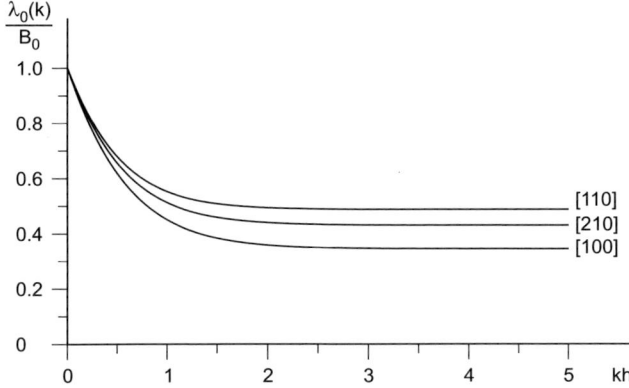

Fig. 4.39. Effective elastic energy associated with fluctuations in alloy composition in the epitaxial film. The minimum eigenvalue $\lambda_0(\boldsymbol{k})$ of the integral operator B_{elast} is plotted for three directions of the wave vector $\boldsymbol{k}$ in the surface plane as a function of kh. It follows from the plot that the minimum value is reached asymptotically for $\boldsymbol{k} \parallel [100]$ as $kh \to \infty$

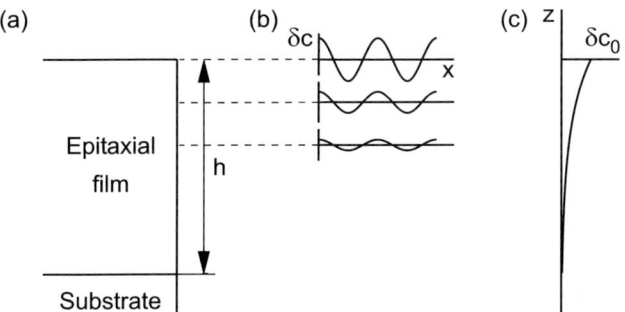

Fig. 4.40. 'Soft' mode of alloy composition fluctuations in an epitaxial film. (**a**) Schematic cross-section of the epitaxial film on a substrate. (**b**) Profile of alloy composition fluctuations in the 'soft' mode at three different depths in the film, according to (4.45a). (**c**) Exponential attenuation of the amplitude of composition fluctuations $\delta c_0(z) \sim \exp[-k(h-z)]$ with depth from the surface

Since the 'soft' mode is localized at the stress-free surface, the effect of stress relaxation at the free surface becomes essential for this mode. This effect manifests itself through the fact that some components of the stress tensor vanish at the surface,

$$\sigma_{xz} = \sigma_{yz} = \sigma_{zz} = 0 \; . \tag{4.46}$$

This leads to a reduction in the effective elastic free energy by a factor of approximately $1/3$ with respect to the elastic energy in the bulk crystal. Correspondingly, the critical temperature of the coherent spinodal decomposition of alloys in the epitaxial film is higher than that in the bulk,

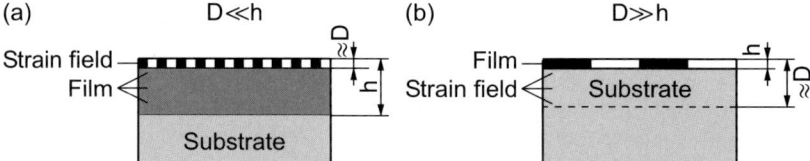

Fig. 4.41. Two different types of alloy composition modulation in an epitaxial film. (**a**) Short-wavelength composition modulation with period $D \ll h$. Composition modulation and strain field exist only in a thin layer of thickness $\sim D$ under the surface. (**b**) Long-wavelength composition modulation with period $D \gg h$. Composition modulation occurs only in the film, but the strain field penetrates into the substrate over a distance $\sim D$. *Black* and *white areas* mark the domain with modulated composition, the *dark grey area* marks a non-decomposed alloy, and the *light grey area* marks the substrate

$$T_c^{\text{film}} = T_c^{(0)} - \frac{c_{11}}{2(c_{11}+c_{12})}\Theta > T_c^{\text{bulk}}. \tag{4.47}$$

Since the effective elastic energy of the 'soft' mode in the film is smaller than the one calculated by Glas [4.85], for a composition profile that is constant in the z direction, the critical temperatures of (4.47) are higher than those calculated in [4.85].

The second part of the theoretical problem of spinodal decomposition in the epitaxial film, as in the bulk, is to obtain the final inhomogeneous structure of the decomposed alloy. This problem was solved in work by Ipatova et al. [4.87] and by Shchukin and Starodubtsev [4.88]. The surface was assumed to remain flat and the equilibrium profile of composition modulation was sought as a linear combination of 'soft' modes (4.45a) and (4.45b). The variation of the total free energy due to alloy decomposition is

$$\Delta F_{\text{total}} = \Delta F_{\text{chem}} + E_{\text{elast}}^{(0)} + \Delta E_{\text{elast}}. \tag{4.48}$$

Here $E_{\text{elast}}^{(0)}$ is the asymptotic value of the elastic energy of the 'soft' mode corresponding to $kh \to \infty$, where all elastic energy is concentrated in a thin layer of thickness $k^{-1} \ll h$, and penetration of strain fields into the substrate may be neglected. ΔE_{elast} is the contribution to the elastic energy due to penetration of the strain field into the substrate.

The optimum period D of composition modulation is determined by the interplay between ΔE_{elast} and $F_{\text{chem}} + E_{\text{elast}}^{(0)}$. For long-wavelength modulations where $D \gg h$ (Fig. 4.41b), strain penetrates into the substrate over a large distance $\sim D$, which increases ΔE_{elast} and is energetically unfavorable. For short-wavelength modulations where $D \ll h$ (Fig. 4.41a), due to the exponential decay of the 'soft' modes, significant alloy decomposition occurs only in a thin layer of thickness $\sim D$ beneath the surface. Then the lowering of the free energy $F_{\text{chem}} + E_{\text{elast}}^{(0)}$ due to decomposition is small ($\sim D$), and this is also energetically unfavorable. Therefore the optimum period is equal to the film thickness, $D_{\text{opt}} \sim h$, up to an order of magnitude.

4.3 Activated Alloy Phase Separation During Overgrowth of Quantum Dots 295

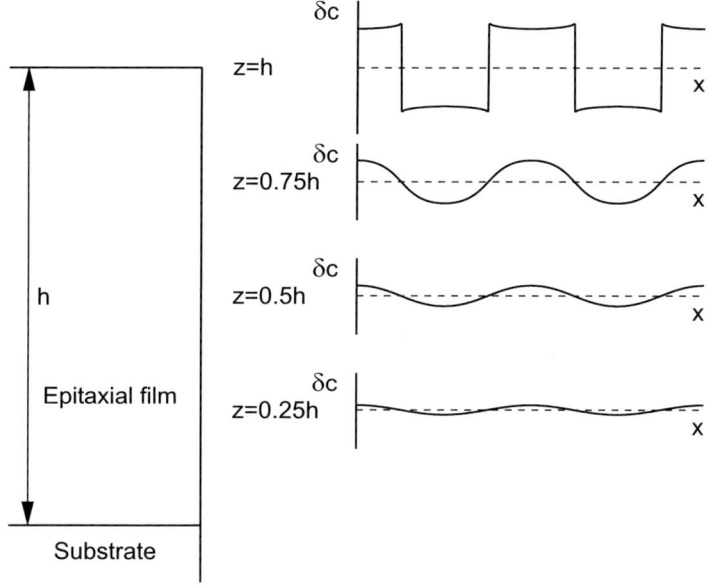

Fig. 4.42. Equilibrium profile of alloy composition modulation in an epitaxial film

Numerical minimization of the free energy (4.48) [4.87, 4.88] shows that the final equilibrium state of a decomposed alloy is a structure with composition modulation in one of the two elastically soft directions, either [100] or [010]. The composition profile has sharp domain boundaries at the surface, while the composition profile in the thickness of the film becomes smoother and the modulation amplitude decreases (Fig. 4.42).

Such an inhomogeneous composition profile decaying into the depth of the epitaxial film away from the surface has been observed by Bert et al. [4.89] in cross-sectional transmission electron microscopy (TEM) of a quaternary InGaAsP alloy film grown by liquid phase epitaxy on an InP(001) substrate.

The equilibrium phase diagram of an alloy epitaxial film containing regions of absolutely unstable, metastable, and stable alloys, was obtained in [4.88].

Generally, the theory of spinodal decomposition of alloys in epitaxial films demonstrates that the stabilizing effect of elastic strain is approximately 3 times weaker than that for the decomposition in the bulk, due to relaxation of elastic stress at the free surface. Correspondingly, critical temperatures of alloy instabilities in epitaxial films are higher than those for bulk alloys. If an unstable alloy film is subject to equilibration, e.g., upon annealing or growth interruption, the decomposition will start at the surface. In the final equilibrium composition-modulated structures, modulation is well pronounced at the surface and decays with depth in the epitaxial film.

Phase Separation of Alloys During Epitaxial Growth. The thermodynamic theory of alloy phase separation applies to closed systems which do not exchange material with the environment and are allowed to evolve towards equilibrium. This may be realized, for example, by annealing a sample, or by a long growth interruption introduced in a growth experiment. In principle, a system can also be regarded as closed during very slow, quasi-equilibrium growth. The evolution of alloy composition occurs via migration of atoms in the gradient of the chemical potential. Such migration is a combination of drift and diffusion. In bulk samples, the kinetic mechanism of alloy phase separation involves bulk diffusion. In films, bulk and surface diffusion may coexist.

Concerning semiconductor alloys, composition-modulated structures are often observed in the as-grown sample. This implies that phase separation occurs during growth. A growing crystal is supplied with material from atomic or molecular beams, gas flow, or melt. Thus the growing crystal is no longer a closed system, but an open one. The epitaxial growth kinetics of semiconductors in open systems is to a large extent determined by the fact that the bulk diffusivity of atoms is typically lower than the surface diffusivity by several orders of magnitude. The flux of atoms which are being incorporated into a growing crystal drives atoms from positions on the surface to positions in the bulk where atoms are no longer mobile. Therefore, in a process of alloy epitaxial growth, fluctuations in alloy composition usually only form at the surface of the growing crystal. Composition fluctuations in the thickness of an already grown epitaxial film are 'frozen'. Similarly to the buried strained islands in multisheet arrays discussed in Sects. 4.1 and 4.2, fluctuations in alloy composition, frozen into the entire thickness of the epitaxial film, create a strain field at the surface. This strain field affects the surface migration kinetics of adatoms and their incorporation into the crystal.

One may then consider the question as to whether composition fluctuations at the advancing surface increase during the process of crystal growth. If such an increase occurs, it implies kinetic instability for the growth of a homogeneous alloy.

'Compositional stress-driven instability' has been found in alloy solidification by Spencer et al. [4.90] and in vapor phase alloy growth by Malyshkin and Shchukin [4.91]. A simple interpretation of such instability may be given within the following simple model [4.91]. The main assumptions are:

- Diffusion of atoms occurs within a thin layer of thickness d beneath the surface. The thickness d exceeds the lattice parameter a but is smaller than all other characteristic lengths.
- Growth is so slow that every surface layer of thickness d reaches a partial equilibrium, i.e., an equilibrium under the constraint of 'frozen' composition profile across the entire thickness of the already grown epitaxial film.

Within this model, one may carry out linear stability analysis. In the first layer of thickness d, let fluctuations $\delta c(x,y)$ occur in the alloy composition.

4.3 Activated Alloy Phase Separation During Overgrowth of Quantum Dots

Then it is possible to obtain a criterion for the situation in which the fluctuation amplitude at the crystal surface increases as crystal growth proceeds.

The variation of the total free energy of the alloy epitaxial film due to fluctuations in composition is given by (4.41). Substituting B_{elast} from (4.42), explicitly writing the contribution from the subsurface layer of thickness d and omitting contributions of order d^2, one obtains

$$\delta F_{\text{total}} = \int \frac{d^2\mathbf{k}}{(2\pi)^2} \left\{ \int_{h-d}^{h} dz \frac{1}{2} \left[\frac{\partial^2 f_{\text{chem}}}{\partial c^2} + B_0 \right] |\widetilde{\delta c}(\mathbf{k}; z)|^2 \right. \tag{4.49}$$

$$+ \frac{1}{2} \int_{h-d}^{h} dz \int_{0}^{h-d} dz' \, \widetilde{\delta c}^*(\mathbf{k}; z)\widetilde{\delta c}(\mathbf{k}; z') \, k \left[\sum_{s=1}^{3} C_s \exp\left[-\alpha_s k|z-z'|\right] \right.$$

$$\left. + \sum_{s=1}^{3} \sum_{s'=1}^{3} D_{ss'} \exp\left[-\alpha_s k(h-z)\right] \exp\left[-\alpha_{s'} k(h-z')\right] \right]$$

$$+ \frac{1}{2} \int_{0}^{h-d} dz \int_{h-d}^{h} dz' \, \widetilde{\delta c}^*(\mathbf{k}; z)\widetilde{\delta c}(\mathbf{k}; z') \, k \left[\sum_{s=1}^{3} C_s \exp\left[-\alpha_s k|z-z'|\right] \right.$$

$$\left. \left. + \sum_{s=1}^{3} \sum_{s'=1}^{3} D_{ss'} \exp\left[-\alpha_s k(h-z)\right] \exp\left[-\alpha_{s'} k(h-z')\right] \right] \right\} + \cdots .$$

As the thickness d of the subsurface layer is the minimum characteristic length, the alloy composition fluctuation δc in this layer may be regarded as independent of z and equal to its value on the surface. The composition fluctuations $\widetilde{\delta c}(k; h)$ forming at the surface are determined by the constrained equilibrium, i.e., by the equilibrium for given values of δc in the thickness of the film, $0 \leq z \leq h-d$. This constrained equilibrium is determined by

$$\frac{\delta(\delta F_{\text{chem}})}{\delta(\widetilde{\delta c}(\mathbf{k}; h))} = 0 \,. \tag{4.50}$$

By keeping only the lowest order terms in d in the variation of the free energy (4.49), and substituting it into (4.50), one obtains an equation that connects composition fluctuations at the surface with the 'frozen' composition fluctuation in the thickness of the film:

$$\left[\frac{\partial^2 f_{\text{chem}}}{\partial c^2} + B_0 \right] \widetilde{\delta c}(\mathbf{k}; h) + \sum_{s=1}^{3} R_s(\varphi) \, k \int_{0}^{h} dz \, \exp\left[-\alpha_s(\varphi)k(h-z)\right] \widetilde{\delta c}(\mathbf{k}; z)$$

$$= 0 \,, \tag{4.51}$$

where

$$R_s(\varphi) = C_s(\varphi) + \sum_{s'=1}^{3} D_{s's}(\varphi) \,. \tag{4.52}$$

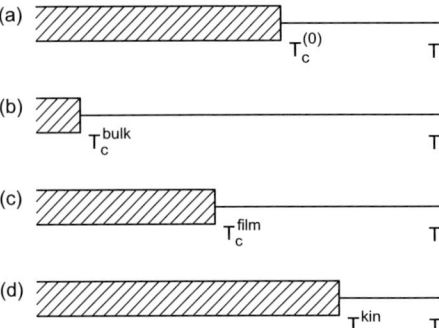

Fig. 4.43. Critical temperatures of alloy instability against fluctuations in composition. (**a**) Critical temperature of thermodynamic instability of a bulk alloy against incoherent phase separation into two phases. The temperature $T_c^{(0)}$ is defined in (4.30). (**b**) Critical temperature of thermodynamic instability of a bulk alloy against coherent phase separation. The temperature T_c^{bulk} is defined in (4.31). (**c**) Critical temperature of thermodynamic instability of an alloy in an epitaxial film bounded by a stress-free surface. The temperature T_c^{film} is defined in (4.47). (**d**) Critical temperature of kinetic instability of alloy epitaxial growth. The temperature T_c^{kin} is defined in (4.55). At temperatures above the critical temperature, alloys are stable for any composition. At temperatures below the critical temperature (*hatched areas*), alloys are unstable in a certain region of composition

By seeking the solution of the integral equation (4.51) in the form $\widetilde{\delta c}(\boldsymbol{k};z) \sim \exp(\gamma k z)$, one obtains the characteristic equation for the amplification increment γ, viz.,

$$\left[\frac{\partial^2 f_{\text{chem}}}{\partial c^2} + B_0\right] + \sum_{s=1}^{3} \frac{R_s(\varphi)}{\alpha_s(\varphi) + \gamma} = 0 \,. \tag{4.53}$$

Equation (4.53) may be transformed to a cubic equation with respect to γ, with three roots γ_1, γ_2, and γ_3. The growth of a homogeneous alloy is kinetically unstable against fluctuations in alloy composition if

$$\max \operatorname{Re} \gamma_{1,2,3} > 0 \,. \tag{4.54}$$

The solution of (4.53) obtained in [4.91] showed that kinetic instability occurs for the first time for composition fluctuations with wave vector $\boldsymbol{k}$ parallel to the elastically soft direction [100] or [010]. The critical temperature of kinetic instability is

$$T_c^{\text{kin}} = T_c^{(0)} + \frac{c_{12}}{c_{11}+c_{12}}\Theta \,. \tag{4.55}$$

In addition, it was shown in [4.91] that, when the kinetic instability occurs for the first time, i.e., when $\operatorname{Re}\gamma = 0$, the imaginary part $\operatorname{Im}\gamma = 0$. Thus instability occurs for the first time as a steady instability, and not as an oscillatory one.

4.3 Activated Alloy Phase Separation During Overgrowth of Quantum Dots   299

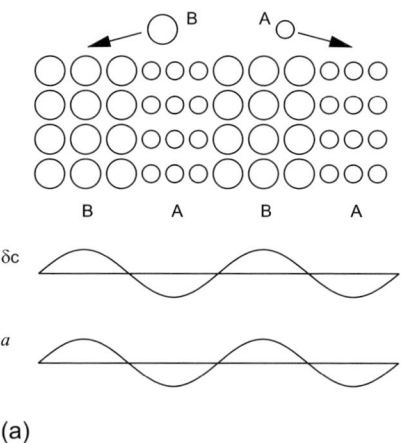

Fig. 4.44. Elastic interaction of adatoms at the surface with a composition-modulated alloy film. Smaller atoms (A) are attracted by domains with a locally smaller lattice parameter, i.e., by domains with an excess of A atoms. Larger atoms (B) are attracted by domains with a locally larger parameter, i.e., by domains with an excess of B atoms

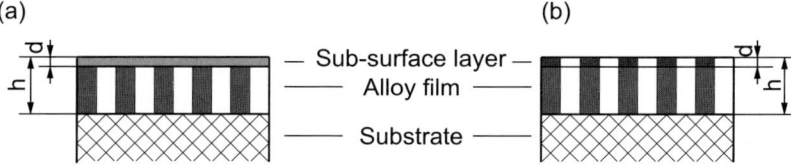

Fig. 4.45. Comparison of elastic energies for two structures. (**a**) The alloy composition in the film is modulated, and the subsurface layer of thickness $d \ll h$ contains a homogeneous alloy. (**b**) The alloy composition in the subsurface layer is modulated exactly as in the film

Figure 4.43 compares critical temperatures for physically distinct processes of phase separation in alloys. It is remarkable that the effect of coherence strain on thermodynamic and kinetic instabilities is opposite! In closed systems, coherence strain hinders alloy decomposition and stabilizes a homogeneous alloy. In open systems, elastic strain favors alloy decomposition and leads to an increase in the critical temperature of instability.

The following comment illustrates this effect (Fig. 4.44). Atoms A and B mixed in an alloy have different atomic radii. Therefore domains with an excess of atoms of smaller radius, say A, have a smaller local lattice parameter than the average. Similarly, domains with an excess of atoms of larger radius, say B, have a larger local lattice parameter than the average. Therefore atoms of larger radius (B), incorporating into the growing crystal, are attracted by surface areas with a larger lattice parameter, i.e., by areas with an excess of B atoms. Similarly, atoms of smaller radius (A) are attracted by surface areas having a smaller lattice parameter, i.e, by areas with an excess of atoms A. This is the elastic driving force to kinetic instability of alloy growth.

To elucidate in more detail the effect of elastic interaction on kinetic instability, we focus on the boundary of the instability region, where $\gamma = 0$. Then the composition profile in the film does not vary with z, i.e., $\widetilde{\delta c}(\boldsymbol{k}; z) = \widetilde{\delta c}(\boldsymbol{k}) = \text{Const.}(z)$. Figure 4.45 shows a thin layer of thickness d deposited on top of a film of thickness $h - d$ and compares two situations. In Fig. 4.45a the alloy composition in the surface layer is homogeneous, i.e., $\delta c = 0$. In Fig. 4.45b, the alloy composition in the subsurface layers is the same as in the underlying film. The difference in the free energy is given by (4.49), into which the z-independent profile of δc should be substituted. This difference is

$$\Delta F \equiv F^{(\mathrm{b})} - F^{(\mathrm{a})}$$
$$= d \int \frac{\mathrm{d}^2 k}{(2\pi)^2} |\widetilde{\delta c}(\boldsymbol{k})|^2 \left[\frac{\partial^2 f_{\mathrm{chem}}}{\partial c^2} + B_0 + \sum_{s=1}^{3} \frac{R_s(\varphi)}{\alpha_s(\varphi)} \right] . \quad (4.56)$$

The first term in the integrand of (4.56) is the contribution of the alloy chemical free energy. Its sign depends on temperature and the average composition of the alloy. The second term is due to the energy of elastic interaction between composition fluctuations in the subsurface layer and the substrate. It is always positive, hinders alloy decomposition, and stabilizes homogeneous alloy growth. The third term is the elastic interaction between composition fluctuations in the subsurface layer and those in the rest of the film thickness. It is negative, favors alloy decomposition, and destabilizes homogeneous alloy growth. The effect of this term was illustrated in Fig. 4.44. The remarkable feature of the elastic interaction is that the destabilizing term in the elastic interaction outweighs the stabilizing term, and the resulting effect of elastic strain on the alloy growth is destabilizing.

Guyer and Voorhees [4.92, 4.93] developed a kinetic theory of alloy compositional and morphological instability during epitaxial growth. Their approach takes into account condensation of atoms from the vapor, evaporation of atoms, migration of adatoms on the surface, and their incorporation into the growing film. In this model, the alloy composition at the film surface is determined by a condition of local equilibrium with a vapor, and the surface kinetics uses a network constraint that couples the diffusion fluxes of A and B atoms. In addition to [4.91], the lattice-matched and lattice-mismatched growth of an alloy are both considered in [4.92, 4.93]. It is shown that, for a lattice-mismatched alloy, compositional instability is always coupled to morphological instability, and the advancing surface of the growing crystal is no longer flat. For the particular case of lattice-matched growth, the result of [4.93] for the critical temperature of kinetic instability coincides in the slow deposition limit with that of (4.55) [4.91].

Ipatova et al. [4.94, 4.95] developed a kinetic theory of compositional instabilities during the step-flow growth of an alloy on a vicinal surface. In this model, both the bulk elastic anisotropy of a cubic crystal and the anisotropy of surface diffusion are taken into account. It has been shown that, due to the interplay between these two anisotropies, the wave vector of the most unstable mode of composition fluctuations can point in any direction differing from that of the elastically soft axis. In addition, an array of steps on the surface has its own anisotropy and thus also affects the direction of $\boldsymbol{k}$ for the most unstable mode.

4.3 Activated Alloy Phase Separation During Overgrowth of Quantum Dots

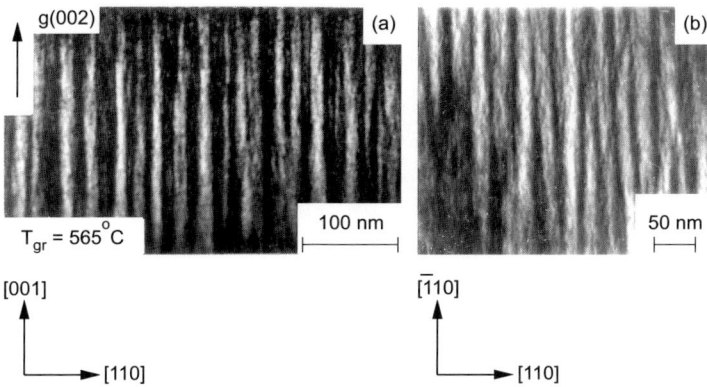

Fig. 4.46. TEM images of $In_{0.52}Al_{0.48}As$ layers grown on an InP substrate by MOCVD at $T_{gr} = 650°C$. (**a**) $[\bar{1}10]$ cross-section (002) TEM dark-field (DF) image. (**b**) [001] plan-view DF image. Both images reveal quasi-periodic contrast in the [110] direction. With kind permission of S.W. Jun et al. [4.96]

In particular, the theory of Ipatova et al. [4.94, 4.95] explains the experimental results of Jun et al. [4.96], where the growth of an InAlAs alloy lattice-matched to an InP substrate results in the formation of a structure with composition modulation in the [110] direction. Transmission electron microscopy (TEM) images obtained in both plan view and cross-section (Fig. 4.46) reveal a quasi-periodic band-like contrast in the [110] lateral direction. Such contrast was observed in samples grown at 565°C and 590°C, and at 615°C, although considerably less pronounced in the latter case. No contrast was observed in samples grown at higher temperatures.

In the theory of alloy growth instability due to Léonard and Desai [4.97–4.99], the surface kinetics is linked to variational derivatives of the total free energy. These authors focused in particular on the way the dependence of elastic moduli on composition affects alloy growth instability.

Venezuela and Tersoff [4.100] developed a model for the instability of alloy growth on a vicinal surface in the step-flow mode. It was shown that a mere difference in surface mobilities of adatoms A and B may lead to a compositional instability in alloy growth.

The theory developed by Spencer et al. [4.101–4.103] involves the derivation of chemical potentials for each alloy component on the surface, and a description of the surface diffusion of each species. The effects of misfit strain, compositional strains, and different mobilities of the alloy components on the morphological stability of strained alloy film growth were considered. In particular, the theory of [4.103] predicts an asymmetry of stability with respect to the sign of the misfit for $Si_{0.5}Ge_{0.5}$ films, due to the different mobilities of Si and Ge adatoms. This provides an explanation for numerous experimental data [4.103].

4.3.2 Steady-State Composition-Modulated Structures in Growing Alloy Films

Most theories dealing with kinetic instability of epitaxial alloy growth have focused on linear stability analysis. Such an analysis gives criteria, e.g., average alloy composition, temperature, misfit, etc., for which the growth of a homogeneous alloy is unstable against fluctuations in alloy composition (compositional instability) and/or fluctuations in surface profile (morphological instability).

Only a few papers have addressed non-linear effects in perturbation amplitude. For example, Léonard and Desai [4.97] obtained theoretically the final inhomogeneous structure of a growing alloy with 1D lateral composition modulation. In another paper [4.98], they took into account the variation of elastic moduli with alloy composition and demonstrated the preferred formation of convex hard particles embedded in an elastically soft matrix.

Barvosa-Carter et al. [4.104] (see also [4.105, 4.106]) emphasized the effect of stress on surface mobility and showed that such a dependence can be either stabilizing or destabilizing, depending on the sign of the stress.

Shchukin and Starodubtsev [4.107, 4.108] considered steady-state composition-modulated structures forming at later stages of alloy epitaxial growth. They considered the epitaxial growth of a binary alloy $A_{1-c}B_c$ lattice-matched on average to the (001) substrate. Both the substrate and the deposit were treated as elastically anisotropic cubic crystals. The elastic moduli of the deposit were assumed composition-independent and equal to those of the substrate.

Figure 4.47 illustrates mass conservation conditions during alloy growth. The net flux of each species A or B into a given surface region is the sum of contributions from the beam and surface migration, and results in the incorporation of atoms into the solid and a corresponding advance of the surface in the direction of the local normal n,

$$N_0 \frac{\partial c_{A,B}}{\partial t} + (V_{gr} \cdot n) N_0 c_{A,B} = -\nabla_S J^S_{A,B} - J^{beam}_{A,B} \cdot n . \tag{4.57}$$

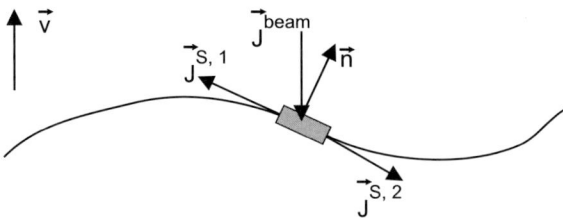

Fig. 4.47. Schematic illustration of the mass conservation condition during alloy growth. The net flux of each species A or B into the shaded surface region is the sum of contributions from the beam and surface migration, causing the surface to advance in the direction of the local normal n

4.3 Activated Alloy Phase Separation During Overgrowth of Quantum Dots

Here the compositions are $c_A = 1 - c$ and $c_B = c$. The surface fluxes of adatoms A and B are given by the Fickian law,

$$\mathbf{J}_A^S = -\frac{D_{AA}}{k_B T}\nabla_S \mu_A - \frac{D_{AB}}{k_B T}\nabla_S \mu_B, \tag{4.58a}$$

$$\mathbf{J}_B^S = -\frac{D_{BA}}{k_B T}\nabla_S \mu_A - \frac{D_{BB}}{k_B T}\nabla_S \mu_B, \tag{4.58b}$$

where the surface diffusivities $D_{AA}, D_{AB}, D_{BA}, D_{BB}$ of adatoms are assumed to be isotropic, and $\mu_{A,B}$ are the chemical potentials of adatoms A and B defined by

$$\mu_A(\mathbf{r}) = \frac{\delta F}{\delta N_A}(\mathbf{r}), \tag{4.59a}$$

$$\mu_B(\mathbf{r}) = \frac{\delta F}{\delta N_B}(\mathbf{r}). \tag{4.59b}$$

The total Helmholtz free energy F is the sum of four contributions,

$$F = F_{\text{chem}} + F_{\text{grad}} + F_{\text{surf}} + F_{\text{elast}}. \tag{4.60}$$

It is convenient to describe the local state of the advancing surface by the deviation of the local surface profile $h(x, y; t)$ from the flat one:

$$h(x, y; t) = V_{\text{gr}} t + \zeta(x, y; t), \tag{4.61}$$

where V_{gr} is the growth velocity related to the average advance of the crystal surface, and $\zeta(x, y; t)$ is the fluctuating component of the surface profile. The basic equations of alloy epitaxial growth (4.57)–(4.59b) can be formulated in terms of the kinetic equations for a joint evolution of the fluctuations in alloy composition $\delta c(x, y, z = h(x, y; t); t)$ at the advancing surface and the surface profile $\zeta(x, y; t)$. We introduce the corresponding kinetic coefficients,

$$D_\zeta = D_{AA} + 2D_{AB} + D_{BB}, \tag{4.62}$$

relating to the evolution of the surface profile, and

$$D_c = c^2 D_{AA} - 2c(1-c)D_{AB} + (1-c)^2 D_{BB}, \tag{4.63}$$

relating to the substitutional diffusion of surface adatoms. Further simplification occurs if one assumes that adatoms A and B have equal diffusivities, $D_{AA} = D_{BB}$, and that the average composition of the growing alloy equals $\bar{c} = 1/2$. Then the joint evolution of the alloy composition and the surface profile obeys the following equations [4.108]:

$$\frac{\partial \zeta}{\partial t} = a \frac{1}{k_B T} \nabla \left[D_\zeta \nabla \left(a \frac{\delta F}{\delta \zeta} - (\delta c) \frac{\delta F}{\delta[\delta c]} \right) \right], \tag{4.64a}$$

$$\frac{\partial \delta c}{\partial t} = \frac{1}{k_B T} \nabla \left(D_c \nabla \frac{\delta F}{\delta[\delta c]} \right) - \left(v + \frac{\partial \zeta}{\partial t} \right) \frac{\delta c}{a}, \tag{4.64b}$$

where a is the lattice parameter in the vertical direction. The last term in (4.64a) relates to the non-linear coupling between substitutional migration of atoms A and B on a surface of a given profile and their joint migration due to changes in the surface profile.

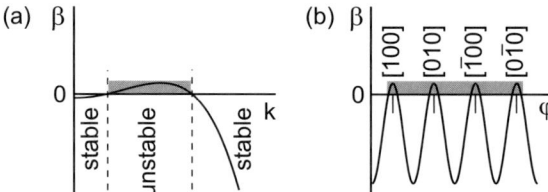
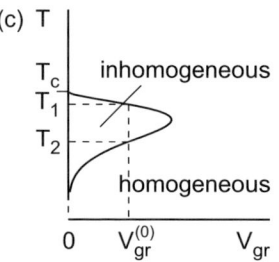

Fig. 4.48. (a) Amplification rate of alloy composition fluctuations vs. absolute value of the wave vector k for a fixed direction of k. The *shaded area* corresponds to an instability. (b) Amplification rate of alloy composition fluctuations vs. direction of k. (c) Linear stability phase diagram for epitaxial alloy growth

Linear Stability Analysis. In order to carry out the linear stability analysis of (4.64a) and (4.64b), one needs to expand the free energy up to quadratic terms in composition fluctuations δc and in surface profile fluctuations ζ, then substitute $\delta c(x,y;t) \sim \exp(\beta t)\exp\left[\mathrm{i}\,(k_x x + k_y y)\right]$ and $\zeta(x,y;t) \sim \exp(\beta t)\exp\left[\mathrm{i}\,(k_x x + k_y y)\right]$ into the linear equations. This yields the following characteristic equation for the amplification rate β,

$$\beta = -\frac{D_c(T)}{k_\mathrm{B}T}k^2\left[\frac{\partial^2 f_\mathrm{chem}(T;c)}{\partial c^2}+\kappa k^2 + B_0 + \sum_{s=1}^{3}\frac{R_s(\varphi)kV_\mathrm{gr}}{\beta+\alpha_s(\varphi)kV_\mathrm{gr}}\right]-\frac{V_\mathrm{gr}}{a}. \tag{4.65}$$

The summation over s includes contributions to the elastic interaction from three acoustic waves (in the static limit), as in (4.53). $R_s(\varphi)$ is the elastic interaction energy and $\alpha_s(\varphi)$ is the attenuation coefficient of the static acoustic wave. The criterion $\mathrm{Re}\,\beta > 0$ implies instability of epitaxial alloy growth against fluctuations in alloy composition (compositional instability) and/or fluctuations in surface profile (morphological instability).

Note that in (4.65), if we use an approximation of fast surface diffusion ($D_c \to \infty$), substitute $\beta = \gamma k V_\mathrm{gr}$, and neglect the term κk^2, then (4.65) reduces to the characteristic equation (4.53) for a simpler model of compositional instability of the alloy growth.

It follows from the linear stability analysis of (4.64a) and (4.64b) that, in the considered situation of lattice-matched growth, compositional and morphological instabilities are not coupled, the surface remains flat, and

4.3 Activated Alloy Phase Separation During Overgrowth of Quantum Dots

only compositional instability may occur. This is similar to the results of [4.93, 4.99]. The main difference between the current analysis and that of [4.93, 4.99] is that we take into account the elastic anisotropy of the substrate and the growing film via the angular dependence of $R_s(\varphi)$ and $\alpha_s(\varphi)$.

To solve (4.65), we use the regular solution approximation for the chemical free energy and the Arrhenius formula for the temperature-dependent diffusion coefficient $D_c(T) = D_c^{(0)} \exp(-E_a/k_B T)$, where E_a is the activation energy. Figure 4.48a displays the amplification rate β vs. the absolute value of the wave vector $\mathbf{k}$ for a fixed direction of $\mathbf{k}$. For small values of k, i.e., for long-wavelength fluctuations in composition, the time interval τ during which surface adatoms are mobile (the deposition time of one monolayer) is not sufficient for migration of adatoms over a long distance $\sim 1/k$, i.e.,

$$\tau = \frac{a}{V_{\mathrm{gr}}} \ll \frac{1}{D_c k^2} . \tag{4.66}$$

For large values of k, the gradient energy $\sim \kappa k^2$ makes fluctuations in alloy composition energetically unfavorable.

Figure 4.48b displays the amplification rate β vs. the direction of $\mathbf{k}$, showing that instability occurs for the first time for composition fluctuations in elastically soft directions $\langle 100 \rangle$.

Figure 4.48c shows the linear stability phase diagram in temperature–growth velocity variables. Instability of alloy growth is governed by competition between surface migration and burial by the incoming flux. Let the temperature T be fixed. Then, if V_{gr} is higher than a certain critical value, fast burial does not allow surface migration to create composition fluctuations and the alloy grows homogeneously. Now let $V_{\mathrm{gr}} = V_{\mathrm{gr}}^{(0)}$ be fixed and T decrease. If $T < T_1(V_{\mathrm{gr}}^{(0)})$, the sum of chemical and elastic driving forces to phase separation outweighs the entropy driving force to mixing, and instability occurs. If $T < T_2(V_{\mathrm{gr}}^{(0)})$, the diffusion coefficient, decreasing with decreasing temperature, becomes too small and instability does not occur.

Non-Linear Coupling between Compositional and Morphological Instabilities. In the linear approximation, the two instabilities are not coupled, i.e., the composition instability does not lead to a morphological one, and the surface remains flat in the early stages of growth. However, the two instabilities are coupled in the non-linear regime, as shown in Fig. 4.49. Let the composition-modulated structure consist of alternating domains with compositions $c_1 < \bar{c}$ (A-rich domains) and $c_2 > \bar{c}$ (B-rich domains) and let atom A have a smaller radius than atom B. Then a bulk alloy $\mathrm{A}_{1-c_1}\mathrm{B}_{c_1}$ has a smaller lattice parameter than bulk $\mathrm{A}_{1-c_2}\mathrm{B}_{c_2}$, i.e., $a_0(c_1) < \bar{a} \equiv a(\bar{c}) < a_0(c_2)$ (Fig. 4.49d). In the composition-modulated structure of Fig. 4.49b, domains are coherently conjugated. Then A-rich domains are stretched with respect to the intrinsic lattice parameter of the bulk alloy with composition c_1, i.e., $a(c_1) > a_0(c_1)$. At the same time, they are compressed with respect to the average lattice parameter of the structure $\bar{a}$. Similarly, B-rich domains are

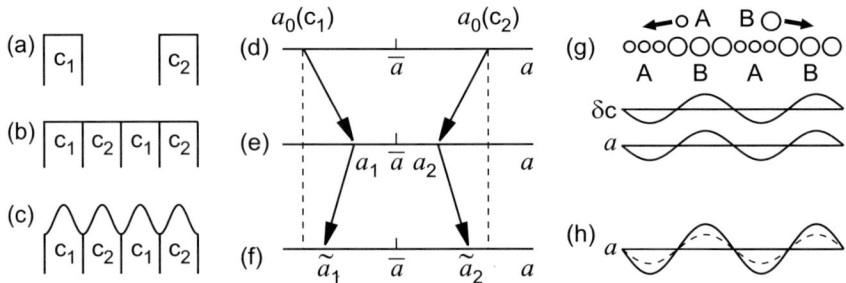

Fig. 4.49. Coupling between composition modulation and surface profile. (**a**) Two separate bulk alloys. (**b**) Composition-modulated structure with a planar surface. (**c**) Composition-modulated structure with a non-planar surface. (**d**) Lattice parameters of bulk alloys with compositions c_1 and c_2. (**e**) Lattice parameters of coherently conjugated composition domains in a stressed system with a planar surface. (**f**) Lattice parameters of coherently conjugated domains in a partially relaxed system with a non-planar surface. (**g**) Elastic driving force to phase separation: atoms A(B) are attracted by A(B)-rich domains. Modulation of composition δc and modulation of the lattice parameter a. (**h**) Modulation of the lattice parameter at the surface. The *dashed line* refers to the planar surface and the *solid line* refers to a non-planar surface

compressed with respect to the intrinsic lattice parameter of bulk $A_{1-c_2}B_{c_2}$, and stretched with respect to $\bar{a}$. Since the surface consists of alternating domains under tensile and compressive stress, it will be unstable against undulations, if the modulation wavelength exceeds a certain critical value. Then the energy gain due to elastic relaxation exceeds the energy cost of a non-planar surface profile, and the surface profile will consist of troughs over domain boundaries and crests in the center of each composition domain (Fig. 4.49c). Due to elastic relaxation, lattice parameters of A-rich and B-rich domains are shifted towards their intrinsic values, $a_0(c_1)$ and $a_0(c_2)$, respectively (Fig. 4.49f).

Figures 4.49g and h illustrate the reaction of surface undulations on phase separation. B atoms with larger atomic radius are incorporated preferentially into surface regions which are stretched with respect to the average lattice parameter of the structure $\bar{a}$, i.e., into B-rich domains (see, for example, [4.94]). Similarly, A atoms prefer to incorporate into A-rich domains. The elastic driving force to phase separation is proportional to the actual variation of the lattice parameter, $a(c_2) - a(c_1)$. Since this variation increases due to surface undulations (Fig. 4.49f), the elastic driving force to phase separation increases too.

Steady-State Modulated Structures. To obtain stable steady-state solutions of the kinetic equations (4.64a) and (4.64b), the equations were solved numerically, and the stability of solutions was checked by integrating the kinetic equation over time in the vicinity of the steady-state solutions [4.108].

4.3 Activated Alloy Phase Separation During Overgrowth of Quantum Dots

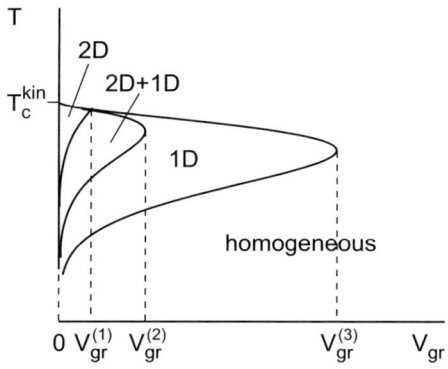

Fig. 4.50. Steady-state phase diagram containing a region with homogeneous growth, a region with growth of a 1D modulated structure, a region with growth of a 2D modulated structure, and a region where the stable growth of both 1D and 2D structures is possible. For a set of material parameters typical of semiconductor alloys, the calculated values of V_{gr} are as follows: $V_{gr}^{(1)} = 1.2 \text{ Å s}^{-1}$, $V_{gr}^{(2)} = 4.0 \text{ Å s}^{-1}$, and $V_{gr}^{(3)} = 96 \text{ Å s}^{-1}$. Thus, a kinetic phase transition between the growth of a 1D structure and the growth of a 2D structure can indeed occur at $V_{gr} < V_{gr}^{(2)}$, i.e., at growth velocities typical in molecular beam epitaxy

The method for obtaining steady-state structures is described in detail elsewhere [4.109].

The calculated steady-state phase diagram of alloy growth on the (001) substrate is shown in Fig. 4.50. The phase diagram was calculated using numerical values for material parameters which are typical for III–V semiconductors. The elastic moduli are $c_{11} = 1.0 \times 10^{12} \text{erg cm}^{-3}$, $c_{12} = c_{44} = 0.5 \times 10^{12} \text{erg cm}^{-3}$. The surface energy is 50 meV Å^{-2}. The lattice mismatch between the final pure components of the alloy is 7%. For the critical temperature of kinetic instability in the slow deposition limit, the value $T_c^{kin} = 1\,000$ K was used. The surface diffusivity of adatoms was taken as $D_c(T) = 1.0 \times 10^{-5} \exp\left(-E_a/k_B T\right) \text{ cm}^2\text{s}^{-1}$, where the activation energy is $E_a = 1.5$ eV.

The steady-state phase diagram of Fig. 4.50 contains:

- a region of homogeneous alloy growth,
- a region where the growth of a structure modulated only in one elastically soft direction ([100] or [010]) is stable,
- a region where the growth of a 2D structure modulated in both [100] and [010] directions is stable,
- a region where the growth of both a 1D modulated structure and a 2D modulated structure is stable.

This type of pattern selection can be explained as follows. The modulation period is known to increase with decreasing growth velocity [4.97]. On the other hand, the interplay between surface and elastic energies favors surface undulations at sufficiently large periods. Thus, surface undulations should occur at sufficiently low growth velocity $V_{\rm gr}$. An essentially non-planar surface favors a 2D structure modulated in both the [100] and [010] elastically soft directions over a 1D structure modulated in only one direction, either [100] or [010]. The effect is similar to the one for static structures of strained islands in lattice-mismatched systems, where 2D structures (pyramids) provide a more efficient elastic relaxation than 1D structures (prisms) [4.110].

In the region of the phase diagram where the growth of both 1D and 2D modulated structures is stable, the actual steady-state structure formed during growth depends on initial conditions. To obtain the structure for any particular initial conditions, e.g., for growth starting from a homogeneous flat substrate, it is necessary to integrate the kinetic equations (4.64a) and (4.64b) over the entire time of evolution. To the best of our knowledge, this problem has not been addressed in the literature.

4.3.3 Alloy Growth on Stressors: Activated Phase Separation

The self-organization phenomena at crystal surfaces and in epitaxial films described so far in this book can be roughly synthesised into two large groups. First, there are phenomena in pure crystals including, for example, periodic surface nanoscale faceting or the formation of single-sheet or multisheet structures of 2D or 3D islands. In these heteroepitaxial systems, the driving force for nanostructure formation is, in general, a difference between the substrate and the adsorbate materials. This may be a difference in lattice parameter, surface energy, surface stress, or something else. Each of the two materials may be a pure crystal. The second group of self-organization phenomena is associated with phase separation in alloys and the formation of composition-modulated structures.

Phenomena in which both the formation of strained islands and an inhomogeneous profile of alloy composition occur together have been considerably less well investigated. Among the few studied effects is an enhanced nucleation of islands in strained alloy systems [4.111]. Islands nucleate at substantially different alloy compositions than the alloy layer, i.e., they are enriched in the alloy constituent which has a larger lattice mismatch with the substrate. For example, for the $In_{1-c}Ga_cAs/GaAs$ system, islands will be enriched in indium. Another recognized effect is the formation of an inhomogeneous composition profile within the islands (see, for example, [4.112]).

Very recently the two branches of self-organization phenomena have indeed merged, providing a new way of controlling geometric parameters and the electronic spectra of nanostructures. Figure 4.51 illustrates the idea of the overgrowth of initial InAs islands by a GaInAs alloy proposed by Maximov et al. [4.113]. Coherent InAs/GaAs islands overgrown by an InGa(Al)As alloy layer serve as a model system. The laterally varying strain at the surface created by the InAs islands affects the overgrowth with an InGa(Al)As alloy

4.3 Activated Alloy Phase Separation During Overgrowth of Quantum Dots

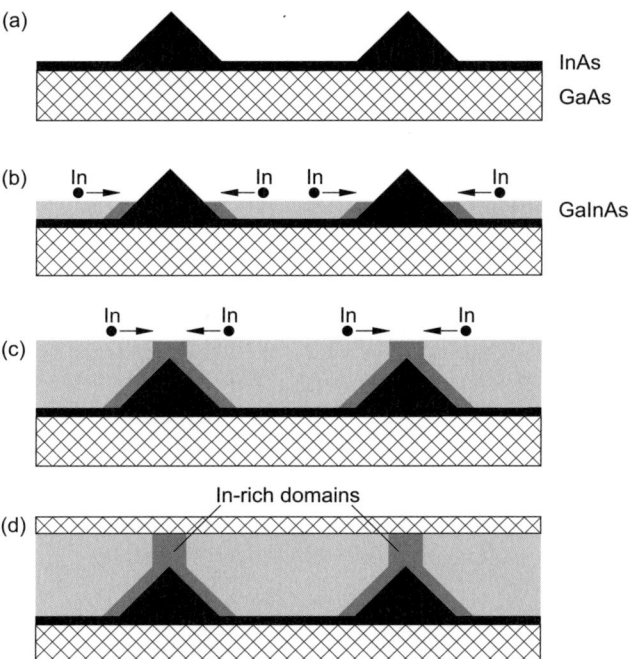

Fig. 4.51. Schematics of the overgrowth of InAs islands by an GaInAs alloy. (**a**) 3D coherently strained InAs islands over an InAs wetting layer on a GaAs substrate. (**b**) Initial stage of overgrowth of InAs islands by the GaInAs alloy. Adatoms of In incorporate preferentially at the facets of initial InAs islands, forming In-rich domains in the capping layer around initial InAs islands. (**c**) Latest stages of overgrowth of InAs islands by the alloy. When InAs islands are completely overgrown, In-rich domains form in the capping layer over the islands. (**d**) Completely overgrown structure. In-rich domains are formed in the vicinity of initial InAs islands, increasing the effective size of quantum dots in both lateral and vertical directions

layer by strain-driven surface migration. One may expect the following qualitative picture of the alloy phase separation activated by stressors. For the conventional overgrowth of InAs QDs by pure GaAs, it has been found [4.4] that gallium atoms prefer to migrate away from QDs towards pseudomorphically strained regions, having an in-plane lattice parameter equal to that of unstrained GaAs. A similar effect should occur during the overgrowth of InAs islands by an InGa(Al)As alloy: indium atoms will accumulate at the IsAs islands, increasing their lateral size. When the islands are completely covered, the tensile strain on top of the QDs will favor indium accumulation from the growing alloy. The latter could eventually increase the height of the islands. We thus expect the activated phase separation of the alloy to increase the indium content in the vicinity of the InAs islands. In other words, the effective lateral size and height of the QDs will increase, providing enhanced

310   4. Engineering of Complex Nanostructures

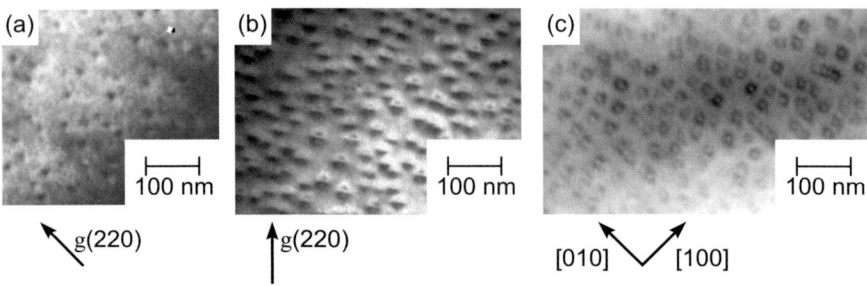

Fig. 4.52. Plan-view TEM images of overgrown InAs/GaAs islands. (**a**) Overgrowth by GaAs. The average lateral size of islands (slightly overestimated) is 12 nm. (**b**) Overgrowth by 5 nm of $Ga_{0.85}In_{0.15}As$ alloy. The average lateral size of islands (slightly overestimated) is 22 nm. (**c**) Overgrowth by 5 nm of $Ga_{0.85}In_{0.15}As$ alloy. The image was taken under different imaging conditions, providing an accurate average lateral size of 18 nm for the islands

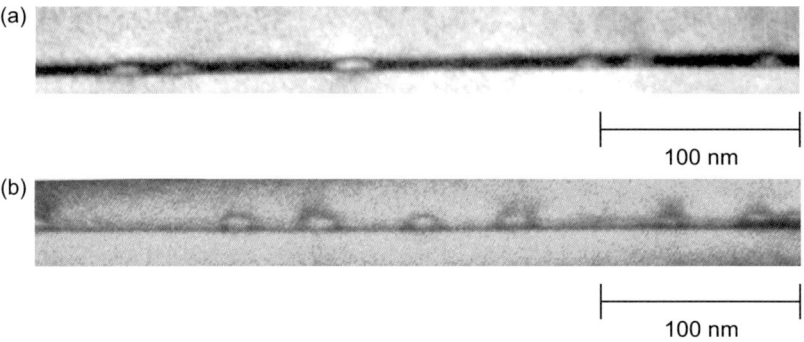

Fig. 4.53. Cross-sectional TEM images of InAs/GaAs islands overgrown by (**a**) $Ga_{0.85}In_{0.15}As$ and (**b**) $Ga_{0.70}Al_{0.15}In_{0.15}As$. Adding Al leads to an increase in the effective island height

localization of electron and holes, and redshifting the photoluminescence (PL) spectrum of the QDs.

The chemical driving force to phase separation in the growing alloy will also contribute to the formation of In-rich domains close to and above InAs islands. The effect will be pronounced in a certain temperature interval, whereas it will be hindered at high temperatures due to entropy effects and at low temperatures due to a decrease in the surface diffusivity of adatoms, similarly to the situation in Fig. 4.48.

Figure 4.52 supports the expected trend, showing a plan-view TEM image of the original InAs islands covered by GaAs (Fig. 4.52a) as well as images of similar islands covered by a 5 nm $In_{0.15}Ga_{0.85}As$ layer (Figs. 4.52b and c). Figures 4.52a and b show a significant contribution from the local strain and Fig. 4.52c, taken under diffraction conditions far removed from an exact Bragg orientation, allows a more precise determination of the island lateral

4.3 Activated Alloy Phase Separation During Overgrowth of Quantum Dots

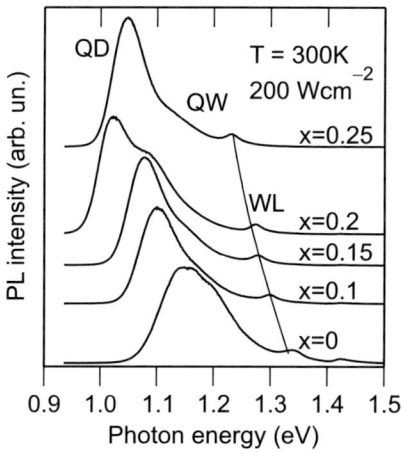

Fig. 4.54. Photoluminescence (PL) spectra of InAs/GaAs quantum dots (QDs) overgrown by $Ga_{1-x}In_xAs$ with different alloy compositions x

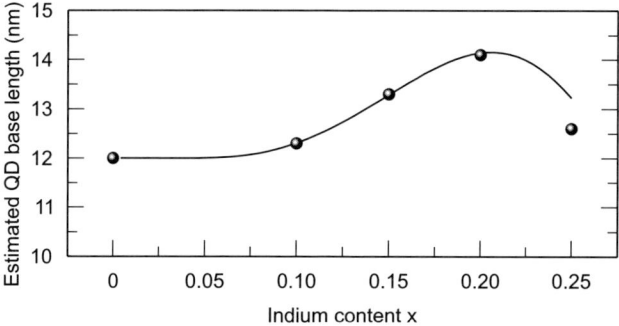

Fig. 4.55. Estimated lateral size of InAs quantum dots (QDs) overgrown by $Ga_{1-x}In_xAs$ vs. indium composition x

size and shape. Figures 4.52b and c show a clear increase in width for islands capped by an InGaAs alloy, from 12 nm to 18 nm.

Cross-sectional TEM images of the alloy-capped sample are shown in Figs. 4.53a and b. Overgrowth by a quaternary $Ga_{0.70}Al_{0.15}In_{0.15}As$ alloy (Fig. 4.53b) leads to a significant increase in island height compared to overgrowth by a ternary $Ga_{0.85}In_{0.15}As$ alloy (Fig. 4.53a). These data indicate that addition of aluminum enhances phase separation.

Figure 4.54 shows the PL spectra of QDs capped by a 4-nm-thick $In_xGa_{1-x}As$ alloy layer with indium composition x varying from 0 to 0.25. As the indium composition increases from 0 to 0.2, the PL peak from QDs shifts towards lower photon energies. This agrees with an increase in QD volume. The same trend may be due to the lowering of the energy bandgap in the alloy cap layer surrounding QDs. However, the bandgap in the layer continues to fall for In content $x > 0.2$, whilst the PL maximum from the QDs exhibits a blueshift.

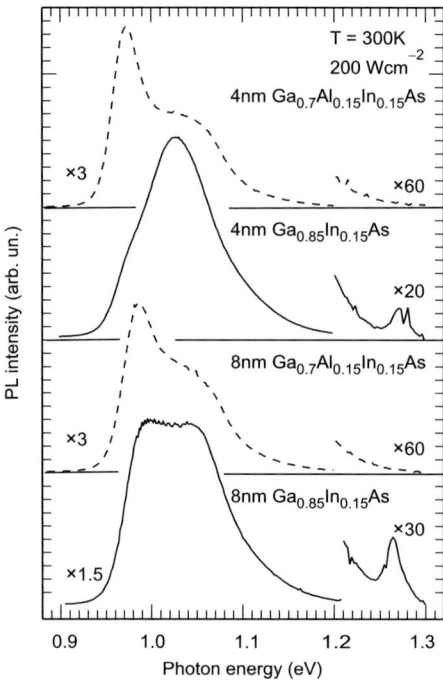

Fig. 4.56. Photoluminescence (PL) spectra of InAs/GaAs quantum dots (QDs) overgrown by $Ga_{0.85}In_{0.15}As$ (*solid lines*) and $Ga_{0.7}Al_{0.15}In_{0.15}As$ (*dashed lines*)

Figure 4.55 shows the dependence of QD lateral size estimated from plan-view TEM on the indium composition x in the cap layer [4.114]. As the indium composition increases from 0 to 0.2, the lateral size of the islands increases too, in accordance with the redshift of the PL spectrum in Fig. 4.54. A further increase in indium composition leads to the onset of dislocation in large islands. Dislocated islands are efficient sinks for indium atoms, extracting them from the remaining coherent islands. This reduces the average size of coherent islands and leads to a blueshift in the PL.

The effect of aluminum on the PL spectra of QDs is shown in Fig. 4.56. Despite an increase in the energy bandgap in the $Ga_{0.7}Al_{0.15}In_{0.15}As$ alloy layer compared to the $Ga_{0.85}In_{0.15}As$ alloy layer, the PL maximum from the QDs exhibits a redshift. This effect can be unambiguously attributed to the enhanced phase separation due to aluminum, in agreement with the cross-sectional TEM data of Fig. 4.53.

Thus, by varying the alloy composition, the thickness of the alloy cap layer, and the amount of initially deposited InAs, it is possible to tune the ground state transition energy of the QDs. For QDs overgrown by $Ga_{0.7}Al_{0.15}In_{0.15}As$, the ground state transition energy is significantly redshifted (up to 200 meV) compared to the original InAs/GaAs QDs. This allows us to reach the technologically important 1.3 μm spectral region, whilst maintaining a high PL efficiency and low defect density. Moreover, it is possible to combine two complex growth modes described in the paper and to fabricate a multisheet array of QD layers, each capped by an alloy material. Detailed description of this growth mode and characteristics of the corresponding QD injection laser are given in [4.113].

Therefore, the overgrowth of coherent strained InAs islands by an InGa(Al)As alloy leads to activated alloy phase separation and the formation of In-rich domains in the vicinity of initial islands. This in turn leads to an increase in the effective lateral size and height of QDs and a significant redshift of the PL spectrum.

To conclude the present chapter, we have demonstrated that by combining self-organization phenomena and carrying out subsequent nano-engineering, nanostructures can be designed to meet practically any requirement on geometrical parameters and electronic spectra. Improving size uniformity and spatial ordering via multi-layered growth of QDs, seeding QDs to obtain independent control over island volume and density, engineering the exciton wave function in order to tune the spectral position and polarization of emitted light, using the effects of activated alloy phase separation, defect-reduction techniques, all these are powerful experimental tools that considerably expand the possible device applications for epitaxial nanostructures.

5. Devices Based on Epitaxial Nanostructures

> Therefore shall they eat of the fruit of their own way, and be filled with their own devices.
>
> *Proverbs 1:31*

The vast effort that has gone into fabricating and studying epitaxial nanostructures like quantum wires, quantum dots, coupled quantum dots, etc., has been to a large extent motivated and fueled by their potentially advantageous and unique device applications. Those in optical communications (lasers, optical amplifiers, photodetectors) and data processing and storage (single-electron devices) are based mainly on the fundamental properties of QD nanostructures, i.e., their discrete energy spectrum. A much lower threshold current density and much higher temperature stability (compared to conventional quantum well systems) up to and above room temperature is one of the remarkable features of the new nano-objects.

From this intense QD research, a novel idea has emerged: to use ensembles of QDs as a new medium. This provides an efficient way of overcoming the problem of thermal evaporation of carriers from individual dots when they are statistically recaptured by neighboring ones. Consequently, arrays of shallow submonolayer quantum dots are currently being used as an active medium for lasers, whilst collective Coulomb blockade may be used in electronic devices, extending their operation to room temperature. So far, optoelectronic applications of QDs as the active medium for semiconductor diode lasers have been the most successful. The major part of this chapter will be concerned with precisely this example of nanostructures at work. We believe that progress in QD photonic devices will be an additional motivation, paving the way to the successful use of new nano-objects in electronic devices as well.

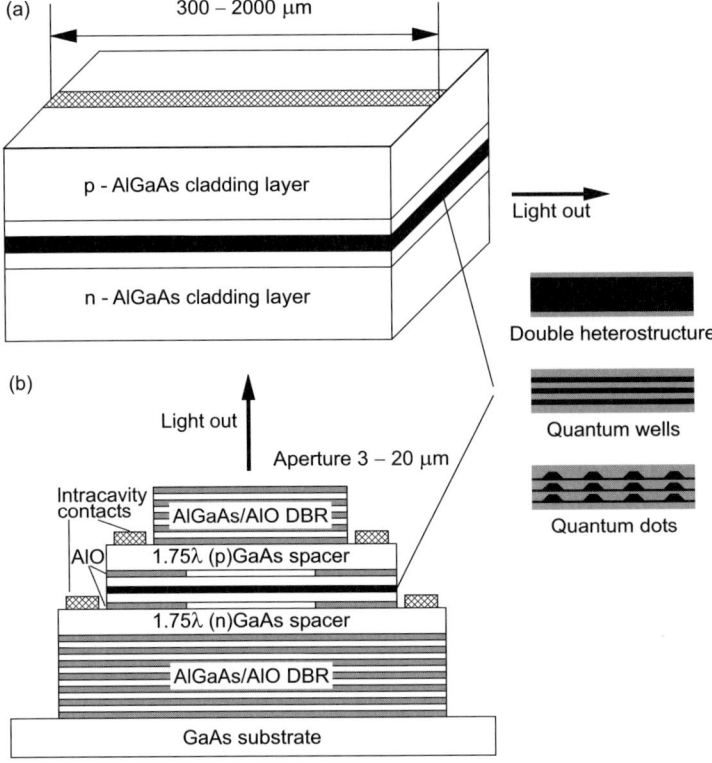

Fig. 5.1. Comparison of (**a**) edge- and (**b**) surface-emitting lasers. In vertical cavity surface-emitting lasers (VCSELs), all-semiconductor (e.g., AlAs–GaAs) or selectively-oxidized AlO–GaAs distributed Bragg reflectors (DBRs) are generally used to achieve high reflectivity. The active region is typically made up of ultra-thin layers or, more recently, stacks of coherent nano-inclusions (quantum dots), as shown in the figure. For single-mode operation, the thickness of the waveguide region in an edge-emitting structure should not generally exceed 0.6–0.8 µm, and for VCSELs the lateral dimensions of the microcavity should lie within 3–5 µm, depending on the specific design

5.1 Quantum Dot Heterostructure Lasers

The lasers currently dominating practical applications can be divided into two types. In edge-emitting devices (Fig. 5.1a), the active medium, for example a thin layer, is placed in a waveguide region having a larger refractive index than the surrounding cladding layers. The laser light is confined in this narrow waveguide. The advantage of the edge-emitting laser is a compact output aperture realized simultaneously with a high light output power. High reflection and antireflection dielectric coatings are usually deposited on the rear and front facets. In a vertical cavity surface-emitting laser (VCSEL) the photons are bounced in the vertical direction in a high finesse cavity

(Fig. 5.1b). The cavity is very short and the gain per cycle is very low. It is thus of key importance to ensure very low losses at each reflection, providing very high reflectivity at the cavity edges, otherwise lasing will not be possible.

5.1.1 Basic Advantages of Heterostructure Lasers

The history of semiconductor lasers has experienced periods of gradual evolution and periods of revolutionary change. The electrically-driven semiconductor laser based on population inversion between ionized impurities and free carriers (unipolar laser) was proposed in 1959 by Basov et al. [5.1]. The idea of a semiconductor laser based on a degenerately doped p^+–n^+ junction was introduced in 1961 [5.2]. Even at that stage, it was noted in passing that lower current densities could be used if the semiconductors forming the p–n junction had forbidden gaps of different widths [5.2]. It was also proposed in [5.2] to use the resonant cavity idea to ensure light output in a direction perpendicular to the plane of the p–n junction, i.e., the concept of the surface-emitting laser was born. However, laser action in semiconductor diodes was first realized in the edge geometry parallel to the plane of the p–n junction by Hall et al. in 1962 [5.3]. It soon became clear that devices based on the degenerately doped p^+–n^+ junction exhibit current densities that are too high to allow relevant practical applications.

The idea of "exploiting quantum effects in heterostructure semiconductor lasers to produce wavelength tunability" and achieving a "lower lasing threshold" via "the change in the density of states which results from reducing the number of translational degrees of freedom of the carriers" was originally introduced by Dingle and Henry in 1976 [5.4]. However, for about a quarter of a century, lasers using structures with carrier confinement in two directions (quantum wires) or all three directions (quantum dots) appeared to be considerably less successful in practical realizations compared with so-called quantum well (QW) devices, where quantum confinement of charge carriers occurs in only one direction. The main advantage in using size-quantized heterostructures in lasers arises from an increase in the density of states for charge carriers near band edges (Fig. 1.3). For the active medium of the laser, this concentrates most injected non-equilibrium carriers in an increasingly narrow energy range near the bottom of the conduction band and/or the top of the valence band. This enhances the maximum material gain, assuming the same homogeneous (or inhomogeneous) broadening and reduces the influence of temperature on device performance. For structures with size quantization in more than one direction, a singularity occurs in the density of states. This is shown in Fig. 1.3 (see also Fig. 5 in [5.4]). The above positive effects are enhanced compared with what happens for quantum wells.

As can be easily understood, the ultimate size quantization in solids is realized in a quantum dot (QD), which represents a semiconductor crystal measuring only a few nanometers across, coherently inserted into a semiconductor matrix of larger bandgap. A QD still has the basic properties of the atom, whilst providing geometrical dimensions that allow practical application of atomic physics to the field of semiconductor devices.

Why is the enhancement of the density of states at the band edge so important for diode lasers? In contrast to the case of a dilute gas of atoms (see Fig. 1.4), the atoms in a crystal are strongly bound to each other. The interactions between closely spaced atoms in a bulk crystal lead unavoidably to a broadening of the electronic spectrum. The absorption band exhibits strong broadening of the order of several electronvolts, in marked contrast to the sharp line absorption spectrum of single atoms. At high temperatures, lattice vibrations (phonons) can easily stimulate transitions of charge carriers in the energy range defined by the lattice temperature and/or scatter the carriers. The 'tails' of the carrier distributions near the bottom of the conduction band and the top of the valence band increase remarkably with temperature. For the same average concentration of injected carriers, a broadening of their energy spectra causes, among other disadvantages, a decreased gain and a degradation in laser performance.

In complete contrast to the above scenario, when the motion of a charge carrier in a crystal is limited to a very small volume, the energy spectrum of the charge carrier is quantized. (It is important that the matrix material should provide a larger bandgap than the QD material and that the potential wells should be attractive for both electrons and holes.) This is similar to the case of electron quantization in the attractive Coulomb potential of an atomic nucleus. In the simplest case of an infinite barrier at the QD–matrix interface, and a 3-dimensional rectangular QD, the size quantization energy is given by

$$E(n_x, n_y, n_z) = \frac{\hbar^2}{2m^*} \left(\frac{\pi^2 n_x^2}{L_x^2} + \frac{\pi^2 n_y^2}{L_y^2} + \frac{\pi^2 n_z^2}{L_z^2} \right), \tag{5.1}$$

where m^* is the effective electron mass, L_x, L_y, and L_z are the box dimensions, and n_x, n_y, n_z are (positive integer) quantum numbers. Electrons in crystals usually have rather small effective masses, and hence relatively large boxes of about 10 nm can result in a large energy separation between electron sublevels (about 100 meV for a GaAs QD). The latter value significantly exceeds the thermal energy at room temperature (26 meV). The population of excited states can thus be avoided and the ultimate temperature-insensitive material gain can potentially be realized.

5.1.2 Development of Heterostructure Lasers

The road to quantum dot heterostructure lasers has been rather long. It is widely recognized now [5.5, 5.6] that the major breakthrough in the field of semiconductor diode lasers occurred when the idea of using a double heterostructure (DHS) as the active region of an injection laser to achieve efficient electron confinement was implemented independently by Alferov and

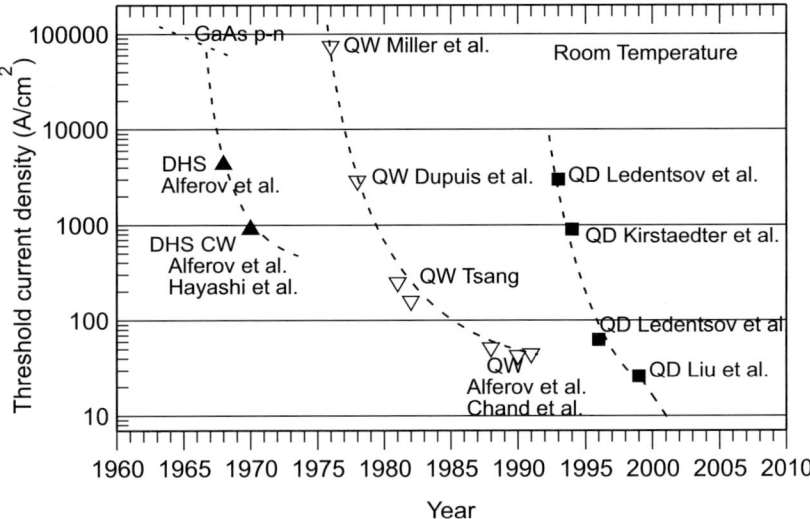

Fig. 5.2. Development of heterostructure lasers. Results marked in figures were reported in original papers. Double heterostructure (DHS) lasers [5.11–5.13]. Quantum well (QW) lasers [5.15, 5.16, 5.18–5.22]. Quantum dot (QD) lasers [5.30, 5.34–5.36]. Results are dated according to the year of submission. The submission date, if different from the publication date, is given in the list of references

Kasarinov [5.7] and Kroemer [5.8]. An important theoretical step was also taken in 1966, when it was realized that optical confinement (wave guiding) was fundamental in order to achieve lasing in a DHS [5.9]. The scientific community was largely skeptical at first about the possibility of finding a heterostructure with sufficiently different bandgaps and good lattice matching. It was widely believed at the time that III–V materials with different bandgap energies must have different lattice constants. Nevertheless, lasing was soon achieved using a lattice-mismatched GaAsP–GaAs DHS [5.10] at low temperature. At the same time, only the discovery of the stability of AlGaAs alloys and subsequent practical realization of a lattice-matched AlGaAs–GaAs sandwich-like DHS made it possible to achieve first low threshold current density pulsed lasing at room temperature [5.11] and, soon after, continuous wave operation [5.12, 5.13] (see Fig. 5.2).

This was a key event for semiconductor diode lasers. In the following years, there was an explosive development of technologies related to diode lasers. Until the middle of the 1980s, DHS lasers with a relatively thick active region dominated the market. Further progress was related to the use of size-quantized heterostructures as active regions in the device [5.4]. As already stressed above, Dingle and Henry [5.4] did not restrict their patent exclusively to quantum wells. They also understood the importance of the Coulomb interaction effects and wrote: "the combination of confinement and Coulomb attraction between the electron and hole increases the strength of the optical absorption and emission processes and hence the laser gain." The authors also wrote: "the interaction of carriers with the lattice, with impurities and with other carriers tends to alter the energetics of band edge transitions" and can "alter the density of states significantly by producing a distribution of bandgap energies." However, they concluded that "neither of these two effects should substantially diminish the change in the density of states which results from reducing the number of translational degrees of freedom of the carriers."

Further efforts in QW development were restricted by the fact that the technology available at the time only permitted research on structures with ultrathin layers. Photo-pumped operation at low temperature was achieved in 1975 [5.14], and soon after that, room-temperature operation was reported [5.15]. The threshold excitation densities were very high (corresponding to current densities of 75–300 kA/cm^2), and most researchers remained skeptical regarding the use of QWs in lasers. Nevertheless, injection lasing was soon demonstrated at room temperature in a quantum well heterostructure (QWHS) laser with threshold current density 2.8 kA/cm^2 [5.16]. It was also predicted and then shown experimentally that the modified density of states in a QWHS laser leads to improved temperature stability of the threshold current compared with conventional DHS lasers [5.17]. However, practical applications of QWHS lasers only came about after a significant reduction in the threshold current density in multiple-QW [5.18] and single-QW separate-confinement heterostructure lasers had been demonstrated [5.19] (see Fig. 5.2). GaAs QWs confined by short-period superlattices achieved threshold current densities of 40–50 A/cm^2 [5.20, 5.21]. Similar thresholds (45 A/cm^2) were reported for AlGaAs lasers with strained InGaAs quantum wells [5.22].

By the beginning of the 1980s, progress in QWHS lasers on the one hand, and experimental advances in crystal growth on misoriented [5.23] or patterned [5.24] substrates on the other, revived interest in heterostructures with size-quantization in more than one direction. The influence of high magnetic fields on the characteristics of DHS lasers was considered experimentally and theoretically [5.25] and results were interpreted as originating from a magnetic-field-induced modification of the density of states. Some reduction of the threshold current at moderate magnetic fields was also observed, in general agreement with theory.

In 1982 Arakawa and Sakaki [5.26] considered theoretically some effects in lasers based on heterostructures with size quantization in one, two, and three directions. They wrote: "Most importantly, the threshold current of such a laser is predicted to be far less temperature sensitive than that of a conventional laser, reflecting the reduced dimensionality of the electronic state." The authors performed experimental studies on a QW laser placed in a high magnetic field directed perpendicular to the QW plane. They demonstrated that the characteristic temperature (T_0) describing the exponential growth of the threshold current with temperature increases in the magnetic field from 144°C to 313°C.

In 1986 Asada et al. [5.27] made a theoretical study of the gain in quantum wire and quantum dot lasers. Assuming the same homogeneous broadening of about 6 meV, the authors demonstrated a significantly increased material gain up to 10^4 cm^{-1} for InGaAs QDs with cubic shape and size 10 nm. This value was much higher than for QWs. They also pointed out that the material gain in a small QD increases in inverse proportion to the cube of the QD size (i.e., a material gain of the order of 10^5 cm^{-1} can be realized in QDs with diameter 5 nm, assuming inhomogeneous broadening of about 20 meV). To realize high modal gain, however, it is important to fabricate dense arrays of small QDs.

It should be noted, however, that certain theoretical predictions were pessimistic. In 1988, Vahala took the inhomogeneity of QDs into consideration [5.28]. The influence of doping on the transparency current was also treated. He concluded that "for high gain operation, a medium composed of quantum boxes does not offer significant advantages over a conventional bulk semiconductor unless quantum box fabrication tolerances are tightly controlled." Benisty et al. [5.29] suggested that the low luminescence efficiency of QDs produced by ion-etching of QW samples results from the lack of matching energies for phonon relaxation of carriers in QDs (a kind of phonon bottleneck effect in the QD).

5.1.3 The Key Breakthrough: Self-Organized Growth

All discussions on QD lasers were of a merely academic nature until the first experimental demonstration of lasing in self-organized QDs [5.30] (see Fig. 5.3). Real applications of QDs to lasers only became possible when techniques of self-organized growth were developed to such a level that dense arrays of uniform QDs could be produced and, simultaneously, defects of all types could be significantly reduced, as described in detail in Chaps. 3 and 4 of the present book.

Self-organized growth represents a cost-effective nanotechnology. Hundreds of billions of nanoscale islands, ordered in size and shape, can be formed on a crystal surface per second and per square centimeter, without making any significant changes in growth equipment and post-growth processing technology.

The effect of spontaneous islanding in the InAs–GaAs system was already known well before the mid-1980s, as mentioned in Sect. 3.3. However, the technique did not attract much attention as a prototype for QD lasers, because it was considered too difficult to produce dislocation-free QDs with sufficiently uniform size and shape using this approach. In 1995, it was still argued in the literature that the luminescence observed in structures with InAs islands was not related to QDs [5.31]. Ga vacancies in GaAs, or "radiative recombination via dislocations, probably those 60°-dislocations formed

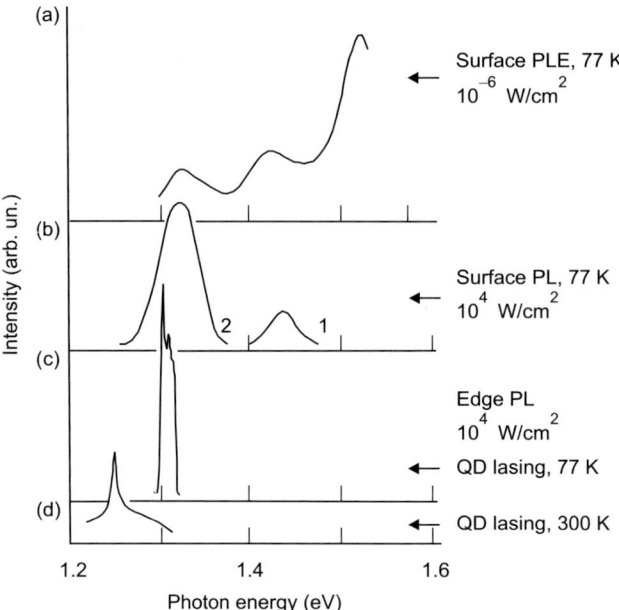

Fig. 5.3. Photo-pumped lasing in quantum dots according to [5.30]

during overgrowth" were put forward as possible sources for long-wavelength emission.

Nevertheless, it is precisely the islanding approach that has led to the breakthrough in QD lasers. In 1993, photopumped lasing in QD heterostructures was demonstrated by Ledentsov et al. [5.30] (see Fig. 5.3). In that paper, InGaAs QDs were inserted in the central part of a GaAs cavity deposited on top of a thick AlGaAs layer and the structure was covered by a thin AlGaAs layer to prevent surface recombination. Photopumped lasing via QDs was realized both at low and room temperature (see Fig. 5.3). It was proposed in [5.30] to stay close to the 2D–3D growth transition revealed in the reflection high-energy electron diffraction (RHEED) pattern (the so-called streaky-to-spotty transition) to reduce the probability of defect formation, and to use relatively low substrate temperatures (as compared to the typical InAs–GaAs growth range [5.32]) resulting in higher area density of islands and lower density of defects.

Low threshold injection lasing from InGaAs QDs was reported in August 1994 by Ledentsov et al. [5.33] and by Kirstaedter et al. [5.34]. The lasing was observed at low and room temperature. At 77 K it occurred at 1.24 eV and 1.31 eV for two different structures, being in both cases clearly within the range of the QD luminescence [5.34]. Most importantly, the threshold current density was remarkably low (< 100 A/cm^2) and remained practically unaffected by temperature up to about 150 K, as expected for QD lasers.

At higher temperatures the threshold current increased due to thermally activated escape of carriers from quantum dots to the surrounding GaAs matrix. Other researchers later confirmed these key features of quantum dot lasers.

The development of QD lasers has led to a better understanding of the basic properties of these devices. The original predictions were generally based on simplified assumptions: infinite barriers, one confined electron level and one confined hole level, bimolecular e–h recombination, ultrafast energy relaxation of injected carriers, an equilibrium carrier distribution, lattice-matched heterostructures, and similar confinement volume for electrons and holes. In realistic devices, these assumptions have had to be replaced over the last few years [5.37–5.40]:

- finite barriers,
- many electron and hole levels,
- monomolecular (excitonic) recombination,
- non-equilibrium carrier distribution,
- strained heterostructures with completely different potential wells for electrons and holes.

This has led to more realistic theoretical predictions [5.37, 5.38, 5.41]. For example, the temperature stability of the threshold current may be either low or high, depending on the particular size, shape, number of electron and hole levels, and density of the QDs.

The use of QDs in diode lasers brings with it several decisive technological advantages:

- Greatly extended tunability of emission wavelength by QD size and composition on a given substrate. Lasing wavelengths are realized in the 1.3–1.4 µm spectral range, and lasing in the 1.4–1.5 and 1.5–1.6 µm spectral ranges is also potentially achievable using GaAs substrates. The latter are important for telecom and free-space applications.
- Very low threshold current densities (< 7 A/cm^2 per QD sheet) and simultaneously very low internal losses (~ 1–3 cm^{-1}) and high quantum efficiencies (> 80–96%) have been demonstrated.
- Carrier confinement in narrow gap QDs placed in a wide gap matrix can prevent nonequilibrium carrier spreading and non-radiative recombination. This improves radiation hardness and suppresses facet overheating, increasing the threshold for catastrophic optical mirror damage (COMD).

5.1.4 State of the Art in Quantum Dot Lasers: Taking an Upper Hand

The development of quantum dot technology and understanding of the fundamental processes underlying self-organized formation of QDs has led to substantial progress in laser fabrication. By combining self-organization phenomena and nanostructural engineering, particularly, size-limited island growth (described in Sect. 3.3), activated alloy phase separation (Sect. 4.3), defect-reduction techniques (Sect. 4.1), and stacking of quantum dots (Sect. 4.1), semiconductor nanostructures have been successfully applied in optoelectronic devices. A detailed description of the present state of the art in QD lasers may be found in reviews by Ledentsov [5.42] and Bimberg et al. [5.43]. A brief survey is given here.

Threshold Current Density. Threshold current density is an important characteristic of a laser. It tells us at which current density the modal gain overcomes overall (internal and external) losses, thereby enabling lasing. External losses can be made very small by depositing dielectric high-reflectivity (HR) coatings. However, as internal losses (scattering, free-carrier absorption, self-absorption, etc.) cannot be made negligibly small (e.g., the need for carrier injection requires p-doped cladding layers), the value of the threshold current density serves as an important quality indicator.

QD lasers have already demonstrated their potential for devices which operate at threshold current densities in the range 10–20 A/cm^2 in long devices with HR facet coatings [5.44]. For conventional devices, suitable for high efficiency operation, HR front facet coatings cannot be applied. Here, the measure of quality is the value of the threshold current density per single quantum well or quantum dot insertion. The best values of 7–10 A/cm^2 per QD insertion are realized in devices with a large number of QD stacks [5.45, 5.54].

A record low transparency current, which is the extrapolated value of the threshold current density for infinite cavity length, of 6 A/cm^2 per dot layer, an internal quantum efficiency of 98%, and an internal loss below 1.5 cm^{-1} have been demonstrated [5.46].

Characteristic Temperature T_0. T_0 is the temperature at which the device must be heated to increase the threshold current by a factor of e = 2.718. The first QD injection lasers already had infinite characteristic temperature in a temperature range below 150–180 K [5.34]. However, at higher temperatures, the threshold current density was considerably less temperature-stable and T_0 was inferior (50–80 K) to values characteristic for commercial GaAs-based quantum well devices.

Several approaches have been proposed to improve T_0 values. These include placing QDs in a quantum well [5.47], increasing the bandgap of the martrix [5.45], and others. However, in some cases where T_0 values were increased (up to 350 K in [5.47]), the penalty for the increase in threshold current density was also high. First, a high characteristic temperature (160 K) at ambient temperature operation (below 40°C) together with low threshold current density operation (70 A/cm^2 for 3-fold-stacked QDs) was realized in [5.48]. As the devices operate in the range 1.26–1.28 μm, traditionally covered by InP-based devices with T_0 values near 50 K, this was a remarkable success. It became possible to build the first high-power contin-

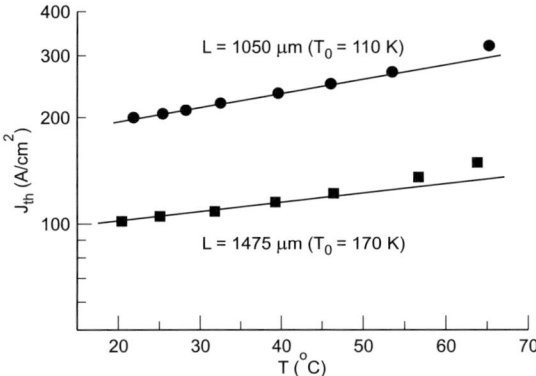

Fig. 5.4. Temperature dependence of the threshold current density of a quantum dot (QD) laser with 5-fold QD stack and uncoated facets

uous wave (CW) QD lasers (3 W in CW operation [5.49]). However, in the CW regime, the T_0 value was already much lower than 160 K [5.49].

The possibility of very high characteristic temperatures in QD lasers (180 K in [5.50] and 230 K in [5.51] up to 80°C) has recently been demonstrated by applying a QD p-modulation doping technique [5.50, 5.51], an approach which has also been used in other publications [5.52]. The penalty for p-doping is a reduced differential efficiency ($< 20\%$ in [5.51]). The advantage is the potential to increase the modulation bandwidth beyond 20 GHz [5.53].

It has recently become possible [5.42, 5.54], using undoped 5-fold-stacked QDs, to increase the T_0 value to 170 K (up to 65°C) without paying the penalty of increased internal losses (see Fig. 5.4). This has made it possible to design 1.3 μm GaAs devices operating at 85% differential efficiency, low threshold current density (100 A/cm^2, 1.5 mm cavity length, uncoated) and high characteristic temperature, all realized in the same device.

Spectral Features of QD Lasers. Even in the early stages of QD laser research it was found that the lasing spectrum of broad stripe QD lasers exhibits laser mode grouping [5.53]. Devices with 40–60 μm stripe widths were studied in [5.53]. Intensity modulation of the lasing spectrum was observed and spectral shifts of the mode groups with temperature were found. It was concluded that the effect may be explained by the formation of a transverse Fabry–Pérot resonator, affecting the gain spectrum in the longitudinal direction. Later, similar spectral features were addressed in a number of publications [5.55–5.59]. In some papers [5.55–5.57], it was concluded that mode grouping is related to leaky modes in the vertical direction arising from the limited thickness of the confinement layers. However, no systematic study on the stripe width or confinement layer thickness has ever been performed.

Another explanation of the observed spectral behavior was given in [5.58]. It was suggested that the spacing between the mode groups can be defined by homogeneous broadening. No detailed temperature studies of the effect were presented. Very recently, systematic studies of the mode grouping effect have been performed on QD lasers with a range of stripe widths and in an extended temperature range [5.59]. It was shown that the spacing between the mode groups is not a function of temperature and is defined by the stripe width, in general agreement with [5.53]. It was also proposed that, by introducing a proper periodic etching profile at the edges of the stripe, it may be possible to suppress or enhance the hole-burning effect in the gain spectra due to transverse cavity modes at particular wavelengths and, potentially, realize wavelength-stabilized, or even single-longitudinal-mode lasing.

Time-Response of QD Lasers. It is known that electron relaxation occurs at a very slow rate in high-purity bulk materials. GaAs exciton photoluminescence (PL) evolution times up to 4 ns have been reported [5.60]. This so-called phonon bottleneck effect for electron energy relaxation is dramatically reduced in QDs, where photoluminescence (PL) evolution times of the order of 10–40 ps have been reported, depending on the particular QD geometry, even at low temperatures and excitation densities [5.45]. This potentially opens the way to devices operating in the 20–100 Gb/s range. Relaxation oscillations in practical devices also indicated a potential for modulation bandwidths larger than 10 GHz [5.45]. Under conditions of high excitation density and lattice temperature, the time response can be even faster [5.52].

However, in order to gain practical advantage from these features, proper device design is essential. At high temperatures, population of matrix states will have an adverse effect on time response [5.61]. It has been shown that, using the idea of resonant tunneling carrier injection into QDs [5.62], devices can be made which operate in the 1 µm wavelength range at 15 GHz and at room temperature [5.63]. Furthermore, p-doping of QDs was proposed to increase the high-frequency response of QD lasers up to 20 GHz and beyond [5.50]. Oxide-confined two-section bistable quantum dot (QD) 1.3 µm lasers with an integrated intracavity QD saturable absorber demonstrate passive mode locking at a repetition time of 7.4 GHz and with a duration of 17 ps [5.64]. These results suggest that a carefully designed QD laser is also a candidate for ultrashort pulse generation.

High-Power Operation. Broad-area GaAs-based QD lasers show a CW output power of ≈ 3–6 W depending on the spectral region. The estimated maximum modal gain for QD ground state lasing is about 14 cm^{-1} for 3-fold-stacked QDs and can be increased up to 20–35 cm^{-1} for 5- to 10-fold-stacked QD active regions [5.54]. The internal losses derived from the slope of the inverse differential efficiency vs. cavity length are as low as 1.5 cm^{-1}.

Figure 5.5 shows the pulsed performance of a QD laser based on 10-fold stacked QDs. The cavity length is 1.45 mm and uncoated facets are used. A differential efficiency as high as 85% and a threshold current density as low as 100 A/cm^2 are achieved. The characteristic temperature is 150 K in the temperature range 0–70°C. All the key values are significantly better than those for commercial InP-based 1.3 µm devices. On the one hand, this

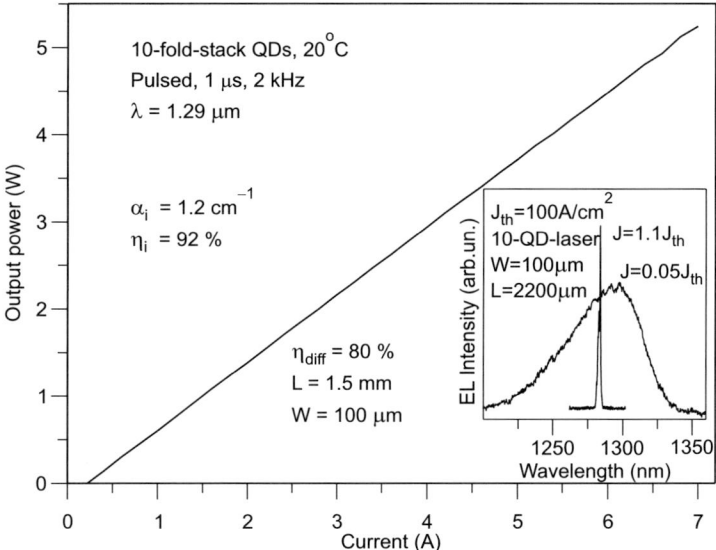

Fig. 5.5. High-power operation of a long-wavelength laser. A 10-fold stack of QDs is used. The *inset* shows electroluminescence (EL) and lasing spectra

QD laser may be used as a standard transmitter, since the ground state modal gain is sufficient to maintain lasing in 300-µm-long devices (with HR-coated rear facet). On the other hand, the device can be used in high-power applications, for example in Raman pumps, where high efficiency is essential at long cavity lengths. Raman pumps in the 1.2–1.3 µm wavelength range can be used in new types of fiber, enabling low losses over the whole 1.25–1.65 µm spectral range. Cost-efficient devices can be very useful for dense wavelength division multiplexing (DWDM) and coarse wavelength division multiplexing (CWDM) systems of the near future. Narrow 7-µm-wide stripes exhibited single-transverse-mode kink-free operation up to 330 mW for uncoated facets (see Fig. 5.6).

Special studies of the laser beam quality have been performed by Ribbat et al. [5.65] for three different types of InGaAs/GaAs laser:

- MOCVD-grown 1.1 µm quantum well lasers,
- MOCVD-grown 1.1 µm QD lasers,
- MBE-grown 1.3 µm QD lasers.

An intrinsic filamentation suppression has been demonstrated for narrow stripe gain guided QD lasers compared to QW lasers with identical waveguide structure and similar lasing wavelength. Filamentation suppression is even stronger for 1.3 µm QD lasers due to the longer emission wavelength. This novel effect can be attributed [5.65] to reduced in-plane carrier diffusion and the reduced α-factor in QD lasers compared to QW lasers.

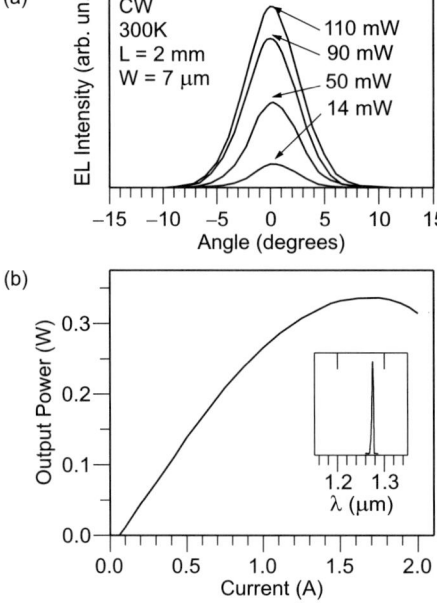

Fig. 5.6. High-power kink-free single-mode operation of a long-wavelength QD laser. (**a**) Far-field pattern in the lateral direction. (**b**) Output power. The lasing spectrum is shown in the *inset*

Operation Lifetime. Quantum dot lasers demonstrate enhanced resistivity in temperature-accelerated degradation tests. Operation lifetimes are in excess of 400 hr CW at 60°C and at output power in excess of 300 mW, without any noticeable degradation, for devices tested without facet protection in humid room ambient. With facet coatings, operation lifetimes in excess of 1 250 hr at 50°C heat sink temperature and output power 1.5 W per output facet have been demonstrated [5.43] (see Fig. 5.7).

The enhanced stability of QD devices was also revealed in intentional degradation tests using high-energy proton bombardment [5.66]. The increase in threshold current after treatment was only half as great in QD devices as in QW lasers, even if the initial characteristics were similar.

Recently, Sellin et al. [5.67] demonstrated operation times of 3 040 hr at 1.0 and 1.5 W output power for MOCVD-grown facet-passivated high-power laser diodes based on six-fold stacks of self-organized InGaAs/GaAs quantum dot layers as gain medium. Neither degradation nor significant changes in the conversion efficiencies were observed during the lifetime measurements. A maximum output power of 4.7 W was obtained in CW mode, 11.7 W was achieved in quasi-CW operation (50 µm pulses and 50 Hz repetition frequency). The output power was limited by catastrophic optical mirror damage, occurring at a power density of about 19.5 MW/cm^2 on the front facet.

Potential Extension of the Wavelength Range. Quantum dots on GaAs substrates could potentially cover the whole telecom wavelength range.

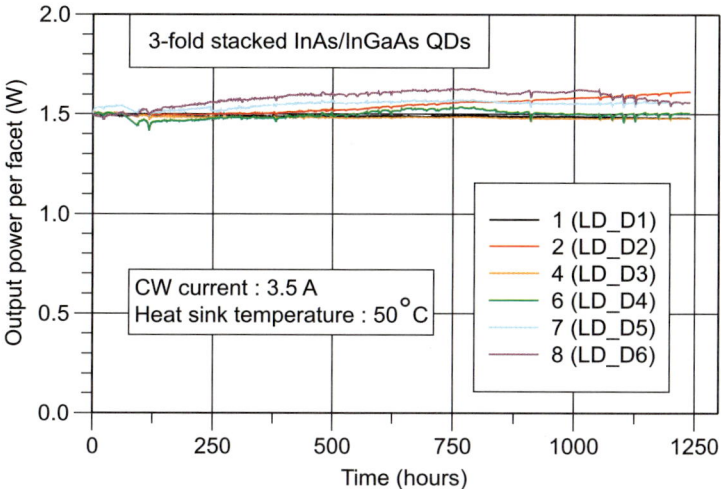

Fig. 5.7. Demonstrating the long operation lifetime of quantum dot lasers

QD PL in the range 1.42–1.7 µm has been reported in a number of publications [5.68–5.70].

Lasers on Submonolayer Quantum Dots. The advantages of submonolayer quantum dots (SML QDs) over 3-dimensional QDs are discussed in detail in Sect. 3.2. The latest results reported by Kovsh et al. [5.71] demonstrate a high output power of 6 W.

Laser structures in which a multi-sheet array of 0.5 ML InAs QDs in a GaAs matrix were used as the active medium were grown by molecular beam epitaxy. The waveguide was formed by a 0.6-µm-thick undoped $Al_{0.15}Ga_{0.85}As$ layer sandwiched between 1.5-µm-thick $Al_{0.3}Ga_{0.7}As$ cladding layers. n- and p-type cladding layers were highly doped ($\approx 1 \times 10^{18}$ cm^{-3}) with Si and Be, respectively. The AlGaAs waveguide contained a GaAs layer 30 nm thick with two sheets of SML QDs. The spacer thickness was 12 nm. The lasers were formed by cleaving into bars of different cavity lengths. Pulsed measurements were conducted on a probe station. Then individual devices were mounted p-side down on a copper heatsink by In solder for continuous wave (CW) operation measurements after high- and low-reflectivity facet coatings. The laser diode with cavity 1 mm long lases at 940 nm.

CW output power and power conversion efficiency vs. drive current are shown in Fig. 5.8. The power of the SML QD laser (solid line) reaches its highest value of 6 W at a current of 6.5 A. At this value the laser fails due to catastrophic optical mirror damage (COMD). The slope efficiency of 1 W/A ($\eta_{\text{diff}} = 76\%$) remains constant up to 6 A and negligibly decreases beyond this value due to thermal rollover. High doping of cladding layers allows one to achieve a very low series resistance of 6.7×10^{-5} Ωcm^{-2}, among the lowest values reported for diode lasers using aluminum. As a result, conversion

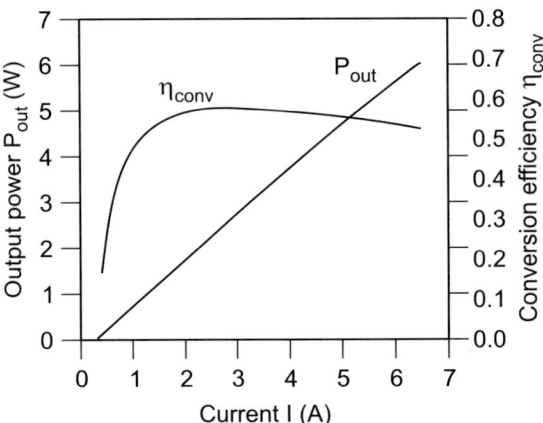

Fig. 5.8. Output power and conversion efficiency vs. drive current for a laser based on a two-fold stack of submonolayer InAs/GaAs quantum dots

efficiency peaks at 58% and remains higher than 50% until the laser fails. The maximum output power and efficiency shown are much higher than the best values reported for lasers based on 3-dimensional (Stranski–Krastanow) QDs (3.5 W and 45%) [5.72] and only slightly worse than QW lasers emitting at 0.98 μm.

The maximum output power of 6 W corresponds to an internal power density at COMD of 12.5 MW/cm^2. We believe that this relatively low value compared with the reported record for AlGaAs-based lasers [5.73, 5.74] can be improved after careful optimization of the facet coating technique.

In this context it is worth noting that low temperature sensitivity of device characteristics is essential to prevent thermal rollover at high drive currents. In the range 10–60°C, the temperature dependence of the threshold current [$\sim \exp(T/T_0)$] for the SML QD laser is described by a characteristic temperature $T_0 = 150$ K. These results demonstrate the potential of the submonolayer technique for forming the active region in high-power lasers.

Quantum Dot Vertical Cavity Surface-Emitting Lasers (VCSELs). Success in conventional edge-emitting QD devices has stimulated interest in the application of QDs to VCSELs. The first surface-emitting QD lasers were realized in 1996 [5.75, 5.76]. In 1997, very low threshold current density (down to 170 A/cm^2) QD VCSELs were fabricated with 10 μm oxide-confined apertures [5.77]. Threshold current densities in VCSELs are usually much higher than in edge-emitters. This is due to the increased spreading of nonequilibrium carriers out of the aperture region and higher injection densities needed to achieve lasing. The wall plug efficiency of the device with 10 μm output aperture approached 16%. It was also shown that, in QD VCSELs, one can go down to very small (submicrometer) oxide apertures and still have a reduction in the threshold current density.

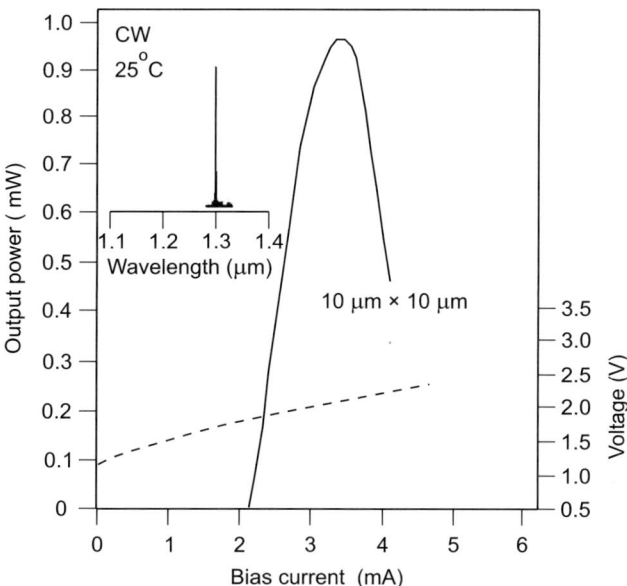

Fig. 5.9. Typical L–I–V curves of 1.3 μm GaAs-based quantum dot VCSELs

News of the first ever GaAs-based 1.3 μm VCSEL was published by Ledentsov et al. [5.78] in March 2000 (see also [5.68, 5.79]). The microcavity was surrounded by p- and n-$Al_{0.98}Ga_{0.02}As$ layers (less than $\lambda/4$ thick) followed by 1λ-thick p- and n-GaAs current spreading/intracavity contact spacer layers, doped to 10^{18} cm^{-3}. Intracavity contacts were used. The spacer layers were followed by DBRs composed of alternating $Al_{0.98}Ga_{0.02}As$ and $\lambda/4$-thick GaAs layers. The $Al_{0.98}Ga_{0.02}As$ layers in the DBR, as well as those surrounding the optical cavity, were selectively oxidized to form Al(Ga)O. QDs were centered in a 1λ-thick GaAs optical microcavity, with edges doped to 10^{17} cm^{-3}. The ends of the microcavity were composed of $Al_xGa_{1-x}As$, linearly graded from $x = 0.02$ up to 0.98.

Currently, the output CW power for the best devices approaches 1.25 mW for 8 μm aperture, and the maximum wall-plug efficiency lies in the range 15–20%. Typical CW light power–current–voltage (L–I–V) characteristics for a QD VCSEL are shown in Fig. 5.9. Operation lifetimes in excess of 5 000 h at 50°C are demonstrated for these 1.3 μm GaAs VCSELs.

Suppression of nonequilibrium carrier surface recombination on side walls of ultrasmall mesa structures has triggered intense research and promises applications in photonic bandgap crystals (PBCs) and microcavities. Hence the realization of practically thresholdless lasers may soon become a reality.

Impact of the QD Density of States on VCSEL Design. There exists a major difference between QD and QW VCSELs related to the discrete nature of QD states. Firstly, the VCSEL resonant cavity wavelength, which is

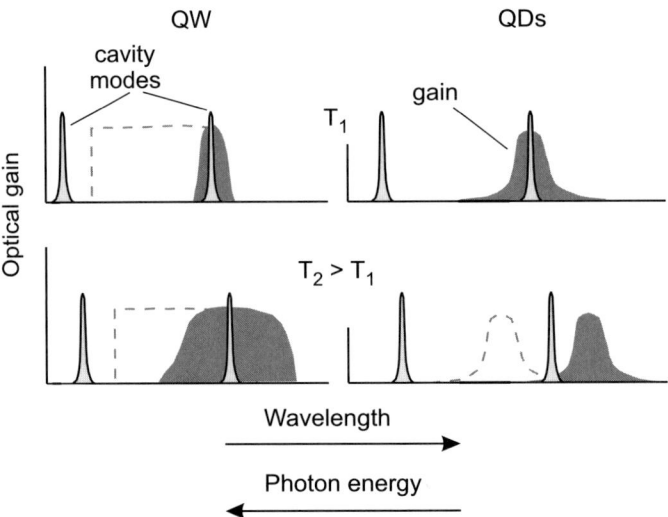

Fig. 5.10. Interconnection between the density of states and the cavity mode for the quantum well (*left*) and quantum dot (*right*) cases. The *dashed line* indicates the density of states which are not filled with injected carriers at the given current density

largely defined by the width of the vertical cavity and its refractive index, is a weak function of temperature. However, the effective bandgap of a semiconductor shrinks with increasing temperature. This causes the gain onset energy to shift away from the VCSEL cavity mode. For most applications the device must nevertheless work in a wide temperature range (typically from $-20°C$ to $+80°C$). Secondly, such a device heats significantly under high current density operation. The temperature-induced detuning between the semiconductor band edge and the cavity mode is not a problem for QW VCSELs, where a continuum of states exists. An increase in injection current causes population of energy states above the gain onset energy and the device continues to operate (see Fig. 5.10).

In a QD there exists no continuum in the density of states. The spread in energy for the QD ground states caused by the QD size and shape dispersion (inhomogeneous width) is typically about 30–40 meV. The separation between the ground and excited states is greater than this value (80–120 meV). Neither injection current nor temperature can significantly broaden the QD spectrum. The full width at half maximum (FWHM) of the photoluminescence spectrum from the QD array does not change up to 200°C (FWHM $\approx$ 30 meV). Once the gain spectrum drifts away from the VCSEL cavity mode, device operation terminates. Thus, QDs with exceptionally large size dispersion (FWHM > 80 meV) are required for broad temperature range and high-power VCSEL applications. The problem may be solved by stacking QDs with different sizes to achieve similar gain spectrum broadening. One

5.2 Quantum Dot Nanostructures for Single-Electron Devices

should mention that in some applications of edge-emitting lasers, for example in fiber grating wavelength-stabilized fiber pump lasers, the need for a broad QD gain spectrum will be of importance.

To conclude, recent results obtained up to the end of the year 2002 for lasers based on both 3-dimensional QDs grown in Stranski–Krastanow mode and 2-dimensional QDs grown in submonolayer mode can surpass conventional quantum well lasers as regards the main parameters.

5.2 Quantum Dot Nanostructures for Single-Electron Devices

The idea of using quantum dots to control electric charges in the system at the ultimate level of one electron is related to the Coulomb blockade phenomenon. The classical electrostatic energy of a quantum dot with capacity C which is capacitively coupled to a gate at a bias voltage V_g is given by

$$E = \frac{Q^2}{2C} - Q\alpha V_g , \qquad (5.2)$$

where α is a dimensionless factor relating the gate voltage to the potential of the island and Q is the charge. Mathematically, the minimum energy is reached for a charge $Q_{\min} = \alpha C V_g$. However, the charge has to be an integer multiple of the elementary charge e, i.e., $Q = Ne$. If V_g has a value such that $Q_{\min}/e = N_{\min}$ is an integer, the charge cannot fluctuate, as long as the temperature is low enough, i.e.,

$$k_B T \ll \frac{e^2}{2C} . \qquad (5.3)$$

Tunneling in or out of the dot is suppressed by the Coulomb barrier $e^2/(2C)$, and conductance is very low. Similarly, the differential capacitance is small. This effect is called Coulomb blockade and has been extensively studied in the literature [5.80–5.83]. Peaks in the tunneling current occur when the gate voltage is such that the energies for N and $N+1$ electrons are degenerate, i.e., $N_{\min} = N + 1/2$. The expected level spacing is

$$e\alpha \Delta V_g = \frac{e^2}{C} + \Delta \varepsilon_N , \qquad (5.4)$$

where $\Delta \varepsilon_N$ denotes the change in lateral (kinetic) quantization energy for the added electron and e^2/C corresponds to the charging energy.

The Coulomb charging energy calculated for self-organized InAs/GaAs QDs is of the same order of magnitude as the energy spacing between individual energy levels. For lens-shaped InAs/GaAs QDs with a radius of 25 nm, a calculated charging energy is $\approx$ 30 meV [5.84]. For pyramid-shaped InAs/GaAs QDs, calculations [5.39] yield a charging energy varying from 25 meV for a pyramid base length of 10 nm to 18 meV for a pyramid base length of 20 nm. One should bear in mind, however, that the localization energy of electrons in InAs/GaAs QDs is only about 200–300 meV at maximum. This corresponds to about $10k_BT$ and the probability of escape to the matrix is unacceptably high. A localization energy of about $100k_BT$ is required for data storage applications.

It follows that single-electron devices operating at room temperature can hardly be constructed on the basis of individual QDs because of the strong thermal evaporation of carriers. Systems based on ensembles of a moderate number of, say, 10 to 100 coupled QDs are more realistic, so that the probability of simultaneous escape of electrons is negligible.

The impact of electronic coupling of dots on the Coulomb blockade of an array of QDs has been analyzed by Stafford and Das Sarma [5.85]. Three zero-temperature phases were identified, depending on the strength of inter-dot tunneling:

- Coulomb blockade of individual dots (regime of small coupling),
- collective Coulomb blockade (intermediate coupling, where at least separate minibands exist),
- breakthrough of Coulomb blockade (strong coupling regime).

We propose that collective Coulomb blockade may be used to construct logic elements operating at room temperature. Let a set of $\mathcal{N}$ electronically coupled dots be capacitively coupled to a gate, and the charging energy of every dot be of the order of the thermal energy, $e^2/(2C) \sim k_BT$. Let the voltage be such that, at $T = 0$, dots are not charged. Then at room temperature the number of dots charged due to thermal fluctuations will be about $\sqrt{\mathcal{N}}$. In order to have all or nearly all dots charged by one electron, a voltage which overcomes the Coulomb barrier must be applied.

Now let the voltage be such that, at $T = 0$, every dot contains one electron. Then the charge of the ensemble of QDs will fluctuate mainly within the interval from $\left(\mathcal{N} - \sqrt{\mathcal{N}}\right)e$ to $\left(\mathcal{N} + \sqrt{\mathcal{N}}\right)e$. With a sufficient number of dots, say a few tens of dots, in the ensemble, it will be possible to ensure distinct switches between logical 0 and logical 1 at room temperature.

6. Conclusion

> When we asked Pooh what the opposite of an Introduction was, he said "The what of a what?" which didn't help us as much as we had hoped, but luckily Owl kept his head and told us that the Opposite of an Introduction, my dear Pooh, was a Contradiction; and, as he is very good at long words, I am sure that that's what it is.
>
> Alan Alexander Milne. *The House at Pooh Corner*

About a decade after quantum dot (QD) and quantum wire research has become the mainstream of semiconductor physics, it is worth asking: has reality fulfilled the hopes and expectations of QDs? Where do we stand now? What is the present state of the art?

There is indeed no contradiction between expectations and reality. Although the horizon is not completely cloudless, there is plenty of sunshine. Quantum dots have arrived! Using the effects of self-organization at crystal surfaces, in a certain window of growth parameters, semiconductor heterostructures can be fabricated with a high density of coherent inclusions in a wide bandgap matrix, displaying a discrete electronic spectrum up to room temperature and above. These structures have a low density of defects and can be fabricated in a massively parallel production-friendly way. They can store or transfer electrons, exhibit bright photoluminescence, and can be used as active media in ultrahigh-quality semiconductor diode lasers.

Having learned to make small and simple QD structures, we now seek to make them larger and much more complex, just by combining the effects of self-organization and further nanogrowth engineering. This is in no way a step backwards. The idea is to use QDs and, more generally, self-organized nanostructures, in a new and intelligent way as building blocks for our cities under the surface of semiconductor wafers. By increasing nanostructure complexity, we gain in tunability and functionality.

In this way, a large number of recipes has emerged for "baking cakes with raisins." A well understood lesson is how to make 'raisins' of different density, volume, shape, and alignment. For example, when growing multisheet arrays, it is possible to improve uniformity in the size and arrangement of nano-insertions. By choosing different materials for QDs in the first and subsequent sheets, one can control the density and size of QDs independently and grow a high density of large dots. By varying the spacer thickness in multisheets, one can choose between vertical columns and a checkerboard arrangement of dots in the cross-section plane. By manipulating the geometry, it is possible to tune between unpolarized and highly polarized surface emission from the dots, and to choose between different edge-emission polarizations. By capping the dots by an alloy, the phase separation in the cap layer can be activated, effectively increasing the size of the QDs and shifting the photoluminescence spectrum towards ultimately longer wavelengths. It is possible to grow large islands, among which a substantial fraction are dislocated, then apply multi-cycle overgrowth and thermal etching, selectively eliminating dislocated islands and keeping coherent ones.

By applying a set of sophisticated nanogrowth and nanoengineering techniques, quantum dot lasers have been developed which surpass conventional quantum well lasers as regards the major parameters: threshold current density, temperature stability of the threshold current density, and differential efficiency. Lasers emitting at 1.3 µm have been grown on GaAs substrates with key parameters far exceeding those of state-of-the-art quantum well lasers on InP substrates. It is quite clear that QD lasers will take over from such conventional quantum well lasers.

One of the branches of nanostructure technology has driven us down to so-called submonolayer islands, which may be regarded as sub-nanostructures. Excellent lasers have also been fabricated using this approach. Further developments in this direction will include submonolayer Ge/Si structures which, due to their intriguing optical properties, are realistic candidates for lasers in the long-wavelength spectral range, using indirect gap materials.

In addition, type-II quantum-wire and QD AlAs/GaAs(311)A nanostructures may be used for lasers in a wavelength range from red to green. A broad interval of wavelengths is covered by InGaN/GaN lasers, which are effectively QD lasers. They will probably solve the problem of efficient ultraviolet, bright blue and green light emitters.

QD nanostructures will also be used in optical amplifiers and photodetectors. The success of quantum dots in the field of optoelectronics will be a powerful stimulating factor for the broader application of fantastic nanoworlds, e.g., in the field of quantum cryptography and quantum computation.

A. Energy of a Strained Disk with Perturbed Shape

In this appendix, we calculate the variation of the energy of a 2-dimensional island, due to a small perturbation in the circular disk shape of the island. The energy of the island is the sum of the energy of the island boundary and the elastic relaxation energy,

$$E_{\text{isl}} = \eta_{\text{b}} L_{\text{b}} - \eta_{\text{d}} I ,\qquad(\text{A.1})$$

where η_{b} is the energy per unit length of the island boundary and L_{b} is the length of the boundary. The quantity η_{d}, which refers to the elastic relaxation energy, has units of energy per unit length and equals

$$\eta_{\text{d}} = \frac{(\Delta\tau)^2 (1-\nu^2)}{\pi Y},\qquad(\text{A.2})$$

where $\Delta\tau$ is the difference in surface stress between the island surface and the substrate, and Y and ν are the Young's modulus and the Poisson ratio of the substrate, respectively. In the isotropic model, the surface stress tensors of both the substrate and the island surface are assumed to be isotropic, $\tau_{\alpha\beta}^{(1,2)} = \tau^{(1,2)} \delta_{\alpha\beta}$. Due to the long-range nature of elastic forces, the energy of elastic relaxation is a non-local functional of the island shape. The quantity I from (A.1), written in the $\boldsymbol{k}$ space, is

$$I = \pi \int \frac{\mathrm{d}^2 \boldsymbol{k}}{(2\pi)^2} \exp(-2ka) \, k \left|\widetilde{\Theta}(\boldsymbol{k})\right|^2 .\qquad(\text{A.3})$$

Consider an island with circular shape of radius R. Let its shape be perturbed

$$r(\varphi) = r + \zeta(\varphi) ,\qquad(\text{A.4})$$

where

$$\zeta(\varphi) = \sum_{m \neq 0} c_m \exp(im\varphi) .\qquad(\text{A.5})$$

Since $r(\varphi)$ is a real function, $c_{-m} = c_m^*$, the island area is

$$A = \int_0^{2\pi} \mathrm{d}\varphi \int_0^{r(\varphi)} \mathrm{d}r' r' .\qquad(\text{A.6})$$

Substituting the expansion (A.4) into the integrand of (A.6), one obtains

$$A = \int_0^{2\pi} d\varphi \, \frac{1}{2} \left[r + \sum_{m \neq 0} c_m \exp(im\varphi) \right] \left[r + \sum_{m' \neq 0} c_{m'} \exp(im'\varphi) \right] . \quad \text{(A.7)}$$

Integration over φ yields

$$A = \pi \left(r^2 + \sum_{m \neq 0} |c_m|^2 \right) . \quad \text{(A.8)}$$

Since the island area equals the area of the unperturbed circle, $A = \pi R^2$, (A.8) implies that

$$r = R \left[1 - \frac{1}{2R^2} \sum_{m \neq 0} |c_m|^2 \right] , \quad \text{(A.9)}$$

which gives the corrected radius of the island to second order in perturbation theory.

A.1 Energy of the Disk Boundary

The length of the island boundary can be written as the integral

$$L_\text{b} = \oint dl = \int_0^{2\pi} d\varphi \sqrt{(dr)^2 + r^2(d\varphi)^2} . \quad \text{(A.10)}$$

By substituting the boundary profile (A.4) into (A.10), one obtains the boundary length as a series sum,

$$L_\text{b} = 2\pi \left(r + \frac{1}{2r} \sum_{m \neq 0} m^2 |c_m|^2 \right) . \quad \text{(A.11)}$$

Substituting the perturbed average radius with second order corrections from (A.9) into (A.11), one obtains

$$L_\text{b} = 2\pi R \left[1 + \frac{1}{2R^2} \sum_{m \neq 0} (m^2 - 1)|c_m|^2 \right] . \quad \text{(A.12)}$$

It should be noted that the terms with $m = \pm 1$ drop out of the summation in (A.12). Such perturbations of the island profile correspond to a pure translation of the island as a whole and obviously do not contribute to the change in boundary length. The boundary energy of the island is therefore

$$E_\text{bound} = 2\pi R \, \eta_\text{b} \left[1 + \frac{1}{2R^2} \sum_{m \neq 0} (m^2 - 1)|c_m|^2 \right] . \quad \text{(A.13)}$$

A.2 Elastic Relaxation Energy of the Disk

The elastic relaxation energy of the island depends on the island shape and is proportional to the integral I from (A.3). To evaluate this integral, let us first calculate the Fourier transform of the shape function $\widetilde{\Theta}(r)$. Let φ be the polar angle of the position vector r, and ψ the polar angle of the wave vector k. Then,

$$\widetilde{\Theta}(\boldsymbol{k}) = \int d^2r \exp\left[-i\boldsymbol{k}\cdot\boldsymbol{r}\right]\Theta(\boldsymbol{r})$$

$$= \int_0^{2\pi} d\varphi \int_0^{r+\zeta(\varphi)} dr\, r\, \exp\left[-ikr\cos(\varphi-\psi)\right]$$

$$= \int_0^{2\pi} d\varphi \int_0^{r} dr'\, r'\, \exp\left[-ikr\cos(\varphi-\psi)\right]$$

$$+ \int_0^{2\pi} d\varphi \int_r^{r+\zeta(\varphi)} dr'\, r'\, \exp\left[-ikr'\cos(\varphi-\psi)\right]. \tag{A.14}$$

Now, in the first term on the right-hand side of (A.14), we integrate over φ, and in the second term, we replace the variables of integration, $r' \to r + \rho$, and expand in powers of ρ up to second order terms. Hence,

$$\widetilde{\Theta}(\boldsymbol{k}) = 2\pi \int_0^r dr'\, r'\, J_0(kr') \tag{A.15}$$

$$+ r \int_0^{2\pi} d\varphi \exp\left[-ikr\cos(\varphi-\psi)\right] \int_0^{\zeta(\varphi)} d\rho \left(1+\frac{\rho}{r}\right)\left[1 - ik\rho\cos(\varphi-\psi)\right].$$

The integral in the first term equals $k^{-2}(kr)J_1(kr)$, where J_m is the Bessel function of order m. After integrating the second term over ρ and retaining terms up to second order in ζ, one obtains

$$\widetilde{\Theta}(\boldsymbol{k}) = 2\pi \frac{(kr)J_1(kr)}{k^2} \tag{A.16}$$

$$+ r \int_0^{2\pi} d\varphi \exp\left[-ikr\cos(\varphi-\psi)\right] \left[\zeta(\varphi) + \left(\frac{1}{r} - ik\cos(\varphi-\psi)\right)\frac{1}{2}\zeta^2(\varphi)\right].$$

Now let us substitute the expansion of the boundary shape (A.5) into (A.16), to obtain

340   A. Energy of a Strained Disk with Perturbed Shape

$$\widetilde{\Theta}(\boldsymbol{k}) = 2\pi \frac{(kr)J_1(kr)}{k^2} \tag{A.17}$$

$$+ r \sum_{m \neq 0} c_m \exp(im\psi) \int_0^{2\pi} d\varphi \exp\left[-ikr\cos(\varphi - \psi)\right] \exp\left[im(\varphi - \psi)\right]$$

$$+ r \sum_{m \neq 0} \sum_{p \neq m} c_{p-m} c_m \exp(ip\psi) \int_0^{2\pi} d\varphi \exp\left[-ikr\cos(\varphi - \psi)\right]$$

$$\times \exp\left[ip(\varphi - \psi)\right] \left[\frac{1}{r} - ik\cos(\varphi - \psi)\right].$$

To proceed, we note again that we are expanding the shape factor $\widetilde{\Theta}(\boldsymbol{k})$ up to second order terms in c_m. In the first term of (A.17), we therefore substitute the island radius r up to second order terms from (A.9). In the second term, we can retain $r \approx R$, and the integral over φ yields $2\pi J_m(kr) \approx 2\pi J_m(kR)$. In the third term, we take into account the fact that this term will later be integrated over ψ. We therefore retain only the term with $p = 0$, which gives a non-vanishing integral, and substitute $r \to R$. Carrying out these approximations, (A.17) reduces to

$$\widetilde{\Theta}(\boldsymbol{k}) = 2\pi \frac{(kR)J_1(kR)}{k^2} - 2\pi \frac{(kR)J_0(kR)}{k^2} (kR) \frac{1}{2R^2} \sum_{m \neq 0} |c_m|^2$$

$$+ 2\pi R \sum_{m \neq 0} c_m \exp[im\psi] J_m(kR)$$

$$+ 2\pi \sum_{m \neq 0} |c_m|^2 \frac{1}{2} \left[1 + \frac{d}{d(kR)}\right] J_0(kR) . \tag{A.18}$$

After simplifying, we obtain

$$\widetilde{\Theta}(\boldsymbol{k}) = 2\pi \left[\frac{(kR)J_1(kR)}{k^2} + R^2 \sum_{m \neq 0} J_m(kR) \frac{c_m}{R} \exp(im\psi) \right. \tag{A.19}$$

$$\left. - \frac{R^2}{2} J_1(kR) \sum_{m \neq 0} \left|\frac{c_m}{R}\right|^2 \right].$$

Now, by substituting the expansion of the shape factor $\widetilde{\Theta}(\boldsymbol{k})$ from (A.19) into (A.3), then integrating over ψ, one obtains the following equation for the integral I:

$$I = \pi(2\pi)^2 \int \frac{d^2\boldsymbol{k}}{(2\pi)^2} \exp[-2ka] \, k \left[\left(\frac{(kR)J_1(kR)}{k^2}\right)^2 \right. \tag{A.20}$$

$$\left. + \frac{R^4}{2} \sum_{m \neq 0} \left[J_m^2(kR) - J_1^2(kR)\right] \left|\frac{c_m}{R}\right|^2 \right].$$

It should be noted that the coefficient of the term with $m = 1$ in (A.20) vanishes, since this type of perturbation is a pure translation of the island as a whole. Introducing dimensionless variables

$$kR = x\,, \qquad \frac{2a}{R} = \lambda\,, \tag{A.21}$$

equation (A.20) reduces to

$$I = I_0(1) + \sum_{m \neq 0} [I_2(m) - I_2(1)] \frac{1}{2} \left|\frac{c_m}{R}\right|^2\,, \tag{A.22}$$

where

$$I_0(m) = \pi R \int_0^\infty dx\, \exp(-\lambda x)\, J_m^2(x)\,, \tag{A.23a}$$

$$I_2(m) = \pi R \int_0^\infty dx\, \exp(-\lambda x)\, x^2\, J_m^2(x) = \frac{d^2 I_0(m)}{d\lambda^2}\,. \tag{A.23b}$$

A.3 Evaluation of Integrals

Although the integrals of (A.23a) and (A.23b) can be found in the literature and can be evaluated by the program Mathematica, the resulting expressions are still too cumbersome and require further simplification. We therefore reproduce the evaluation of the integrals in the present section. To evaluate the integrals (A.23a) and (A.23b), it is convenient to express the Bessel functions in terms of the confluent hypergeometric function,

$$J_m(z) = \frac{1}{2^m \Gamma(m+1)} z^m \exp(-iz)\, {}_1F_1\left(m + \frac{1}{2}, 2m+1, 2iz\right)$$

$$= \frac{1}{2^m \Gamma(m+1)} z^m \exp(iz)\, {}_1F_1\left(m + \frac{1}{2}, 2m+1, -2iz\right)\,. \tag{A.24}$$

Substituting (A.24) into the integrand of (A.23a), the latter reduces to

$$I_0(m) = \pi R \frac{1}{2^{2m} [\Gamma(m+1)]^2} \int_0^\infty dx\, \exp(-\lambda x)\, x^{2m} \tag{A.25}$$

$$\times {}_1F_1\left(m + \frac{1}{2}, 2m+1, -2ix\right) {}_1F_1\left(m + \frac{1}{2}, 2m+1, 2ix\right)\,.$$

Integrals of this type are evaluated analytically in the Mathematical Appendices of L.D. Landau and E.M. Lifshitz: *Quantum Mechanics*, Pergamon, Oxford, 1980; equations (f.9) and (f.10). The result is

A. Energy of a Strained Disk with Perturbed Shape

$$J = \int_0^\infty dz \exp(-\lambda z) \, z^{\gamma-1} \, {}_1F_1(\alpha, \gamma, kz) \, {}_1F_1(\alpha, \gamma, k'z) \quad (A.26)$$

$$= \Gamma(\gamma) \, \lambda^{\alpha+\alpha'-\gamma} (\lambda - k)^{-\alpha} (\lambda - k')^{-\alpha'} \, {}_1F_2\left(\alpha, \alpha', \gamma, \frac{kk'}{(\lambda-k)(\lambda-k')}\right).$$

The straightforward substitution of α, α', γ, k, k' from (A.25) into (A.26) would give rise to uncertainties at $\lambda \to 0$. To resolve these uncertainties, we introduce a small quantity ε and substitute

$$\alpha = \alpha' = m + \frac{1}{2}, \quad \gamma = 2m + 1 + \varepsilon, \quad k = 2\mathrm{i}, \quad k' = -2\mathrm{i}, \quad (A.27)$$

into (A.26). We then obtain

$$I_0(m) = \pi R \frac{1}{2^{2m}[\Gamma(m+1)]^2} \Gamma(2m+1+\varepsilon) \lambda^{-\varepsilon} (\lambda - 2\mathrm{i})^{-(m+1/2)}$$

$$\times (\lambda + 2\mathrm{i})^{-(m+1/2)} \, {}_1F_2\left(m+\frac{1}{2}, m+\frac{1}{2}, 2m+1+\varepsilon, \frac{4}{\lambda^2+4}\right)$$

$$= \pi R \frac{1}{2^{2m}[\Gamma(m+1)]^2} \Gamma(2m+1+\varepsilon) \lambda^{-\varepsilon} \frac{1}{2^{2m+1}} \frac{1}{(1+\lambda^2/4)^{m+1/2}}$$

$$\times {}_1F_2\left(m+\frac{1}{2}, m+\frac{1}{2}, 2m+1+\varepsilon, \frac{1}{1+\lambda^2/4}\right). \quad (A.28)$$

In order to regularize equation (A.28), we use formula (e.7) from the Mathematical Appendices of *Quantum Mechanics*, relating hypergeometric functions of the arguments z and $(1-z)$,

$${}_1F_2(\alpha, \beta, \gamma, 1-z) = \frac{\Gamma(\gamma)\Gamma(\gamma-\alpha-\beta)}{\Gamma(\gamma-\alpha)\Gamma(\gamma-\beta)} {}_1F_2(\alpha, \beta, \alpha+\beta+1-\gamma, z)$$

$$+ \frac{\Gamma(\gamma)\Gamma(\alpha+\beta-\gamma)}{\Gamma(\alpha)\Gamma(\beta)} z^{\gamma-\alpha-\beta} \, {}_1F_2(\gamma-\alpha, \gamma-\beta, \gamma+1-\alpha-\beta, z). \quad (A.29)$$

Substituting

$$\alpha = \beta = m + \frac{1}{2}, \quad \gamma = 2m + 1 + \varepsilon, \quad (A.30)$$

into (A.29), we obtain

$${}_1F_2\left(m+\frac{1}{2}, m+\frac{1}{2}, 2m+1+\varepsilon, 1-z\right) = \quad (A.31)$$

$$\frac{\Gamma(2m+1+\varepsilon)\Gamma(\varepsilon)}{\Gamma(m+1/2+\varepsilon)\Gamma(m+1/2+\varepsilon)} {}_1F_2\left(m+\frac{1}{2}, m+\frac{1}{2}, 1-\varepsilon, z\right)$$

$$+ \frac{\Gamma(2m+1+\varepsilon)\Gamma(-\varepsilon)}{\Gamma(m+1/2)\Gamma(m+1/2)} z^\varepsilon \, {}_1F_2\left(m+\frac{1}{2}+\varepsilon, m+\frac{1}{2}+\varepsilon, 1+\varepsilon, z\right).$$

As $\varepsilon \to 0$, (A.31) contains an uncertainty of the type $0 \times \infty$. To resolve this uncertainty, we transform

A.3 Evaluation of Integrals   343

$$\Gamma(\varepsilon) = \frac{\Gamma(1+\varepsilon)}{\varepsilon}, \quad \Gamma(-\varepsilon) = -\frac{\Gamma(1-\varepsilon)}{\varepsilon}. \tag{A.32}$$

Then (A.31) reduces to

$$_1F_2\left(m+\frac{1}{2}, m+\frac{1}{2}, 2m+1+\varepsilon, 1-z\right) = \Gamma(2m+1+\varepsilon)z^{\varepsilon/2}\frac{1}{\varepsilon} \tag{A.33}$$

$$\times \left[\frac{\Gamma(1+\varepsilon)}{\Gamma(m+1/2+\varepsilon)\Gamma(m+1/2+\varepsilon)} z^{-\varepsilon/2} {}_1F_2\left(m+\frac{1}{2}, m+\frac{1}{2}, 1-\varepsilon, z\right) \right.$$

$$\left. - \frac{\Gamma(1-\varepsilon)}{\Gamma(m+1/2)\Gamma(m+1/2)} z^{\varepsilon/2} {}_1F_2\left(m+\frac{1}{2}+\varepsilon, m+\frac{1}{2}+\varepsilon, 1+\varepsilon, z\right) \right].$$

Now, introducing the expansion of the hypergeometric function up to linear terms in z,

$$_1F_2(\alpha, \beta, \gamma, z) = 1 + \frac{\alpha\beta}{\gamma}z + \cdots, \tag{A.34}$$

into (A.33), we obtain

$$_1F_2\left(m+\frac{1}{2}, m+\frac{1}{2}, 2m+1+\varepsilon, 1-z\right) = \Gamma(2m+1+\varepsilon)z^{\varepsilon/2}\frac{1}{\varepsilon} \tag{A.35}$$

$$\times \left\{ \frac{\Gamma(1+\varepsilon)}{\Gamma(m+1/2+\varepsilon)\Gamma(m+1/2+\varepsilon)} z^{-\varepsilon/2} \left[1 + \frac{(m+1/2)^2}{1-\varepsilon} z \right] \right.$$

$$\left. - \frac{\Gamma(1-\varepsilon)}{\Gamma(m+1/2)\Gamma(m+1/2)} z^{\varepsilon/2} \left[1 + \frac{(m+1/2+\varepsilon)^2}{1+\varepsilon} z \right] \right\}.$$

We now set $\varepsilon \to 0$. To do this, we expand the expression in square brackets up to linear terms in ε,

$$\lim_{\varepsilon \to 0} {}_1F_2\left(m+\frac{1}{2}, m+\frac{1}{2}, 2m+1+\varepsilon, 1-z\right) \tag{A.36}$$

$$= \Gamma(2m+1)\, z^0\, \frac{1}{\varepsilon}\, \frac{\Gamma(1)}{\Gamma(m+1/2)\Gamma(m+1/2)}$$

$$\times \left\{ [1+\psi(1)\varepsilon] \left[1+\psi\left(m+\frac{1}{2}\right)\varepsilon\right]^{-2} \left(1 - \frac{1}{2}\ln z\, \varepsilon\right) \right.$$

$$\times \left[1 + \frac{(m+1/2)^2}{1}z + \left(m+\frac{1}{2}\right)^2 z\varepsilon \right]$$

$$- [1-\psi(1)\varepsilon]\left(1 + \frac{1}{2}\ln z\,\varepsilon\right)$$

$$\left. \times \left[1 + \frac{(m+1/2)^2}{1}z + \left(m+\frac{1}{2}\right)^2\left(\frac{2}{m+1/2}-1\right)z\varepsilon \right] \right\},$$

where $\psi(z)$ is the logarithmic derivative of the Gamma function, that is, $\psi(z) \equiv [\ln \Gamma(z)]'$. Simplifying (A.36) yields

$$\lim_{\varepsilon \to 0} {}_1F_2\left(m + \frac{1}{2}, m + \frac{1}{2}, 2m + 1 + \varepsilon, 1 - z\right) \tag{A.37}$$

$$= \frac{\Gamma(2m+1)}{\Gamma(m+1/2)\Gamma(m+1/2)} \left[1 + \left(m + \frac{1}{2}\right)^2 z\right]$$

$$\times \left[-\ln z + 2\psi(1) - 2\psi\left(m + \frac{1}{2}\right) + 2\left(m^2 - \frac{1}{4}\right) z\right].$$

Now note that $(1 + \lambda^2/4)^{-1} = 1 - z$. After substituting (A.37) into (A.28), we obtain

$$I_0(m) = \pi R \frac{1}{2^{2m} \left[\Gamma(m+1)\right]^2} \Gamma(2m + 1 + \varepsilon) \lambda^{-\varepsilon} \frac{1}{2^{2m+1}} (1-z)^{m+1/2}$$

$$\times \frac{\Gamma(2m+1)}{\Gamma(m+1/2)\Gamma(m+1/2)} \left[1 + \left(m + \frac{1}{2}\right)^2 z\right]$$

$$\times \left[-\ln z + 2\psi(1) - 2\psi\left(m + \frac{1}{2}\right) + 2\left(m^2 - \frac{1}{4}\right) z\right]. \tag{A.38}$$

We may now set $\varepsilon = 0$ and note that

$$\frac{1}{2^{4m}} \left[\frac{\Gamma(2m+1)}{\Gamma(m+1)\Gamma(m+1/2)}\right]^2 \tag{A.39}$$

$$= \frac{1}{2^{2m}} \left[\frac{1 \cdot 2 \cdot 3 \cdot \ldots \cdot (2m-1)}{1 \cdot 2 \cdot \ldots \cdot m \cdot \Gamma\left(\frac{1}{2}\right) \cdot \frac{1}{2} \cdot \frac{3}{2} \cdot \ldots \cdot \left(m - \frac{1}{2}\right)}\right]^2$$

$$= \frac{1}{[\Gamma(1/2)]^2} \left[\frac{1 \cdot 2 \cdot \ldots \cdot (2m-1)}{2 \cdot 4 \cdot \ldots \cdot 2m \cdot 1 \cdot 3 \cdot \ldots \cdot (2m-1)}\right]^2 = \frac{1}{\pi}.$$

Hence, substituting (A.39) into (A.38) and retaining terms up to order z, we obtain

$$I_0(m) = \pi R \frac{1}{2\pi} \left\{\ln\left(\frac{1}{z}\right) + 2\left[\psi(1) - \psi\left(m + \frac{1}{2}\right)\right] \right. \tag{A.40}$$

$$\left. + \left(m^2 - \frac{1}{4}\right) z \ln\left(\frac{1}{z}\right) + 2\left[1 + \psi(1) - \psi\left(m + \frac{1}{2}\right)\right] \left(m^2 - \frac{1}{4}\right) z\right\}.$$

We now substitute

$$z = \frac{\lambda^2/4}{1 + \lambda^2/4}, \tag{A.41}$$

into (A.40) to obtain

$$I_0(m) = R\left\{\ln\left(\frac{2}{\lambda}\right) + \frac{\lambda^2}{8} + \left[\psi(1) - \psi\left(m + \frac{1}{2}\right)\right]\right.$$
$$\left. + \left(m^2 - \frac{1}{4}\right)\frac{\lambda^2}{4}\ln\left(\frac{2}{\lambda}\right) + \left[1 + \psi(1) - \psi\left(m + \frac{1}{2}\right)\right]\left(m^2 - \frac{1}{4}\right)\frac{\lambda^2}{4}\right\}. \quad (A.42)$$

To further simplify (A.42), we carry out the transformation

$$\frac{1}{m^2 - \frac{1}{4}} - 2\psi\left(m + \frac{1}{2}\right) = \frac{1}{m - \frac{1}{2}} - \frac{1}{m + \frac{1}{2}} - \psi\left(m + \frac{1}{2}\right) - \psi\left(m + \frac{1}{2}\right)$$
$$= -\psi\left(m - \frac{1}{2}\right) - \psi\left(m + \frac{3}{2}\right). \quad (A.43)$$

Now by substituting (A.43) into (A.42), we obtain

$$I_0(m) = R\left\{\ln\left(\frac{1}{\lambda}\right) + \left[\ln 2 + \psi(1) - \psi\left(m + \frac{1}{2}\right)\right] + \left(m^2 - \frac{1}{4}\right)\frac{\lambda^2}{4}\ln\left(\frac{1}{\lambda}\right)\right.$$
$$\left. + \left[\ln 4 + 2 + 2\psi(1) - \psi\left(m - \frac{1}{2}\right) - \psi\left(m + \frac{3}{2}\right)\right]\left(m^2 - \frac{1}{4}\right)\frac{\lambda^2}{8}\right\}.$$
$$(A.44)$$

To obtain the quantity $I_2(m)$, we differentiate (A.44) twice with respect to λ. This yields

$$I_2(m) = \frac{d^2}{d\lambda^2}I_0(m) = R\left\{-\frac{1}{\lambda^2} + \left(m^2 - \frac{1}{4}\right)\frac{1}{4}\left[2\ln\left(\frac{1}{\lambda}\right) - 1 + \ln 4\right.\right.$$
$$\left.\left. + 2\psi(1) - \psi\left(m - \frac{1}{2}\right) - \psi\left(m + \frac{3}{2}\right)\right]\right\}. \quad (A.45)$$

When we calculate $I_2(m) - I_2(1)$, which enters the quadratic coefficient in the energy expansion (A.20), the most divergent term λ^{-2} in (A.45) vanishes and

$$I_2(m) - I_2(1) = R\left\{\left(\frac{m^2 - 1}{2}\right)\left[\ln\left(\frac{1}{\lambda}\right) - \frac{1}{2} + \ln 2\right]\right.$$
$$\left. - \frac{1}{4}\left(m^2 - \frac{1}{4}\right)\left[-2\psi(1) + \psi\left(m - \frac{1}{2}\right) + \psi\left(m + \frac{3}{2}\right)\right]\right.$$
$$\left. + \frac{3}{16}\left[-2\psi(1) + \psi\left(\frac{1}{2}\right) + \psi\left(\frac{5}{2}\right)\right]\right\}. \quad (A.46)$$

A.4 Stiffness of the Disk against Shape Perturbations

After we have evaluated all the necessary integrals (A.23a) and (A.23b), we can write down the energy of an island with the shape of a perturbed disk. Let us consider $I_0(1)$ and set $\lambda = 0$ except in the logarithmic term. Then, since

$$\psi(1) = -\gamma, \quad \psi\left(\frac{1}{2}\right) = -\gamma - 2\ln 2, \quad \psi\left(\frac{3}{2}\right) = -\gamma - 2\ln 2 + 2, \quad \text{(A.47)}$$

where $\gamma = 0.5772$ is the Euler constant, we obtain

$$I_0(1) = R \ln\left(\frac{8}{e^2 \lambda}\right). \tag{A.48}$$

We now collect all terms in the island energy. Substituting (A.46) and (A.48) into (A.22), (A.20), and (A.3), and adding the boundary energy from (A.13), we obtain the total energy of a perturbed island up to quadratic terms in the perturbation amplitudes:

$$\Delta E = 2\pi R \eta_b \left[1 + \sum_{m \neq 0} \frac{m^2 - 1}{2} \left|\frac{c_m}{R}\right|^2 \right] + 2\pi R \eta_d \left[-\ln\left(\frac{4R}{e^2 a}\right) \right.$$

$$+ \sum_{m \neq 0} \left\{ -\left(\frac{m^2 - 1}{2}\right) \left[\ln\left(\frac{R}{a}\right) - \frac{1}{2}\right] \right.$$

$$+ \left(m^2 - \frac{1}{4}\right) \frac{1}{4} \left[-2\psi(1) + \psi\left(m - \frac{1}{2}\right) + \psi\left(m + \frac{3}{2}\right)\right]$$

$$\left.\left. - \frac{3}{16} \left[-2\psi(1) + \psi\left(\frac{1}{2}\right) + \psi\left(\frac{5}{2}\right)\right]\right\} \left|\frac{c_m}{R}\right|^2 \right], \tag{A.49}$$

where η_b is defined above as the energy of the island boundary per unit length, and η_d can be interpreted as the elastic relaxation energy per unit length of the boundary. By rearranging terms in (A.49), we can write it in the more convenient form,

$$\Delta E = 2\pi R \eta_d \left[\frac{\eta_b}{\eta_d} - \ln\left(\frac{4R}{e^2 a}\right)\right]$$

$$+ 2\pi R \eta_d \sum_{m \neq 0} \left\{\frac{m^2 - 1}{2} \left[\frac{\eta_b}{\eta_d} - \ln\left(\frac{R}{a}\right) + \frac{1}{2}\right]\right.$$

$$+ \left(m^2 - \frac{1}{4}\right) \frac{1}{4} \left[-2\psi(1) + \psi\left(m - \frac{1}{2}\right) + \psi\left(m + \frac{3}{2}\right)\right]$$

$$\left. - \frac{3}{16} \left[-2\psi(1) + \psi\left(\frac{1}{2}\right) + \psi\left(\frac{5}{2}\right)\right]\right\} \left|\frac{c_m}{R}\right|^2. \tag{A.50}$$

A.4 Stiffness of the Disk against Shape Perturbations

This expansion gives us both the energy of the unperturbed circular disk and the quadratic expansion of the energy in terms of perturbation amplitudes. From the equation for the optimum radius of a disk, we obtain

$$\frac{\eta_b}{\eta_d} = \ln\left(\frac{R_{\text{opt}}}{a}\right) + 2\ln 2 - 3 . \tag{A.51}$$

Substituting (A.51) into (A.50), we obtain the expansion of the energy of a disk in terms of perturbation amplitudes as a function of the actual disk radius R and the optimum radius R_{opt},

$$\Delta E = 2\pi\eta_d R\left[-\ln\left(\frac{R}{R_{\text{opt}}}\right) - 1\right] + 2\pi\eta_d R \sum_{m\neq 0} \Lambda_m \left|\frac{c_m}{R}\right|^2 , \tag{A.52}$$

where the stiffness Λ_m of a circular island against the mth perturbation harmonic is

$$\Lambda_m = -\frac{m^2-1}{2}\ln\left(\frac{R}{R_{\text{opt}}}\right)$$
$$+ \frac{4m^2-1}{16}\left[4\ln 2 - 5 - 2\psi(1) + \psi\left(m-\frac{1}{2}\right) + \psi\left(m+\frac{3}{2}\right)\right]$$
$$- \frac{3}{16}\left[4\ln 2 - 5 - 2\psi(1) + \psi\left(\frac{1}{2}\right) + \psi\left(\frac{5}{2}\right)\right] . \tag{A.53}$$

To calculate the partition function of an array of islands in Sect. 2.2, the perturbed profile of the island boundary is expanded in cosines and sines (3.82) rather than in complex exponents (A.5). The two expansions are equivalent, if we set

$$c_m = \begin{cases} \frac{a_m - ib_m}{2} & \text{if } m > 0 , \\ c^*_{-m} & \text{if } m < 0 . \end{cases} \tag{A.54}$$

Furthermore, it is convenient to express the energy of an island in terms of the number of atoms N rather than the radius ρ. Substituting $\rho = a\sqrt{N}/\sqrt{\pi}$ and (A.54) into (A.52), we obtain the following expansion of the island energy:

$$\Delta E = -2\sqrt{\pi}\eta_d a\sqrt{N}\ln\left(e\sqrt{\frac{N}{N_{\text{opt}}}}\right) + \frac{\pi^{3/2}\eta_d}{a\sqrt{N}}\sum_{m=1} \Lambda_m (a_m^2 + b_m^2) , \tag{A.55}$$

which is used in Sect. 3.2 to evaluate the entropy of shape fluctuations.

It should be noted that, for $m = \pm 1$, the stiffness Λ_m from (A.53) vanishes, since $m = \pm 1$ corresponds to a pure translation which does not affect the island energy. For $R = R_{\text{opt}}$, and $|m| \geq 2$, the stiffness $\Lambda_m > 0$, which means that circular islands are stable. However, as the island radius increases, Λ_m becomes negative. Instability occurs for the first time for $m = \pm 2$ and $R = e^{1/3}R_{\text{opt}}$. Fluctuations with $m = \pm 2$ correspond to an elliptical distortion of

A. Energy of a Strained Disk with Perturbed Shape

a circular island. This instability is similar to the instability of square islands against transformation into rectangular islands, found by Tersoff and Tromp [J. Tersoff, R.M. Tromp: Phys. Rev. Lett. **70**, 2782 (1993)].

B. Elastic Interaction of Two Strained Disks

The elastic interaction energy between strained 2-dimensional islands is given by the double integral over the two surfaces,

$$E_{\text{inter}} = \frac{(1-\nu^2)(\Delta\tau)^2}{\pi Y} \int_{A_1} d^2r \int_{A_2} d^2r' \frac{1}{|r-r'|^3} , \tag{B.1}$$

where A_1 and A_2 are the surface areas of the two islands. If the distance between the two islands is much larger than the lateral size of the islands, one can pull the term $|r-r'|^{-3}$ out of the integral. The remaining integrals will then give the product of the surface areas A_1 and A_2 of the two islands,

$$E_{\text{inter}} = \frac{(1-\nu^2)(\Delta\tau)^2}{\pi Y} \frac{A_1 A_2}{R^3} , \tag{B.2}$$

where R is the distance between the two islands. However, for a dense array of islands, it is important to know the exact equation for the interaction energy of two nearby islands. To evaluate the interaction energy of two circular islands (disks), it is convenient to introduce polar coordinates, r_1, φ_1 for the first island, and r_2, φ_2 for the second island. Then the interaction energy (B.1) reduces to

$$E_{\text{inter}} = \frac{(1-\nu^2)(\Delta\tau)^2}{\pi Y} \int_0^{\rho_1} dr_1 r_1 \int_0^{2\pi} d\varphi_1 \int_0^{\rho_2} dr_2 r_2 \int_0^{2\pi} d\varphi_2 \frac{1}{|R+r_2-r_1|^3} , \tag{B.3}$$

where ρ_1 and ρ_2 are the radii of the two disks. The integrand of (B.3) is proportional to the interaction energy between the two points depicted in Fig. B.1a. If we integrate first over φ_1 and φ_2, this provides the interaction energies between the two rings. Further integration over r_1 and r_2 gives the interaction energy between the two disks.

To carry out the integration over polar angles φ_1 and φ_2, we follow the approach of Rickman and Srolovitz [J.M. Rickman, D.J. Srolovitz: Surf. Sci. **284**, 211 (1993)] and use the expansion

$$\frac{1}{|r-r'|} = \sum_{m=-\infty}^{\infty} \int_0^{\infty} dk \, J_m(kr) J_m(kr') \exp[im(\varphi - \varphi')] , \tag{B.4}$$

B. Elastic Interaction of Two Strained Disks

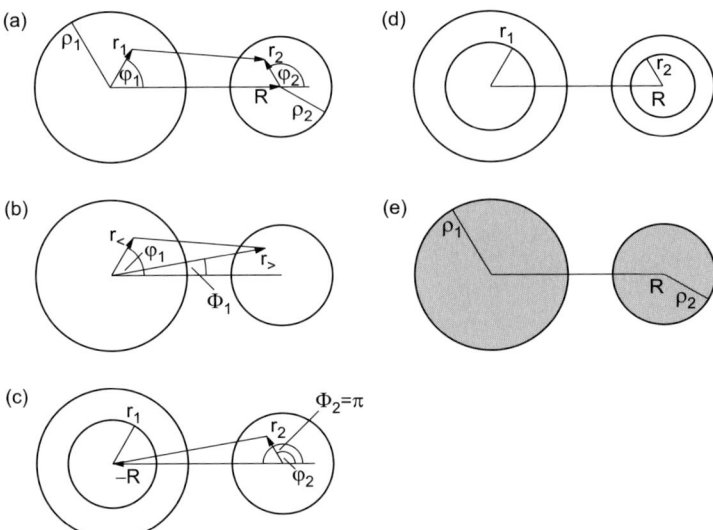

Fig. B.1. Successive stages in the calculation of the elastic interaction energy between two circular islands with radii ρ_1 and ρ_2, separated by distance R. (**a**) The integrand of (B.3) is proportional to the interaction between the two points r_1 and r_2. (**b**) To integrate the interaction energy over the ring of the first island, we choose the origin in the center of the first island and apply (B.8). (**c**) The interaction energy, integrated over φ_1 and given by (B.9), corresponds to the interaction between the ring of radius r_1 of the first island and the point r_2 of the second island. To integrate the interaction energy over the ring of the second island, we choose the origin in the center of the second island and apply (B.11) for the integration. (**d**) The interaction energy, integrated over φ_1 and φ_2 and given in (B.13) corresponds to the interaction between the ring of radius r_1 of the first island and the ring of radius r_2 of the second island. (**e**) The interaction energy, integrated over φ_1, φ_2, r_1, and r_2 and given by (B.15) corresponds to the interaction between the two islands

where r, r' are the radial coordinates and φ, φ' the polar angles of the two points. J_m is a Bessel function of order m. The integral in (B.4) has been evaluated by Rickman and Srolovitz and equals

$$\frac{1}{|\boldsymbol{r}-\boldsymbol{r'}|} = \mathrm{Re} \sum_{m=0}^{\infty} \frac{(2-\delta_{m0})\Gamma(m+1/2)}{\Gamma(m+1)\Gamma(1/2)} \exp[im(\varphi-\varphi')] \left(\frac{r_<^m}{r_>^{m+1}}\right)$$
$$\times {}_2F_1\left(m+\frac{1}{2},\frac{1}{2};m+1;(r_</r_>)^2\right), \tag{B.5}$$

where $r_<$ is the smaller of r and r', $r_>$ is the larger of r and r', the function ${}_2F_1\left(m+\frac{1}{2},\frac{1}{2};m+1;[r_</r_>]^2\right)$ is a hypergeometric function, and Re denotes the real part.

To evaluate the integral of (B.3) with $|\boldsymbol{r}-\boldsymbol{r'}|^{-3}$, it is useful to consider an expansion of $|\boldsymbol{r}-\boldsymbol{r'}|^{-3}$ similar to (B.5). We note that

$$\frac{1}{|\boldsymbol{r}-\boldsymbol{r}'|^3} = \nabla^2 \frac{1}{|\boldsymbol{r}-\boldsymbol{r}'|}, \tag{B.6}$$

and write down the Laplacian in polar coordinates,

$$\nabla^2 = \frac{1}{r_>}\frac{\partial}{\partial r_>}\left(r_>\frac{\partial}{\partial r_>}\right) + \frac{1}{r_>^2}\frac{\partial^2}{\partial \varphi^2}. \tag{B.7}$$

Expanding the hypergeometric function $_2F_1\left(m+\frac{1}{2},\frac{1}{2};m+1;[r_</r_>]^2\right)$ as a power series in $(r_</r_>)^2$, substituting the expansion into (B.5), and applying the Laplacian from (B.7), we obtain the formula

$$\frac{1}{|\boldsymbol{r}-\boldsymbol{r}'|^3} = 4\left[\Gamma(1/2)\right]^2 \sum_{p=0}^{\infty}\left[\frac{\Gamma(3/2+p)}{\Gamma(1/2)\Gamma(1+p)}\right]^2 \left(\frac{r_<}{r_>}\right)^{2p}$$

$$+ \sum_{m\neq 0} \exp\left[im(\varphi-\varphi')\right] \ldots . \tag{B.8}$$

Let us note here that we have to substitute the expansion (B.8) into the integrand of (B.3) and integrate over φ_1. Putting the origin in the center of the first disk we find that, in (B.8), the smaller of the two radii $r_<$ is the radius of the ring r_1, and the larger $r_>$ is the distance between the center of the first ring and the current point on the second ring, i.e., $r_> = |\boldsymbol{R}+\boldsymbol{r}_2|$, the angle $\varphi = \varphi_1$, and the angle $\varphi' = \Phi_1$ (see Fig. B.1b). After integrating over φ_1, only the axially symmetric part (with $m=0$) of the expansion (B.8) remains. This integral equals

$$I_1 = \int_0^{2\pi} d\varphi_1 \frac{1}{|\boldsymbol{r}-\boldsymbol{r}'|^3} = 2\pi \sum_{p=0}^{\infty}\left[\frac{\Gamma(3/2+p)}{\Gamma(3/2)\Gamma(1+p)}\right]^2 \frac{r_1^{2p}}{r_>^{2p+3}}. \tag{B.9}$$

The quantity I_1 from (B.9) is proportional to the interaction energy between the ring of radius r_1 of the first disk and the particular point of the second disk. To perform the second integration over the position of the point on the second ring in (B.3), i.e., over φ_2, we put the origin at the center of the second island. Then we expand $1/r_>^{2p+3}$ in each term of the series in (B.9), in a similar way to (B.5) and (B.8).

To carry out such an expansion, we note that $r_> = |\boldsymbol{R}-\boldsymbol{r}_2|$, where $\boldsymbol{R}$ is the vector connecting the centers of the two disks, and $\boldsymbol{r}_2$ is the vector connecting the center of the second disk with the current point on the disk. Then we use the identity

$$\frac{1}{|\boldsymbol{R}+\boldsymbol{r}_2|^{3+2p}} = \frac{1}{3^2 \cdot 5^2 \cdot \ldots \cdot (2p+1)^2}(\nabla^2)^p \frac{1}{|\boldsymbol{R}+\boldsymbol{r}_2|^3}$$

$$= \frac{1}{2^{2p}}\left[\frac{\Gamma(3/2)}{\Gamma(3/2+p)}\right]^2 (\nabla^2)^p \frac{1}{|\boldsymbol{R}+\boldsymbol{r}_2|^3}. \tag{B.10}$$

To use the expansion of $|\boldsymbol{R}-\boldsymbol{r}_2|^{-3}$ from (B.8) in (B.10), we replace $r_< \to r_2$ and $r_> \to R$, then substitute the expansion in (B.10). Hence,

$$\frac{1}{|\boldsymbol{R}+\boldsymbol{r}_2|^{3+2p}} = \frac{1}{2^{2p}} \left[\frac{\Gamma(3/2)}{\Gamma(3/2+p)}\right]^2 (\nabla_{\boldsymbol{R}}^2)^p \tag{B.11}$$

$$\left\{\frac{4}{R^3} \sum_{q=0}^{\infty} \left[\frac{\Gamma(3/2+q)}{\Gamma(1/2)\Gamma(1+q)}\right]^2 \left(\frac{r_2}{R}\right)^{2q} + \sum_{m\neq 0} \exp\left[im(\Phi_2 - \varphi_2)\right]\ldots\right\},$$

where $\Phi_2 = \pi$ is the polar angle of the vector $\boldsymbol{R}$ and φ_2 is the polar angle corresponding to the current point on the ring (see Fig. B.1). By applying the Laplacian p times to (B.11), we obtain

$$\frac{1}{|\boldsymbol{R}+\boldsymbol{r}_2|^{3+2p}} = \left[\frac{\Gamma(3/2)}{\Gamma(3/2+p)}\right]^2 \sum_{q=0}^{\infty} \left[\frac{\Gamma(q+p+3/2)}{\Gamma(3/2)\Gamma(1+q)}\right]^2 \frac{r_2^{2q}}{R^{2q+2p+3}}$$

$$+ \sum_{m\neq 0} \exp\left[im(\Phi_2 - \varphi_2)\right]\ldots. \tag{B.12}$$

We now substitute the expansion (B.12) into (B.9) and integrate over φ_2. The term with $m = 0$ will be multiplied by 2π and the other terms will vanish. Hence,

$$I_2 = \int_0^{2\pi} d\varphi_2 \int_0^{2\pi} d\varphi_1 \frac{1}{|\boldsymbol{r}-\boldsymbol{r}'|^3} \tag{B.13}$$

$$= (2\pi)^2 \sum_{p=0}^{\infty} \sum_{q=0}^{\infty} \left[\frac{\Gamma(3/2+p+q)}{\Gamma(3/2)\Gamma(p+1)\Gamma(q+1)}\right]^2 \frac{r_1^{2p} r_2^{2q}}{R^{2q+2p+3}}.$$

It is convenient now to change the order of summation and introduce a new variable $s = p + q$. Then the outer summation will be carried out over s, from 0 to ∞, and the inner summation will be carried out over p, from 0 to s. Hence,

$$I_2 = (2\pi)^2 \sum_{s=0}^{\infty} \left\{\sum_{p=0}^{s} \left[\frac{r_1^p r_2^{s-p}}{\Gamma(p+1)\Gamma(s-p+1)}\right]^2\right\} \left[\frac{\Gamma(3/2+s)}{\Gamma(3/2)}\right]^2 \frac{1}{R^{3+2s}}.$$
$$\tag{B.14}$$

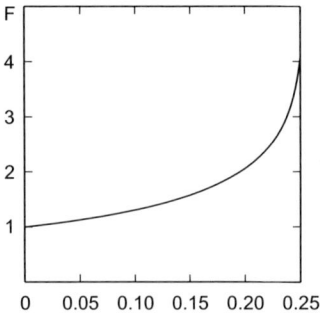

Fig. B.2. Correction factor $F(\xi_1, \xi_2)$ in the elastic interaction energy of two equally sized circular islands ($\xi_2 = \xi_1 = \xi$), defined by (B.16). $\xi = (\rho/R)^2$, where ρ is the radius of the islands and R is the distance between the centers of the islands. When $\xi \to 0$, the islands are remote, and for $\xi = 0.25$, the islands touch each other

B. Elastic Interaction of Two Strained Disks

The quantity I_2 is proportional to the elastic interaction energy between two rings (see Fig. B.1). Now, to obtain the interaction energy between the two disks, we substitute I_2 from (B.14) into (B.3) and integrate over the radii r_1 and r_2. This yields

$$E_{\text{inter}} = \frac{(1-\nu^2)(\Delta\tau)^2}{\pi Y} (\pi\rho_1^2)(\pi\rho_2^2) \frac{1}{R^3} F\left[\left(\frac{\rho_1}{R}\right)^2, \left(\frac{\rho_2}{R}\right)^2\right]. \tag{B.15}$$

Here $F(\xi_1, \xi_2)$ gives the correction factor due to the interaction energy of two remote islands from (B.2). This correction factor is

$$F(\xi_1, \xi_2) = \sum_{s=0}^{\infty} \left[\frac{\Gamma(3/2+s)}{\Gamma(3/2)}\right]^2 \left\{\sum_{p=0}^{s} \xi_1^p \xi_2^{s-p} \left[\frac{1}{\Gamma(p+1)\Gamma(s-p+1)}\right]^2 \frac{1}{(p+1)(s-p+1)}\right\}. \tag{B.16}$$

The correction factor equals 1 for remote islands, where $\xi_1 \to 0$ and $\xi_2 \to 0$. Figure B.2 depicts the value of F for equally sized islands ($\xi_2 = \xi_1 = \xi$). The correction factor increases rapidly as the islands come closer together and touch. The factor $F(\xi, \xi)$ reaches the value ≈ 4.075 when the two equally sized islands touch.

C. Stiffness of a Hexagonal Array of Interacting Strained Disks

The total energy of an array of interacting strained 2-dimensional circular islands (disks) is

$$E_{\text{total}} = -\eta_{\text{d}} a \sum_p \sqrt{N(p)} \ln\left(e\sqrt{\frac{N(p)}{N_{\text{opt}}}}\right)$$
$$+ \frac{\eta_{\text{d}} a^4}{2} \sum_p \sum_{q \neq p} U(p,q)\left[N(p), N(q), R(p,q)\right] , \quad \text{(C.1)}$$

where p and q label islands, $N(p)$ is the number of atoms in the pth island, and $R(p,q) = |\mathbf{R}(p) - \mathbf{R}(q)|$ is the distance between the centers of the pth and qth islands. The coefficient η_{d} is the constant of elastic interaction,

$$\eta_{\text{d}} = \frac{(1-\nu^2)(\Delta\tau)^2}{\pi Y} , \quad \text{(C.2)}$$

and $U(p,q)\left[N(p), N(q), R(p,q)\right]$ is the elastic interaction energy between the two islands calculated in (B.15). In terms of the numbers of atoms in islands, the interaction energy is

$$U(p,q)\left[N(p), N(q), R(p,q)\right] = \frac{N(p)N(q)}{R(p,q)^3} F\left[\frac{a^2 N(p)}{\pi R(p,q)^2}, \frac{a^2 N(q)}{\pi R(p,q)^2}\right] , \quad \text{(C.3)}$$

where $F(\xi_1, \xi_2)$ is the correction factor accounting for the contribution of the higher elastic multipoles to the interaction energy and given by (B.16). For remote islands, the interaction is purely dipole–dipole, and $F \to 1$.

The structure of an array of islands at low temperatures is close to that at $T = 0$. The number of atoms in the islands exhibits small fluctuations around the average value and the positions of the islands fluctuate around positions in the ideal hexagonal superlattice. The partition function of the array of islands is determined by these small fluctuations and can be calculated as a Gaussian integral near the maximum of the exponent. To evaluate this integral, we expand the total energy (C.1) in powers of the fluctuation amplitudes $\delta N(p)$, $\delta X(p)/a$, and $\delta Y(p)/a$. The first order corrections vanish and we expand up to second order terms,

C. Stiffness of a Hexagonal Array of Interacting Strained Disks

$$E_{\text{total}} = E_{\text{total}}^{(0)} + \frac{1}{2}\sum_{p,q}\Phi_{00}(p,q)\delta N(p)\delta N(q)$$

$$+ \frac{1}{2}\sum_{\alpha=1}^{2}\sum_{p,q}\Phi_{\alpha 0}(p,q)\frac{\delta R_\alpha(p)}{a}\delta N(q)$$

$$+ \frac{1}{2}\sum_{\beta=1}^{2}\sum_{p,q}\Phi_{0\beta}(p,q)\delta N(q)\frac{\delta R_\beta(q)}{a}$$

$$+ \frac{1}{2}\sum_{\alpha,\beta=1}^{2}\sum_{p,q}\Phi_{\alpha\beta}(p,q)\frac{\delta R_\alpha(p)}{a}\frac{\delta R_\beta(q)}{a}, \tag{C.4}$$

where $\alpha, \beta = 1, 2$ label in-plane Cartesian coordinates, $R_1 = X$, and $R_2 = Y$. The coefficients Φ form the $3\mathcal{N}\times 3\mathcal{N}$ matrix of second derivatives of the total energy, where $\mathcal{N}$ is the number of islands in the system, and show a certain similarity with the force constants in the dynamics of a crystal lattice. In this comparison, fluctuations in island positions correspond to acoustic vibrations, and fluctuations in the number of atoms in the islands correspond to optical vibrations. Fluctuations in island shape, considered for each island separately in Appendix B, correspond to intramolecular vibrations.

Since all stiffness coefficients $\Phi_{ij}(p,q)$ are defined for an ideal periodic array of equally sized and equally shaped islands, it is convenient to use the Fourier representation and expand all fluctuations in plane waves,

$$\delta N(p) = \frac{1}{\mathcal{N}}\sum_{k\in\text{BZ}}\widetilde{\delta N}(k)\exp[i\mathbf{k}\cdot\mathbf{R}(p)], \tag{C.5a}$$

$$\delta R_\alpha(p) = \frac{1}{\mathcal{N}}\sum_{k\in\text{BZ}}\widetilde{\delta R_\alpha}(k)\exp[i\mathbf{k}\cdot\mathbf{R}(p)]. \tag{C.5b}$$

By substituting the Fourier expansions of the fluctuations from (C.5a) and (C.5b) into (C.4), we obtain the equation for the total energy variation as a sum over the Brillouin zone,

$$E_{\text{total}} = E_{\text{total}}^{(0)} + \frac{1}{2\mathcal{N}}\sum_{k\in\text{BZ}}\left\{D_{00}(k)|\widetilde{\delta N}(k)|^2 + \frac{1}{2}\sum_{\alpha=1}^{2}D_{\alpha 0}(k)\frac{\widetilde{\delta R}_\alpha^*(k)}{a}\widetilde{\delta N}(k)\right.$$

$$\left.+ \frac{1}{2}\sum_{\beta=1}^{2}D_{0\beta}(k)\widetilde{\delta N}^*(k)\frac{\widetilde{\delta R}_\beta(k)}{a} + \frac{1}{2}\sum_{\alpha,\beta=1}^{2}D_{\alpha\beta}(k)\frac{\widetilde{\delta R}_\alpha^*(k)}{a}\frac{\widetilde{\delta R}_\beta(k)}{a}\right\}. \tag{C.6}$$

For practical calculation of the 3×3 matrix $D_{ij}(k)$, it is convenient to use the relationship between the distance R_0 separating the centers of nearest-neighboring circular islands in the hexagonal superlattice, the number of atoms in the islands N, and the surface coverage q. The coverage q is the ratio of the island area to the area of the superlattice unit cell,

C. Stiffness of a Hexagonal Array of Interacting Strained Disks

$$q = \frac{Na^2}{\frac{\sqrt{3}}{2}R_0^2} . \tag{C.7}$$

To obtain the matrix $D_{ij}(\boldsymbol{k})$, we proceed as follows. First, we expand the total energy up to second order terms in the fluctuations $\delta N(p)$ and $\delta R_\alpha(p)$. Then we substitute the Fourier expansions of the fluctuations from (C.5a) and (C.5b) and obtain the expansion of the total energy in the form (C.6). As a result, we obtain the stiffness matrix $D_{ij}(\boldsymbol{k})$ in the form

$$D_{00}(\boldsymbol{k}) = \frac{\eta_d a}{4N^{3/2}} \left\{ \ln\left(e\sqrt{\frac{N}{N_{\text{opt}}}}\right) \right. \tag{C.8a}$$

$$+ \frac{4}{2\sqrt{\pi}} \left(\frac{\sqrt{3}}{2}\right)^{3/2} q^{3/2} \left[\sum_{r(j)\neq 0} \frac{1}{r(j)^3} \left([2\xi F_{10}(\xi) + \xi^2 F_{20}(\xi)] \right. \right.$$

$$\left. \left. + \exp\left[-i\boldsymbol{k} \cdot \boldsymbol{r}(j)\right] \left[F_{00}(\xi) + 2\xi F_{10}(\xi) + \xi^2 F_{11}(\xi)\right] \right) \right] \right\},$$

$$D_{\beta 0}(\boldsymbol{k}) = \frac{\eta_d a}{4N^{3/2}} \times \frac{4}{2\sqrt{\pi}} \left(\frac{\sqrt{3}}{2}\right)^{3/2} q^{3/2} \times \left(\frac{\sqrt{3}}{2}qN\right)^{1/2} \tag{C.8b}$$

$$\times \sum_{r(j)\neq 0} \exp\left[-i\boldsymbol{k} \cdot \boldsymbol{r}(j)\right] \frac{m_\beta(j)}{r(j)^4}$$

$$\times \left[-3F_{00}(\xi) - 9\xi F_{10}(\xi) - 2\xi^2 \left[F_{20}(\xi) + F_{11}(\xi)\right]\right] ,$$

$$D_{0\alpha}(\boldsymbol{k}) = [D_{\alpha 0}(\boldsymbol{k})]^* , \tag{C.8c}$$

$$D_{\alpha\beta}(\boldsymbol{k}) = \frac{\eta_d a}{4N^{3/2}} \times \frac{4}{2\sqrt{\pi}} \left(\frac{\sqrt{3}}{2}\right)^{3/2} q^{3/2} \times \left(\frac{\sqrt{3}}{2}qN\right) \tag{C.8d}$$

$$\times \sum_{r(j)\neq 0} \frac{1 - \exp\left[-i\boldsymbol{k} \cdot \boldsymbol{r}(j)\right]}{r(j)^5} \times \left\{ [-3\delta_{\alpha\beta} + 15m_\alpha(j)m_\beta(j)] \right.$$

$$\times F_{00}(\xi) + 4\left[10m_\alpha(j)m_\beta(j) - \delta_{\alpha\beta}\right]\xi F_{10}(\xi)$$

$$\left. + 8m_\alpha(j)m_\beta(j)\xi^2 \left[F_{20}(\xi) + F_{11}(\xi)\right] \right\} .$$

Summations are carried out over all islands except the given 0th island, $j \neq 0$. The position vector $\boldsymbol{r}(j)$ of the jth island is given in units of R_0. The vector $m_\alpha(j)$ is the unit vector directed from the 0th island to the jth island,

$$m_\alpha(j) = \frac{r_\alpha(j)}{r(j)} . \tag{C.9}$$

The quantities ξ depend on the distance between the islands R_j,

$$\xi(j) = \frac{\sqrt{3}}{2\pi} q \frac{1}{r(j)^2} . \tag{C.10}$$

The quantities $F_{ij}(\xi)$ are related to the correction factor F, which describes the deviation of the exact interaction energy between two islands from the dipole–dipole interaction and is given by (B.16). The quantities $F_{ij}(\xi)$ are defined by

$$F_{00}(\xi) = F(\xi_1, \xi_2)\Big|_{\xi_1=\xi_2=\xi} , \tag{C.11a}$$

$$F_{10}(\xi) = \frac{\partial F(\xi_1, \xi_2)}{\partial \xi_1}\Big|_{\xi_1=\xi_2=\xi} , \tag{C.11b}$$

$$F_{20}(\xi) = \frac{\partial^2 F(\xi_1, \xi_2)}{\partial \xi_1^2}\Big|_{\xi_1=\xi_2=\xi} , \tag{C.11c}$$

$$F_{11}(\xi) = \frac{\partial^2 F(\xi_1, \xi_2)}{\partial \xi_1 \partial \xi_2}\Big|_{\xi_1=\xi_2=\xi} . \tag{C.11d}$$

Equations (C.8a)–(C.8d) are used in Sect. 3.2 to evaluate the root mean square fluctuations in the number of atoms in islands, the distance between nearest-neighboring islands, and the entropy of the island fluctuations in volumes and positions.

References

Chapter 1

1.1 http://nobelprizes.com/
1.2 S. Luryi, J. Xu, A. Zaslavsky (Eds.): *Future Trends in Microelectronics. The Road Ahead* (Wiley, New York 1999)
1.3 The data has been collected and synthesized from numerous on-line sources, i.e., http://www.siliconstrategies.com, http://www.news.com, http://www.imia.com
1.4 N.N. Ledentsov: In *Proceedings of the 23rd International Conference on Physics of Semiconductors*, Berlin, Germany, July 22–27, 1996, eds. M. Scheffler and R. Zimmermann (World Scientific, Singapore 1996) Vol. **1**, p. 21
1.5 D. Bimberg, M. Grundmann, N.N. Ledentsov: *Quantum Dot Heterostructures* (Wiley, Chichester 1998)
1.6 S.Y. Chou, P.R. Krauss, L. Long: J. Appl. Phys. **79**, 6101 (1996)
1.7 IBM J. Res. Develop. **37**, 288 (1993). Special issue on X-ray lithography
1.8 R.L. Kubena, F.P. Stratton, J.W. Ward, G.M. Atkinson, R.J. Joyce: J. Vac. Sci. Technol. B **7**, 1798 (1989)
1.9 U. Drodofsky, J. Stuhler, T. Schulze, M. Drewsen, B. Brezger, T. Pfau, J. Mlynek: Appl. Phys. B **65**, 755 (1997)
1.10 D.M. Eigler, E.K. Schweizer: Nature **344**, 524 (1990)
1.11 A.D. Kent, D.M. Shaw, S.V. Molnar, D.D. Awschalom: Science **262**, 1249 (1993)
1.12 S.C. Minne, Ph. Flueckiger, H.T. Soh, C.F. Quate: J. Vac. Sci. Technol. B **13**, 1380 (1995)
1.13 S.C. Minne, J.D. Adams, G. Yaralioglu, S.R. Manalis, A. Atalar, C.F. Quate: Appl. Phys. Lett. **73**, 1742 (1998)
1.14 M. Despont, J. Brugger, U. Drechsler, U. Dürig, W. Häberle, M. Lutwyche, H. Rothuizen, R. Stutz, R. Widmer, H. Rohrer, G. Binnig, P. Vettiger: in *Technical Digest of Twelfth IEEE International Micro Electro Mechanical Systems Conference EMS 99* (IEEE, Piscataway 1999) p. 564
1.15 H. Haken: *Synergetics, an Introduction: Nonequilibrium Phase Transitions and Self-Organization in Physics, Chemistry, and Biology*, 3rd edn. (Springer, Berlin 1983)
1.16 C. Teichert: Phys. Rep. **365**, 335 (2002)
1.17 S.K. Scott: *Oscillations, Waves, Chaos in Chemical Kinetics* (Oxford University Press, Oxford 1994)
1.18 H. Nishimori, N. Ouchi: Phys. Rev. Lett. **71**, 197 (1993)
1.19 H. Haken, J. Portugali: Environ. Plann. B Plann. Des. **22**, 35 (1993)
1.20 D. Helbing, P. Molnar: in *Self-Organization of Complex Structures. From Individual to Collective Dynamics* (Gordon and Breach, London 1997), p. 569
1.21 B.S. Kerner: Phys. Rev. Lett. **81**, 3797 (1998)

1.22 N.N. Ledentsov, M. Grundmann, N. Kirstaedter, O. Schmidt, R. Heitz, J. Böhrer, D. Bimberg, V.M. Ustinov, V.A. Shchukin, P.S. Kop'ev, Zh.I. Alferov, S.S. Ruvimov, A.O. Kosogov, P. Werner, U. Richter, U. Gösele, J. Heydenreich: In *Proceedings of the 7th International Conference on Modulated Semiconductor Structures*, Madrid, Spain, July 1995, Solid State Electron. **40**, 785 (1996)

1.23 F. Hatami, N.N. Ledentsov, M. Grundmann, J. Böhrer, F. Heinrichsdorff, M. Beer, D. Bimberg, S.S. Ruvimov, P. Werner, U. Gösele, J. Heydenreich, S.V. Ivanov, B.Ya. Meltser, P.S. Kop'ev, Zh.I. Alferov: Appl. Phys. Lett. **67**, 656 (1995)

1.24 Yu.G. Musikhin, D. Gethsen, D.A. Bedarev, N.A. Bert, W.V. Lundin, A.F. Tsatsul'nikov, A.V. Sakharov, A.S. Usikov, Zh.I. Alferov, I.L. Krestnikov, N.N. Ledentsov, A. Hoffmann, D. Bimberg: Appl. Phys. Lett. **80**, 2099 (2002)

1.25 M. Strassburg, V. Kutzer, U.W. Pohl, A. Hoffmann, I. Broser, N.N. Ledentsov, D. Bimberg, A. Rosenauer, U. Fischer, D. Gerthsen, I.L. Krestnikov, M.V. Maximov, P.S. Kop'ev, Zh.I. Alferov: Appl. Phys. Lett. **72**, 942 (1998)

1.26 A.G. Khachaturyan: *Theory of Phase Transformations and the Structure of Solid Solutions* (Nauka, Moscow 1974), in Russian

1.27 A.G. Khachaturyan: *Theory of Structural Transformations in Solids* (Wiley, New York 1983)

1.28 A. Pimpinelli, J. Villain: *Physics of Crystal Growth* (Cambridge University Press, Cambridge 1998)

1.29 N.N. Ledenstov: *Growth Processes and Surface Phase Equilibria in Molecular Beam Epitaxy*, Springer Tracts in Modern Physics **156** (Springer, Berlin 1999)

1.30 M. Zinke-Allmang: Thin Solid Films **346**, 1 (1999)

1.31 J.L. Merz, A.-L. Barabási, J.K. Furdyna, R.S. Williams: in *Future Trends in Microelectronics. The Road Ahead*, ed. by S. Luryi, J. Xu, and A. Zaslavsky (Wiley, New York 1999) p. 237

1.32 V.A. Shchukin, D. Bimberg: Rev. Mod. Phys. **71**, 1125 (1999)

1.33 P. Politi, G. Grenet, A. Marty, A. Ponchet, J. Villain: Phys. Rep. **324**, 271 (2000)

1.34 I.L. Krestnikov, N.N. Ledentsov, A. Hoffmann, D. Bimberg: phys. stat. sol. (a) **183**, 207 (2001)

Chapter 2

2.1 D.M. Eigler, E.K. Schweizer: Nature **344**, 524 (1990)

2.2 R. Heckingbottom: In *Molecular Beam Epitaxy and Heterostructures*, ed. by L.L. Chang and K. Ploog (Nijhoff, Dordrecht 1985)

2.3 J.G. Amar, F. Family: Phys. Rev. Lett. **74**, 2066 (1995)

2.4 V. Bressler-Hill, S. Varma, A. Lorke, B.Z. Nosho, P.M. Petroff, W.H. Weinberg: Phys. Rev. Lett. **74**, 3209 (1995)

2.5 S. Liu, L. Bönig, J. Detch, H. Metiu: Phys. Rev. Lett. **74**, 4495 (1995)

2.6 G.S. Bales, D.C. Chrzan: Phys. Rev. Lett. **74**, 4879 (1995)

2.7 A. Madhukar, P. Chen, Q. Xie, A. Konkar, T.R. Ramachandran, N.P. Kobayashi, R. Viswanathan: In *Low Dimensional Structures Prepared by Epitaxial Growth or Regrowth on Patterned Substrates*, Proceedings of the NATO Advanced Workshop, February 20–24, 1995, Ringberg Castle, Germany, ed. by K. Eberl, P. Petroff, and P. Demeester (Kluwer, Dordrecht 1995) p. 19

2.8 K.M. Chen, D.E. Jesson, S.J. Pennycook, T. Thundat, R.J. Warmack: Proc. Mat. Res. Soc. Symp. **399**, 271 (1995)
2.9 N. Kobayashi, T.R. Ramachandran, P. Chen, A. Madhukar: Appl. Phys. Lett. **68**, 3299 (1996)
2.10 Y. Chen, J. Washburn: Phys. Rev. Lett. **77**, 4046 (1996)
2.11 A.-L. Barabási: Appl. Phys. Lett. **70**, 2565 (1997)
2.12 H.T. Dobbs, D.D. Vvedensky, A. Zangwill, J. Johansson, N. Carlsson, W. Seifert: Phys. Rev. Lett. **79**, 897 (1997)
2.13 D.E. Jesson, G. Chen, K.M. Chen, S.J. Pennycook: Phys. Rev. Lett. **80**, 5156 (1998)
2.14 A.F. Andreev: Pis'ma Zh. Eksp. Teor. Fiz. **32**, 654 (1980) [JETP Letters **32**, 640 (1980)]
2.15 A.F. Andreev: Zh. Eksp. Teor. Fiz. **80**, 2042 (1980) [Sov. Phys. JETP **53**, 1063 (1981)]
2.16 V.I. Marchenko: Zh. Eksp. Teor. Fiz. **81**, 1141 (1981) [Sov. Phys. JETP **54**, 605 (1981)]
2.17 V.I. Marchenko: Pis'ma Zh. Eksp. Teor. Fiz. **33**, 397 (1981) [JETP. Lett. **33**, 381 (1981)]
2.18 V.I. Marchenko: Pis'ma Zh. Eksp. Teor. Fiz. **55**, 72 (1992) [JETP Lett. **55**, 73 (1992)]
2.19 D. Vanderbilt: Surf. Sci. **268**, L300 (1992)
2.20 J. Tersoff, R.M. Tromp: Phys. Rev. Lett. **70**, 2782 (1993)
2.21 V.A. Shchukin, A.I. Borovkov, N.N. Ledentsov, D. Bimberg: Phys. Rev. B **51**, 10104 (1995)
2.22 V.A. Shchukin, A.I. Borovkov, N.N. Ledentsov, P.S. Kop'ev: Phys. Rev. B **51**, 17767 (1995)
2.23 K.-O. Ng, D. Vanderbilt: Phys. Rev. B **52**, 2177 (1995)
2.24 V.A. Shchukin, N.N. Ledentsov, P.S. Kop'ev, D. Bimberg: Phys. Rev. Lett. **75**, 2968 (1995)
2.25 N.N. Ledentsov, M.V. Maximov, P.S. Kop'ev, V.M. Ustinov, M.V. Belousov, B.Ya. Meltser, S.V. Ivanov, V.A. Shchukin, Zh.I. Alferov, M. Grundmann, D. Bimberg, S.S. Ruvimov, W. Richter, P. Werner, U. Gösele, U. Heidenreich, P.D. Wang, C.M. Sotomayor Torres: Microelectronics Journal **26**, 871 (1995)
2.26 N.N. Ledentsov, M. Grundmann, N. Kirstaedter, O. Schmidt, R. Heitz, J. Böhrer, D. Bimberg, V.M. Ustinov, V.A. Shchukin, A.Yu. Egorov, A.E. Zhukov, S. Zaitsev, P.S. Kop'ev, Zh.I. Alferov, S.S. Ruvimov, A.O. Kosogov, P. Werner, U. Gösele, J. Heydenreich: Solid State Electron. **40**, 785 (1996)
2.27 I. Daruka, A.-L. Barabási: Phys. Rev. Lett. **79**, 3708 (1997)
2.28 A.Y. Cho, J.R. Arthur Jr.: Progr. Solid State Chem. **10**, 157 (1975)
2.29 C.T. Foxon: Acta Electron. **21**, 139 (1978)
2.30 E.G. Scott, D.A. Andrews, G.J. Davies: J. Vac. Sci. Technol. B **4**, 534 (1986)
2.31 C.E.C. Wood, C.R. Stanley, G.W. Wicks, M.B. Esi: J. Appl. Phys. **54**, 1868 (1983)
2.32 A. Madhukar: Surf. Sci. **132**, 344 (1983)
2.33 S. Clarke, D.D. Vvedensky: Appl. Phys. Lett. **51**, 340 (1987)
2.34 R. Heckingbottom, C.J. Todd, G.J. Davies: J. Electrochem. Soc. **127**, 444 (1980)
2.35 R. Heckingbottom, G. Davies: J. Cryst. Growth **50**, 644 (1980)
2.36 R. Heckingbottom, G.J. Davies, K.A. Prior: Surf. Sci. **132**, 375 (1983)
2.37 R. Heckingbottom: In *Molecular Beam Epitaxy and Heterostructures*, ed. by L.L. Chang and K. Ploog (Nijhoff, Dordrecht 1985)
2.38 H. Seki, A. Koukitu: J. Cryst. Growth **78**, 342 (1986)

2.39 P.S. Kop'ev, N.N. Ledentsov: Sov. Phys. Semicond. **22**, 1093 (1988)
2.40 S.V. Ivanov, P.S. Kop'ev, N.N. Ledentsov: J. Cryst. Growth **104**, 345 (1990)
2.41 S.V. Ivanov, P.S. Kop'ev, N.N. Ledentsov: J. Cryst. Growth **111**, 151 (1991)
2.42 P.S. Kop'ev, S.V. Ivanov, A.Yu. Egorov, D.Yu. Uglov: J. Cryst. Growth **96**, 533 (1989)
2.43 J. Shen, C. Chatillon: J. Cryst. Growth **106**, 543 (1990)
2.44 S.Yu. Karpov, Yu.V. Kovalchuk, V.E. Myachin, Yu.V. Pogorelski: J. Crys. Growth **129**, 563 (1993)
2.45 F.A. Kröger: *The Chemistry of Imperfect Crystals* (Wiley, New York 1964)
2.46 C.T. Foxon, J.J. Harris, D. Hilton, J. Hewett, C. Roberts: Semicond. Sci. Technol. **4**, 582 (1989)
2.47 Zh.I. Alferov, S.V. Ivanov, P.S. Kop'ev, N.N. Ledentsov, B.Ya. Mel'tser, M.I. Nemenov, V.M. Ustinov: Fiz. Tekh. Poluprovodn. **24**, 152 (1989) [Sov. Phys. Semicond. **24**, 92 (1989)]
2.48 Y. Horikoshi, M. Kawashima, H. Yamaguchi: Jap. J. Appl. Phys. **27**, 169 (1988)
2.49 L.D. Landau, E.M. Lifshitz: *Statistical Physics*. Part I (Pergamon, New York 1980)
2.50 D.T.J. Hurle: J. Phys. Chem. Solids **40**, 613 (1979)
2.51 M. Yoshida, K. Watanabe: J. Electrochem. Soc. **132**, 1733 (1985)
2.52 J.A. van Vechten: J. Electrochem. Soc. **122**, 419 (1975)
2.53 M.B. Panish, M. Ilegems: In *Progress in Solid State Chemistry*, Vol. 7, ed. by H. Reiss and J. McCaldin (Pergamon, Oxford 1972)
2.54 M. Tmar, A. Gabriel, C. Chatillon, I. Ansara: J. Cryst. Growth **68**, 557 (1984)
2.55 M. Tmar, A. Gabriel, C. Chatillon, I. Ansara: J. Cryst. Growth **69**, 421 (1984)
2.56 G.B. Stringfellow: J. Cryst. Growth **27**, 21 (1974)
2.57 N.N. Ledentsov: *Growth Processes and Surface Phase Equilibria in Molecular Beam Epitaxy*, Springer Tracts in Modern Physics **156** (Springer, Berlin 1999)
2.58 F. Turco, J. Massies, G.P. Contour: Rev. Phys. Appl. **22**, 827 (1987)
2.59 T. Nakagawa, S. Gonda, S. Emura, S. Shimazu: J. Cryst. Growth **87**, 276 (1988)
2.60 D. Dutartre, M. Gavand: J. Cryst. Growth **66**, 647 (1984)
2.61 G.B. Stringfellow: *Organometallic Vapor-Phase Epitaxy: Theory and Practice*, 2nd edn. (Academic Press, San Diego 1999)
2.62 H.-J. Güntherodt, R. Wiesendanger (Eds.): *Scanning Tunneling Microscopy I*, Springer Series in Surface Sciences **20** (Springer, Berlin 1994)
2.63 C.I. Chen: *Introduction to Scanning Tunneling Microscopy* (Oxford University Press, Oxford 1993)
2.64 P. Hirsch, A. Howie, R. Nicolson, D. Pashley, M. Whelan: *Electron Microscopy of Thin Crystals* (Butterworths, Washington 1965)
2.65 J.C.H. Spencer: *Experimental High-Resolution Electron Microscopy* (Clarendon Press, Oxford 1980)
2.66 L. Reimer: *Transmission Electron Microscopy*, Springer Series in Optical Sciences **36** (Springer, Berlin 1984)
2.67 H. Cerva, H. Oppolzer: Progr. Cryst. Growth and Charact. **20**, 231 (1990)
2.68 A. Ourmazd, F.H. Baumann, M. Bode, Y. Kim: Ultramicroscopy **34**, 237 (1990)
2.69 D. Bimberg, F. Heinrichsdorff, R.K. Bauer, D. Gerthsen, D. Stenkamp, D.E. Mars, J.N. Miller: J. Vac. Sci. Technol. B **10**, 1793 (1992)

2.70 W. Neumann, H. Hofmeister, D. Conrad, K. Scheerschmidt, S. Ruvimov: Z. Kristallogr. **211**, 147 (1996)
2.71 B.A. Joyce, J.H. Neave, P.J. Dobson, P.K. Larsen: Phys. Rev. B **29**, 814 (1984)
2.72 P.K. Larsen, P.J. Dobson (Eds.): *Reflection High Energy Electron Diffraction and Electron Imaging of Surfaces*, NATO ASI Series B: Physics **188** (Plenum Press, New York 1988)
2.73 D.E. Aspnes: J. Vac. Sci. Technol. B **3**, 1498 (1985)
2.74 W. Richter, D.R.T. Zanh: In *Optical Characterization of Epitaxial Semiconductor Layers*, ed. by G. Bauer and W. Richter (Springer, Berlin 1996) p. 12.
2.75 W. Richter: Appl. Phys. A **75**, 129 (2002)
2.76 G.B. Bokii, V.B. Lazarev (Eds.): *Crystallochemistry Problems in Semiconductor Materials Research* (Nauka, Moscow 1975) in Russian
2.77 W.J. Bartels, W. Nijman: J. Cryst. Growth **44**, 518 (1978)
2.78 L. Tapfer, K. Ploog: Phys. Rev. B **33**, 5565 (1986)
2.79 A. Segmüller, I.C. Noyan, V.S. Speriosu: Progr. Cryst, Growth and Charact. **18**, 21 (1989)
2.80 A. Krost, F. Heinrichsdorff, D. Bimberg, A. Darhuber, G. Bauer: Appl. Phys. Lett. **68**, 785 (1996)
2.81 C. Teichert: Phys. Rep. **365**, 335 (2002)
2.82 J.M. Elson, J.M. Bennett: J. Opt. Soc. Am. **69**, 31 (1979)
2.83 J.D. Kiely, D.A. Bonnell: J. Vac. Sci. Technol. B **15**, 1483 (1997)
2.84 R. Guckenberger: In *Procedures of Scanning Probe Microscopies*, ed. by R.J. Colto, A. Engel, J.E. Frommer, H.E. Gaub, A.A. Gewirth, R. Guckenberger, J. Rabe, W.M. Heckel, B. Parkinson (Wiley, Chichester 1998) p. 24.
2.85 M.A. Lutz, R.M. Feenstra, P.M. Mooney, J. Tersoff, J.O. Chu: Surf. Sci. **316**, L1075 (1994)
2.86 J.S. Villarrubia: J. Res. Natl. Inst. Stand. Technol. **102**, 425 (1997)
2.87 R. Leon, J. Wellman, X.Z. Liao, J. Zuo, D.J.H. Cockayne: Appl. Phys. Lett. **76**, 1558 (2000)
2.88 V.A. Shchukin, N.N. Ledentsov, V.M. Ustinov, Yu.G. Musikhin, V.B. Volovik, A. Schliwa, O. Stier, R. Heitz, D. Bimberg: In *Morphological and Compositional Evolution of Heteroepitaxial Semiconductor Thin Films*, ed. by A.-L. Barabási, E. Jones, and J. Mirecki Millunchick, Mat. Res. Soc. Symp. Proc. **618** (Pittsburgh 2000) p. 79
2.89 N.N. Ledentsov, V.A. Shchukin, D. Bimberg, V.M. Ustinov, N.A. Cherkashin, Yu.G. Musikhin, B.V. Volovik, G.E. Cirlin, Zh.I. Alferov: Semicond. Sci. and Technol. **16**, 502 (2001)
2.90 Y.-W. Mo, D.E. Savage, B.S. Swartzentruber, M.G. Lagally: Phys. Rev. Lett. **65**, 1020 (1990)
2.91 W. Wu, J.R. Tucker, G.S. Solomon, J.S. Harris Jr.: Appl. Phys. Lett. **71**, 1083 (1997)
2.92 H. Eisele, O. Flebbe, T. Kalka, C. Preinesberger, F. Heinrichsdorff, A. Krost, D. Bimberg, M. Dähne-Prietsch: Appl. Phys. Lett. **75**, 106 (1999)
2.93 N. Liu, J. Tersoff, O. Baklenov, A.L. Holmes Jr., C.K. Shih: Phys. Rev. Lett. **84**, 334 (2000)
2.94 J. Tersoff: Phys. Rev. Lett. **52**, 465 (1984)
2.95 H. Eisele: *Cross-Sectional Scanning Tunneling Microscopy of InAs/GaAs Quantum Dots*, Berlin Studies in Solid State Physics **10** (Wissenschaft und Technik Verlag, Berlin 2002)
2.96 A. Rosenauer: Habilitationsschrift, Universität Karlsruhe (2001)
2.97 P. Doyle, P. Turner: Acta Cryst. **A24**, 390 (1968)
2.98 G. Smith, R. Burge: Acta Cryst. **15**, 182 (1962)

2.99 P.M. Petroff: J. Vac. Sci. Technol. **14**, 973 (1977)
2.100 S. Guha, A. Madhukar, K.C. Rajkumar: Appl. Phys. Lett. **57**, 2110 (1990)
2.101 S.S. Ruvimov, P. Werner, K. Scheerschmidt, U. Richter, U. Gösele, J. Heydenreich, N.N. Ledentsov, M. Grundmann, D. Bimberg, V.M. Ustinov, A.Yu. Egorov, P.S. Kop'ev, Zh.I. Alferov: Phys. Rev. B **51**, 14766 (1995)
2.102 G. Möllenstedt, H. Düker: Z. Physik **145**, 377 (1965)
2.103 J.M. Gibson, R. Hull, J.C. Bean, M.M.J. Treacy: Appl. Phys. Lett. **46**, 649 (1985)
2.104 M.M.J. Treacy, J.M. Gibson: J. Vac. Sci. Technol. **4**, 1458 (1986)
2.105 S.S. Ruvimov, K. Scheerschmidt: phys. stat. sol. (a) **150**, 471 (1995)
2.106 Q. Xie, P. Chen, A. Madhukar: Appl. Phys. Lett. **65**, 2051 (1994)
2.107 K. Muraki, S. Fukatsu, Y. Shiraki, R. Ito: Appl. Phys. Lett. **61**, 557 (1992)
2.108 U. Woggon, W. Langbein, J.M. Hvam, A. Rosenauer, T. Remmele, D. Gerthsen: Appl. Phys. Lett. **71**, 377 (1997)
2.109 A. Rosenauer, W. Oberst, D. Litvinov, D. Gerthsen, A. Förster, R. Schmidt: Phys. Rev. B **61**, 8276 (2000)
2.110 J.M. Garcia, G. Medeiros-Ribeiro, K. Schmidt, T. Ngo, J.L. Feng, A. Lorke, J. Kotthaus, P.M. Petroff: Appl. Phys. Lett. **71**, 2014 (1997)
2.111 G.D. Lian, J. Yuan, L.M. Brown, G.H. Kim, D.A. Ritchie: Appl. Phys. Lett. **73**, 49 (1998)
2.112 O. Brandt, K. Ploog, R. Bierwolf, M. Hohenstein: Phys. Rev. Lett. **68**, 1339 (1992)
2.113 R. Bierwolf, M. Hohenstein, F. Phillipp, O. Brandt, G. Crook, K. Ploog: Ultramicroscopy **49**, 273 (1993)
2.114 S. Paciornik, R. Kilaas, U. Dahmen: Ultramicroscopy **50**, 255 (1993)
2.115 P.H. Jouneau, A. Tardot, G. Feulliet, H. Marietta, J. Gibert: J. Appl. Phys. **75**, 7310 (1994)
2.116 H. Seitz, M. Seibt, F. Baumann, K. Ahlborn, W. Schröter: phys. stat. sol. (a) **150**, 625 (1995)
2.117 A. Rosenauer, S. Kaiser, T. Reisinger, J. Zweck, W. Gebhardt, D. Gerthsen: Optik (Stuttgart) **102**, 63 (1996)
2.118 D. Gerthsen, E. Hahn, B. Neubauer, A. Rosenauer, O. Schön, M. Heuken, A. Rizzi: phys. stat. sol. (a) **177**, 145 (2000)
2.119 W. Press, B. Flannery, S. Teukolsky, W. Vetterling: *Numerical Recipes in C. The Art of Scientific Computing* (Cambridge University Press, Cambridge 1988)
2.120 L.D. Marks: Ultramicroscopy **62**, 43 (1996)
2.121 M. Strassburg, V. Kutzer, U.W. Pohl, A. Hoffmann, I. Broser, N.N. Ledentsov, D. Bimberg, A. Rosenauer, U. Fischer, D. Gerthsen, I.L. Krestnikov, M.V. Maximov, P.S. Kop'ev, and Zh.I. Alferov: Appl. Phys. Lett. **72**, 942 (1998)
2.122 T.I. Kamins, E.C. Carr, R.S. Williams, S.J. Rosner: J. Appl. Phys. **81**, 211 (1997)
2.123 R.L. Park, M.G. Lagally (Eds.): *Methods of Experimental Physics*, Vol. 22: *Surfaces* (Academic Press, Orlando 1985) Chap. 5
2.124 M.G. Lagally, D.E. Savage, M.C. Tringides: in *Reflection High-Energy Electron Diffraction and Reflection Electron Imaging of Surfaces*, ed. by P.K. Larsen and P.J. Dobson, NATO ASI Series, Series B: Physics **188**, 139 (1988)
2.125 K. Hingerl, D.E. Aspnes, I. Kamiya, L.T. Florez: Appl. Phys. Lett. **63**, 885 (1993)
2.126 G. Bauer, W. Richter (Eds.): *Optical Characterization of Epitaxial Semiconductor Layers* (Springer, Berlin, Heidelberg 1996)

2.127 J.-T. Zettler: Progr. Cryst. Growth Charact. Mater. **35**, 27 (1997)
2.128 D.E. Aspnes: Mater. Sci. Eng. B **30**, 109 (1995)
2.129 I. Kamiya, D.E. Aspnes, H. Tanaka, L.T. Florez, J.P. Harbison, R. Bhat: Phys. Rev. Lett. **68**, 627 (1992)
2.130 F. Reinhardt, W. Richter, A.B. Müller, D. Gursche, P. Kurpas, K. Ploska, K.C. Rose, M. Zorn: J. Vac. Sci. Technol. B **11**, 1427 (1993)
2.131 J.M. Olson, A. Kibbler: J. Cryst. Growth **77**, 182 (1986)
2.132 A.A. Darhuber, V. Holy, J. Stangl, G. Bauer, A. Krost, F. Heinrichsdorff, M. Grundmann, D. Bimberg, V.M. Ustinov, P.S. Kop'ev, A.O. Kosogov, P. Werner: Appl. Phys. Lett. **70**, 955 (1997)
2.133 A.A. Darhuber, P. Schittenhelm, V. Holy, J. Stangl, G. Bauer, G. Abstreiter: Phys. Rev. B **55**, 15652 (1997)
2.134 D. Bimberg, M. Grundmann, N.N. Ledentsov: *Quantum Dot Heterostructures* (Wiley, Chichester 1998)

Chapter 3

3.1 H. Haken: *Synergetics. An Introduction: Nonequilibrium Phase Transitions and Self-Organization in Physics, Chemistry and Biology*, 3rd edn. (Springer, Berlin 1983)
3.2 H. Haken: *Advanced Synergetics. Instability Hierarchies of Self-Organizing Systems and Devices*, 2nd edn. (Springer, Berlin 1987)
3.3 A. Madhukar: Surf. Sci. **132**, 344 (1983)
3.4 S. Mukherjee, E. Pehlke, J. Tersoff: Phys. Rev. B **49**, 1919 (1994)
3.5 E.D. Williams, R.J. Phaneuf, J. Wei, N.C. Bartelt, T.L. Einstein: Surf. Sci. **294**, 219 (1993); Surf. Sci. **310** 451 (1994)
3.6 D.-J. Liu, J.D. Weeks, M.D. Johnson, E.D. Williams: Phys. Rev. B **55**, 7653 (1997)
3.7 C. Teichert: Phys. Rep. **365**, 335 (2002)
3.8 A. Konkar, A. Madhukar, P. Chen: Appl. Phys. Lett. **72**, 220 (1998)
3.9 G. Jin, J.L. Liu, S.G. Thomas, Y.H. Luo, K.L. Wang, B.Y. Nguyen: Appl. Phys. Lett. **75**, 2752 (1999)
3.10 J.H. Li, V. Holý, M. Meduna, S.C. Moss, A.G. Norman, A. Mascarenhas, J.L. Reno: Phys. Rev. B **66**, 115312 (2002)
3.11 G. Wulff: Z. Kristallogr. Mineral. **34**, 449 (1901)
3.12 C. Herring: Phys. Rev. **82**, 87 (1951)
3.13 A.A. Chernov: Uspekhi Fiz. Nauk **73**, 277 (1961) [Sov. Phys. Uspekhi **4**, 116 (1961)]
3.14 W.W. Mullins: In *Metal Surfaces: Structure, Energetics and Kinetics* (American Society for Metals, Metals Park 1963) p. 17
3.15 C. Rotman, M. Wortis: Phys. Reports **103**, 59 (1984)
3.16 D.J. Eaglesham, A.E. White, L.C. Feldman, N. Moriya, D.C. Jacobson: Phys. Rev. Lett. **70**, 1643 (1993)
3.17 N.C. Bartelt, T.L. Einstein, C. Rottman: Phys. Rev. Lett. **66**, 961 (1991)
3.18 J.E. Métois, J.C. Heyraud: Ultramicroscopy **31**, 73 (1989)
3.19 H.M. van Pinxteren, J.W.M. Frenken: Europhys. Lett. **21**, 43 (1993)
3.20 J.W. Gibbs: *Collected Works*, Vol. 1, *Thermodynamics* (Longmans, London 1928)
3.21 V.I. Marchenko, A.Ya. Parshin: Zh. Eksp. Teor. Fiz. **79**, 257 (1980) [Sov. Phys. JETP **52**, 129 (1980)]
3.22 R.J. Needs: Phys. Rev. Lett. **58**, 53 (1987)

3.23 R. Shuttleworth: Proc. Phys. Soc. London, Sect. A **63**, 444 (1950)
3.24 C. Herring: *The Physics of Powder Metallurgy*, ed. by W.E. Kingston (McGraw-Hill, New York 1951)
3.25 V. Fiorentini, M. Methfessel, M. Scheffler: Phys. Rev. Lett. **71**, 1051 (1993)
3.26 A. Garcia, J.E. Northrup: Phys. Rev. B **71**, 17350 (1993)
3.27 J. Dabrowski, E. Pehlke, M. Scheffler: Phys. Rev. B **49** 4790 (1994)
3.28 A.F. Andreev, Yu.A. Kosevich: Zh. Eksp. Teor. Fiz. **81**, 1435 (1981) [Sov. Phys. JETP **54**, 761 (1981)]
3.29 P. Nozières, D.E. Wolf: Z. Phys. B **70**, 399 (1988); Z. Phys. B **70**, 507 (1988)
3.30 Yu.A. Kosevich: Progr. Surf. Sci. **55**, 1 (1997)
3.31 L.D. Landau, E.M. Lifshitz: *Theory of Elasticity* (Pergamon, New York 1959)
3.32 Y. Saito, H. Uemura, M. Uwaha: Phys. Rev. B **63**, 045422 (2001)
3.33 D.J. Cheng, R.F. Wallis, L. Dobrzynsky: Surf. Sci. **43**, 400 (1974)
3.34 A.F. Andreev: Pis'ma Zh. Eksp. Teor. Fiz. **32**, 654 (1980) [JETP Letters **32**, 640 (1980)]
3.35 A.F. Andreev: Zh. Eksp. Teor. Fiz. **80**, 2042 (1980) [Sov. Phys. JETP **53**, 1063 (1981)]
3.36 V.I. Marchenko: Zh. Eksp. Teor. Fiz. **81**, 1141 (1981) [Sov. Phys. JETP **54**, 605 (1981)]
3.37 R. Nötzel, N.N. Ledentsov, L. Däweritz, M. Hohenstein, K. Ploog: Phys. Rev. Lett. **67**, 3812 (1991)
3.38 R. Nötzel, N.N. Ledentsov, L. Däweritz, K. Ploog, M. Hohenstein: Phys. Rev. B **45**, 3507 (1992)
3.39 Zh.I. Alferov, A.Yu. Egorov, A.E. Zhukov, S.V. Ivanov, P.S. Kop'ev, N.N. Ledentsov, B.Ya. Mel'tser, V.M. Ustinov: Fiz. Tekh. Poluprovodn. **26**, 1715 (1992) [Sov. Phys. Semicond. **26**, 959 (1992)]
3.40 V.A. Shchukin, A.I. Borovkov, N.N. Ledentsov, P.S. Kop'ev: Phys. Rev. B **51**, 17767 (1995)
3.41 M. Kasu, N. Kobayashi: Appl. Phys. Lett. **62**, 1262 (1993)
3.42 V.A. Shchukin, A.I. Borovkov, N.N. Ledentsov, D. Bimberg: Phys. Rev. B **51**, 10104 (1995)
3.43 Z.V. Popovic, M.B. Vukmirovic, Y.S. Raptis, E. Anastassakis, R. Nötzel, K. Ploog: Phys. Rev. B **52**, 5789 (1995)
3.44 A.J. Shields, R. Nötzel, M. Cardona, L. Däweritz, K. Ploog: Appl. Phys. Lett. **60**, 2537 (1992)
3.45 Z.V. Popovic, E. Richter, J. Spitzer, M. Cardona, A.J. Shields, R. Nötzel, K. Ploog: Phys. Rev. B **49**, 7577 (1994)
3.46 E. Tournié, R. Nötzel, K.H. Ploog: Phys. Rev. B **49**, 11053 (1994)
3.47 M. Ilg, R. Nötzel, K. Ploog: Appl. Phys. Lett. **62**, 1472 (1993)
3.48 R. Nötzel, H.-P. Schönherr, Z. Niu, L. Däweritz, K. Ploog: J. Cryst. Growth **201/202**, 814 (1999)
3.49 R. Nötzel, D. Essler, M. Hohenstein, K. Ploog: J. Appl. Phys. **74**, 431 (1993)
3.50 M. Wassermeier, J. Sudijono, M.D. Johnson, K.T. Leung, B.C. Orr, L. Däweritz, K. Ploog: Phys. Rev. B **51**, 14721 (1995)
3.51 P. Moriarty, Y.-R. Ma, A.W. Dunn, P.H. Beton, M. Henini, C. McGinley, E. McLoughlin, A.A. Cafolla, G. Hughes, S. Dowes, D. Teehan, B. Murphy: Phys. Rev. B **55**, 15397 (1997)
3.52 L. Geelhaar, J. Márquez, K. Jacobi: Phys. Rev. B **60**, 15890 (1999)
3.53 P. Castrillo, G. Armelles, L. González, P.S. Domínguez, L. Colombo: Phys. Rev. B **51**, 1647 (1995)
3.54 S.W. da Silva, Yu.A. Pusep, J.C. Galzerani, M.A. Pimenta, D.I. Lubyshev, P.P. Gonzalez Borrero, P. Basmaji: Phys. Rev. B **53**, 1927 (1996)

3.55 D. Lüerßen, A. Dinger, H. Kalt, W. Braun, R. Nötzel, K. Ploog, J. Tümmler, J. Geurts: Phys. Rev. **57**, 1631 (1998)
3.56 G. Armelles, P. Castrillo, P.S. Dominguez, L. Gonzälez, A. Ruiz, D.A. Contreras-Solorio, V.R. Velasco, F.García-Moliner: Phys. Rev. B **49**, 14020 (1994)
3.57 W. Langbein, D. Lüerßen, H. Kalt, J.M. Hvam, W. Braun, K. Ploog: Phys. Rev. B **54**, 10784 (1996)
3.58 V.A. Volodin, M.D. Efremov, V.V. Preobrazhenskii, B.P. Semyagin, V.V. Bolotov, V.A. Sachkov: Semiconductors **34**, 62 (2000)
3.59 G. Armelles, P. Castrillo, P.D. Wang, C.M. Sotomayor Torres, N.N. Ledentsov, N.A. Bert: Solid State Commun. **94**, 613 (1995)
3.60 A.B. Vorob'ev, A.K. Gutakovsky, V.Ya. Printz, M.A. Putyato: Appl. Phys. Lett. **77**, 2976 (2000)
3.61 Y. Hsu, W.L. Wang, T.S. Kuan: Phys. Rev. B **50**, 4973 (1994)
3.62 O. Brandt, K. Kanamoto, Y. Tokuda, N. Tsukada, O. Wada, J. Tanimura: Phys. Rev. B **48**, 17599 (1993)
3.63 H. Kalt, W. Langbein, D. Lüerssen, A. Dinger, J. Tümmler, J. Geurts, W. Braun, R. Nötzel, K. Ploog: In *Proceedings of the 23rd International Conference on the Physics of Semiconductors*, Berlin, Germany, July 21–26, 1996, ed. by M. Scheffler and R. Zimmermann (World Scientific, Singapore 1996) Vol. **3**, p. 1747.
3.64 N.N. Ledentsov, D. Litvinov, D. Gerthsen, I.P. Soshnikov, V.A. Shchukin, V.M. Ustinov, A.Yu. Egorov, A.E. Zhukov, V.A. Volodin, M.D. Efremov, V.V. Preobrazhenskii, B.P. Semyagin, D. Bimberg, Zh.I. Alferov: J. Electr. Mater. **30**, 463 (2001)
3.65 P.S. Kop'ev, N.N. Ledentsov: In *Abstracts of Invited Lectures and Contributed Papers. The International Symposium on Nanostructures: Physics and Technology*, ed. by P.S.Kop'ev (St. Petersburg, Russia, June 13–18, 1993) p. 26
3.66 M.G. Lagally, D.E. Savage, M.C. Tringides: In *Reflection High-Energy Electron Diffraction and Reflecting Electron Imaging of Surfaces* **188**, NATO Advanced Study Institute, Series B, ed. by P.K. Larsen and P.J. Dobson (Plenum, New York 1988) p.139.
3.67 H.-J. Klaar, C.-A. Huang: Prakt. Metallogr. **31**, 290 (1994)
3.68 A. Rosenauer, S. Kaiser, T. Reisinger, J. Zweck, W. Gebhardt, D. Gerthsen: Optik (Stuttgart) **102**, 63 (1996)
3.69 D. Litvinov, A. Rosenauer, D. Gerthsen, N.N. Ledentsov, D. Bimberg, G.A. Ljubas, V.V. Bolotov, V.A. Volodin, M.D. Efremov, V.V. Preobrazhenskii, B.R. Semyagin, I.P. Soshnikov: Appl. Phys. Lett. **8**, 1080 (2002)
3.70 A.A. Chernov: *Modern Crystallography III* (Springer, Berlin 1984)
3.71 A.O. Golubok, G.M. Gur'yanov, V.N. Petrov, Yu.B. Samsonenko, S.Ya. Tipisev, G.E. Tsyrlin, N.N. Ledentsov: Fiz. Tekh. Poluprovodn. **28**, 516 (1994) [Semiconductors **28**, 317 (1994)]
3.72 N.N. Ledentsov, G.M. Gurianov, G.E. Tsyrlin, V.N. Petrov, Yu.B. Samsonenko, A.O. Golubok, S.Ya. Tipisev: Fiz. Tekh. Poluprovodn. **28**, 903 (1994) [Semiconductors **28**, 526 (1994)]
3.73 J.-K. Zuo, R.J. Warmack, D.M. Zehner, J.F. Wendelken: Phys. Rev. B **47**, 10743 (1993)
3.74 J.-K. Zuo, D.M. Zehner, J.F. Wendelken, R.J. Warmack, H.-N. Yang: Surf. Sci. **301**, 233 (1994)
3.75 R. Koch, M. Borbonus, O. Haase, K.H. Rieder: Phys. Rev. Lett. **67**, 3416 (1991)
3.76 R. Nötzel, L. Däweritz, K. Ploog: Phys. Rev. B **46**, 4736 (1992)

3.77 M. Higashiwaki, M. Yamamoto, T. Higuchi, S. Shimomura, A. Adachi, Y. Okamoto, N. Sano, S. Hiyamizu: Jap. J. Appl. Phys. **35**, L606 (1996)
3.78 R. Nötzel, N.N. Ledentsov, L. Däweritz, K. Ploog: *Method of Fabricating a Compositional Semiconductor Device*, US patent US 5714765, issued 3.02.1998, priority 29.01.1991
3.79 N.N. Ledentsov: to be published
3.80 E. Bauer: Z. Krist. **110**, 372 (1958)
3.81 F.C. Frank, J.H. van der Merwe: Proc. Roy. Soc. London A **198**, 205 (1949)
3.82 M. Volmer, A. Weber: Z. Physik. Chem. **119**, 277 (1926)
3.83 I.N. Stranski, L. Krastanow: Sitzungsberichte der Akademie der Wissenschaften in Wien, Mathematisch-Naturwissenschaftliche Klasse **146**, 797 (1937)
3.84 V.I. Marchenko: Pis'ma Zh. Eksp. Teor. Fiz. **33**, 397 (1981) [JETP. Lett. **33**, 381 (1981)]
3.85 O.L. Alerhand, D. Vanderbilt, R.D. Meade, J.D. Joannopoulos: Phys. Rev. Lett. **61**, 1973 (1988)
3.86 D. Vanderbilt: Surf. Sci. **268**, L300 (1992)
3.87 K. Kern, H. Niehus, A. Schatz, P. Zeppenfeld, J. George, G. Comsa: Phys. Rev. Lett. **67**, 855 (1991)
3.88 P.D. Wang, N.N. Ledentsov, C.M. Sotomayor Torres, P.S. Kop'ev, V.M. Ustinov: Appl. Phys. Lett. **64**, 1526 (1994)
3.89 V. Bressler-Hill, A. Lorke, S. Varma, K. Pond, P.M. Petroff, W.H. Weinberg: Phys. Rev. B **50**, 8479 (1994)
3.90 N.N. Ledentsov, P.D. Wang, C.M. Sotomayor Torres, A.Yu. Egorov, M.V. Maximov, V.M. Ustinov, A.E. Zhukov, P.S. Kop'ev: Phys. Rev. B **50**, 12171 (1994)
3.91 M.V. Belousov, N.N. Ledenstov, M.V. Maximov, P.D. Wang, I.N. Yassievich, N.N. Faleev, I.A. Kozin, V.M. Ustinov, P.S. Kop'ev, C.M. Sotomayor Torres: Phys. Rev. B **51**, 14346 (1995)
3.92 N.N. Ledenstov, I.L. Krestnikov, M.V. Maximov, S.V. Ivanov, S.L. Sorokin, P.S. Kop'ev, Zh.I. Alferov, D. Bimberg, C.M. Sotomayor Torres: Appl. Phys. Lett. **69**, 1343 (1996)
3.93 N.N. Ledenstov, I.L. Krestnikov, M.V. Maximov, S.V. Ivanov, S.L. Sorokin, P.S. Kop'ev, Zh.I. Alferov, D. Bimberg, C.M. Sotomayor Torres: Appl. Phys. Lett. **70**, 2766 (1997)
3.94 R. Engelhardt, V. Türck, U.W. Pohl, D. Bimberg: J. Cryst. growth **184/185**, 311 (1998)
3.95 T. Kümmell, R. Weigand, G. Bacher, A. Forchel: Appl. Phys. Lett. **73**, 3106 (1998)
3.96 V. Türck, S. Rodt, O. Stier, R. Heitz, U.W. Pohl, D. Bimberg, R. Steingruber: Phys. Rev. B **61**, 9944 (2000)
3.97 I.M. Lifshits, V.V. Slyozov: Zh. Eksp. Teor. Fiz. **35**, 479 (1958) [Sov. Phys. JETP **8**, 331 (1959)]
3.98 I.M. Lifshits, V.V. Slyozov: J. Phys. Chem. Solids **19**, 35 (1961)
3.99 C. Wagner: Z. Electrochem. **65**, 581 (1961)
3.100 F.K. Men, W.E. Packard, M.B. Webb: Phys. Rev. Lett. **61**, 2469 (1988)
3.101 O.L. Alerhand, A.N. Berker, J.D. Joannopoulos, D. Vanderbilt, R.J. Hamers, J.E. Demuth: Phys. Rev. Lett. **64**, 2406 (1990)
3.102 T.W. Poon, S. Yip, P.S. Ho, F.F. Abraham: Phys. Rev. Let. **65**, 2161 (1990)
3.103 G. Boishin, L.D. Sun, M. Hohage, P. Zeppenfeld: Surf. Sci. **512**, 185 (2002)
3.104 V.I. Marchenko: Pis'ma Zh. Eksp. Teor. Fiz. **55**, 72 (1992) [JETP Lett. **55**, 73 (1992)]
3.105 K.-O. Ng, D. Vanderbilt: Phys. Rev. B **52**, 2177 (1995)

3.106 R. Plass, J.A. Last, N.C. Bartelt, G.L. Kellogg: Nature **412**, 875 (2001)
3.107 P. Zeppenfeld, M. Krzyzowski, C. Romainczuk, G. Comsa, M.G. Lagally: Phys. Rev. Lett. **72**, 2737 (1994)
3.108 A.A. Maradudin, X. Huang, A.P. Mayer: J. Appl. Phys. **70**, 53 (1991)
3.109 J. Tersoff, R.M. Tromp: Phys. Rev. Lett. **70**, 2782 (1993)
3.110 A. Grossmann, W. Erley, J.B. Hannon, H. Ibach: Phys. Rev. Lett. **77**, 127 (1996).
3.111 V.A. Shchukin, N.N. Ledentsov, D. Bimberg: In *Self-Organized Processes in Semiconductor Alloys: Spontaneous Ordering, Composition Modulation, and 3D Islanding*, ed. by D.M. Follstaedt, B.A. Joyce, A. Mascarenhas, T. Suzuki, Mat. Res. Soc. Symp. Proc. V. **583** (Pittsburgh, USA 2000) p. 23
3.112 L.G. Wang, P. Kratzer, M. Scheffler, N. Moll: Phys. Rev. Lett. **82**, 4042 (1999)
3.113 T. Shitara, D.D. Vvedensky, J.H. Neave, B.A. Joyce: Mat. Res. Soc. Proc. **312**, 267 (1993)
3.114 C. Ratch, A. Zangwill, P. Smilauer: Surf. Sci. **314**, L937 (1994); C. Ratch, P. Smilauer, A. Zangwill, D. Vvedensky: J. Phys. I **6**, 575 (1996)
3.115 J.A. Stroscio, D.T. Pierce: Phys. Rev. B **49**, 8522 (1994)
3.116 M. Meixner, E. Schöll, V.A. Shchukin, D. Bimberg: Phys. Rev. Lett. **87**, 236101 (2001)
3.117 A.B. Bortz, M.H. Kalos, J.L. Lebowitz: J. Comp. Phys. **17**, 10 (1975)
3.118 V.A. Shchukin, N.N. Ledentsov, A. Hoffmann, D. Bimberg, I.P. Soshnikov, B.V. Volovik, V.M. Ustinov, D. Litvinov, D. Gerthsen: phys. stat. sol. (b), **224**, 503 (2001)
3.119 Zh.I. Alferov, S.V. Ivanov, P.S. Kop'ev, A.V. Lebedev, N.N. Ledentsov, M.V. Maximov, I.V. Sedova, T.V. Shubina, A.A. Toropov: Superlattices Microstruct. **15**, 65 (1994)
3.120 C. Benoit à la Guillaume, J.M. Denber, F. Salvan: Phys. Rev. **177**, 567 (1969)
3.121 I.L. Krestnikov, M.V. Maximov, A.V. Sakharov, P.S. Kop'ev, Zh.I. Alferov, N.N. Ledentsov, D. Bimberg, C.M. Sotomayor Torres: J. Cryst. Growth **184/185**, 545 (1998)
3.122 N.N. Ledentsov, A.F. Tsatsul'nikov, A.Yu. Egorov, P.S. Kop'ev, A.R. Kovsh, M.V. Maximov, V.M. Ustinov, B.V. Volovik, A.E. Zhukov, Zh.I. Alferov, I.L. Krestnikov, D. Bimberg, A. Hoffmann: Appl. Phys. Lett. **74**, 161 (1999)
3.123 I.L. Krestnikov, M. Strassburg, M. Caesar, A. Hoffmann, U.W. Pohl, D. Bimberg, N.N. Ledentsov, P.S. Kop'ev, Zh.I. Alferov, D. Litvinov, A. Rosenauer, D. Gerthsen: Phys. Rev B **60**, 8696 (1999)
3.124 L.C. Lenchyshyn, M.L.W. Thewalt, D.C. Houghton, J.-P. Noel, N.L. Rowell, J.C. Sturn, X. Xiao: Phys. Rev. B **47**, 16655 (1993)
3.125 T. Baier, U. Mantz, K. Thonke, R. Sauer, F. Schäffler, H.J. Herzog: *Proc. 22nd International Conference on the Physics of Semiconductors*, ed. by D.J. Lockwood, Vancouver, Canada, August 15–19, 1994, Vol. 2 (World Scientific, Singapore 1995) p. 1568
3.126 O.P. Pchelyakov, Yu.B. Bolokhvityaniov, A.V. Dvurechenskii, L.V. Sokolov, A.I. Nikiforov, A.I. Yakimov: Fiz. Tekhn. Poluprovodn. **34**, 1281 (2000) [Semiconductors **34** (2000)]
3.127 M.W. Dashiel, U. Denker, C. Müller, G. Costantini, C. Manzano, K. Kern, O.G. Schmidt: Appl. Phys. Lett. **80**, 1279 (2002)
3.128 N.N. Ledentsov, M. Grundmann, F. Heinrichsdorff, D. Bimberg, V.M. Ustinov, A.E. Zhukov, M.V. Maximov, Zh.I. Alferov, J.A. Lott: IEEE J. Sel. Top. Quantum Electron. **6**, 439 (2000)
3.129 I.L. Krestnikov, N.N. Ledentsov, A. Hoffmann, D. Bimberg: phys. stat. sol. (a) **183**, 207 (2001)

3.130 N.N. Ledentsov, J. Böhrer, M. Beer, F. Heinrichsdorff, M. Grundmann, D. Bimberg, S.V. Ivanov, B.Ya. Meltser, I.N. Yassievich, N.N. Faleev, P.S. Kop'ev, Zh.I. Alferov: Phys. Rev. B **52**, 14058 (1995)

3.131 A. Makarov, N.N. Ledentsov, A.F Tsatsul'nikov, G.E. Tsyrlin, V.A. Egorov, V.M. Ustinov, N.D. Zakharov, P. Werner: Fiz. Tekhn. Poluprovodn. **36** (2002) [Semiconductors **36** (2002)]

3.132 N.D. Zakharov, P. Werner, U. Gösele, G. Gerth, G. Cirlin, V.A. Egorov, B.V. Volovik: Mat. Sci. Engineering B **87**, 92 (2001)

3.133 N.D. Zakharov, G.E. Cirlin, P. Werner, U. Gösele, G. Gerth, B.V. Volovik, N.N. Ledentsov, V.M. Ustinov: In *Proceedings of the 9th International Symposium on Nanostructures: Physics and Technology 2001* (St. Petersburg, Russia, June 18–22, 2001) p. 21

3.134 L.P. Rokinson, D.C. Tsui, J.L. Benton, Y.-H. Xie: Appl. Phys. Lett. **75**, 2413 (1999)

3.135 A.R. Kovsh, A.E. Zhukov, N.A. Maleev, S.S. Mikhrin, A.V. Vasil'ev, Yu.M. Shernyakov, D.A. Livshits, M.V. Maximov, D.S. Sizov, N.V. Kryzhanovskaya, N.A. Pikhtin, V.A. Kapitonov, I.S. Tarasov, N.N. Ledentsov, V.M. Ustinov, J.S. Wang, L.Wei, G. Lin, J.Y. Chi: In *Proc. 10th Int. Symp. on Nanostructures: Physics and Technology* (St. Petersburg, Russia, June 17–21, 2002) p. 395

3.136 A.V. Sakharov, W.V. Lundin, I.L. Krestnikov, V.A. Semenov, A.S. Usikov, A.F. Tsatsul'nikov, Yu.G. Musikhin, M.V. Baidakova, Zh.I. Alferov, N.N. Ledentsov, J. Holst, A. Hoffmann, D. Bimberg, I.P. Soshnikov, D. Gerthsen: phys. stat. sol. (b) **216**, 435 (1999)

3.137 A.F. Tsatsul'nikov, I.L. Krestnikov, W.V. Lundin, A.V. Sakharov, A.P. Kartashova, A.S. Usikov, Zh.I. Alfeorv, N.N. Ledentsov, A. Strittmatter, A. Hoffmann, D. Bimberg, I.P. Soshnikov, D. Litvinov, A.P. Rosenauer, D. Gerthsen, A. Plaut: Semicond. Sci. Technol. **15**, 766 (2000)

3.138 W.J. Schaffer, M.D. Lind, S.P. Kowalczyk, R.W. Grant: J. Vac. Sci. Technol. B **1**, 688 (1983)

3.139 B.F. Lewis, F.J. Grunthaner, A. Madhukar, R. Fernandez, J. Maserjian: J. Vac. Sci. Technol. B **2**, 419 (1984)

3.140 F.J. Grunthaner, M.Y. Yen, T.C. Lee, A. Madhukar, B.F. Lewis: Appl. Phys. Lett. **46**, 983 (1985)

3.141 M.Y. Yen, A. Madhukar, B.F. Lewis, R. Fernandez, L. Eng, F.J. Grunthaner: Surf. Sci. **174**, 606 (1986) (submitted September 1, 1985)

3.142 J.M. Gibson, R. Hull, J.C. Bean, M.M.J. Treacy: Appl. Phys. Lett. **46**, 649 (1985)

3.143 L. Goldstein, F. Glas, J.Y. Marzin, M.N. Charasse, G. Le Roux: Appl. Phys. Lett. **47**, 1099 (1985) (submitted July 26, 1985; published November 15, 1985)

3.144 F. Glas, C. Guille, P. Hénoc, F. Houzay: Inst. Phys. Conf. Ser. **87**, Sect. 2, p. 71 (1987)

3.145 S. Guha, A. Madhukar, K.C. Rajkumar: Appl. Phys. Lett. **57**, 2110 (1990)

3.146 M. Tabuchi, S. Noda, A. Sasaki: In *Science and Technology of Mesoscopic Structures*, ed. by S. Namba, C. Hamaguchi, and T. Ando (Springer, Tokyo 1992) p. 379

3.147 J.M. Moison, F. Houzay, F. Barthe, L. Leprince, E. André, O. Vatel: Appl. Phys. Lett. **64**, 196 (1994)

3.148 D. Leonard, M. Krishnamurthy, C.M. Reaves, S.P. Denbaars, P.M. Petroff: Appl. Phys. Lett. **63**, 3203 (1993)

3.149 N.N. Ledentsov, M. Grundmann, N. Kirstaedter, J. Christen, R. Heitz, J. Böhrer, F. Heinrichsdorff, D. Bimberg, S.S. Ruvimov, P. Werner, U. Richter, U. Gösele, J. Heydenreich, V.M. Ustinov, A.Yu. Egorov, M.V. Maximov, P.S. Kop'ev, Zh.I. Alferov: In *Proceedings of the 22nd International Conference on Physics of Semiconductors*, ed. by D.J. Lockwood, Vancouver, Canada, August 1994 (World Scientific, Singapore 1994) Vol. **3**, p. 1855

3.150 J.-Y. Marzin, G.M. Gérard, A. Izraël, G. Barrier, D. Bastard: Phys. Rev. Lett. **73**, 716 (1994)

3.151 M. Grundmann, J. Christen, N.N. Ledentsov, J. Böhrer, D. Bimberg, S.S. Ruvimov, P. Werner, U. Richter, U. Gösele, J. Heydenreich, V.M. Ustinov, A.Yu. Egorov, A.E. Zhukov, P.S. Kop'ev, Zh.I. Alferov: Phys. Rev. Lett. **74**, 4043 (1995)

3.152 N.N. Ledentsov, M. Grundmann, N. Kirstaedter, O. Schmidt, R. Heitz, J. Böhrer, D. Bimberg, V.M. Ustinov, V.A. Shchukin, P.S. Kop'ev, Zh.I. Alferov, S.S. Ruvimov, A.O. Kosogov, P. Werner, U. Richter, U. Gösele, J. Heydenreich: In *Proceedings of the 7th International Conference on Modulated Semiconductor Structures*, Madrid, Spain, July 1995, Solid State Electron. **40**, 785 (1996)

3.153 D. Vanderbilt, L.K. Wickham: In *Evolution of Thin-Film and Surface Microstructures*, ed. by C.V. Thompson, J.Y. Tsao, and D.J. Srolovitz, Mat. Res. Soc. Symp. Proc. **202**, 555 (MRS, Pittsburgh 1991)

3.154 C. Ratsch, A. Zangwill: Surf. Sci. **293**, 123 (1993)

3.155 V.A. Shchukin, N.N. Ledentsov, P.S. Kop'ev, D. Bimberg: Phys. Rev. Lett. **75**, 2968 (1995)

3.156 I. Daruka, A.-L. Barabási: Phys. Rev. Lett. **79**, 3708 (1997)

3.157 V.A. Shchukin, D. Bimberg: Rev. Mod. Phys. **71**, 1125 (1999)

3.158 C. Priester, M. Lannoo: Phys. Rev. Lett. **75**, 73 (1995)

3.159 A. Madhukar, P. Chen, Q. Xie, A. Konkar, T.R. Ramachandran, N.P. Kobayashi, R. Viswanathan: In *Low Dimensional Structures Prepared by Epitaxial Growth or Regrowth on Patterned Substrates*, Proceedings of the NATO Advanced Workshop, February 20–24, 1995, Ringberg Castle, Germany, ed. by K. Eberl, P. Petroff, and P. Demeester (Kluwer, Dordrecht 1995) p. 19

3.160 N. Kobayashi, T.R. Ramachandran, P. Chen, A. Madhukar: Appl. Phys. Lett. **68**, 3299 (1996)

3.161 K.M. Chen, D.E. Jesson, S.J. Pennycook, T. Thundat, R.J. Warmack: Proc. Mat. Res. Soc. Symp. **399**, 271 (1995)

3.162 D.E. Jesson, G. Chen, K.M. Chen, S.J. Pennycook: Phys. Rev. Lett. **80**, 5156 (1998)

3.163 V.A. Shchukin, N.N. Ledentsov, V.M. Ustinov, Yu.G. Musikhin, V.B. Volovik, A. Schliwa, O. Stier, R. Heitz, D. Bimberg: In *Morphological and Compositional Evolution of Heteroepitaxial Semiconductor Thin Films*, ed. by A.-L. Barabási, E. Jones, and J. Mirecki Millunchick, Mat. Res. Soc. Symp. Proc. **618**, 79–90 (MRS, Pittsburgh 2000)

3.164 N.N. Ledentsov, V.A. Shchukin, D. Bimberg, V.M. Ustinov, N.A. Cherkashin, Yu.G. Musikhin, B.V. Volovik, G.E. Cirlin, Zh.I. Alferov: Semicond. Sci. and Technol. **16**, 502 (2001)

3.165 R. Leon, J. Wellman, X.Z. Liao, J. Zuo, D.J.H. Cockayne: Appl. Phys. Lett. **76**, 1558 (2000)

3.166 D.J. Eaglesham, M. Cerullo: Phys. Rev. Lett. **64**, 1943 (1990)

3.167 Y.-W. Mo, D.E. Savage, B.S. Swartzentruber, M.G. Lagally: Phys. Rev. Lett. **65**, 1020 (1990)

3.168 R.J. Asaro, W.A. Tiller: Metall. Trans. **3**, 1789 (1972)
3.169 M.A. Grinfel'd: Dokl. Akad. Nauk SSSR **290**, 1358 (1986) [Sov. Phys. Dokl. **31**, 831 (1986)]
3.170 D. Srolovitz: Acta Metall. **37**, 621 (1989)
3.171 B.J. Spencer, P.W. Voorhees, S.H. Davis: Phys. Rev. Lett. **67**, 3696 (1991)
3.172 D.E. Jesson, S.J. Pennycook, J.-M. Baribeau, D.C. Houghton: Phys. Rev. Lett. **71**, 1744 (1993)
3.173 W.H. Yang, D.J. Srolovitz: Phys. Rev. Lett. **71**, 1593 (1993)
3.174 E. Pehlke, N. Moll, A. Kley, M. Scheffler: Appl. Phys. A **65**, 525 (1997)
3.175 B.K. Chakraverty: J. Phys. Chem. Solids **28**, 2401 (1967)
3.176 J. Drucker: Phys. Rev. B **48**, 18203 (1993)
3.177 A. Madhukar, Q. Xie, P. Chen, A. Koknar: Appl. Phys. Lett. **64**, 2727 (1994)
3.178 D. Leonard, K. Pond, P.M. Petroff: Phys. Rev. B **50**, 11687 (1994)
3.179 F. Hatami, N.N. Ledentsov, M. Grundmann, J. Böhrer, F. Heinrichsdorff, M. Beer, D. Bimberg, S.S. Ruvimov, P. Werner, U. Gösele, J. Heydenreich, S.V. Ivanov, B.Ya. Meltser, P.S. Kop'ev, Zh.I. Alferov: Appl. Phys. Lett. **67**, 656 (1995)
3.180 A. Ponchet, A. Le Corre, H. L'Haridon, B. Lambert, S. Salaün: Appl. Phys. Lett. **67**, 1850 (1995)
3.181 R. Leon, S. Fafard, D. Leonard, J.L. Merz, P.M. Petroff: Appl. Phys. Lett. **67**, 521 (1995)
3.182 R. Apetz, L. Vescan, A. Hartmann, C. Dieker, H. Lüth: Appl. Phys. Lett. **66**, 445 (1995)
3.183 P. Schittenhelm, M. Gail, J. Brunner, J.F. Nützel, G. Abstreiter: Appl. Phys. Lett. **67**, 1292 (1995)
3.184 D.E. Jesson, K.M. Chen, S.J. Pennycook: MRS Bulletin **21**, 31 (1996)
3.185 S.H. Xin, P.D. Wang, A. Yin, C. Kim, M. Dobrowolska, J.L. Merz, J.K. Furdyna: Appl. Phys. Lett. **69**, 3884 (1996)
3.186 V.M. Ustinov, E.R. Weber, S. Ruvimov, Z. Liliental-Weber, A.E. Zhukov, A.Yu. Egorov, A.R. Kovsh, A.F. Tsatsul'nikov, P.S. Kop'ev: Appl. Phys. Lett. **72**, 362 (1998)
3.187 G.E. Cirlin, V.G. Dubrovskii, V.N. Petroff, N.K. Polyakov, N.P. Korneeva, V.N. Demidov, A.O. Golubok, S.A. Masalov, D.V. Kurochkin, O.M. Gorbenko, N.I. Komyak, N.N. Ledentsov, Zh.I. Alferov, D. Bimberg: In *Proceedings of the 6th International Symposium on Nanostructures: Physics and Technology*, St. Petersburg, Russia, June 22–26, 1998
3.188 N.N. Ledentsov, V.M. Ustinov, A.Yu.Egorov, A.E. Zhukov, M.V. Maximov, I.G. Tabatadze, P.S. Kop'ev: Fiz. Tekh. Poluprovodn. **28**, 1484 (1994) [Semiconductors **28**, 832 (1994)] (submitted December 29, 1993)
3.189 N. Kirstaedter, N.N. Ledentsov, M. Grundmann, D. Bimberg, V.M. Ustinov, S.S. Ruvimov, M.V. Maximov, P.S. Kop'ev, Zh.I. Alferov, U. Richter, P. Werner, U. Gösele, J. Heydenreich: Electron. Lett. **30**, 1416 (1994)
3.190 D. Bimberg, N. Kirstaedter, N.N. Ledentsov, Zh.I. Alferov, P.S. Kop'ev, V.M. Ustinov: IEEE Journal of Selected Topics in Quantum Electronics, **3**, 196 (1997)
3.191 D. Bimberg, M. Grundmann, N.N. Ledentsov, S.S. Ruvimov, P. Werner, U. Richter, J. Heydenreich, V.M. Ustinov, P.S. Kop'ev, Zh.I. Alferov: Thin Solid Films **267**, 32 (1995)
3.192 G. Cirlin, G.M. Guryanov, A.O. Golubok, S.Ya. Tipissev, N.N. Ledentsov, P.S. Kop'ev, M. Grundmann, D. Bimberg: Appl. Phys. Lett. **67**, 97 (1995)

3.193 M. Grundmann, N.N. Ledentsov, R. Heitz, L. Eckey, J. Christen, J. Böhrer, D. Bimberg, S.S. Ruvimov, P. Werner, U. Richter, U. Gösele, J. Heydenreich, V.M. Ustinov, A.Yu. Egorov, A.E. Zhukov, P.S. Kop'ev, Zh.I. Alferov: phys. stat. sol. (b) **188**, 249 (1995)
3.194 V.A. Shchukin, N.N. Ledentsov, M. Grundmann, P.S. Kop'ev, D. Bimberg: Surf. Sci. **352–354**, 117 (1996)
3.195 V.A. Shchukin, N.N. Ledentsov, M. Grundmann, D. Bimberg: In *Optical Properties of Low Dimensional Semiconductors*, NATO ASI Series, Series E: Applied Physics, ed. by G. Abstreiter, A. Aydinli, and J.-P. Leburton (Kluwer, Dordrecht 1997) p. 257
3.196 A.Yu. Kaminski, R.A. Suris: In *Proceedings of the 23rd International Conference on Physics of Semiconductors*, Berlin, Germany, July 22–27, 1996, ed. by M. Scheffler and R. Zimmermann (World Scientific, Singapore 1996), Vol. **2**, p. 1337
3.197 K.M. Chen, D.E. Jesson, S.J. Pennycook, T. Thundat, R.J. Warmack: Phys. Rev. B **56**, R1700 (1997)
3.198 C. Duport, C. Priester, J. Villain: in *Morphological Organization in Epitaxial Growth and Removal*, ed. by Z. Zhang and M. Lagally (World Scientific, Singapore 1997)
3.199 E. Pehlke, N. Moll, M. Scheffler: In *Proceedings of the 23rd International Conference on Physics of Semiconductors*, Berlin, Germany, July 22–27, 1996, ed. by M. Scheffler and R. Zimmermann (World Scientific, Singapore 1996), Vol. **2**, p. 1301
3.200 B.J. Spencer, J. Tersoff: Phys. Rev. Lett. **79**, 4858 (1997)
3.201 A.G. Khachaturyan: *Theory of Phase Transformations and the Structure of Solid Solutions* (Nauka, Moscow 1974) in Russian
3.202 A.G. Khachaturyan: *Theory of Structural Transformations in Solids* (Wiley, New York 1983)
3.203 A.L. Roitburd: Phys. Status Solidi (a) **37**, 329 (1976)
3.204 D. Wolf: Phys. Rev. Lett. **70**, 627 (1993)
3.205 E. Steimetz, F. Scheinle, J.-T. Zettler, W. Richter: J. Cryst. Growth **170**, 208 (1997)
3.206 K. Georgsson, N. Carlsson, L. Samuelson, W. Seifert, L.R. Wallenberg: Appl. Phys. Lett. **67**, 2981 (1995)
3.207 S.S. Ruvimov, P. Werner, K. Scheerschmidt, U. Richter, U. Gösele, J. Heydenreich, N.N. Ledentsov, M. Grundmann, D. Bimberg, V.M. Ustinov, A.Yu. Egorov, P.S. Kop'ev, Zh.I. Alferov: Phys. Rev. B **51**, 14766 (1995)
3.208 S.S. Ruvimov, K. Scheerschmidt: phys. stat. sol. (a) **150**, 471 (1995)
3.209 I.P. Ipatova, V.G. Malyshkin, V.A. Shchukin: J. Appl. Phys. **74**, 7198 (1993)
3.210 I.P. Ipatova, V.G. Malyshkin, V.A. Shchukin: Phil. Mag. B **70**, 557 (1994)
3.211 M. Grundmann, O. Stier, D. Bimberg: Phys. Rev. B, **52**, 11969 (1995)
3.212 K. Portz, A.A. Maradudin: Phys. Rev. B **16**, 3535 (1977)
3.213 V.A. Shchukin, D. Bimberg: Appl. Phys. A **67**, 687 (1998)
3.214 N.N. Ledentsov, D. Bimberg, Yu.M. Shernyakov, V. Kochnev, M.V. Maximov, A.V. Sakharov, I.L. Krestnikov, A.Yu. Egorov, A.F. Tsatsul'nikov, B.V. Volovik, V.M. Ustinov, P.S. Kop'ev, Zh.I. Alferov, A.O. Kosogov, P. Werner: Appl. Phys. Lett. **70**, 2888 (1997)
3.215 G.-X. Qian, R.M. Martin, D.J. Chadi: Phys. Rev. B **38**, 7649 (1998)
3.216 N. Moll, A. Kley, E. Pehlke, M. Scheffler: Phys. Rev. B **54**, 8844 (1996)
3.217 J. Tersoff: Phys. Rev. B. **43**, 9377 (1991)
3.218 C. Roland, G.H. Gilmer: Phys. Rev. B **47**, 16286 (1993)
3.219 J.A. Floro, G.A. Lucadamo, E. Chason, L.B. Freund, M. Sinclair, R.D. Twesten, R.Q. Hwang: Phys. Rev. Lett. **80**, 4717 (1998)

3.220 J. Tersoff, Y.H. Phang, Zh. Zhang, M.G. Lagally: Phys. Rev. Lett. **75**, 2730 (1995)
3.221 J. Tersoff: Phys. Rev. Lett. **77**, 2017 (1996)
3.222 J.E. Guyer, P.W. Voorhees: Phys. Rev. B **54**, 11710 (1996)
3.223 F. Léonard, R.F. Desai: Phys. Rev. B **57**, 4805 (1998)
3.224 I.P. Ipatova, V.G. Malyshkin, A.A. Maradudin, V.A. Shchukin, R.F. Wallis: Phys. Rev. B **57**, 12969 (1998)
3.225 B.J. Spencer, P.W. Voorhees, J. Tersoff: Phys. Rev. B **64**, 235318 (2001)
3.226 S.V. Ghasias, A. Madhukar: in *Proceedings of the SPIE Symposium on Growth of Advanced Semiconductor Structures* (SPIE, Bellingham, Washington, USA 1988) Vol. **944**, p. 16
3.227 S.V. Ghasias, A. Madhukar: J. Vac. Sci. Technol. B **7**, 264 (1989)
3.228 Y. Chen, J. Washburn: Phys. Rev. Lett. **77**, 4046 (1996)
3.229 A.-L. Barabási: Appl. Phys. Lett. **70**, 2565 (1997)
3.230 A. Polimene, A. Patané, M. Capizzi, F. Martelli, L. Nasi, G. Salviatti: Phys. Rev. B **53**, R4213 (1996)
3.231 M. Berti, A.V. Drigo, G. Rossetto, G. Torzo: J. Vac. Sci. Technol. B **15**, 1794 (1997)
3.232 D.J. Bottomley: Appl. Phys. Lett. **72**, 783 (1998)
3.233 K. Ozasa, Y. Aoyagi, Y.J. Park, L. Samuelson: Appl. Phys. Lett. **71**, 797 (1997)
3.234 F. Heinrichsdorff, A. Krost, D. Bimberg, A.O. Kosogov, P. Werner: Appl. Surf. Sci. **123/124**, 725 (1998)
3.235 C. Li, C. Jagadish: Appl. Phys. Lett. **69**, 2551 (1996)
3.236 K. Muraki, S. Fukatsu, Y. Shiraki, R. Ito: Appl. Phys. Lett. **61**, 557 (1992)
3.237 U. Woggon, W. Langbein, J.M. Hvam, A. Rosenauer, T. Remmele, D. Gerthsen: Appl. Phys. Lett. **71**, 377 (1997)
3.238 A. Rosenauer, W. Oberst, D. Litvinov, D. Gerthsen, A. Förster, R. Schmidt: Phys. Rev. B **61**, 8276 (2000)
3.239 M.V. Maximov, A.F. Tsatsul'nikov, B.V. Volovik, D.A. Bedarev, A.Yu. Egorov, A.E. Zhukov, A.R. Kovsh, N.A. Bert, V.M. Ustinov, P.S. Kop'ev, Zh.I. Alferov, N.N. Ledentsov, D. Bimberg, I.P. Soshnikov, P. Werner: Appl. Phys. Lett. **75**, 2347 (1999)
3.240 T.I. Kamins, E.C. Carr, R.S. Williams, S.J. Rosner: J. Appl. Phys. **81**, 211 (1997)
3.241 T.I. Kamins, G. Medeiros-Ribeiro, D.A.A. Ohlberg, R. Stanley Williams: Appl. Phys. A: **67**, 1 (1998)
3.242 J. Tersoff, B.J. Spencer, A. Rastelli, H. von Känel: Phys. Rev. Lett. **89**, 196104 (2002)
3.243 N.N. Ledentsov: In *Proceedings of the 23rd International Conference on Physics of Semiconductors*, Berlin, Germany, July 22–27, 1996, ed. by M. Scheffler and R. Zimmermann (World Scientific, Singapore 1996) Vol. **1**, p. 21
3.244 G.E. Cirlin, V.G. Dubrovskii, V.N. Petrov, N.K. Polyakov, N.P. Korneeva, V.N. Demidov, A.O. Golubok, S.A. Masalov, D.V. Kurochkin, O.M. Gorbenko, N.I. Komyak, V.M. Ustinov, A.Yu. Egorov, A.R. Kovsh, M.V. Maximov, A.F. Tsatsul'nikov, B.V. Volovik, A.E. Zhukov, P.S. Kop'ev, Zh.I. Alferov, N.N. Ledentsov, M. Grundmann, D. Bimberg: Semicond. Sci. Technol. **13**, 1262 (1998)
3.245 T. Mano, H. Fujioka, K. Ono, Y. Watanabe, M. Oshima: Appl. Surf. Sci. **130–132**, 768 (1998)

3.246 N.D. Zakharov, P. Werner, V.M. Ustinov, G.E. Cirlin, O.V. Smolski, D.V. Denisov, Zh.I. Alferov, N.N. Ledentsov, R. Heitz, D. Bimberg: *Proc. 7th Int. Symp. Nanostructures: Physics and Technology* (St. Petersburg, Russia, 1999) p. 216

3.247 N.D. Zakharov, P. Werner, R. Heitz, D. Bimberg, N.N. Ledentsov, V.M. Ustinov, D.V. Denisov, Zh.I. Alferov, G.E. Cirlin: Mat. Res. Soc. Symp. Proc. **571**, 247 (1999), ed. by H. Lee, S. Moss, D. Norris, and D. Ila (Pittsburgh 1999)

3.248 R. Heitz, N.N. Ledentsov, D. Bimberg, A.Yu. Egorov, M.V. Maximov, V.M. Ustinov, A.E. Zhukov, Zh.I. Alferov, G.E. Cirlin, I.P. Soshnikov, N.D. Zakharov, P. Werner, U. Gösele: Appl. Phys. Lett. **74**, 1701 (1999)

3.249 V.N. Petrov, N.K. Polyakov, V.A. Egorov, G.E. Cirlin, N.D. Zakharov, P. Werner, V.M. Ustinov, D.V. Denisov, N.N. Ledentsov, Zh.I. Alferpv: Fiz. Tekh. Poluprovodn. **34**, 838 (2000) [Semiconductors, **34**, 841 (2000)]

3.250 S. Nakamura, T. Mukai, M. Senoh: Jpn. J. Appl. Phys. Part 2, **32**, L8 (1993)

3.251 S. Nakamura, M. Senoh, S. Nagahama, N. Iwasa, T. Matsushita, H. Kioku, Y. Sugimoto: Jpn. J. Appl. Phys. Part 2, **35**, L74 (1996)

3.252 I. Akasaki, S. Sota, H. Sakai, T. Tanaka, M. Koike, H. Amano: Electron. Lett. **32**, 1105 (1996)

3.253 I.L. Krestnikov, W.V. Lundin, A.V. Sakharov, V.A. Semenov, A.S. Usikov, A.F. Tsatsul'nikov, Z.I. Alferov, N.N. Ledentsov, A. Hoffmann, D. Bimberg: Appl. Phys. Lett. **75**, 1192 (1999)

3.254 H. Morkoc: *Nitride Semiconductors and Devices* (Springer, Berlin 1999)

3.255 S.C. Jain, M. Willander, J. Narayana, R. Van Overstraeten: J. Appl. Phys. **87**, 965 (2000)

3.256 S. Chichibu, T. Azuhata, T. Sota, S. Nakamura: Appl. Phys. Lett. **69**, 4188 (1996)

3.257 Y. Narukawa, Y. Kawakami, M. Funato, S. Fujita, S. Fujita, S. Nakamura: Appl. Phys. Lett. **70**, 981 (1997)

3.258 Y. Narukawa, K. Sawada, Y. Kawakami, S. Fujita, S. Fujita, S. Nakamura: J. Cryst. Growth **189**, 606 (1998)

3.259 N.N. Ledentsov, Z.I. Alferov, I.L. Krestnikov, W.V. Lundin, A.V. Sakharov, I.P. Soshnikov, A.F. Tsatsul'nikov, D. Bimberg, A. Hoffmann: Compound Semicond. **5**, 61 (1999)

3.260 Yu.G. Musikhin, D. Gethsen, D.A. Bedarev, N.A. Bert, W.V. Lundin, A.F. Tsatsul'nikov, A.V. Sakharov, A.S. Usikov, Zh.I. Alferov, I.L. Krestnikov, N.N. Ledentsov, A. Hoffmann, D. Bimberg: Appl. Phys. Lett. **80**, 2099 (2002)

3.261 D. Gerthsen, E. Hahn, B. Neubauer, A. Rosenauer, O. Schön, M. Heuken, A. Rizzi: phys. stat. sol. (b) **177**, 145 (2000)

Chapter 4

4.1 L. Goldstein, F. Glas, J.Y. Marzin, M.N. Charasse, G. Le Roux: Appl. Phys. Lett. **47**, 1099 (1985)

4.2 T. Kuan, S.S. Iyer: Appl. Phys. Lett. **59**, 2242 (1991)

4.3 J.Y. Yao, T.G. Andersson, G.L. Dunlop: J. Appl. Phys. **69**, 2224 (1991)

4.4 Q. Xie, P. Chen, A. Madhukar: Appl. Phys. Lett. **65**, 2051 (1994)

4.5 Q. Xie, A. Madhukar, P. Chen, N. Kobayashi: Phys. Rev. Lett. **75**, 2542 (1995)

4.6 N.N. Ledentsov, J. Böhrer, D. Bimberg, I.V. Kochnev, M.V. Maximov, P.S. Kop'ev, Zh.I. Alferov, A.O. Kosogov, S.S. Ruvimov, P. Werner, U. Gösele: Appl. Phys. Lett. **69**, 1095 (1996)

4.7 N.N. Ledentsov, M. Grundmann, N. Kirstaedter, O. Schmidt, R. Heitz, J. Böhrer, D. Bimberg, V.M. Ustinov, V.A. Shchukin, P.S. Kop'ev, Zh.I. Alferov, S.S. Ruvimov, A.O. Kosogov, P. Werner, U. Richter, U. Gösele, J. Heydenreich: In *Proceedings of the 7th International Conference on Modulated Semiconductor Structures*, Madrid, Spain, July 1995, Solid State Electron. **40**, 785 (1996)

4.8 N.N. Ledentsov, V.A. Shchukin, M. Grundmann, N. Kirstaedter, J. Böhrer, O. Schmidt, D. Bimberg, V.M. Ustinov, A.Yu. Egorov, A.E. Zhukov, P.S. Kop'ev, S.V. Zaitsev, N.Yu. Gordeev, Zh.I. Alferov, A.I. Borovkov, A.O. Kosogov, S.S. Ruvimov, P. Werner, U. Gösele, J. Heydenreich: Phys. Rev. B **54**, 8743 (1996)

4.9 M.S. Miller, J. Malm, M. Pistol, S. Jeppesen, B. Kowalski, K. Georgsson, L. Samuelson: J. Appl. Phys. **80**, 3360 (1996)

4.10 G.S. Solomon, J.A. Trezza, A.F. Marshall, J.S. Harris: Phys. Rev. Lett. **76**, 952 (1996)

4.11 A.A. Maradudin, R.F. Wallis: Surf. Sci. **91**, 423 (1980)

4.12 D. Srolovitz: Acta Metall. **37**, 621 (1989)

4.13 W.W. Mullins: J. Appl. Phys. **28**, 333 (1957)

4.14 J. Tersoff, C. Teichert, M.G. Lagally: Phys. Rev. Lett. **76**, 1675 (1996)

4.15 F. Liu, S.E. Davenport, H.M. Evans, M.G. Lagally: Phys. Rev. Lett. **82**, 2528 (1999)

4.16 G.S. Solomon, S. Komarov, J.S. Harris Jr., Y. Yamamoto: J. Cryst. Growth **175–176**, 707 (1997)

4.17 Y. Nakata, Y. Sugiyama, T. Futatsugi, N. Yokoyama: J. Cryst. Growth **175–176**, 713 (1997)

4.18 M.K. Zundel, P. Specht, K. Eberl, N.Y. Jin-Phillipp, F. Phillipp: Appl. Phys. Lett. **71**, 2972 (1997)

4.19 R. Heitz, A. Kalburg, Q. Xie, M. Grundmann, P. Chen, A. Hoffmann, A. Madhukar, D. Bimberg: Phys. Rev. B **57**, 9050 (1998)

4.20 C. Teichert, M.G. Lagally, L.J. Peticolas, J.C. Bean, J. Tersoff: Phys. Rev. B **53**, 16334 (1996)

4.21 E. Mateeva, P. Sutter, J.C. Bean, M.G. Lagally: Appl. Phys. Lett. **71**, 3233 (1997)

4.22 P. Schittenhelm, G. Abstreiter, A. Darhuber, G. Bauer, P. Werner, A. Kosogov: Thin Solid Films **294**, 291 (1997)

4.23 F. Heinrichsdorff, A. Krost, M. Grundmann, D. Bimberg, A. Kosogov, P. Werner, F. Bertram, J. Christen: In *Proceedings of the 23rd International Conference on Physics of Semiconductors*, Berlin, Germany, July 22–27, 1996, ed. by M. Scheffler and R. Zimmermann (World Scientific, Singapore 1996), Vol. 2, p. 1321; F. Heinrichsdorff, A. Krost, M. Grundmann, D. Bimberg, F. Bertram, J. Christen, A. Kosogov, P. Werner, J. Heydenreich, U. Gösele: J. Cryst. Growth **170**, 568 (1997)

4.24 V.M. Ustinov, A.Yu. Egorov, A.R. Kovsh, A.E. Zhukov, M.V. Maximov, A.F. Tsatsul'nikov, N.Yu. Gordeev, S.V. Zaitsev, Yu.M. Shernyakov, N.A. Bert, P.S. Kop'ev, Zh.I. Alferov, N.N. Ledentsov, J. Böhrer, D. Bimberg, A.O. Kosogov, P. Werner, U. Gösele: J. Cryst. Growth **175/176**, 689 (1997)

4.25 D. Bimberg, N. Kirstaedter, N.N. Ledentsov, Zh.I. Alferov, P.S. Kop'ev, V.M. Ustinov: IEEE Journal of Selected Topics in Quantum Electronics **3**, 196 (1997)

4.26 L.V. Asryan, R.A. Suris: IEEE Journal of Selected Topics on Quantum Electronics **3**, 148 (1997)
4.27 M.V. Maximov, D.A. Bedarev, A.Yu. Egorov, P.S. Kop'ev, A.R. Kovsh, A.V. Lunev, Yu.G. Musikhin, Yu.M. Shernyakov, A.F. Tsatsul'nikov, V.M. Ustinov, B.V. Volovik, A.E. Zhukov, Zh.I. Alferov, N.N. Ledentsov, D. Bimberg: In *Proceedings of the 24th International Conference on Physics of Semiconductors*, Jerusalem, Israel, August 2–7, 1998, ed. by D. Gershoni (World Scientific, Singapore 1998) CDROM
4.28 I. Mukhametzhanov, R. Heitz, J. Zeng, P. Chen, A. Madhukar: Appl. Phys. Lett. **73**, 1841 (1998)
4.29 A.F. Tsatsul'nikov, A.Yu. Egorov, A.E. Zhukov, A.R. Kovsh, V.M. Ustinov, N.N. Ledentsov, M.V. Maksimov, A.V. Sakharov, A.A. Suvorova, P.S. Kop'ev, Zh.I. Alferov, D. Bimberg: Fiz. Tekh. Poluprovodn. **31**, 109 (1997) [Semiconductors **31**, 88 (1997)]
4.30 O. Stier, M. Grundmann, D. Bimberg: Phys. Rev. B **59**, 5688 (1999)
4.31 P. Yu, W. Langbein, K. Leosson, J.M. Hvam, N.N. Ledentsov, D. Bimberg, V.M. Ustinov, A.Yu. Egorov, A.E. Zhukov, A.F. Tsatsul'nikov, Yu.G. Musikhin: Phys. Rev. B **60** 16680 (1999)
4.32 A.A. Darhuber, V. Holy, J. Stangl, G. Bauer, A. Krost, F. Heinrichsdorff, M. Grundmann, D. Bimberg, V.M. Ustinov, P.S. Kop'ev, A.O. Kosogov, P. Werner: Appl. Phys. Lett. **70**, 955 (1997)
4.33 J.M. Garcia, G. Medeiros-Ribeiro, K. Schmidt, T. Ngo, J.L. Feng, A. Lorke, J. Kotthaus, P.M. Petroff: Appl. Phys. Lett. **71**, 2014 (1997)
4.34 A. Lorke, R.J. Luyken, A.O. Govorov, J.P. Kotthaus, J.M. Garcia, P.M. Petroff: Phys. Rev. Lett. **84**, 2223 (2000)
4.35 W.T. Tsang: Appl. Phys. Lett. **38**, 661 (1981)
4.36 N.N. Ledentsov: to be published
4.37 R.K. Tsui: U.S. Patent 5,075,744, December 24 1991
4.38 J.L. Jewell, H. Temkin: U.S. Patent 5,719,894, February 17 1998
4.39 J.L. Jewell, H. Temkin: U.S. Patent 5,960,018, September 28 1999
4.40 T. Inoue, T. Eshita: U.S. Patent 5,019,874, May 28 1991
4.41 J.C. Bean, G.S. Higashi, R. Hull, J.L. Peticolas: U.S. Patent 5,091,767, February 25 1992
4.42 J. Narayan, J.C.C. Fan: U.S. Patent 5,208,182, May 4 1993
4.43 Y. Chen, R.P. Schneider Jr., Sh.-Y. Wang: U.S. Patent 5,927,995, July 27 1999
4.44 S. Luryi: U.S. Patent 4,806,996, February 21 1989
4.45 E.A. Fitzgerald Jr., D.G. Ast: U.S. Patent 5,156,995, October 20 1992
4.46 J.L. Jewell, H. Temkin: U.S. Patent 5,960,018, September 28 1999
4.47 M. Strassburg, V. Kutzer, U.W. Pohl, A. Hoffmann, I. Broser, N.N. Ledentsov, D. Bimberg, A. Rosenauer, U. Fischer, D. Gerthsen, I.L. Krestnikov, M.V. Maximov, P.S. Kop'ev, Zh.I. Alferov: Appl. Phys. Lett. **72**, 942 (1998)
4.48 V.A. Shchukin, D. Bimberg, V.G. Malyshkin, N.N. Ledentsov: Phys. Rev. B **57**, 12262 (1998)
4.49 G.W. Farnell: *Properties of Elastic Surface Waves* Vol. 6: *Physical Acoustics*, ed. by W.P. Mason and R.N. Thurston (Academic Press, New York 1970)
4.50 A.M. Kosevich, Yu.A. Kosevich, E.S. Syrkin: Zh. Eksp. Teor. Fiz. **88**, 1089 (1985) [Sov. Phys. JETP **61**, 639 (1985)]
4.51 K. Portz, A.A. Maradudin: Phys. Rev. B **16**, 3535 (1977)
4.52 A.F. Andreev: Pis'ma Zh. Eksp. Teor. Fiz. **32**, 654 (1980) [JETP Letters **32**, 640 (1980)]

4.53 V.I. Marchenko: Pis'ma Zh. Eksp. Teor. Fiz. **33**, 397 (1981) [JETP. Lett. **33**, 381 (1981)]
4.54 O.L. Alerhand, D. Vanderbilt, R.D. Meade, J.D. Joannopoulos: Phys. Rev. Lett. **61**, 1973 (1988)
4.55 D. Vanderbilt: Surf. Sci. **268**, L300 (1992)
4.56 K.-O. Ng, D. Vanderbilt: Phys. Rev. B **52**, 2177 (1995)
4.57 J.D. Eshelby: In *Solid State Physics*, Vol. 3, ed. by F. Seitz and D. Turnbull (Academic Press, New York 1956) p. 79
4.58 M. Strassburg, R. Heitz, V. Türck, S. Rodt, U.W. Pohl, A. Hoffmann, D. Bimberg, I.L. Krestnikov, V.A. Shchukin, N.N. Ledentsov, Zh.I. Alferov, D. Litvinov, A. Rosenauer, D. Gerthsen: J. Electron. Mater. **28**, 506 (1999)
4.59 I.L. Krestnikov, M. Straßburg, M. Caesar, A. Hoffmann, U.W. Pohl, D. Bimberg, N.N. Ledentsov, P.S. Kop'ev, Zh.I. Alferov, D. Litvinov, A. Rosenauer, D. Gerthsen: Phys. Rev. B **60**, 8696 (1999)
4.60 I.L. Krestnikov, M.V. Maximov, A.V. Sakharov, P.S. Kop'ev, Zh.I. Alferov, N.N. Ledentsov, D. Bimberg, C.M. Sotomayor Torres: J. Cryst. Growth **184/185**, 545 (1998)
4.61 G. Springholz, V. Holy, M. Pinczolits, G. Bauer: Science **282**, 1675 (1998)
4.62 V. Holý, G. Springholz, M. Pinczolits, G. Bauer: Phys. Rev. Lett. **83**, 356 (1999)
4.63 G. Springholz, M. Pinczolits, P. Mayer, V. Holy, G. Bauer, H.H. Kang, L. Salamanca-Riba: Phys. Rev. Lett. **84**, 4669 (2000)
4.64 H. Li, J. Wu, Z. Wang, T. Daniels-Race: Appl. Phys. Lett. **75**, 1173 (1999)
4.65 H. Li, T. Daniles-Race, M.-A. Hasan: J. Vac. Sci. Technol. B **19**, 1471 (2001)
4.66 H. Li, J. Wu, Zh. Wang, T. Daniels-Race: Appl. Phys. Lett. **80**, 1367 (2002)
4.67 N. Liu, C.-K. Shih, O. Baklenov, A.L. Holmes Jr.: In *Abstracts of the 1999 Fall Meeting of the Materials Research Society* (Boston, November 29– December 3, 1999) p. 167
4.68 X.-D. Wang, N. Liu, S. Govindaraju, A.L. Holmes Jr., C.-K. Shih: unpublished
4.69 A. Zunger, S. Mahajan: in *Handbook on Semiconductors*, ed. by T.S. Moss. Vol. **3**, ed. by S. Mahajan (Elsevier Science, Amsterdam 1994) p. 1399
4.70 J.W. Cahn: Acta Met. **9**, 795 (1961)
4.71 J.W. Cahn: Acta Met. **10**, 179 (1962)
4.72 J.W. Cahn: Trans. Met. Soc. **242**, 166 (1968)
4.73 A.G. Khachaturyan: phys. stat. sol. **35**, 119 (1969)
4.74 A.G. Khachaturyan: *Theory of Phase Transformations and Structure of Solid Solutions* (Nauka, Moscow 1974) in Russian
4.75 A.G. Khachaturyan: *Theory of Structural Transformations in Solids* (Wiley, New York 1983)
4.76 B. de Cremoux, P. Hirtz, J. Ricciardi: in *GaAs and Related Compounds 1980*, ed. by H.W. Thim, Inst. Phys. Conf. Ser. **56**, 115 (Institute of Physics, London 1981)
4.77 B. de Cremoux: J. Phys. (Paris) **43**, C5–19 (1982)
4.78 G.B. Stringfellow: J. Cryst. Growth **58**, 194 (1982)
4.79 K. Onabe: Jap. J. Appl. Phys. **21**, L323 (1982)
4.80 M. Ilegems, M.B. Panish: J. Phys. Chem. Solids **35**, 409 (1974)
4.81 A.S. Jordan, M. Ilegems: J. Phys. Chem. Solids **36**, 329 (1975)
4.82 A.G. Khachaturyan, R.A. Suris: Kristallografiya **13**, 83 (1968) [Sov. Phys. Crystallogr. **13**, 63 (1968)]
4.83 G.B. Stringfellow: J. Electr. Mater. **11**, 903 (1982)
4.84 G.B. Stringfellow: J. Cryst. Growth. **65**, 454 (1983)
4.85 F. Glas: J. Appl. Phys. **62**, 3201 (1987)

4.86 I.P. Ipatova, V.G. Malyshkin, V.A. Shchukin: J. Appl. Phys. **74**, 7198 (1993)
4.87 I.P. Ipatova, V.G. Malyshkin, V.A. Shchukin: Phil. Mag. B **70**, 557 (1994)
4.88 V.A. Shchukin, A.N. Starodubtsev: Phys. Low Dim. Structures **11/12**, 203 (1998)
4.89 N.A. Bert, L.S. Vavilova, I.P. Ipatova, V.A. Kapitonov, A.V. Murashova, N.A. Pikhtin, A.A. Sitnikova, I.S. Tarasov, V.A. Shchukin: Fiz. Tekhn. Poluprovodn. **33**, 544 (1999) [Semiconductors **33**, 510 (1999)]
4.90 B.J. Spencer, P.W. Voorhees, S.H. Davis, G.B. McFadden: Acta Metall. Mater. **40**, 1599 (1992)
4.91 V.G. Malyshkin, V.A. Shchukin: Fiz. Tekh. Poluprovodn. **27**, 1932 (1993) [Semiconductors **27**, 1062 (1993)]
4.92 J.E. Guyer, P.W. Voorhees: Phys. Rev. Lett. **74**, 4031 (1995)
4.93 J.E. Guyer, P.W. Voorhees: Phys. Rev. B **54**, 11710 (1996)
4.94 I.P. Ipatova, V.G. Malyshkin, A.A. Maradudin, V.A. Shchukin, R.F. Wallis: *Proceedings of the 23rd International Symposium on Compound Semiconductors*, St. Petersburg, Russia, September 23–27, 1996; Inst. Phys. Conf. Series **155**, 323 (1997)
4.95 I.P. Ipatova, V.G. Malyshkin, A.A. Maradudin, V.A. Shchukin, R.F. Wallis: Phys. Rev. B **57**, 12969 (1998)
4.96 S.W. Jun, T.-Y. Seong, J.H. Lee, B. Lee: Appl. Phys. Lett. **68**, 3443 (1996)
4.97 F. Léonard, R.F. Desai: Phys. Rev. B **55**, 9990 (1997)
4.98 F. Léonard, R.F. Desai: Phys. Rev. B **56**, 4955 (1997)
4.99 F. Léonard, R.F. Desai: Phys. Rev. B **57**, 4805 (1998)
4.100 P. Venezuela, J. Tersoff: Phys. Rev. B **58** 10871 (1998)
4.101 B.J. Spencer, P.W. Voorhees, J. Tersoff: Phys. Rev. Lett. **84**, 2499 (2000)
4.102 B.J. Spencer, P.W. Voorhees, J. Tersoff: Appl. Rev. Lett. **76**, 3022 (2000)
4.103 B.J. Spencer, P.W. Voorhees, J. Tersoff: Phys. Rev. B **64**, 235318 (2001)
4.104 W. Barvosa-Carter, M.J. Aziz, L.J. Gray, T. Kaplan: Phys. Rev. Lett. **81**, 1445 (1998)
4.105 J.F. Sage, W. Barvosa-Carter, M.J. Aziz: Appl. Phys. Lett. **77**, 516 (2000)
4.106 M. Aziz: In *Morphological and Compositional Evolution of Heteroepitaxial Semiconductor Thin Films*, ed. by A.-L. Barabási, E. Jones, and J. Mirecki Millunchick, Mat. Res. Soc. Symp. Proc. **618**, 233 (MRS, Pittsburgh 2000)
4.107 V.A. Shchukin, A.N. Starodubtsev: In *Proc. 26th Int. Symp. on Compound Semiconductors*, Berlin, Germany, August 21–25, 1999, Am. Inst. Phys. Conf. Ser. **166**, 231 (New York 1999)
4.108 V.A. Shchukin, A.N. Starodubtsev: In *Self-Organized Processes in Semiconductor Alloys: Spontaneous Ordering, Composition Modulation, and 3D Islanding*, ed. by D.M. Follstaedt, B.A. Joyce, A. Mascarenhas, and T. Suzuki, Proc. Mat. Res. Soc. Symp. **583**, 327 (MRS, Pittsburgh 2000)
4.109 A.N. Starodubtsev: PhD Thesis, Abraham Ioffe Physical Technical Institute of the Russian Academy of Sciences, St. Petersburg (2000) in Russian
4.110 V.A. Shchukin, N.N. Ledentsov, P.S. Kop'ev, D. Bimberg: Phys. Rev. Lett. **75**, 2968 (1995)
4.111 J. Tersoff: Phys. Rev. Lett. **81**, 3183 (1998)
4.112 N. Liu, J. Tersoff, O. Baklenov, A.L. Holmes Jr., C.K. Shih: Phys. Rev. Lett. **84**, 334 (2000)
4.113 M.V. Maximov, A.F. Tsatsul'nikov, B.V. Volovik, D.S. Sizov, Yu.M. Shernyakov, I.N. Kaiander, A.E. Zhukov, A.R. Kovsh, S.S. Mikhrin, V.M. Ustinov, Zh.I. Alferov, R. Heitz, V.A. Shchukin, N.N. Ledentsov, D. Bimberg, Yu.G. Musikhin, W. Neumann: Phys. Rev. B **62**, 16671 (2000)
4.114 F. Guffarth, R. Heitz, A. Schliwa, O. Stier, N.N. Ledentsov, A.R. Kovsh, V.M. Ustinov, D. Bimberg: Phys. Rev. B **64**, 085305 (2001)

Chapter 5

5.1 N.G. Basov, B.M. Vul', Yu.M. Popov: Zh. Eksp. Teor. Fiz. **37**, 416 (1959) [Sov. Phys. JETP]
5.2 N.G. Basov, O.N. Krokhin, Yu.M. Popov: Zh. Eksp. Teor. Fiz. **40**, 1320 (1961) [Sov. Phys. JETP]
5.3 R.N. Hall, G.E. Fenner, J.D. Kingsley, T.J. Soltys, R.O. Carlson: Phys. Rev. Lett. **9**, 366 (1962) (submitted September 24, 1962)
5.4 R. Dingle, C.H. Henry: *Quantum Effects in Heterostructure Lasers*, U.S. Patent No. 3982207 (September 21, 1976)
5.5 Zh.I. Alferov: In *Proc. 99th Nobel Symposium*, Arild, June 4–8, 1996, Physica Scripta **68**, 32 (1996)
5.6 I. Hayashi: IEEE Trans. Electron. Devices **31**, 1630–1645 (1984)
5.7 Zh.I. Alferov, R.F. Kazarinov: *Double Heterostructure Laser*, Authors Certificate No 27448, Application No 950840 with a priority from March 30, 1963
5.8 H. Kroemer: *A Proposed Class of Heterojunction Injection Lasers*, Proc. IEEE **51**, 1782 (1963) submitted October 14, 1963
5.9 Zh.I. Alferov: Fiz. Tverd. Tela **8**, 3102 (1966) [Sov. Phys. Solid. State **8**, 10, 2480 (1967)]
5.10 Zh.I. Alferov, D.Z. Garbuzov, V.S. Grigor'eva, Yu.V. Zhilyaev, L.V. Kradinova, V.I. Korol'kov, E.P. Morozov, O.A. Ninua, E.L. Portnoi, V.D. Prochukhan, M.K. Trukan: Fiz. Tverd. Tela **9**, 279 (1967) (submitted July 15, 1966) [Sov. Phys. Solid State **9**, 208 (1967)]
5.11 Zh.I. Alferov, V.M. Andreev, E.L. Portnoi, M.K. Trukan: Fiz. Tekn. Poluprovodn. **3**, 1328 (1969) submitted December 30, 1968 [Sov. Phys. Semicond. **3**, 1107 (1970)]
5.12 Zh.I. Alferov, V.M. Andreev, D.Z. Garbuzov, Yu.V. Zhilyaev, E.P. Morozov, E.L. Portnoi, V.G. Trofim: Fiz. Tekh. Poluprovodn. **4**, 1826 (1970) [Sov. Phys. Semicond. **4**, 1573 (1970)] submitted May 6, 1970
5.13 I. Hayashi, M.B. Panish, P.W. Foy, S. Sumski: Appl. Phys. Lett. **17**, 109 (1970) submitted June 8, 1970
5.14 J.P. van der Ziel, R. Dingle, R.C. Miller, W. Wiegmann, W.A. Nordland Jr.: Appl. Phys. Lett. **26**, 463 (1975)
5.15 R.C. Miller, R. Dingle, A.C. Gossard, R.A. Logan, W.A. Nordland Jr., W. Wiegmann: J. Appl. Phys. **47**, 4509 (1976)
5.16 R.D. Dupuis, P.D. Dapkus, N. Holonyak Jr., E.A. Rezek, R. Chin: Appl. Phys. Lett. **32**, 295 (1978)
5.17 N. Holonyak Jr., R. Kolbas, R.D. Dupuis, P.D. Dapkus: IEEE J. Quantum. Electron. **16**, 170 (1980)
5.18 W.T. Tsang: Appl. Phys. Lett. **39**, 786 (1981)
5.19 W.T. Tsang: Appl. Phys. Lett. **40**, 217 (1982)
5.20 Zh.I. Alferov, A.M. Vasiliev, S.V. Ivanov, P.S. Kop'ev, N.N. Ledentsov, B.Ya. Meltser, V.M. Ustinov: Pis'ma Zh. Tekn. Fiz. **14**, 1803 (1988) [Sov. Phys. Techn. Phys. Lett. **14**, 782 (1988)]
5.21 Zh.I. Alferov, S.V. Ivanov, P.S. Kop'ev, N.N. Ledentsov, M.E. Lutsenko, M.I. Nemenov, B.Ya. Meltser, V.M. Ustinov, S.V. Shaposhnikov: Fiz. Tekn. Poluprovodn. **24**, 152 (1990) [Sov. Phys. Semicond. **24**, 92 (1990)]
5.22 N. Chand, E.E. Becker, J.P. van der Ziel, S.N.Gu. Chu, N.K. Dutta: Appl. Phys. Lett. **58**, 1704 (1991)
5.23 P.M. Petroff, A.C. Gossard, R.A. Logan, W. Weigman: Appl. Phys. Lett. **41**, 635 (1982)
5.24 E. Kapon, M.C. Tamargo, D.M. Hwang: Appl. Phys. Lett. **50**, 347 (1987)
5.25 H.A. Bluyssen, L.J. Ruyven: IEEE J. Quantum Electron. **17**, 2191 (1981)

5.26 Y. Arakawa, H. Sakaki: Appl. Phys. Lett. **40**, 939 (1982)
5.27 M. Asada, M. Miyamoto, Y. Suematsu: IEEE J. Quantum Electron. **22**, 1915 (1986)
5.28 K.J. Vahala: IEEE J. Quantum. Electron. **24**, 523 (1988)
5.29 H. Benisty, C.M. Sotomayor Torres, C. Weisbuch: Phys. Rev. B **44**, 10945 (1991)
5.30 N.N. Ledentsov, V.M. Ustinov, A.Yu. Egorov, A.E. Zhukov, M.V. Maximov, I.G. Tabatadze, P.S. Kop'ev: Fiz. Tekh. Poluprovodn. **28**, 1484 (1984) submitted December 29, 1983 [Semiconductors **28**, 832 (1994)]
5.31 O. Brandt, M. Ilg, K. Ploog: Microelectronics Journal **26**, 861 (1995)
5.32 D. Leonard, M. Krishnamurthy, C.M. Reaves, S.P. Denbaars, P.M. Petroff: Appl. Phys. Lett. **63**, 3203 (1993)
5.33 N.N. Ledentsov, M. Grundmann, N. Kirstaedter, J. Christen, R. Heitz, J. Böhrer, F. Heinrichsdorff, D. Bimberg, S.S. Ruvimov, P. Werner, U. Richter, U. Gösele, J. Heydenreich, V.M. Ustinov, A.Yu. Egorov, M.V. Maximov, P.S. Kop'ev, Zh.I. Alferov: In *Proceedings of the 22nd International Conference on Physics of Semiconductors*, Vancouver, Canada, August 1994, ed. by D.J. Lockwood (World Scientific, Singapore 1994) Vol. **3**, p. 1855
5.34 N. Kirstaedter, N.N. Ledentsov, M. Grundmann, D. Bimberg, V.M. Ustinov, S.S. Ruvimov, M.V. Maximov, P.S. Kop'ev, Zh.I. Alferov, U. Richter, P. Werner, U. Gösele, J. Heydenreich: Electron. Lett. **30**, 1416 (1994)
5.35 N.N. Ledentsov, J. Böhrer, D. Bimberg, S.V. Zaitsev, V.M. Ustinov, A.Yu. Egorov, A.E. Zhukov, M.V. Maximov, P.S. Kop'ev, Zh.I. Alferov, A.O. Kosogov, U. Gösele, S.S. Ruvimov: Mat. Res. Soc. Symp. Proc. **421**, 133, ed. by R.J. Shul, S.J. Pearton, F. Ren, and C.-S. Wu (Pittsburgh 1996)
5.36 G.T. Liu, A. Stintz, H. Li, K.J. Malloy, L.F. Lester: Electron. Lett. **35**, 1163 (1999)
5.37 For a review, see D. Bimberg, M. Grundmann, N.N. Ledentsov: *Quantum Dot Heterostructures* (Wiley, Chichester 1998)
5.38 M. Grundmann, D. Bimberg: Jpn. J. Appl. Phys. **36**, 4181 (1997)
5.39 O. Stier, M. Grundmann, D. Bimberg: Phys. Rev. B **59**, 5688 (1999)
5.40 A. Zunger: MRS Bulletin **23**, 35 (1998)
5.41 L.V. Asryan, M. Grundmann, N.N. Ledentsov, O. Stier, R.A. Suris, D. Bimberg: IEEE J. Quantum Electron. **37**, 418 (2001)
5.42 N.N. Ledentsov: J. Select. Topics Quantum Electron. **8**, 1015 (2002)
5.43 D. Bimberg, N.N. Ledentsov, J.A. Lott: MRS Bulletin **27**, 531 (2002)
5.44 P.G. Eliseev, H. Li, A. Stinz, G.T. Liu, T.C. Newell, K.J. Malloy, L.F. Lester: Appl. Phys. Lett. **77**, 262 (2000)
5.45 N.N. Ledentsov, V.A. Shchukin, M. Grundmann, N. Kirstaedter, J. Böhrer, O. Schmidt, D. Bimberg, V.M. Ustinov, A.Yu. Egorov, A.E. Zhukov, P.S. Kop'ev, S.V. Zaitsev, N.Yu. Gordeev, Zh.I. Alferov, A.I. Borovkov, A.O. Kosogov, S.S. Ruvimov, P. Werner, U. Gösele, J. Heydenreich: Phys. Rev. B **54**, 8743 (1996)
5.46 R. Sellin, Ch. Ribbat, M. Grundmann, N.N. Ledentsov, D. Bimberg: Appl. Phys. Lett. **78**, 1207 (2001)
5.47 M.V. Maximov, I.V. Kochnev, Yu.M. Shernyakov, S.V. Zaitsev, N.Yu. Gordeev, A.F. Tsatsul'nikov, A.V. Sakharov, I.L. Krestnikov, P.S. Kop'ev, Zh.I. Alferov, N.N. Ledentsov, D. Bimberg, A.O. Kosogov, P. Werner, U. Gösele: Jap. J. Appl. Phys. **36**, Part 1, 4221 (1997)
5.48 Yu.M. Shernyakov, D.A. Bedarev, E.Yu. Kondrat'eva, P.S. Kop'ev, A.R. Kovsh, N.A. Maleev, M.V. Maximov, V.M. Ustinov, B.V. Volovik, A.E. Zhukov, Zh.I. Alferov, N.N. Ledentsov, D. Bimberg: Electron. Lett. **35**, 898 (1999)

5.49 A.E. Zhukov, A.R. Kovsh, V.M. Ustinov, S.S. Mikhrin, N.A. Maleev, E.Yu. Kondrat'eva, D.A. Livshits, M.V. Maximov, B.V. Volovik, D.A. bedarev, Yu.G. Musikhin, N.N. Ledentsov, P.S. Kop'ev, D. Bimberg: IEEE Photonics Technol. Lett. **11**, 1345 (1999)

5.50 O.B. Shchekin, D.G. Deppe: Appl. Phys. Lett. **80**, 3277 (2002)

5.51 O.B. Shchekin, J. Ahn, D.G. Deppe: to be published

5.52 P. Borri, S. Schneider, W. Langbein, U. Woggon, A.E. Zhukov, V.M. Ustinov, Zh.I. Alferov, D. Ouyang, D. Bimberg: Appl. Phys. Lett. **79**, 2633 (2001)

5.53 S.V. Zaitsev, N.Yu. Gordeev, V.M. Ustinov, A.E. Zhukov, A.Yu. Egorov, M.V. Maximov, A.F. Tsatsul'nikov, N.N. Ledentsov, P.S. Kop'ev, Zh.I. Alferov, D. Bimberg: Fiz. Tekh. Poluprovodn. **31** 539 (1997) [Semiconductors **31**, 455 (1997)]

5.54 A.R. Kovsh, N.A. Maleev, A.E. Zhukov, S.S. Mikhrin, A.P. Vasil'ev, Yu.M. Shernyakov, M.V. Maximov, D.A. Livshits, V.M. Ustinov, Zh.I. Alferov, N.N. Ledentsov, D. Bimberg: Electron. Lett. **38**, 1104 (2002)

5.55 L. Harris, D.J. Mowbray, M.S. Skolnick, M. Hopkinson, G. Hill: Appl. Phys. Lett. **73**, 969 (1998)

5.56 D. Bhattacharya, E.A. Avrutin, A.C. Bryce, J.H. Marsh, B. Bimberg, F. Heinrichsdorff, V.M. Ustinov, S.V. Zaitsev, N.N. Ledentsov, P.S. Kop'ev, Zh.I. Alferov, A.I. Onishchenko, E.P. O'Reilly: IEEE J. Select. Topics Quantum Electron. **5**, 648 (1999)

5.57 L. Harris, A.D. Ashmore, D.J. Mowbray, M.S. Skolnick, M. Hopkinson, G. Hill, J. Clark: Appl. Phys. Lett. **75**, 3512 (1999)

5.58 M. Sugawara, K. Mukai, Y. Nakata, H. Ishikawa, A. Sakamoto: Phys. Rev. B **61**, 7595 (2000)

5.59 D. Ouyang, R. Heitz, N.N. Ledentsov, S. Bognár, R.L. Sellin, Ch. Ribbat, D. Bimberg: Appl. Phys. Lett. **81** (2002)

5.60 J. Aaviksoo, I. Reimand, V.V. Rossin, V.V. Travnikov: Phys. Rev. B **45**, 1473 (1992)

5.61 D. Bimberg, M. Grundmann, N.N. Ledentsov, M.H. Mao, Ch. Ribbat, R.L. Sellin, V.M. Ustinov, A.E. Zhukov, Zh.I. Alferov, J.A. Lott: phys. stat. sol. (b) **224**, 787 (2001)

5.62 L.V. Asryan, S. Luryi: IEEE J. Quantum Electron. **37**, 905 (2001)

5.63 P. Bhattacharya, S. Ghosh: Appl. Phys. Lett. **80**, 3482 (2002)

5.64 X. Huang, A. Stintz, H. Li, L.F. Lester, J. Cheng, K.J. Malloy: Appl. Phys. Lett. **78**, 2825 (2001)

5.65 C. Ribbat, R. Sellin, I. Kainder, F. Hopfer, N.N. Ledentsov, D. Bimberg, A.R. Kovsh, V.M. Ustinov, A.E. Zhukov, M.V. Maximov: to be published

5.66 C. Ribbat, R. Sellin, M. Grundmann, D. Bimberg, N.A. Sobolev, M.C. Carmo: Electron. Lett. **37**, 174 (2001)

5.67 R.L. Sellin, C. Ribbat, D. Bimberg, F. Rinner, H. Konstanzer, M.T. Kelemen, M. Mikula: Electron. Lett. **38**, 883 (2002)

5.68 N.N. Ledentsov, M. Grundmann, F. Heinrichsdorff, D. Bimberg, V.M. Ustinov, A.E. Zhukov, M.V. Maximov, Zh.I. Alferov, J.A. Lott: IEEE J. Sel. Top. Quantum Electron. **6**, 439 (2000)

5.69 B.V. Volovik, D.S. Sizov, A.F. Tsatsul'nikov, Yu.G. Musikhin, N.N. Leentsov, V.M. Ustinov, V.A. Egorov, V.N. Petrov, N.K. Polyakov, G.E. Tsyrlin: Fiz. Tekh. Poluprovodn. **34** (2000) [Semiconductors **34**, 1316 (2000)]

5.70 J. Tatebayashi, M. Nishioka, Y. Arakawa: Appl. Phys. Lett. **78**, 174 (2001)

5.71 A.R. Kovsh, A.E. Zhukov, N.A. Maleev, S.S. Mikhrin, A.V. Vasil'ev, Yu.M. Shernyakov, D.A. Livshits, M.V. Maximov, D.S. Sizov, N.V. Kryzhanovskaya, N.A. Pikhtin, V.A. Kapitonov, I.S. Tarasov, N.N. Ledentsov, V.M. Ustinov, J.S. Wang, L. Wei, G. Lin, J.Y. Chi: In *Proc.*

10th Int. Symp. Nanostructures: Physics and Technology (St. Petersburg, Russia, June 17–21, 2002) p. 395

5.72 A.R. Kovsh, A.E. Zhukov, D.A. Lifshits, A.Yu. Egorov, V.M. Ustinov, M.V. Maximov, Yu.G. Musikhin, N.N. Ledentsov, P.S. Kop'ev, Zh.I. Alferov, D. Bimberg: Elecron. Lett. **35**, 1161 (1999); A.E. Zhukov, A.R. Kovsh, S.S. Mikhrin, N.A. Maleev, V.M. Ustinov, D.A. Lifshits, I.S. Tarasov, D.A. Bedarev, M.V. Maximov, A.F. Tsatsul'nikov, I.P. Soshnikov, P.S. Kop'ev, Zh.I. Alferov, N.N. Ledentsov, D. Bimberg: Electron. Lett. **35**, 1845 (1999)

5.73 X. He, S. Srinivasan, S. Wilson, C. Mitchell, R. Patel: Electron. Lett. **34**, 2126 (1998)

5.74 D.A. Livshits, I.V. Kochnev, V.M. Lantratov, N.N. Ledentsov, T.A. Nalyot, I.S. Tarasov, Z.I. Alferov: Electron. Lett. **36**, 1848 (2000)

5.75 H. Saito, K. Nishi, I. Ogura, S. Sugou, Y. Sugimoto: Appl. Phys. Lett. **69**, 3140 (1996)

5.76 R. Schur, F. Sogawa, M. Nishioka, S. Ishida, Y. Arakawa: In *Proc. 15th IEEE International Semiconductor Laser Conference*, Haifa, Israel, October 13–18, 1997, paper PDP3, p. 6 (1997)

5.77 J.A. Lott, N.N. Ledentsov, V.M. Ustinov, A.Yu. Egorov, A.E. Zhukov, P.S. Kop'ev, Zh.I. Alferov, D. Bimberg: Electron. Lett. **33**, 1150 (1997)

5.78 N.N. Ledentsov, D. Bimberg, V.M. Ustinov, J.A. Lott, Zh.I. Alferov: Memoirs of the Institute of Scientific and Industrial Research, Osaka **57**, 80 (2000)

5.79 J.A. Lott, N.N. Ledentsov, V.M. Ustinov, N.A. Maleev, A.E. Zhukov, A.R. Kovsh, M.V. Maximov, B.V. Volovik, Zh.I. Alferov, D. Bimberg: Electron. Lett. **36**, 1384 (2000)

5.80 H. Grabert, H. Horner (Eds.): Z. Phys. B **85**, 317 (1991)

5.81 H. Fukuyama, T. Ando: *Transport Phenomena in Mesoscopic Systems* (Springer, Berlin 1992)

5.82 H. Grabert, M.H. Devoret (Eds.): *Single Charge Tunnelling: Coulomb Blockade Phenomena* (Plenum Press, New York 1992)

5.83 M.A. Kastner: Physics Today **1**, 25 (1993)

5.84 A. Wojs, P. Hawrylak: Phys. Rev. B **53**, 10841 (1996)

5.85 C.A. Stafford, S. Das Sarma: Phys. Rev. Lett. **72**, 3590 (1994)

Index

activated alloy phase separation 262, 309, 310
adatom 138, 215
- condensation 215
- surface diffusion 296
- surface diffusion and drift 239, 300
- surface migration 239, 300
alloy
- absolutely unstable 288
- chemical free energy 283
- critical temperature of instability 286, 293, 298
- elastic energy 284
- formation enthalpy 284
- gradient energy 284
- metastable 288
- mixing entropy 284
- soft mode of composition fluctuations 293
- stable 288
- thermodynamic instability 288
alloy growth
- compositional instability 296, 306
- critical temperature of kinetic instability 298
- morphological instability 300, 306
annealing 59, 99, 168, 206, 257, 260–262
arsenic vapor pressure 209, 212
Asaro–Tiller–Grinfel'd instability 166
atomic force microscopy (AFM) 36, 46, 99, 278

binodal curve 288
bulk diffusion 296

capillarity 71, 72
cathodoluminescence (CL) 107, 162
closed system 58, 168
coherent island 159, 160, 162, 167, 168, 211, 212, 255, 256, 258, 259, 262
composition-modulated structure 301

configuration entropy 125, 134
- entropy of fluctuations
-- in island positions 128, 133
-- in island shape 126
-- in island volume 128, 133
corrugated superlattice 78, 87, 104
Coulomb blockade 333
cross-sectional scanning tunneling microscopy (XSTM) 39

defect reduction 255
digital analysis of lattice images (DALI) 10, 45, 47, 143, 144, 230, 263, 270
disk-to-stripe transition 116, 119, 120
dislocated island 159, 160, 162, 167, 168, 211, 212, 255, 258, 259, 262
distribution of island sizes 138, 139
double heterostructure (DHS) laser 319

edge-emitting laser 316
elastic anisotropy 184, 264, 277
elastic force dipole 98, 269
elastic force monopole 74, 98, 111
elastic Green function 184, 265, 267, 291
elastic relaxation
- at crystal edges 74, 176
- at domain boundaries 111
- volume elastic relaxation 176, 179, 194
elastically hard direction 277
elastically soft direction 184, 187, 277, 285, 293, 304
electronic density of states 4, 317, 332
energy of edges 176
engineering of exciton wave function 249, 250, 274
equilibrium
- constrained 59
- global 59
- local 59, 204, 300

– partial   59, 238
equilibrium constant   23, 27
equilibrium crystal shape (ECS)   61
equilibrium vapor pressure   25, 28, 31–33
evaporation   260, 262
exciton oscillator strength   106, 146

faceting   64, 65
Frank–van der Merwe (FM) growth mode   105, 193

Gibbs phase rule   22
growth interruption   59, 168, 206–208, 213, 220

high resolution transmission electron microscopy (HRTEM)   10, 43, 47, 87, 94, 143, 144, 151, 160, 228, 230, 244, 263, 270
– processing   84, 88, 89
hut cluster   165

impurity segregation   30
irreversible phase transition   211

kinetic Monte Carlo simulations   141

Laplace pressure   72
lateral association of quantum dots (LAQD)   224, 261
liquidus   23, 25, 26, 28, 31–33
low energy electron diffraction (LEED)   49

mass action law   23, 29
metalorganic chemical vapor deposition (MOCVD)   33, 34, 51, 213, 229
metalorganic vapor phase epitaxy (MOVPE)   33
miscibility curve   288
molecular beam epitaxy (MBE)   17, 207, 209
Moore's law   1

nanostructure   2
non-linear growth stage   306

open system   58, 299
optical reflectance   273
oscillator strength   106, 146
Ostwald ripening   163, 169, 181, 189, 193, 211, 212

overgrowth   255, 257, 260–262, 309, 310

periodically faceted surface   74
– AlAs vicinal (001)   100
– AlAs(311)   76
– GaAs vicinal (001)   99, 100
– GaAs(311)   76
– Si vicinal (111)   98
– TaC(110)   101
phase
– equilibrium   22, 26
– separation   64, 65
photoluminescence (PL)   54, 104, 145, 153–155, 158, 162, 208, 217, 221, 226, 247, 248, 272, 273, 311, 312
– blueshift   221, 311
– edge PL   249, 274, 275
– polarization   249, 250, 274
– redshift   221, 311
– surface PL   250
platelet   200
precursor   33

quantum dot   4
quantum dot (QD) laser   319, 323
– characteristic temperature   324, 325
– high power operation   327
– injection laser   322
– laser on submonolayer QDs   330
– operation lifetime   328, 329
– photo-pumped laser   322
– threshold current density   324, 325
– vertical cavity surface-emitting laser (QD VCSEL)   330, 331
quantum dot array
– CdSe/ZnSe submonolayer   107, 272
– electronically coupled QDs   243
– Ge/Si(001) submonolayer   152
– InAs/GaAs   162, 207, 209, 216, 221
– InAs/GaAs submonolayer   145
– InGaAs/GaAs   162
– PbSe/PbEuTe   278
quantum well   4, 229
quantum well (QW) laser   319, 323, 324
quantum well wire (QWW) superlattice   78, 100
quantum wire   4, 78, 281
– GaAs/AlAs vicinal (001)   100
– GaAs/AlAs(311)   78
– InAs/InAlAs   281

Rayleigh wave

– generalized   264
– ordinary   264
reflectance anisotropy spectroscopy (RAS)   51, 53
reflection high energy electron diffraction (RHEED)   49, 50, 52, 76
resonant waveguiding   147
reversible phase transition   206, 209

scanning tunneling microscopy (STM)   37, 46, 98, 101, 114, 115, 165
seeding of quantum dots   247, 248
self-assembly   58
self-organization   58, 321
semiconductor alloy   28
single-electron device   333
size-limited island growth   163, 181, 182, 189, 207
solidus   22, 31
solubility curve   288
spectroscopic ellipsometry   51
spinodal curve   288
steady-state modulated structure   307
stiffness of an array of disks   131
strain-induced barrier
– for the attachment of an adatom to a strained island   201
– for the nucleation of a new atomic layer on a facet of a strained island   203
Stranski–Krastanow (SK) growth mode   105, 167, 193
stressor   248, 309
striped domain structure   112
sub-nanostructure   336
sublimation   23, 27
submonolayer island array   106, 143
– CdSe/ZnSe(001)   147, 263, 270, 272–275
– Ge/Si(001)   151, 153–155
– InAs/GaAs(001)   143–145
– Pb/Cu(111)   120
surface domain structure
– CuO/Cu(110)   114, 115
– Pb/Cu(111)   120
surface energy   61, 67, 175, 179, 194
– strain-induced renormalization   67, 177, 182

surface excess elastic modulus   67, 175
surface stress   65, 67, 175
– discontinuity
– – at crystal edges   71, 176
– – at domain boundaries   110, 114
surface-emitting laser (VCSEL)   316

temperature ramping   218, 222
thermal etching   255
three-dimensional island array
– Ge/Si(001)   165
– InAlAs/GaAlAs(001)   247
– InAs/GaAs(001)   157, 159, 160, 162, 179, 207–209, 212, 216–218, 221, 222, 225, 226, 244, 245
– – overgrowth by GaAs   252
– – overgrowth by InGaAlAs   310, 312
– – overgrowth by InGaAs   310–312
– InAs/Si(001)   228
– InGaAs/GaAlAs(001)   247
– InGaAs/GaAs(001)   250
– InGaN/GaN(0001)   230
– PbSe/PbEuTe(111)   278, 280
transmission electron microscopy (TEM)   40, 41, 46
– cross section   100, 157, 222, 248, 252, 256, 259, 262, 280, 281, 301, 310
– modeling   43, 46, 228
– plan view   10, 91–93, 159, 162, 187, 208, 218, 222, 225, 244, 247, 248, 250, 256, 257, 259, 261, 281, 301, 310

vertically anticorrelated growth   263, 268–270, 274, 280, 281
vertically correlated growth   268–270, 274, 280
vicinal surface   97–99, 300
Volmer–Weber (VW) growth mode   105, 193

wetting layer   177, 182, 191, 193, 205, 221
Wiener filtering   45, 83

X-ray diffraction   54

Young's modulus   184

Printing: Druckhaus Berlin-Mitte GmbH
Binding: Buchbinderei Stein & Lehmann, Berlin